Remote Sensing in Archaeology

INTERDISCIPLINARY CONTRIBUTIONS TO ARCHAEOLOGY

Series Editor: **Michael Jochim**, *University of California, Santa Barbara, California*
Founding Editor: **Roy S. Dickens, Jr.**, *Late of University of North Carolina, Chapel Hill, North Carolina*

THE ARCHAEOLOGIST'S LABORATORY
The Analysis of Archaeological Data
E.B. Banning

AURIGNACIAN LITHIC ECONOMY
Ecological Perspectives from Southwestern France
Brooke S. Blades

EARLIEST ITALY
An Overview of the Italian Paleolithic and Mesolithic
Margherita Mussi

EMPIRE AND DOMESTIC ECONOMY
Terence N. D'Altroy and Christine A. Hastorf

EUROPEAN PREHISTORY: A SURVEY
Edited by Saurunas Miliasuskas

THE EVOLUTION OF COMPLEX HUNTER-GATHERERS
Archaeological Evidence from the North Pacific
Ben Fitzhugh

FAUNAL EXTINCTION IN AN ISLAND SOCIETY
Pygmy Hippotamus Hunters of Cyprus
Alan H. Simmons

A HUNTER-GATHERER LANDSCAPE
Southwest Germany in the Late Paleolithic and Neolithic
Michael A. Jochim

MISSISSIPPIAN COMMUNITY ORGANIZATION
The Powers Phase in Southeastern Missouri
Michael J. O'Brien

NEW PERSPECTIVES ON HUMAN SACRIFICE AND RITUAL BODY TREATMENTS
IN ANCIENT MAYA SOCIETY
Edited by Vera Tiesler and Andrea Cucina

REMOTE SENSING IN ARCHAEOLOGY
Edited by James Wiseman and Farouk El-Baz

THE TAKING AND DISPLAYING OF HUMAN BODY PARTS AS TROPHIES BY AMERINDIANS
Edited by Richard J. Chacon and David H. Dye

A Continuation Order Plan is available for this series. A continuation order will bring delivery of each new volume immediately upon publication. Volumes are billed only upon actual shipment. For further information please contact the publisher.

Remote Sensing in Archaeology

Edited by

JAMES WISEMAN

Center for Archaeological Studies, Boston University
Boston, MA, USA

and

FAROUK EL-BAZ

Center for Remote Sensing, Boston University
Boston, MA, USA

 Springer

James R. Wiseman
Center for Archaeological Studies
Boston University
725 Commonwealth Avenue
Boston, MA, 02215
jimw@bu.edu

Farouk El-Baz
Center for Remote Sensing
Boston University
675 Commonwealth Avenue
Boston, MA 02215
Farouk@bu.edu

Cover Illustration:
SIR-C 1994 image of Angkor Wat, Cambodia, and a mosaic detail of Angkor Wat and
Kapilapura mound. See Figure 1, p. 186, and Figure 14, p. 211, for images and
detailed explanations.

Library of Congress Control Number: 2001012345

Additional material to this book can be downloaded from http://extras.springer.com.

HB ISBN 10 0-387-44453-X   e-ISBN 10 0-387-44455-6   SB ISBN 10 0-387-44615-X
HB ISBN 13 978-0-387-44453-6   e-ISBN 13 978-0-387-44455-0

9 8 7 6 5 4 3 2 1

springer.com

Contributors

Turki S.M. Al-Saud. King Abdulaziz City for Science and Technology, Riyadh, Saudi Arabia.
Email: talsaud@kacst.edu.sa

Jennifer R. Bales. 3345 Grand Ave., Billings, MT 59102

Robert D. Ballard. Institute for Archaeological Oceanography, Graduate School of Oceanography, University of Rhode Island, South Ferry Road, Narragansett, RI 02882.
Email: bballard@gso.uri.edu

Deborah Blom. Department of Anthropology, University of Vermont, Williams Hall 508, 72 University Place, Burlington, VT 05405-0168.
Email: deborah.blom@uvm.edu

Ronald G. Blom. NASA Jet Propulsion Laboratory, California Institute of Technology, Pasadena, CA 91109.
Email: ronald.blom@jpl.nasa.gov

Stefano Campana. Landscape Archaeology, Department of Archaeology and History of Arts, University of Siena at Grosseto Convento delle Clarisse, Via Vinzaglio 28, 58100 Grosseto, Italy.
Email: campana@unisi.it

Nicholas Clapp. Thomas Road Productions, 1551 S. Robertson Blvd. Los Angeles, CA 90035.
Email: nicholasclapp@gmail.com

Douglas C. Comer. Cultural Site Research and Management, Inc., 4303 N. Charles St., Baltimore, MD 21218.
Email: dcomer@culturalsite.com

Lawrence B. Conyers. Department of Anthropology, University of Denver, 2000 E. Asbury Street, Denver, CO 80208.
Email: lconyers@du.edu

Nicole Couture. Department of Anthropology, McGill University, Stephen Leacock Building, 855 Sherbrooke Street West, Montreal, QC, H3A 2T7, Canada.
Email: nicole.couture@mcgill.ca

Robert Crippen. NASA Jet Propulsion Laboratory, California Institute of Technology, Pasadena, CA 91109.
Email: robert.crippen@jpl.nasa.gov

Farouk El-Baz. Center for Remote Sensing, Boston University, 725 Commonwealth Avenue, Boston, MA 02215.
Email: farouk@crsa.bu.edu

Charles Elachi. NASA Jet Propulsion Laboratory, California Institute of Technology, Pasadena, CA 91109.
Email: charles.elachi@jpl.nasa.gov

Francisco Estrada-Belli. Department of Anthropology, Vanderbilt University, Box 6050, Nashville, TN 37235.
Email: Francisco.Estrada-Belli@vanderbilt.edu

Diane L. Evans. NASA Jet Propulsion Laboratory, California Institute of Technology, Pasadena, CA 91109.
Email: Diane.L.Evans@jpl.nasa.gov

Tom G. Farr. NASA Jet Propulsion Laboratory, California Institute of Technology, Pasadena, CA 91109.
Email: Tom.G.Farr@jpl.nasa.gov

William R. Fowler, Jr. Department of Anthropology, Vanderbilt University, Box 6050-B, Nashville, TN 37205.
Email: William.R.Fowler@vanderbilt.edu

Riccardo Francovich. Medieval Archaeology, Department of Archaeology and History of Arts, University of Siena, Via Roma 56, 53100 Siena, Italy.
Email: francovich@unisi.it

Tony Freeman. NASA Jet Propulsion Laboratory, California Institute of Technology, Pasadena, CA 91109.
Email: tony.freeman@jpl.nasa.gov

Thomas G. Garrison. Department of Anthropology, Harvard University, 11 Divinity Avenue, Cambridge, MA 02138.
Email: garrison@fas.harvard.edu

Dean Goodman. Geophysical Archaeometry Laboratory, 20014 Gypsy Lane, Woodland Hills, CA 91364.
Email: gal_usa_goodman@msn.com, www.GPR-SURVEY.com

George R. Hedges. Quinn Emanuel Urquhart Oliver & Hedges, LLP, 865 South Figueroa Street, 10th Floor, Los Angeles, CA 90017.
Email: georgehedges@quinnemanuel.com

Scott Hensley. NASA Jet Propulsion Laboratory, California Institute of Technology, Pasadena, CA 91109.
Email: scott.hensley@jpl.nasa.gov

Derrold W. Holcomb. Advanced Sensor Software, Leica Geosystems, GIS & Mapping Division, 2801 Buford Highway, Suite 300, Atlanta, GA 30329. Email: derrold.holcomb@gis.leica-geosystems.com

Burgess F. Howell. Universities Space Research Association, NASA Global Hydrology and Climate Center, Marshall Space Flight Center, Huntsville AL 35805. Email: burgess.howell@mfsc.nasa.gov

Carrie Hritz. Department of Near Eastern Languages and Civilizations, University of Chicago, Oriental Institute, Chicago, IL 60637.

Daniel E. Irwin. NASA Global Hydrology and Climate Center, Marshall Space Flight Center, Huntsville, AL 35805. Email: daniel.irwin@nasa.gov

Magaly Koch. Center for Remote Sensing, Boston University, 725 Commonwealth Avenue, Boston, MA 02215. Email: mkoch@bu.edu

Kenneth L. Kvamme. Department of Anthropology and Archeo-Imaging Laboratory, Old Main 330, University of Arkansas, Fayetteville, AR 72701. Email: kkvamme@uark.edu

Katherine L.R. McKee. Independent Researcher, Oxnard, CA 93036. Email: krmckee@zdelpia.met

David A. Mindell. Massachusetts Institute of Technology, Building E51-194A, Cambridge, MA 02138. E-mail: Mindell@mit.edu

Elizabeth Moore. School of Oriental and African Studies (SOAS), University of London, Thornhaugh Street, Russel Square London WC1H OXG United Kingdom. Email: em4@soas.ac.uk

Yasushi Nishimura. Cultural Heritage Protection Cooperation Office, Asia/Pacific Cultural Centre for UNESCO(ACCU), Nara-shi, 636-8113 Japan. Email: nyasushi@accu.or.jp

Agamemnon G. Pantel. Pantel, del Cueto & Associates Torrimar, #11 Calle Valencia Guaynabo, Puerto Rico 00966-3011. Email: pantel@prtc.net

Salvatore Piro. ITABC-CNR, Institute for Technologies Applied to Cultural Heritage, P.O. Box 10, 00016 Monterotondo Sc. (Roma) Italy. Email: salvatore.piro@itabc.cnr.it

Matthew D. Reynolds. Department of Anthropology & Center for Advanced Spatial Technologies, Old Main 330, University of Arkansas, Fayetteville, AR 72701. Email: mdr01@uark.edu

Benjamin F. Richason III. Department of Geography, St Cloud State University, St Cloud, MN 56301.
Email: bfrichason@stcloudstate.edu

Cordula A. Robinson. Center for Remote Sensing, Boston University, 725 commonwealth Avenue; Boston, MA 02215.
Email: cordula@crsa.bu.edu

A. C. Roosevelt. Department of Anthropology, University of Illinois at Chicago, 1007 West Harrison Street, Chicago, IL 60607.
Email: annaroosevelt@gmail.com

William Saturno. Department of Anthropology, University of New Hampshire, Huddleston Hall, 73 Main Street, Durham, NH 03824-3532.
Email: wsaturno@cisunix.unh.edu

Kent Schneider. USDA Forest Service Heritage Program, 1720 Peachtree Rd NW, Atlanta, GA 30309.
Email: kaschneider@fs.fed.us

Thomas L. Sever. NASA Global Hydrology and Climate Center, Marshall Space Flight Center, Huntsville, AL 35805.
Email: tom.sever@nasa.gov

Irina Lita Shingiray. Department of Archaeology, Boston University, Boston, MA 02215.
Email: iharris@bu.edu

Payson Sheets. Department of Anthropology, University of Colorado, Boulder, CO 80309-0233.
Email: sheetsp@colorado.edu

John H. Stubbs. World Monuments Fund, 949 Park Avenue, New York, NY 10028.
Email: jstubbs@wmf.org

Patrick Ryan Williams. Field Museum of Natural History, Department of Anthropology, 1400 Lake Shore Drive, Chicago, IL 60605.
Email: rwilliams@fieldmuseum.org

James Wiseman. Center for Archaeological Studies, Boston University, 675 Commonwealth Avenue, Boston, MA 02215.
Email: jimw@bu.edu

Juris Zarins. Department of Sociology and Anthropology, Southwest Missouri State University, Springfield, MO 65804.
Email: zarins@missouristate.edu

Ezra B. W. Zubrow. State University of New York, Department of Anthropology, P.O. Box 610005, Buffalo, NY 14260.
Email: zubrow@buffalo.edu

Foreword

Over the last few decades, a revolution took place in our ability to observe and "explore" our home planet using spaceborne and airborne remote sensing instruments. This revolution resulted from the new capability of observing on a large, regional and global, scale surface patterns and features, and from using multispectral instruments to observe surface signatures not visible to the human eye, and to penetrate below the surface cover using microwave radiation.

Over the last decade, this new technology is being brought to bear in the field of archaeology. As nicely and comprehensively described in this book, remote sensing techniques are bringing new powerful tools to help archaeologists in their quest of discovery and exploration. Even though this field is still in its infancy, the different articles in this book give us a glimpse of the things to come and the great potential of remote sensing in archaeology.

The editors have brought an excellent sampling of authors that illustrate how remote sensing techniques are being used in the real world of archeological exploration. A number of chapters illustrate how spaceborne and airborne remote sensing instruments are being used to decipher surface morphological features in arid (Egypt, Arabia), semi-arid (Greece, Ethiopia, Italy), as well as tropical regions (Costa Rica, Guatemala, Cambodia) to help in archeological and paleontological exploration. They illustrate the use of surface-cover penetration with radars, high-resolution multispectral imaging on a regional basis, as well as topographic signatures acquired with spaceborne and airborne sensors.

A number of articles also illustrate the innovative use of ground penetrating radar systems which capitalize on recent developments in signal processing and pattern recognition, as well as the use of visualization and classification techniques to help extract certain patterns associated with man-made structures from natural signatures. In addition, the emerging field of maritime archaeology is addressed where more precision navigation tools combined with visual and acoustic sensors are enabling new capabilities in mapping sea-bottom surfaces.

This book provides an excellent and diverse overview of the emerging capability of remote sensing archaeology and is a very valuable and important text for archaeologists in their quest to use advanced technology to help in their studies of exploration, and for remote sensing technologists and scientists by giving them a good understanding of the challenges that archaeologists find in their endeavors.

The editors are to be applauded for bringing together such an excellent collection of authors and articles to cover this important emerging field.

CHARLES ELACHI, Director
NASA Jet Propulsion Laboratory
Pasadena, California

Acknowledgments

Grants from the J. M. Kaplan Fund made possible the series of conferences on Remote Sensing in Archaeology at Boston University, as well as sessions and workshops at the Archaeological Institute of America (AIA) and the Society for American Archaeology (SAA), that are mentioned in the Introduction. The J. M. Kaplan Fund also helped cover some of the editorial costs of this volume. The National Aeronautics and Space Administration's (NASA) Jet Propulsion Laboratory (JPL) and Global Hydrology and Climate Center contributed to the sponsorship of these activities by supporting the participation of the NASA personnel. The National Research Council, Institute of Technologies Applied to Cultural Heritage (Rome, Italy) contributed to the 2001 conference by supporting expenses of participants from Italy. Marta Ostovich of the Center for Archaeological Studies at Boston University was particularly helpful to the coeditors as an editorial assistant. Ashley McIntosh and Rebecca Anderson, also of the Center, were helpful with proofreading. Several colleagues conscientiously provided peer reviews of the many manuscripts; Cordula Robinson, in the Center for Remote Sensing, was particularly helpful in this regard. We extend our sincere thanks to all the individuals and institutions for their assistance.

Contents

II. AERIAL PHOTOGRAPHY AND FRACTALS

III. GEOGRAPHIC INFORMATION SYSTEMS

IV. GEOPHYSICAL PROSPECTING AND ANALYTICAL PRESENTATIONS

V. MARITIME SETTING APPLICATIONS

VI. CULTURAL RESOURCES AND HERITAGE MANAGEMENT

Introduction

JAMES WISEMAN AND FAROUK EL-BAZ

Remote sensing is studying an object or a phenomenon from a distance. Starting nearly a century ago, archaeologists began to utilize the earliest tool of remote sensing, aerial photography, to gain a better understanding of the regional setting of ancient sites and to detect traces of actual sites (see, e.g., Deuel, 1969; Wilson, 1982). The field of spaceborne remote sensing for civilian uses, however, is much more recent: it began to develop in the 1960s with the advent of the Apollo era of lunar exploration by the National Aeronautics and Space Administration (NASA). At the time, it was necessary to image the moon in enough detail to allow the proper selection of landing sites for the Apollo astronauts (Kosofsky and El-Baz, 1970). In addition, it was required to employ sensing instruments from lunar orbit to allow the extrapolation of findings from the landing sites to larger areas of the moon's surface (Masursky et al., 1978).

Upon the termination of the Apollo program in 1972, NASA initiated remote sensing of the surface of the earth from orbit by the launch of the first Landsat. Multispectral images of the earth's features became available to the scientific community at large, and were soon applied in various fields of research and study (as illustrated in the thorough survey by Lillesand et al., 2004). During the past two decades in particular, archaeologists have used the various data obtained from earth-orbiting satellites to help resolve investigation problems or to discover hidden sites beneath soils and sands (El-Baz, 1990, 1997).

In 1984 NASA held the first-ever Conference on Remote Sensing in Archaeology at the Earth Resources Laboratory of the National Space Technology Laboratories, later renamed the John C. Stennis Space Center, in Bay St. Louis, Mississippi. This conference, funded by NASA, the National Science Foundation, and the National Geographic Society, brought together 22 professional archaeologists to learn about recent advances in the technology at NASA and to discuss possible future applications in archaeology (Sever and Wiseman, 1985). The editors of this volume organized a follow-up workshop in 1986, and international conferences in 1998 and 2001 at Boston University. The 1998 conference, "Remote Sensing in Archaeology from Spacecraft, Aircraft, on Land, and in the Deep Sea," was accompanied by colloquia

1

and workshops by several of the conference participants at the Annual Meetings of the Archaeological Institute of America (1997) and the Society for American Archaeology (1999). The 2001 conference, for which Maurizio Forte (Rome) and P. Ryan Williams, now at the Field Museum, Chicago, joined us as co-organizers, was an Italy-United States Workshop: "The Reconstruction of Archaeological Landscapes through Digital Technologies" (Forte et al., 2003).

Research results presented during these initiatives were well received by the international archaeological community. What is more, undergraduate and graduate courses in applications of remote sensing in archaeology were developed at Boston University beginning in the mid 1990s. It is because of this increasing interest in the topic that we decided to put together this volume for use by both students and researchers.

The volume is divided into six sections, grouping chapters dealing with similar topics related to: radar and satellite imagery, aerial photography and fractals, geographic information systems (GIS), geophysical prospecting at close-range, applications in a maritime setting, and cultural resources and heritage management. Many of the papers in one group, however, also touch on research applications considered in one or more of the other sections, and several deal also with issues of visualization.

RADAR AND SATELLITE IMAGES

The most recent development in the applications of remote sensing in archaeological investigations is the use of radar imagery. For this reason, the first contribution in the first section introduces the reader to the topic. Holcomb and Shingiray survey the various ways researchers have used radar images in a historical perspective. They describe the progression from the early use of single-band, black-and-white radar images to the utilization of recently acquired multi-frequency, multi-polarization radar data. They also address the complex nature of radar images and the importance of correcting their radiometric and geometric distortions through image processing.

In the next chapter El-Baz et al. discuss the use of radar data to establish the geoarchaeological setting of vast regions in the eastern Sahara of North Africa. Radar transmissions have the ability to penetrate dry, fine-grained sand to reveal the underlying topography. Thus, radar images obtained by the U.S. Space Shuttle as well as by Radarsat were used to unveil ancient river courses in otherwise flat and sand-covered plains. This discovery explained the preponderance of prehistoric artifacts in tracts of the Great Sahara in Egypt and adjacent areas in Libya and Sudan. The importance of this finding to the location of groundwater resources, which accumulated during wet climates in the past, is underlined.

Blom et al. used the Space Shuttle radar data to map a segment of a caravan route in the Sultanate of Oman. This research was carried out in the course of searching for Ubar, a legendary site of human habitation along the frankincense trade route from southern Arabia to its northern parts. Radar data and supporting Landsat images were used to locate an ancient fortress, serving also as a caravanserai, which must have collapsed into a sinkhole, perhaps as a result of extraction of much groundwater to irrigate the surrounding oasis.

The most recent radar data were obtained by the Shuttle Radar Topography Mission (SRTM), which employed Interferometric Synthetic Aperture Radar (InSAR). Unlike the imaging radar, InSAR used radar interferometry to produce a near-global elevation model. Evans and Farr describe the technology and the processing of the data. Such data, which can include centimeter-scale changes in topography, have proved to be particularly useful in the study of regional topography and in the assessment of both natural and human-induced hazards. The authors provide several examples of the application of the technology to assist field workers in establishing the local geographic setting of archaeological sites worldwide, and in monitoring cultural-heritage sites.

The next five chapters deal with the application of remotely sensed multispectral images to archaeological investigations, some in concert with radar data.

Comer and Blom discuss the use of eight protocols of SAR data, some in conjunction with multispectral data from other aerial and satellite instruments, which they developed in a study of archaeological sites on San Clemente Island, California. They successfully established statistically based site signatures, which they tested in a variety of ways. The protocols and the reflexive methodology (involving refining earlier results from one protocol on the basis of ground verification and/or the results of another protocol) are applicable to research in other parts of the world. The authors also highlight the implications of their research for cultural resource management, especially the use of SAR in the relatively inexpensive evaluation of archaeological potential of large land areas.

In the following chapter, Saturno et al. used high-resolution data to study ancient Maya landscape that is covered by high, thick vegetation. Images from IKONOS, Quickbird, and EO-1 satellites were utilized to identify the location of ancient settlements in northern Guatemala, which have long been veiled by a dense forest canopy. Micro-environmental characteristics of settlements built with limestone and making use of limestone plaster allowed their recognition as distinct from the surrounding land by changes in the characteristics, variety, and density of vegetation. Such a study, which is ongoing, can facilitate the process of locating archaeological sites in the heavily vegetated areas, as assisted by GPS measurements. The authors discuss in detail various techniques of pan sharpening, a data-fusion process that improves resolution of data, which has been of major utility in their ongoing research.

Similarly, Sheets and Sever used remotely sensed data within a jungle environment in research that led to a hypothesis about the origins of monumental sunken entrances to sites of the chiefdoms in Costa Rica. In the Arenal area, satellite images combined with aerial photographs and ground verification revealed straight, sunken entrances to cemeteries and villages dating back to 500 B.C. These entrances were the terminal points of footpaths, many kilometers long, that were inadvertently entrenched some 2 or 3 m below the level of the surrounding terrain through continued use by generations of people. Their entrenched nature allowed their mapping in the remotely sensed data. The authors argue that it was this culturally recognized entryway to a special place that was adopted later in chiefdom centers such as Cutris on the Atlantic coast.

Close-range remote sensing was applied by Moore et al. to define the archaeological context of the temples of Angkor, Cambodia. In addition to spaceborne radar data from the Space Shuttle, an Airborne Synthetic Aperture Radar (AIRSAR) was used in 1996 to image the area. The researchers discuss the technical data from both polarimetric and interferometric analysis. The importance of the region's hydrology was recognized in antiquity, so that water management became a major feature of ancient land use, which the radar could detect. A number of new sites were identified in radar imagery and later verified on the ground, especially circular, moated prehistoric sites. The authors also draw archaeological inferences regarding the transition from prehistoric times to the Khmer achievement of Hindu-Buddhist cities of Angkor of the 9th–13th centuries A.D.

AERIAL PHOTOGRAPHY AND FRACTALS

Zubrow, who in 1985 was the first scholar to apply fractals to archaeological investigations, uses fractals in this paper to investigate the organizing principles of cultural settlements and their landscapes. The methodology involved measuring fractals of all natural and cultural phenomena in aerial photographs of six selected pueblos and their countryside of the Colorado Plateau and the Rio Grande Valley, and then determining the degree of similarity of patterning between each settlement and its countryside. The author considers this effort a prolegomenon to the use of fractals to determine how prehistoric settlements fit their landscapes.

GEOGRAPHIC INFORMATION SYSTEMS

This section includes three papers that highlight the valuable contributions of geographic information systems (GIS) and their methodologies. The first, by Estrada-Belli and Koch, shows how combining remote sensing and GIS can help reveal an archaeological setting in a dense tropical forest. The case in question is the Holmul region of northeastern Guatemala. Landforms and their vegetation cover are first identified by Landsat, IKONOS and IF-SAR data. The resulting geomorphological map is used to identify settlement sites using GIS modeling techniques. The case serves as a model for future studies in similar environments.

Similarly, Richardson and Hritz utilized multiple sets of remote sensing images with GIS techniques to study a maze of ancient settlements and canals in central Mesopotamia (Iraq). Multispectral images of Landsat, ASTER, and SPOT were used in conjunction with radar data from Radarsat and black-and-white photographs from Corona. The data were used to identify features and detect ancient sites in the regions of Nippur and Lake Dalmaj. Extensive GIS databases were created from the variety of remotely sensed data to clarify the nature and relationships of the settlements and canals.

Another instance of the use of multiple data sources in archaeo-logical work is provided by Campana and Francovich in their study of Late Roman and Medieval landscapes in southern Tuscany (Italy). They discuss the results of combined use of IKONOS and Quickbird data along with microdigital terrain modeling using differential GPS and digital photogram-metry. This procedure was coupled with magnetic-sensing data to study pervious settlement patterns in the area and their evolution through time. They stress the importance of mobile instrumentation in the conduct of field work.

GEOPHYSICAL PROSPECTING AND ANALYTICAL PRESENTATIONS

Remote sensing methodologies are not limited to spaceborne or airborne sensors. Instruments that can be used at or close to the surface of the earth to probe its shallow interior may provide valuable data to archaeologists that cannot be obtained in any other way. Such instruments include the ground-penetrating radar (GPR) and various magnetometers and gravimeters in addition to resistivity meters. On land, researchers have also used seismometers, and beneath water bodies they used sonar instruments. All such sensors have an ability to detect and record specific characteristics of the subsurface in ways that are meaningful to archaeologists.

The most commonly used instrument of geophysical prospection in archaeological investigations is the ground-penetrating radar (GPR). This type of instrument sends radar waves toward and into the ground, and records the returned echoes. From these returns, archaeologists can generate a picture in three-dimensions of the subsurface without the use of a shovel. Conyers provides an overview of GPR technology and methods of processing the reflection profiles to create three-dimensional images and amplitude slice-maps of buried features. He also provides illustrative examples of the use of GPR on archaeological projects at historical sites in Albany, New York, and at sites in the California Sierra Nevada Mountains; Denver, Colorado; and in Petra, Jordan.

Kvamme discusses the use of multiple geophysical sensors and computer methods of handling spatial data. Various instruments might yield data that would confirm, complement, or add to information from others. Comparison of such data side-by-side, however, limited the potential correlations. GIS and computer-based methods allow the superposition of such data for complete correlation and "data fusion." He illustrates the utility of this approach with data from the commercial center of Army City in central Kansas.

Goodman et al. applied advances in visualization and processing of GPR data to a variety of sites dating from the Roman era in Italy (Forum Novum in the Sabine Hills) to a 16th-century church in Puerto Rico. Using 3-D iso-surface rendering and visualization in all cases allowed the informative and accurate depiction of the sites and provided an informed guide to determining where to excavate. Complete topographic corrections to GPR datasets were applied in the study of Japanese burial mounds. In addition, the authors discuss the use of GPS navigation in conducting GPR surveys at Native American cemeteries.

Fowler et al. employed magnetic gradiometry, electrical conductivity, and magnetic susceptibility to study the archaeological site of Ciudad Vieja, the ruins of the first villa of San Salvador, which was the first Spanish-conquest city in El Salvador. The surveys revealed locations of probable buried stone-wall foundations. These and other stone structures were easily detectable by magnetic susceptibility because of the high iron-oxide content of the volcanic stones. Information gained from the geophysical surveys successfully guided the archaeologists in their selection of areas for excavation.

At Tiwanaku in the Andean Altiplano, Williams et al. apply geophysical techniques to resolve questions of characteristics of the polity. They used several techniques, including magnetometry, electrical resistivity, and GPR to study the development of the urban center at the site during the Middle Horizon (approximately 600-100 A.D.). They outline the methodology of recreating the urban structure of the 6-km^2 capital city.

In her article reviewing the use of geophysical survey on the lower Amazon during more than two decades, Anna Roosevelt discusses research methodology, including the fundamental techniques of geophysical remote sensing, and stresses the importance of combining traditional field surveys and excavations with geophysical methods. She provides several case studies that have led to major changes in the scholarly understanding of the nature and evolution of human occupation of the tropical rainforest. Her discussion clearly illustrates the potential uses and problems of geophysical instruments in assisting researchers to understand better archaeological sites in such an environment.

MARITIME SETTING APPLICATIONS

Geophysical instruments are just as useful underwater as underground. They are particularly significant because: (1) coastlines worldwide change in space and time, depending on changes to ocean currents, incoming sediments, and the rise and fall of global seal levels; and (2) much trade was conducted in the past through waterways, and because of accidents, inclement weather, or lack of knowledge, numerous ships lie beneath water bodies worldwide. Much can be learned from the ships and their construction as well as from the artifacts they contain. Thus, instruments that help explore beneath the surface of water are of great benefit to archaeologists who seek information on coastal changes or of sunken vessel sites. Both articles in this section discuss the issues and technologies involved, especially in the deep seas.

Ballard confirms the emergence of archaeological oceanography as a new field of research. He reviews the history of the field of research in recent years, including the evolution of remotely operated vehicle systems (ROVs). He also discusses the challenges to future endeavors that require gaining access to sites and artifacts at great depths.

Mindell describes methods of surveying and documenting archaeological sites in marine environments with great precision. Particularly during the last decade, precision navigation enabled closed-loop control of underwater vehicles for high data-density, multiple-pass surveys. Mindell also discusses the use of acoustical and optical data to produce cm-scale, 3-D maps of archaeological sites.

CULTURAL RESOURCES AND HERITAGE MANAGEMENT

Remotely sensed data can serve as a baseline for documenting human activities at heritage sites. Stubbs and McKee provide an argument for correlation of these data in a GIS for use by resource managers, historians, planners, and engineers to catalog and assess organizational and structural patterns of heritage sites. They also foresee the use of these methods in creating a modern urban dynamic model that will enable conservation and proper management of these non-renewable cultural resources. In describing the increasing use of remote sensing for cultural-heritage purposes, they present case studies from many parts of the world, with special emphasis on developments at the World Heritage Site of Angkor, Cambodia.

These 21 chapters cover a wide range of methods of applying newly developed methodologies to the understanding as well as the preservation of archaeological sites worldwide. The arguments are supported by case studies of topics from around the world. There is no doubt that many new initiatives will result in the availability of new sensors in great variety and with more capabilities in the future. It is with all this in mind that we have prepared this volume to fill the need for documenting the history of remote sensing in archaeological investigations, but in anticipation of new and even more valuable technologies and procedures for additional applications.

It is our hope that this book will be used in classrooms of both undergraduate and graduate students. It is also our intent to encourage professional archaeologists to familiarize themselves with the various uses of remotely sensed data based on the case studies presented in the articles. Employing information gathered by remote sensors can sometimes convert the planning of fieldwork into a high-yield activity: that is, the information can sometimes accurately guide the researcher in a quest for information, patterns, or remains of the past. Such data provide a reliable and fast way to better understand the setting of archaeological sites in all types of environments. Data from remote sensors should be utilized more extensively, and more often, than the worldwide archaeological community uses them today. It is our hope that this book will assist in reaching that objective.

JAMES WISEMAN, Director
Center for Archaeological Studies
Boston University
Boston, MA

FAROUK EL-BAZ, Director
Center for Remote Sensing
Boston University
Boston, MA

REFERENCES

Deuel, L., 1969, *Flights into Yesterday: the Story of Aerial Archaeology*. St. Martin's Press, New York.

El-Baz, F. (1990) Remote sensing in archaeology. In *1991 Yearbook of Science and the Future*, pp. 144–161. Encyclopedia Britannica, Chicago IL.

El-Baz, F. (1997) Space age archaeology. *Scientific American* 277(2):40–45.

Forte, M., Williams, P.R., Wiseman, J., and El-Baz, F., 2003, *The Reconstruction of Archaeological Landscapes through Digital Technologies*. BAR International Series 1151, Oxford.

Kosofsky, L.J., and El-Baz, F., 1970, *The Moon as Viewed by Lunar Orbiter*. National Aeronautics and Space Administration (NASA). Washington, D.C., NASA SP-200.

Lillesand, T. M., Kiefer, R. W., and Chipman, J.W., 2004, *Remote Sensing and Image Interpretation*. Fifth Edition. John Wiley, New York.

Masursky, H., Colton, G.W., and El-Baz, F., 1978, *Apollo Over the Moon: A View from Orbit*. National Aeronautics and Space Administration (NASA). Washington, D.C., NASA SP-362.

Sever, T. L., and Wiseman, J., 1985, *Remote Sensing in Archaeology: Potential for the Future. Report on a Conference, March 1-2, 1984*. Earth Resources Laboratory, NSTL, Mississippi.

Wilson, D. R., 1982, *Air Photo Interpretation for Archaeologists*. B. T. Batsford, London.

Section I

Radar and Satellite Images

Chapter 1

Imaging Radar
in Archaeological Investigations:
An Image Processing Perspective

DERROLD W. HOLCOMB AND IRINA LITA SHINGIRAY

Abstract: This paper provides a survey of the various ways researchers have used radar imagery in their particular studies. The early studies were usually limited to single-band (gray-scale) imagery because little else was available in the public domain. In the past few years, multi-polar, multi-frequency radar imagery has become available on an experimental basis (SIR-C and AirSAR). This development has increased the information content of the available data, but extracting and understanding this information content is not simple. In addition, radar images have significant radiometric and geometric distortions that must be addressed, but often are not.

1. INTRODUCTION

The applications of imaging radar for archaeological investigations are manifold. Radar images are used for archaeological prospecting in regional surveys to detect cultural and natural features and sites, for paleo-landscape reconstruction and ecosystem studies, for cultural heritage monitoring, and for navigation in unknown terrain. This relatively recent remote sensing technique, while still in its experimental stage, has accounted for a number of important archaeological discoveries, and holds great potential for impending archaeological investigations. The unique ability to image through cloud cover, haze, light atmospheric precipitation, and smoke make the radar system an "all weather" sensor (Verstappen, 1977; Lillesand and Kiefer, 2000). This capability of radar sensors is valuable for mapping and studying tropical and subtropical territories, where cloud cover poses a major impediment to optical imaging. In arid environments, a radar beam is able to penetrate up to two meters of dry fine-grained sediments and image rough subsurface features (Elachi et al., 1984; El-Baz, 1997; Robinson, 1998; Abdelsalam et al., 2000). Imaging radar has been critical in detecting previously unknown paleo-drainage channels and related Paleolithic and Neolithic sites in the eastern Sahara Desert (McCauley et al., 1982 and 1986; McHugh et al., 1988 and 1989).

Archaeologists have made use of radar images to reveal differences in texture, roughness, moisture content, topography, and geometry of features and surfaces; to investigate and monitor the ruins of Angkor in Cambodia (Moore, et al., 2004); to identify and study extensive agricultural and drainage canals of Maya Lowlands (Adams, 1980; Pope and Dahlin, 1989; Sever, 1998; Sever and Irwin, 2003); to detect locations of Paleolithic sites in the Gobi Desert of Mongolia (Holcomb, 2001); to identify sites associated with Pleistocene/Holocene human occupation in the deserts of Egypt and Sudan (McHugh et al., 1989); and to help in locating the lost city of Ubar (Blom et al., 1997). As a result of radar data analysis, archaeological sites were detected directly or predicted indirectly in locales where human occupation and activity were likely to occur. These included areas with available water sources, access to arable land, traces of remaining secondary vegetation, sources of raw materials, and nexus of transportation and communication networks.

A sensitivity of radar systems to various directional features—such as roads, trails, canals, fences, and valleys—makes radar images highly suitable for navigation purposes during archaeological field surveys. This method has proved to be especially useful in remote, featureless, and inaccessible terrain (Holcomb, 2001), areas with a dynamic eolian landscape prone to rapid topographic change, and in densely vegetated terrains (Adams, 1982).

The applications of imaging radar in archaeology are still limited by a number of factors. First, low-cost, high-quality radar images, such as the SIR-C experimental data, have a limited spatial coverage and are not available for many areas of archaeological interest. At the same time, the relatively high-cost of commercial radar data can limit their use to large-budget archaeological projects. In addition, the spatial resolution of spaceborne radar images that are available in the public domain is rarely sufficient for the direct detection of archaeological features.

Moreover, since a radar image differs fundamentally from an image acquired with optical remote sensors, analysis of radar data requires specialized training and image processing software in order to achieve adequate interpretation results. The radar data limitations can be overcome if an analyst employs a range of methods of systematic radar data interpretation, multi-sensor images, and image processing techniques in order to detect, enhance, extract, and analyze archaeological information from the radar image.

2. RADAR IMAGING

2.1. Distinctive Properties of Radar

Radar is an active microwave sensing system, which means that it supplies its own source of illumination energy, as opposed to passive microwave radiometer or visible-infrared multispectral sensors, which sense energy that is naturally emitted/reflected from imaged terrain (Lillesand and Kiefer, 2000). Precisely because it is an active sensor, the radar can image through cloud cover (which is transparent to microwave energy beams), precipitation, and even arid sand-sheet under certain conditions.

Radar images present a distorted appearance because of side-looking geometry, which is an outcome of imaging with SAR (Synthetic Aperture Radar) systems. SAR, as opposed to real aperture radar, increases the effective resolution of the image by utilizing the motion of the sensor to simulate an increase in the size of the antenna (Robinson, 1998). The resulting image provides a distinctive view of the environment, one that is far removed from the view produced by optical (sunlight) sensors (Lillesand and Kiefer, 2000).

One distinctive advantage of radar images is sensitivity to topography (especially in hilly areas where the slopes are 20° or more) that emphasizes the geometrical properties of the terrain through patterns and contrasts in image tone and texture (Ford et al., 1983). For archaeological applications these properties can be crucial for direct identification of man-made features such as fortresses, mounds, earthworks, walls, causeways, or bridges, which often possess steep slopes and right-angle or round geometry. These features are often hard to distinguish on conventional optical images, such as LANDSAT data, where indications of vertical geometry and texture are not present or emphasized.

A radar image interpreter has to keep in mind that the pronounced contrast of features, deep shadows, and image foreshortening can create psychological relief illusion, relief distortion (Verstappen, 1977), or topographic inversion (Sabins, 1997). In many cases, however, the relief is indeed displaced on radar images; the result of imaging with a small incidence angle (Ford et al., 1989). Imaging of high relief with a small incidence angle produces a layover effect, which results in hilltops overlaying hill bases in the radar image (Ford et al., 1989). This phenomenon is the extreme case of foreshortening. Layover occurs because radar is a ranging (distancing) sensor. In steep terrain, it is possible that areas of high elevation are closer to the sensor than adjoining low areas. The radar return signal (backscatter) from the higher elevations reaches the sensor sooner and is thus recorded as being geographically closer. Besides this geometric artifact, which can be corrected by orthorectification (Robinson, 1998), layover also produces a radiometric artifact, the summing of the return signal from areas of differing elevation, for which correction is not possible.

Radar data taken with larger incidence angles and lower frequencies is used for detecting the texture and roughness of various types of terrain, vegetation cover, lithological and geological composition, and other information used in regional archaeology studies. This type of mapping of geology, hydrology, and vegetation is possible because the radar signal responds to changes in the micro-geometry of the surfaces imaged, resulting in differences in the amounts of radar energy scattered and returned back to the sensor.

Radar waves are very responsive to feature orientation, especially when the feature is perpendicular to the radar beam—the direction of illumination. Because of this responsiveness, such topographically flat or near-flat features such as roads, trails, canals, fences, and pipes can be easily detected, even when they are somewhat obscured by vegetation or dry sand. Moreover, radar images are superb for mapping of geological structures (Roth and Elachi, 1975), which, in turn, is used to predict groundwater aquifers and locate specific raw materials that have been exploited by humans at different time periods (Wendorf et al., 1987; McHugh et al., 1988; El-Baz, 1998; Abdelsalam et al., 2000).

Another property of a radar sensor, which is potentially important for archaeological detection, is a strong response of the radar beam to objects composed of conductive material with high dielectric properties. These include, for instance, objects of metal and igneous rock, salt surfaces, and soils and vegetation with high moisture content. Thus a wet ground surface will scatter a radar signal more strongly than the same surface when it is dry (Fung and Ulaby, 1983). It should be noted that this scatter results in a loss of signal available for subsurface penetration and for the reflection of surface geometry. The ability to determine soil and plant moisture is important for wetland studies, whereas mapping of salt-flats is used for studies of arid environments. Such information, in turn, is useful for regional archaeological studies, which attempt to predict optimal areas for human occupation, based on indirect detection of economic and social activities of human groups in the past.

Radar images are better suited for the indirect detection of archaeological features, as opposed to direct detection, because of the (relatively) low spatial resolution available from spaceborne sensors. A standard pixel size for such images is 12–25 meters, a resolution too coarse for adequate imaging of many archaeological sites. In some cases, features can be detected on the sub-pixel level, such as roads, railroads, pipelines, fences, bridges, cars, roofs, etc. This kind of detection occurs when the geometry and target-material properties result in a strong return of the radar signal, the high value of which dominates the entire pixel on the digital image (Holcomb, 1998a, 1998b). The use of this effect for direct feature detection in a radar image can be limited or precluded by speckle noise (Dainty, 1980), a radiometric artifact inherent to radar data. Speckle occurs because the SAR image pixels are made up of an incoherent sum of the returns from multiple scatterers within each pixel (Ford et al., 1989). This can limit the discrimination and interpretation of objects on radar images. In response to this problem, a suite of speckle reduction techniques and algorithms has been developed (Lee et al., 1994). There is always a tradeoff, however, between speckle noise reduction and a loss of resolution and detail that can result in the elimination of important data from the radar image (Holcomb, 1998a).

The distinctive character of radar images, together with the imperfections of the human eye-brain system and limits of visual discrimination capacities, often make intuitive interpretation of a radar image unsatisfactory (Verstappen, 1977). For instance, the human eye is adept at connecting discontinuous and often unrelated features (Holcomb, 1998b), which can account for various errors when interpreting radar images. Therefore, the unique capabilities of radar imaging can sometimes only be appreciated in comparison or contrast with other types of data such as panchromatic, multispectral, etc. (Ford et al., 1983).

2.2. Orthocorrection

In the past decade the need for orthocorrection, using a sufficiently accurate digital elevation model (DEM), has become widely recognized in the remote sensing community. While certainly appropriate to pixel-scale georeferencing of optical images, this correction is a near requirement when working with radar images. The reasons for the distortions in radar images are related to the extreme off-nadir viewing and the ranging nature of the sensor; these are discussed in detail in standard radar texts. What is important to recognize is the potential scale of the geometric distortion. It has been demonstrated that, in the incidence-angle range of 22–47 degrees, the positional offset of a feature in a radar image varies from 1 to 2.4 meters for every meter of elevation inaccuracy (Holcomb, 1999b). Thus, in areas of varying topography, the relative position of features (pixels) can be extremely displaced. The magnitude of this distortion is a function of incidence angle, and selection of the most optimum available incidence angle can minimize image distortion thereby increasing interpretability (Singhroy and St. Jean, 1999). In general, larger incidence angles produce less distortion.

The interplay between coregistration (discussed below) and orthocorrection can greatly confound interpretation of multilayer radar datasets. In recent work with airborne multifrequency datasets, Holcomb noted feature displacements between channels. The interband displacement within this orthocorrected radar dataset was as large as 21 meters in longitude and 7 meters in latitude. Clearly such a large interband misregistration would preclude frequency-related comparisons or feature location on the ground.

After an evaluation of the data history and processing sequences it was determined that the initial interchannel mismatch had been much smaller. Because of the extreme topography of the study area, however, the initially small misregistration was greatly amplified during orthocorrection. This example also highlights the importance of precise image-DEM registration to support later fieldwork.

2.3. Radar DEMs

Digital elevation models (DEMs) are important to many remote sensing studies and are, arguably, the base layer for any geographical information system. Their role in orthocorrection is discussed above. DEMS can be generated from radar images by two very different techniques: stereo intersection and interferometry. The stereo intersection technique is identical to that used with optical images (e.g., SPOT stereo-pairs). Because of speckle noise and topographic layover, a radar stereo-pair can never produce as accurate a DEM as a comparable optical stereo-pair. (Stereo radar does have a role to play, however, in areas where cloud or fog precludes the use of optical sensors.) More interesting is the use of radar interferometric image-pairs to create a DEM. This technique and its limitations are discussed elsewhere (Madsen and Zebker, 1998). A DEM of medium resolution, covering much of the world, has been derived by interferometry using a radar sensor system mounted in the U.S. Space Shuttle (SRTM Mission). This DEM could provide an acceptable base layer for orthocorrection. What is significant for archaeology is the very high elevation precision that is obtainable using interferometry, especially from an airborne system. (Note: Recent advances in lidar imaging provide a second source of high precision DEMs).

Once a high-precision DEM is available in the GIS, enhancement operations like shaded relief and image draping can be used to powerful effect in visual interpretation of the full dataset. By amplifying the elevation axis and artificially illuminating the DEM with a low "sun" angle, termed artificial shaded relief, subtle variations in elevation can be enhanced (Note that this shaded-relief technique can be used to emphasize detail in any dataset, such as a SPOT Panchromatic image, not just a DEM). Sever and Irwin (2003) make use of a high-precision DEM generated by the STAR-3i system to develop 3-D visualization of their study area in the Peten, Guatemala. Using Virtual GIS software (Leica, 2003) they are able to "fly" through their dataset looking for areas of interest (Figure 1). In their case, they were able to locate a number of elevated sites, termed bajos. Over 70 of these sites were visited in the field and all contained archaeological materials. Another approach is to drape a multi-band image on top of the (artificial) shaded-relief DEM. The resultant draped image conveys elevation, thematic content, and the correlation between the two in one view. This technique was used by Moore and colleagues in their study of Angkor (below).

Figure 1. Star3-I interferometric DEM shown in shaded relief with exaggerated Z (elevation) axis. This technique makes it possible to highlight small changes in elevation. In many geographic areas, including floodplains, forests, and deserts, man has chosen these slightly elevated spots to occupy. [Image courtesy of Tom Sever (Sever and Irwin, 2003)].

3. EARLY ARCHAEOLOGICAL APPLICATIONS OF IMAGING RADAR

3.1. Wetlands

Early scientific emphasis of radar imaging had been mainly on geological, hydrological, and vegetation mapping (Lillesand and Kiefer, 2000). Nevertheless, when the first experimental images became available to researchers, the advantages of using this technology for archaeological investigations were immediately evident. These early applications of imaging radar resulted in a number of important archaeological discoveries, which came initially as a surprise, and opened new vistas for archaeological remote sensing. Among the first such studies was the investigation of the Maya Lowlands of the Yucatán Peninsula (modern day Mexico, Belize, and Guatemala) by Mesoamerican archaeologists led by Adams (1980; 1982; Adams et al., 1981). Adams used radar images to survey the wetlands and detected a large network of agricultural and drainage canals, evidence of intensive wetland cultivation by Maya civilization, which dominated that region from about the 4th to the 10th century A. D. Previously, evidence for the existence of these canals was obtained with optical aerial photography (Siemens and Puleston, 1972) and some fieldwork, limited by the extremely dense vegetation cover of the Yucatan wetlands. This initial evidence, however, was called into question by the authors themselves, including Puleston, who argued that the extant grid pattern they had imaged was a result of natural processes, not human actions (see Adams et al., 1981).

Adams and his team addressed this issue by carrying out a large-scale survey employing radar (SAR) images; they detected a vast extension of such canal networks throughout the entire region of the Maya Lowlands. The network of canals and raised fields, called *chinampas*, showed up as gray line patterns created by slight differences in the elevation of vegetation (Adams et al., 1981). As a result of the side-looking geometry of the radar sensor, the difference between the taller trees growing between canals on raised fields and the smaller trees or absence of them in the canals was emphasized. Archaeological fieldwork in accessible areas of the wetlands confirmed the man-made nature of these extensive canals.

In addition to these prominent canal systems, the radar successfully delineated the edges of extensive paved surfaces. Two types of signatures seemed to correlate with archaeological sites, which appeared as large irregular spots of light (resulting from reflections from the casings of large buildings) and as distinct conical shadows cast by large mounds (Adams et al., 1981). Direct detection of archaeological features was limited by the low spatial and radiometric resolution of the radar images used for this project.

This discovery helped to resolve a dilemma in the field of Mayan archaeology. The evidence for a large Mayan population was inconsistent with the assumption that their economic subsistence was based on an ineffective slash-and-burn agricultural system (Adams, 1980). When archaeologists detected the *chinampas* network on the radar images, they were able to show that the use of this sophisticated agricultural subsistence pattern required considerable management and centralized control provided by a political organization even more complex than was previously thought (Adams, 1980). Consequently, the curious correlation between the most successful Maya urban centers and their proximity to the largest swamps was also elucidated, supporting the hypothesis of the usefulness of the wetlands for agricultural production (Adams, 1980).

An attempt to contest this hypothesis about the extensive canal system led another team of researchers to investigate Maya agricultural activities by means of environmental studies (Pope and Dahlin, 1989 and 1993). By using a combination of SAR (SEASAT) and optical (Landsat TM) images, they examined the distribution of canal systems, and their relation to wetland hydrology and vegetation. They also detected and mapped vegetation patterns and surface water, and identified five types of wetlands. Only three types of swamps proved suitable for sustaining agriculture, and only these perennial types, according to these researchers, supported the Maya canal systems. They also noted that several important Maya urban centers were associated not with perennial, but seasonal swamps, which could not have been used to construct *chinampas* or grow crops. Therefore, the authors proposed that the noted urban centers were situated near the swamps for reasons unrelated to agricultural practice, but rather in order to extract plentiful natural resources from the wetland ecosystem, for which canals also could have been utilized. In spite of the ongoing debate regarding this issue, the use of radar images in detecting these canal systems has been crucial in revealing the Mayans' sophisticated use of wetlands and suggesting a complex political organization.

3.2. Deserts

Concurrently, on the other side of the globe, in the extremely dry environment of the eastern Sahara Desert, experimental Shuttle Imaging Radar (SIR-A) produced subsurface images of previously unknown valleys and paleo-channels beneath the sand-sheet of Egypt and the Sudan (McCauley et al., 1982). These features, called radar rivers, are not visible on optical remote sensing images or in the field. They were revealed by radar imaging because of its ability to penetrate through a layer of dry sand, and map the structure underneath it. The radar sensor mapped the dendritic and braided dark channels formed by past fluvial action at a depth of up to two meters or more (Elachi et al., 1984:383). The radar rivers virtually absorbed the radar signal, because they are filled with fine-grained alluvial sand and silt, returning an energy value of very low intensity. Therefore the channels appeared dark and stood out on the radar image, in contrast with the brighter intensity return from the bedrock and coarser material of the surroundings (Schaber et al., 1986).

The presence of paleo-drainage systems buried under the eolian sand deposits revealed the existence of multiple playas and oases, which were periodically inhabited by prehistoric humans during the pluvial phases corresponding to the Paleolithic and Neolithic Periods (McCauley et al., 1986; Wendorf et al., 1987; McHugh et al., 1988 and 1989). The consequent fieldwork in the eastern Sahara resulted in finds of multiple Acheulean, Mousterian, Aterian, and Neolithic artifacts and sites "meaningfully located" in relation to the paleo-riverbeds and their terraces (McCauley et al., 1982).

These new finds have led to a fundamental reassessment of the prehistoric archaeology of North Africa (McHugh et al., 1988 and 1989). Before this discovery, there was a lack of geoarchaeological explanation as to why multiple Paleolithic tools were found *in situ* in the middle of the desert, far from any playa paleo-ecosystems (McCauley et al., 1986). The discovery of the radar rivers showed that Acheulean handaxes and other tools (dated to about 0.3–0.5 million years ago) were associated with the riverine environments which previously existed in the region and which, with the advent of hyperaridity, became buried by eolian sand (McHugh et al., 1988). During the next pluvial (50,000 BP) following the arid period, the pebbly banks of the radar rivers were occupied once again—this time by the Middle Paleolithic cultures, which left their Mousterian and Aterian tools.

The latter period was followed by a hiatus in human occupation, until the environmental conditions of the desert changed to those of subhumid and semiarid savannah. Then, beginning at about 9,000 BP, Neolithic inhabitants left ample remains of their material culture not only along the shores of the radar rivers, but also in connection with the inter-dunal basins and uplands (McCauley et al., 1986; Wendorf et al., 1987; McHugh et al., 1988 and 1989).

In conjunction with the SIR-A experiment and the discovery of radar rivers in Egypt, several groups of archaeologists undertook a survey in that region to investigate archaeological site distribution in relation to the buried channels, to determine the age of the channels, and to evaluate their potential as reservoirs of Pleistocene/Holocene rainfall that could have been exploited as groundwater resources (Wendorf et al., 1987; McHugh et al., 1988 and 1989). The researchers carried out a systematic archaeological survey of selected parts of the channels and evaluated their lithostratigraphy. As a result of the survey, hundreds of archaeological sites were found, the majority of which were identified as belonging to the Middle and Late Neolithic Periods. The archaeologists concluded that the radar rivers did not contain water during the Holocene pluvial period (as they did during the Pleistocene), but water was probably trapped in the bedrock and inter-dunal basins in the vicinity of the channels during or after the rains (Wendorf et al., 1987). Eventually, it was postulated that the water that drained from the radar river interfluves ran off or seeped through sand and rock, and collected in the radar river valleys—this attracted humans, as the large number of archaeological sites testifies (McHugh et al., 1988 and 1989). After the Neolithic Period, however, severely dry conditions prevailed again, and the alluvium of the radar rivers was only sporadically used for caravan traffic, which left behind negligible traces of their passage (McCauley et al., 1986).

The discovery of the radar rivers opened a new chapter in paleo-environmental and archaeological studies of arid regions using imaging radar data. Radar images (from SIR-A and B) also revealed subsurface features in the Mojave Desert in California (Blom et al., 1984), in the Al Labbah Plateau in Saudi Arabia (Berlin et al., 1986), and in the Taklamakan (Holcomb, 1992) and Badain Jaran (Walker et al., 1987) Deserts in China. As a result of these studies, researchers evaluated the main parameters of a radar system which allows radar pulse to penetrate through the desert-sand overburden, and concluded that a wavelength of 23 cm(L-band) falls within the narrow estimated optimum range of 1–20 GHz (Schaber et al., 1986). The HH (horizontal transmit-horizontal receive) polarization and a 47 degree incident angle were also found to be favorable. For effective imaging, the desert surface must be smooth, very dry sand with low clay content in order for the radar beam to be able to penetrate, and the subsurface features must possess a radar signature different from that of the surface to contrast with the latter (Schaber et al., 1986; Robinson, 1998).

These early radar images were limited to one band (gray-scale) obtained with a single wavelength and polarization. In order to extract a maximum of information from this data, researchers developed a range of computer image processing techniques to enhance the features of interest and improve radar image analysis.

3.3. The Lost Cities

The discovery of sand-buried features in SIR-A and SIR-B images inspired researchers to apply these data to detect historical archaeological sites in arid deserts. Information about many such sites existed in written sources, but their locations had not been detected or confirmed by archaeological fieldwork investigations, which were limited by the severe environmental conditions of the hyperarid deserts. It was hoped that radar images could help to locate these monuments.

One of these sites was the Arabian "lost city of Ubar", a desert caravansary which once supported the lucrative transcontinental trade in frankincense from 2000 BC to about 300 AD. According to the Koran, the ancient city was destroyed by divine will because of the wickedness and arrogance of its residents (Blom et al., 1997). Ubar was described as a multi-columned fortress located amidst a beautiful oasis, with a large central well—all of which was destroyed as a result of a single catastrophic event and subsequently swallowed by the sands of the Rub al-Khali, the Empty Quarter (Williams, 1992). A team of researchers (Blom, 1992; Blom et al., 1997), compiled a corpus of geographic information including historical maps and remotely sensed images—optical (LANDSAT TM, SPOT, LFC) and imaging radar (SIR-A and B)—and attempted to locate the lost city in the modern country of Oman. Although the radar data revealed prominent fluvial channels (evidence of past pluvial conditions) and some surface tracks incised into the desert rock, the lack of regional coverage of the area under investigation prevented a systematic detection of the caravan trails (Blom et al., 1997; Holcomb, 1998b). In order to overcome this problem, analysis of the optical images was undertaken in order to complement the radar data.

All images in this project were digitally processed using contrast and edge enhancement, spatial filtering, ratioing of bands, and other standard procedures (Blom et al., 1997; Gillespie, 1980; Moik, 1980). Moreover, the images were subjected to more unusual "directed band ratioing" (Crippen et al., 1988), "four component" processing techniques (Crippen, 1997), and merging images from different sensors, which were crucial for detecting and mapping even small desert trails (Blom et al., 1997). Thus, digital image processing and analysis helped to eliminate large portions of the desert and focus concentration on the areas where major tracks converged, revealing the best possible candidate for the lost Ubar—the modern-day village of Shisr. Subsequent fieldwork in Oman—which consisted of an investigation of the old ruins of Shisr and three seasons of archaeological excavations— indicated that the lost city of Ubar had likely been found. The researchers showed that the advantageous location and the water-well were central to the thriving existence and functioning of Ubar as a desert caravansary (Blom et al., 1997). Therefore, after the fortress collapsed in the sinkhole (around the time specified in the written sources), and the well was buried under the rubble or sand, the city ceased to exist until a new well was dug—perhaps many years later, when the legend of Ubar had been erased from the folk memory.

Another study of SIR data emphasized the importance of digital processing for the interpretation of radar images. Holcomb (1992; 1996) used SIR-A data of the Taklamakan Desert to find places that had the potential to contain archaeological sites along the famous Silk Route, which periodically connected Eastern and Western civilizations through trade and cultural exchange. Chinese archaeologists have excavated a number of sites in the area of Lop Nor in the Taklamakan Desert, unearthing burial sites, watch towers, habitation sites—all of which were correlated with nearby watercourses (Wang, 1985; Holcomb, 1992). Holcomb used radar images in an attempt to locate ancient waterways in the desert and evaluate places with potential for past human habitation. The rationale for this approach was based on previous applications of radar images in the eastern Sahara described above, and it was expected that the radar signal would detect old rivers (that had fed Lake Lop) through the overlaying sand sheet and reveal the "fine morphologic structure beneath" (Holcomb, 1992). Optical panchromatic LFC (Large Format Camera) coverage was also used in this study for comparison. The radar rivers were indeed imaged by the SIR-A because the river alluvium was denser in contrast with the less consolidated sand overburden, and thus yielded a distinctive radar return. The braided river courses were terminated by the conspicuous evaporation rings of the Lop Nor, whose level depended on the drainage of the old rivers. The latter, as known from historical sources, have periodically carried large volumes of water (Hedin, 1940)—for instance, when Hedin searched for the ancient Loulan along one of them—but currently the rivers have dried up. As a result, the lake displayed extensive salt-encrusted surfaces, which produce a very bright radar return to the level of saturation. Such playas are fully revealed on the radar images, even when they are covered by a layer of dry sand (Walker et al., 1987).

The radar rivers which were detected in the studies of the Sahara were very large (on a scale of hundreds of meters) and were visible without special enhancements. The radar rivers of the Taklamakan, however, were on the scale of tens of meters and required a suite of digital image enhancements before they became clearly visible (Holcomb, 1992, 1998). The regimens developed for this study were based on the effects that each operation had on each pixel, for the features of interest often were confined to just one pixel in width (25 m) in the radar images. The resolution limitations were confounded by speckle, the filtering of which could lead to the loss of the narrow river channels. It was found that "an accentuating operation followed by a smoothing operation was an effective cycle and that sequences of two or three such cycles yielded the best results" when working close to the limits of detection (Holcomb, 1992). Therefore, the digital radar image processing enhanced and revealed the linear features of one or two pixels in width. A similar procedure was performed on radar images from the Niya River region in Central Asia, where a potential man-made canal was detected from the image (Holcomb, 1996). Besides the fluvial features, such sub-linear objects as walls, roads, causeways, etc. can also be elucidated by digital radar image enhancement, and the result can be used not only for mapping the features, but also for navigation purposes during an archaeological survey in unknown or difficult terrains—as in the case below.

3.4. Paleohydrologic Reconstruction

An imaging radar (RADARSAT) mosaic was used in an archaeological survey in the Gobi Desert of Mongolia to help locate Paleolithic sites and navigate through the Gobi. Prior to the fieldtrip, the radar images were digitally enhanced, in order to reduce speckle noise and emphasize lineaments, topographic, and hydrographic features. The images were then used to help direct ground exploration (Holcomb, 2001). Since most Paleolithic sites were associated with ancient watercourses, the enhanced hydrographic features of the radar maps were very informative during that survey. Since radar was able to image paleohydrology, such as ancient lakes and playas, better than the optical sensors, the radar maps were of great use in the Gobi's Valley of Lakes. This place is renowned for its abundance of archaeological artifacts, since the expeditions of Roy Chapman Andrews in 1920 (Andrews, 1929). In contrast to the survey by Andrews, who concentrated on collecting artifacts only from the contemporary lakeshores, the 2001 reconnaissance was carried out along multiple lake strand-lines, currently covered by sand and were visible only on the radar image, which demarcated the changing lakeshores in the past. In this environment, and in association with other old playas, multiple scattered archaeological artifacts were found. The radar images, therefore, helped in the selection of potential sites for archaeological survey, where time and resources were crucial to the success of the expedition (Holcomb, 2001). Radar's ability to image physical properties of the terrain—namely its topography, texture, moisture, roughness—and to reflect target geometry, thus enhancing linear features, makes it optimal for cross-country navigation in remote and difficult regions.

4. MULTI-CHANNEL RADAR

4.1. Data

The preceding work was done largely through the analysis of single-band (gray-scale) radar images; a restriction largely resulting from the fact that at the time most radar systems, and all commercial systems, operated at a single frequency and polarization. This data limitation is analogous to the use of panchromatic optical images. The analogy is limited because panchromatic optical images tend to be very broadband and contain all polarizations while single-band radar has a very narrow bandwidth and polarized transmit and receive orientations. These parametric differences are not, however, the causal factor in the limited enhancement options.

Mathematically, enhancement of single band radar images is limited to operations such as despeckle, smooth, edge-enhance, texture analysis, or contrast manipulation, as was done in the research discussed above. The more sophisticated image analysis operations such as band ratios, classification, principal components, or other space-transforms require multi-band datasets.

Analytically, a single-band dataset is greatly under-defined in relation to the environment being imaged. In a typical archaeologically interesting area the scene may contain man-made features (ancient and/or modern), natural landscape components (rivers, terrain variations of several scales, surface material) and vegetation (natural and agricultural plants). Various combinations of these scatterers moderate the radar return signal, while the analyst is interested only in the ancient man-made features. The model is under-determined.

Recent commercial space-borne radar systems are:

System	Frequency	Polarization	Wavelength	Incidence Angle
ERS-1, -2	C	VV	5.6 cm	23 degrees
JERS-1	L	HH	23.5 cm	35 degrees
RADARSAT	C	HH	5.6 cm	20–59 d.
Envisat	C	Quad Polar	5.6 cm	14–45 d.

To create a multi-band radar dataset several parameters can be varied; frequency, polarization, and incidence angle. Given the above sensors, one could create a multi-frequency dataset by purchasing separate C- and L-band images and coregistering them. Or one could coregister ERS and RADARSAT images to create a C-band multipolar dataset. Or, obviously, images from all three sensors could be coregistered. Aside from coregistration problems (discussed below), however, there could also be incidence-angle differences or temporal changes complicating the interpretation of such a dataset. Conversely, both temporal changes and incidence-angle differences between the radar images can reveal information of interest to the analyst (Singhroy and St. Jean, 1999)

The recent launch of the ERS follow-on satellite, ENVISAT, offers the option of space-borne multi-polar, multi-incidence angle C-bands datasets. Several more such satellite systems (e.g., RADARSAT II) are planned. In addition, there are experimental radar sensors that generate multi-polar multi-frequency datasets. In the civilian sector, these are the NASA /JPL AIRSAR system and the related NASA/JPL Shuttle-based SIR-C sensor, the Canadian Convair SAR system, and the German X-SAR.

The AIRSAR and SIR-C sensors operate at 3 frequencies and are both polarimetric. Both sensors have C- and L-band. AIRSAR also has P-band and SIR-C also has X-band. Being fully polarimetric means that an image can be created corresponding to any combination of transmit and receive orientations (Evans et al., 1988), not just the standard HH and VV. While X-, C- and L-band sensors have become the workhorses of radar remote sensing, P-band is uncommon. This is in part because P-band is currently only deployed on aircraft, since the radiation cannot easily penetrate the ionosphere. Nevertheless, it is of great interest to archaeologists because of its ability to penetrate both vegetation and some ground surfaces. Recent efforts are directed toward the use of circularly polarized P-band for a spaceborne system.

4.2. Coregistration

Image enhancement regimens based on interband (interchannel) relationships (e.g., band ratios) are predicated on the assumption that the bands are precisely coregistered. For analyses of large features, such as crop delineation, problems resulting from pixel-scale misregistration will occur only at the feature boundaries and may be tolerable. Archaeological analysis, however, is often at the pixel level. For analyses of such finely detailed features, band misregistration precludes interband analysis or enhancement. Misregistration greater than pixel scale is, in general, unacceptable for any but the grossest work. While the following comments are directed toward radar-radar coregistration, they apply to any band registration problem (e.g., radar-optical).

If the bands are not already precisely coregistered, a greatly over-defined polynomial transform should be used to resample one image to the other. The order of the transform (1st, 2nd, or 3rd) is determined by the nature of the misregistration. If it were a simple shift, a first-order transform would suffice. More commonly, there will be non-linear shifts because of different imaging geometry and/or look direction. In extreme cases, a third-order transform may be necessary. Great care must be taken to have tiepoints uniformly distributed over the entire image to prevent introduction of distortions. By over-defining the transform (that is, by having far more than the minimum number of tiepoints) it is possible to reduce the random root-mean-square (rms) error to the subpixel level and to distribute the error uniformly over the entire image. This is easily accomplished by using software with point-prediction capability (Leica, 2003). In practice, well-distributed tiepoints are collected until the "predicted" point consistently falls exactly where it should. At that time, the transform must be correct. This may require 30–60 tiepoints rather than the 6-tiepoint minimum (Holcomb, 2001). Clearly, if one is trying simultaneously to georeference the images, some of the tiepoints must be ground control points (GCPs).

When doing the coregistration, it is generally preferable to register the lower-resolution image to the higher-resolution image, i.e., the high-resolution image is used as the Reference Image. This procedure will allow the greatest accuracy of registration. However, if the lower resolution image has georeferencing that is to be retained, it may be desirable to use it as the Reference Image. A larger number of tiepoints and more attention to precise work would then be required to attain the same registration accuracy. Evaluation of the x- and y-residual and the rms error statistics will indicate the accuracy of registration (Leica, 2003). After creating the coregistered images, they should be co-displayed in a viewer window. Then fade, flicker, and swipe tools can be used to evaluate visually the precision of the coregistration (Leica, 2003).

The multiple bands in commercial datasets are generally considered to be inherently coregistered. For optical sensors, utilizing beam-splitters or diffraction gratings, this is true. The bands in a multi-polar radar dataset are, similarly, assumed coregistered. But problems can arise in multi-frequency radar datasets, as discussed above (orthocorrection), where the initial inter-channel misregistration was amplified during orthocorrection because of large topographic variation in the imaged area.

4.3. Frequency

The two parameters of the radar wave, frequency (wavelength) and polarization, largely determine the nature of the interaction with the target surface. It is, in general, true that the radar signal is most sensitive to, or most affected by, surface character on the scale of the wavelength itself (Evans et al., 1994). The scattering of the incident radar beam is affected, to the first order, by the wavelength relative to the size of the scattering surface "roughness". A mathematical model to quantify the radar response to surface roughness was suggested by Sabins (1997). Citing the work of Peake and Oliver (1971), Sabins developed formulae to define "smooth," "intermediate," and "rough" in terms of wavelength and incidence angle. Because there are three wavelengths of SIR-C or AIRSAR, each with its own smooth, intermediate, and rough components, it becomes possible to describe an even larger range of surface roughness. It also becomes possible to quantify the roughness for a specific spot (pixel). For example, given a 25-degree incidence angle, if a particular pixel is classed as rough in C-band but intermediate in L-band, it would be expected to have a roughness between 1.5 and 5.9 cm based on the given formula. By quantifying 3 (or more) roughness levels, a multi-frequency radar image can be reconstructed as an RGB roughness image (Holcomb, 1999a).

Subsurface imaging through dry sand is also wavelength dependent (Elachi et al., 1984). In general, any radar will penetrate several wavelengths. Thus while X-band will penetrate only a few centimeters, P-band can penetrate several meters. Radar waves will penetrate very dry sand because of its low dielectric constant. Addition of even small amounts of moisture greatly increases the soil dielectric constant and stops penetration (Elachi et al., 1984). For this reason, subsurface archaeological detection using remote sensing (as opposed to ground penetrating radar) has been limited to hyper-arid areas (see 3.2. Early Archaeological Applications of Imaging Radar in Deserts, above).

This relationship extends to penetration through vegetation cover; in general, the longer wavelengths will penetrate deeper into the vegetation. Radar waves are scattered by features on the size scale of the radar wavelength. Thus, the X-band (5.77–2.75 cm) return is generally top-of-canopy scattering by leaves and small branches. C-band (3.8–7.6 cm) will penetrate to some extent through sparse or short vegetation. L-band (7.7–19.3 cm) penetrates through reeds and grasses while P-band (40–77 cm) can often penetrate totally through a canopy to give a return signal from the large branches, tree trunk, and ground. Under these circumstances of deep penetration, the return signal will be a combination of returns from all these target components. Scattering models attempt to devolve this mixed return into its component parts, i.e., branches, trunk, and ground.

4.4. Multi-polarization Radar

Optical images traditionally rely on solar radiation as the incident (input) illumination. Passive radar systems rely on the distribution of naturally emitted microwaves. These both result in a mix of frequencies and polarizations over which the analyst has little or no control. Some experiments have been done with laser remote sensing, and airborne lidar systems now routinely capture a return amplitude image. Controlling the polarization of the illuminating radiation does not (currently) have an equivalent in optical remote sensing. Conversely, the emitted incident illumination from an active radar sensor is totally controlled by the sensor designer/operator. The latter facility allows control of both the frequency and the polarization, horizontal (H) or vertical (V), of the emitted radar pulse. Similarly, the detecting antenna can receive either polarization. All four of the original commercial satellites had one fixed transmit-receive orientation. The next generation sensors will allow several polarization options.

As discussed, the first-order sensitivity of a particular frequency is highly correlated with variations in size or roughness of the target surface. The nature of the polarization effect may be less straightforward to predict (Ulaby et al., 1987). Most sites of archaeological interest will have some amount of vegetation cover. The archaeologist may want to wait until the end of a long dry spell or dormant period to acquire the radar images. Preferable imaging conditions would be leaf-off, dry, senescent vegetation and low soil moisture. A common scenario would then be dry grasses, straw-like vegetation, or dry reeds. Consider how radar waves from the above three commercial radar satellites might interact with this standing, reed-like, senescent vegetation. Unless the vegetation is fairly dense, the L-band wave is likely to "penetrate" through and the return signal will be dominated by a ground interaction. The C-band radiation, having a shorter wavelength, will interact more with the vegetation and the return will be wholly or partially dominated by a vegetation scatter return. Research has shown that vertical standing vegetation has a larger scattering effect if the incident radar signal is vertically polarized (Ulaby et al., 1987). The order of preference for maximum penetration, therefore, might be L-band, then C-HH. C-VV might only be appropriate for low sparse vegetation cover.

4.5. Polarimetric Radar Data

Polarimetric radar sensors capture all orthogonal polarization options "simultaneously." In operation, the radar transmitter emits alternating horizontal (H) and vertical (V) pulses. Both H and V return radiation is received for all transmissions. Evans et al. (1988) and Boerner et al. (1998) provide an overview of this technology, termed radar polarimetry. Given this complete dataset, the analyst can generate any fixed transmit/receive image desired (e.g. HH). In addition, interchannel relationships such as the phase difference between the polarization channels, cross-section ratios (e.g. HH/VV), or HH-VV correlation coefficient can be investigated for enhancing features of interest.

Two NASA/JPL sensors currently provide polarimetric radar data to archaeologists: AIRSAR and SIR-C. The SIR-C sensor, an adaptation of the AIRSAR developed for use from the Space Shuttle, flew two synoptic missions in 1994. A number of archaeologically interesting areas were imaged and can be viewed at the JPL/NASA web site (http://www.jpl.nasa.gov/radar/sircxsar/ archaeology.html). An example of the sites imaged is seen in Figure 2. Note that picture P-45156 is from the work of Moore and Freeman (1998), picture P-44534 is from the work of Holcomb (1995, 1996) and P-45302 is from the work of Blom et al. (1997).

Interpretation of such radar datasets is not as inherently obvious to the human interpreter as it is with multi-band optical datasets. The physical conditions controlling intensity and phase of the radar return signal are still being researched, both within and outside the field of archaeology. Optimal data processing and display regimens are, to date, site and application specific. Analysts must adapt their methodology to the ecological conditions of the specific study area. Thus, a model that works well in heavily vegetated areas (Freeman and Durden, 1998; Moore and Freeman, 1996) might prove inappropriate in an arid region (Cloude and Pottier, 1996).

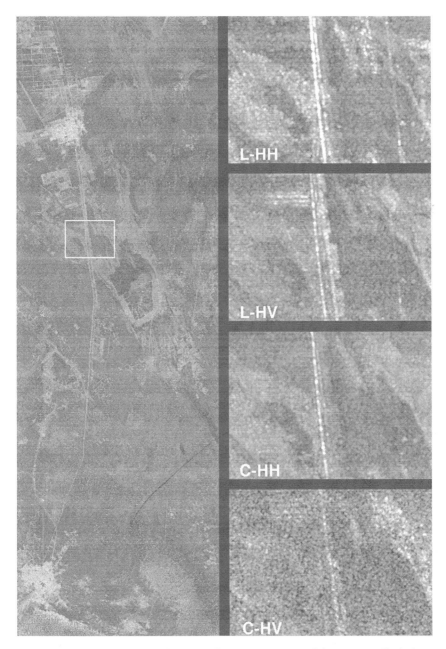

Figure 2. Spaceborne SIR-C radar image showing a segment of the Great Wall of China in north-central China. The wall appears as a thin orange band, running from the top to the bottom of the color image on the left. The black and white images on the right correspond to the area outlined by the box and are the four frequency-polarization channels. (Image courtesy of JPL).

In practice, the issues of multi-frequency and multi-polarization have been intermixed because the available datasets (AIRSAR and SIR-C) provide both. Scientists, logically, look for those bands that emphasize the features of interest to them, with a lesser regard for underlying phenomenological reasons. Moore and Freeman (1995, 1996), working with SIR-C images, find that an RGB= LHH, LHV, CHV color-composite image enables them to interpret their Angkor study area. This image can be viewed on the JPL website (http//:www.jpl.nasa.gov/releases/98/angkor98.html).

4.6. Radar Scattering

A more phenomenological way to interpret radar images is through the use of scattering models (Van Zyl, 1989; Freeman and Durden, 1992; Freeman and Durden, 1998; Cloude and Pottier, 1996; Cloude and Pottier, 1997). This type of analysis attempts to express the scattering matrix, as quantified by the polarimetric signature of each pixel, as a sum of independent elements and to associate each independent element with a physical scattering mechanism (Cloude and Pottier, 1996). For example, intrachannel phase difference helps separate scattering behavior resulting from different wave-target interaction ("bounce") mechanisms. The image can then be displayed as a dataset wherein the layers are a quantification of different bounce mechanisms.

The interaction of the radar wave with the target surface can be visualized as a reflection of the incident radiation (Freeman and Durden, 1992, Cloude and Pottier, 1996). A number of important radar interaction mechanisms are shown in Figure 3. A single (specular) reflection results in the radiation being largely reflected away from the sensor because of the large incidence angle of imaging radar. This mechanism accounts for the dark (or black) pixels of playas, roads, or smooth-water bodies. When the incident radiation reflects off a moderately rough surface, a Bragg scattering model is used. Again, in this context, moderately rough is defined in relation to the radar wavelength. A surface that is moderately rough to X-band (2–3 cm) could appear smooth to, say, L-band (25 cm). This difference was exploited by Holcomb (1999a) to classify riverine deltas in Mongolia. A third interaction is a "double-bounce" from two orthogonal surfaces. This is the mechanism associated with man-made objects such as walls (Holcomb, 1995, Figure 4). This model could also apply to a ground-tree trunk reflection.

A fourth mechanism, volume scattering, is the result of multiple reflections from a cloud of randomly oriented surfaces. This is often termed "canopy" scatter because a vegetative canopy is the typical scenario for this mechanism to apply. As discussed, the ability to penetrate vegetation increases with wavelength. Thus, for certain vegetation types and densities, one might find the C-band return to result largely from volume scattering in the vegetation (canopy) while the L- or P-band return is dominated by a ground-tree trunk, double-bounce mechanism. Such returns are exactly

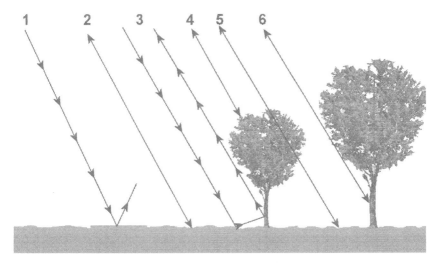

Figure 3. Radar-target interaction mechanisms. 1) Specular reflection, 2) Bragg backscatter, 3) double-bounce, 4) volume scattering, 5) Bragg backscatter after propagation through random medium, and 6) backscatter from anisotropic structure. (adapted from Cloude and Pottier, 1996).

the sort of discrimination and analysis made possible by multi-band radar images.

The fifth mechanism depicted involves propagation of the radar waves through a "transparent" medium; in this case a tree canopy. This mechanism is possible if the vegetation were sufficiently sparse (relative to the wavelength!) to allow canopy penetration with minimal attenuation. The final possible mechanism is a direct scatter back from an oriented dipole. Not all mechanisms are relevant to every imaging scenario. Thus, Freeman and Durden (1992) limit their scatter model to the dominant mechanisms in their area of study.

Archaeological (or other) interpretation of these scattering-model images is not straightforward. A particular target can give several types of return. For example, a tree could give a canopy (volume scatter) return, a ground-tree double-bounce, a ground scatter after propagation through the canopy, a single scatter directly from the trunk, etc. And any given pixel can be the result of contributions from several target types, a mixed pixel. The tree return could be mixed with scatter from low shrubs and bare earth to form one pixel. Clearly, this variety cannot be easily devolved into a few simple scatter categories. Additionally, it is not possible to invert the scatter model back into real-world physical features. Even if it is known that two pixels have exactly the same scatter model (e.g. 1/3 Bragg, 1/3 double-bounce, and 1/3 volume), it still is not possible to say the areas they represent are identical.

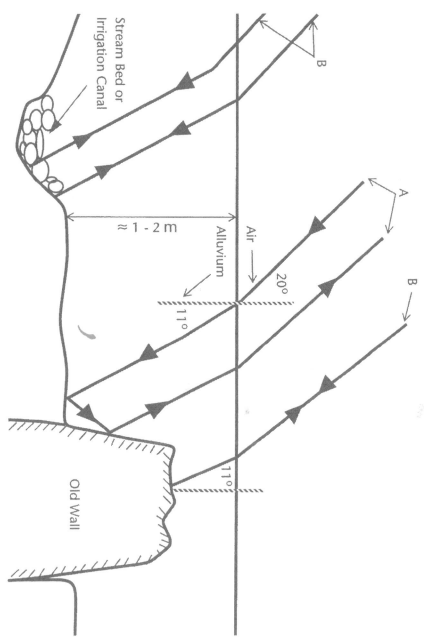

Figure 4. Modeled geometry of wave propagation through alluvium (dry sand) and interaction with buried wall foundation and irrigation canal. (adapted from Holcomb, 1998a).

5. APPLICATIONS OF POLARIMETRIC RADAR IMAGES

5.1. Analytical Approaches

As with the single-band work, site detection has been largely indirect and analysis of these datasets has been logically broken into vegetation-based and geology-based approaches. In both domains, paleo-hydrologic mapping plays a significant role (Moore and Freeman 1995, Holcomb 2001), and is reflective of the importance of water on both sides of the radar-archaeology equation. Availability of water resources has always greatly influenced where man chooses to live. Simultaneously, the presence of moisture or water greatly attenuates the radar signal. This effect is so dominant that the quantitative mapping of soil moisture has been proposed as an operational application for radar images. Dubois et al. (1995) propose a soil moisture computation using the L-band HH/VV backscatter ratio. Freeman, Hensley, and Moore (Freeman et al., 1998) comment on the use of a P-band HH/VV ratio image for delineating old moats from the drier earthworks.

The vegetation-based approach has emphasized the use of polarimetric, multi-frequency datasets for better vegetation discrimination, which would then provide indirect evidence of human modification of the landscape. One of the first experiments directed toward this approach was in 1990 using AIRSAR data of an area in Belize, Guatemala and Mexico (Durden et al, 1991). This flight, not coincidentally, covered many of the areas and ecotypes that had been so extensively studied and debated by archaeologists (Adams, 1980, 1982; Pope and Dahlin, 1989, 1993). The experiment, code-named TREE (Tropical Rainforest Ecology Experiment), clearly had archaeological research as a subtext. In an article in *National Geographic*, O'Neill (1993) mentions that; "Sponsors of the TREE project were curious whether the airborne sensing devices could detect undiscovered ruins under the canopy." In that same article, one of the researchers commented on the apparent correlation between radar images and the archaeological nature of the corresponding ground feature.

In support of this experiment, Freeman, Durden, and colleagues (Durden et al., 1991; Freeman et al., 1992) began to develop frequency-polarization relationships to elucidate the various sorts of vegetation in the scene. They conclude, "Polarimetric radar appears able to identify most of the vegetation types present in the study area" (Durden et al., 1991). From this understanding, a 3-component scattering model was developed as a means to classify vegetation. It is important to recognize that this scattering model was optimized for their environment (tropical forest) and is not necessarily appropriate to other environs where different scatter mechanisms are dominant. Pope et al. (1994) use a totally different set of indices to model this same dataset.

The geology-based approach has been largely directed toward semi-quantitative analysis of terrain evolution time-sequences. Distribution pattern maps of dated deposits are a record of regional climatic history. These data can be used in archaeological studies that seek to understand climatological and landscape evolution and man's role in that landscape and process. It has been shown (Owens, 1997) that on-site geologic mapping of alluvial fans can reveal details of the climatic history through which the fan had evolved and/or been exposed. Such fieldwork is, of course, arduous and expensive. In an initial attempt to extract such information by remote sensing, Farr and Chadwick (1996) investigated the use of SIR-C images to compare geomorphic processes in alluvial fans in China and the U.S. Using both L-HV radar (from SIR-C) and panchromatic SPOT images, they were able to dissect fans in both regions into young, intermediate, and old classes. Differences in source lithologies and regional weathering processes are reflected in the remote sensing signatures. Based on this work they conclude that the signatures are "consistent enough in each of these regions to be used for mapping fan units over large areas".

5.2. Angkor

A comprehensive study using multi-parameter radar images (3-frequency polarimetric SAR) for archaeological investigations centered on the Angkor region in Cambodia. The project was initiated with data collected by the SIR-C sensor flown on Space Shuttle Endeavor in 1994 (Evans, 1994; Moore and Freeman, 1995). In the initial work, Moore eloquently combines both physical and metaphysical observations in the archaeological interpretation of the multi-band radar images. She notes that, as a result of the dominance of textural effects in the image; "One of the most striking visual differences between previous satellite images...and the SIR-C is in the perception of the terrain." (Moore and Freeman, 1995). To quantify these perceptions, they turn to decomposition and modeling of the polarimetric SIR-C dataset.

The use of fully polar SAR increases the number of input data parameters and allows the possibility of modeling the complicated composition of the imaged surface. A 3-component scattering model (double-bounce, volume scatter, and Bragg scatter) was chosen for their study site (Freeman and Durden, 1992, 1998). The three derived scatter components for each band (C and L) were then displayed as red-green-blue (R, G, B) images that were visually interpreted (Moore and Freeman 1996). Extensive field-testing of this scattering model had been undertaken in previous work in Belize (and other areas) where it was found to work well in discriminating various vegetation cover types (Durden et al., 1991; Freeman and Durden, 1992). The similarities of the general vegetation cover between Belize and Cambodia suggested the model might also be appropriate there. Moore and Freeman (1996) note: "The same potential is seen in applying the model to Angkor where classification of landcover is fundamental to understanding terrain and hydrological preferences of prehistoric and historic settlements and water management structures. Decomposition of the radar signatures or measurements into different scattering mechanisms allows the observer to differentiate between different landforms." The authors take a fundamentally correct approach to their data; they frame their questions and direct their efforts to issues that best exploit the information content of the dataset. Firstly, they have wide area and (relatively) low-resolution images, thus they direct their efforts to regional features. Secondly, and importantly, recognizing that radar backscatter is largely moderated by vegetation, moisture and water, they attempt to develop an understanding of the human-landscape evolution through elucidation of landscape remnants. This approach proved especially fruitful in an area to the NW of the main Angkor complex. In this rural area, 11 circular sites are identified consisting of mounds with moats and/or earthworks. Often the associated moats are totally vestigial and are not seen on the ground or in optical (visible) images; their form is revealed in slight vegetation and moisture differences (Moore et al., 2004).

The above analyses indicated that advanced radar images might have a significant role to play in understanding the evolution of the complex Angkor region. In 1996 the NASA/JPL AIRSAR/TopSAR sensor was deployed over Cambodia as part of the PacRim mission (JPL, 1998). This airborne SAR system operates in three frequencies; C-, L-, and P-band. AIRSAR brought three major advances to the dataset. Being airborne allowed the system to be flown at lower altitude that resulted in smaller pixels (5–10M vs 25M). Use of P-band (68 cm) radar, which is not currently operational on a space-borne sensor, allowed deeper vegetation penetration than is possible with L- or C-band. (P-band can penetrate the ionosphere, but the propagation and polarization are affected.) And, given the very slight topographic variations in this area and the overarching importance of water management, the high precision of the TOPSAR DEM was critical.

Analysis of this new dataset again emphasized the use of the (3-component) scattering model to quantify the variations within the images. As a focus of the research was on the role of water management as an under-lying fundamental of human activity in the area, incorporating the DEM into the display was important (Moore et al., 1999; Moore, 2000). The circular sites discussed above were clearly mapped by the P-band, presumably because of radar return from tree-trunks and large branches on the earthworks (Freeman et al., 1998). One of these sites is seen in the upper left quadrant of Figure 5. Another prominent circular site, Kapilapura, was noticed in the NE corner of the Angkor Wat moat. This small circular feature predated but seemed to have been known and purposefully preserved when the rectangular temple was built in the 12th century AD (Moore et al., 1998). This site was identified primarily through examination of the high-resolution digital elevation (topography) data produced by TOPSAR at C-band. The elevation data mapped the top of the vegetation canopy in a densely wooded area and the Kapilapura feature was noticed to have a canopy ~ 1 meter higher than the surrounding forest. Continued analysis of the images in multifarious display modes combined with

Figure 5. Northwest corner of the West Baray, Angkor, Cambodia. Image is from the NASA/JPL AIRSAR sensor, plotted here with elevation encoded as color and C-VV backscatter amplitude as pixel intensity. (Image courtesy of Freeman, Hensley and Moore, 1998).

historical data and site visits enabled the authors to develop an interpretation of Angkor in the context of the spatial arrangement of natural and man-made features. Hydrological engineering, a fundamental precept of Angkorian development, was a focal point of the interpretation (Moore et al., 1999; Moore, 2000). Eventually, analysis of the radar data allowed the authors to interpret the landscape evolution from animistic habitation to the monumental Hindu cities of Angkor (Moore et al., 1998, 2004)

Parts of these data are posted at the Royal Angkor Foundation website (http://www.angkor.iff.hu/index.html).

5.3. Mongolia

The advent of multi-band radar images allows the application of image-analysis techniques such as classification. Several model-based methods have been advanced in the literature (Van Zyl, 1989; Cloude and Pottier, 1997) and their application discussed above (Freeman and Durden, 1998; Moore et al., 1996). A second approach is to use the traditional parametric statistics-based classifiers such as Isodata, Maximum Likelihood, and Mahalanobis distance. With these techniques, the data are clustered in N-dimensional space where N is the number of input bands. Classification accuracies of 70–85% are routinely achieved with six-band Landsat images, thus 4 to 8 bands of radar images might be expected to produce similar accuracies.

In the mid-1990s, Holcomb was working with the Joint Mongolia Russia America Archaeological Expedition (JMRAAE) to investigate possible roles for radar remote sensing (Olsen, 1998). Datasets for this project included RADARSAT, SPOT and SIR-C (Holcomb, 2001). Because the JMRAAE team was focused on the Paleolithic, direct detection of artifacts (stone scatters, habitations, or workshops) was impossible. Efforts were instead directed toward developing a paleohydrological model for the study area (Gobi Desert) to be used in guiding the required fieldwork. The climatic history is recorded in the alluvial fans and, because of the paucity of rainfall in Recent times, this climatic record is preserved. Building on the work of Sabins (1997), Owen et al. (1997), and Farr and Chadwick (1996), a multi-band dataset, L-HH, L-HV, C-HH, C-HV and X-VV, was constructed and classified.

Testing on selected subsets indicated that the best classification results are obtained when the image has been pre-filtered to suppress speckle. Use of a 3X3 Gamma-Maximum a Posteriori (Gamma-MAP) speckle-reduction filter followed by a 3X3 Local Regions segmenting filter (Leica, 2003) retains resolution while providing a superior classification result. In addition, various frequency/polarization layers were removed and the classification repeated. This procedure allowed an evaluation of the relative value of the various bands. Not surprisingly, X-band was found to contribute little to the roughness map; everything was rough to this wavelength (3 cm). Figure 6 show the results of classification using L-HH, L-HV, C-HH and C-HV. Note that only the area below the mountain slopes is classified; areas of high slope are masked out of the analysis. As discussed above (section 2.2. Orthocorrection), the intensity of the radar return is greatly affected by the incidence angle. Since no DEM was available for this area and it could not be radiometrically corrected for slope effects, the classification had to be restricted to areas of low uniform slope. Farr and Chadwick (1996) make a similar decision, restricting their classification to slopes between 4 and 8 degrees.

The classified radar image was found to map, very precisely, the surface structure of extensive alluvial fans in the reconnaissance area (Holcomb, 1998c). Within the area of the SIR-C radar mosaic, twenty roughness classes were identified. These classes were selected based on color differences in various frequency/polarization images and the classification image (Figure 6). For each of the twenty selected classes, representative pixels were located. The intensity of these pixels in each of the various frequency/polarization layers was extracted from the full 5-band SIR-C dataset and used to make plots of the "signature" of each pixel. A number of these "representative pixels" were visited in the field. As described elsewhere

Figure 6. Classified SIR-C multi-polar, multi-frequency radar image. Colors correspond to variations in surface roughness (rock sizes) clearly mapping the structure of the alluvial fans. Gray areas were not classified. Inset boxes delineate areas seen in Figure 7.

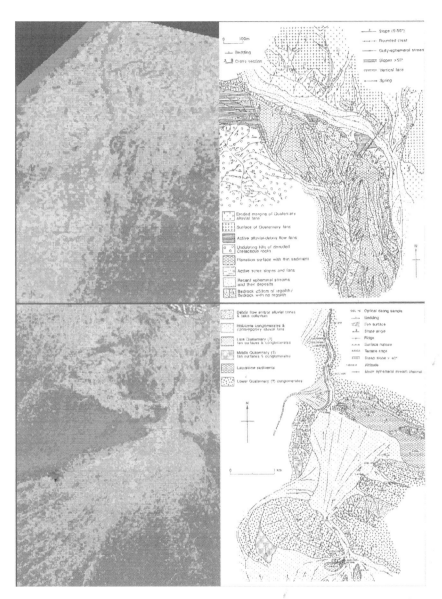

Figure 7. Detail of two alluvial fans from the classified SIR-C image and alluvial fans of similar structure mapped nearby using traditional geological techniques. Mapped fans are approximately 400 km NE of the classified radar fans. The similarity of structure suggests that the multi-channel radar is mapping the geomorphological evolution of the fans structure consistent with field mapping. (Fan maps adapted from Owen et al. 1997).

(Holcomb, 2001), the field team became quite adept at navigating through the radar mosaics using GPS devices. The radar mosaics were carefully prepared and georeferenced in advance and were continually "tweaked" at identifiable locations such as river junctions. In this fashion, the team became confident in the ability to navigate to specific pixels (25M). It was routinely possible to predict, in advance and from kilometers away, the rock-size distribution of an intended destination pixel.

In addition, the alluvial fan structures as mapped by the classified radar images (Figures 6 and 7) compare favorably with geomorphological maps of similar alluvial fans in the same region. The fans shown in the images and sketch-maps of Figure 7 are not the exact same locations; the 2 study areas were several hundred kilometers apart (Holcomb, 2001; Owen et al., 1997). The two classified radar fans were selected from the many seen in Figure 6 because they revealed structures similar to those detailed in the study of Owen and colleagues. This suggests that the radar sensor is mapping the same geological/environmental factors as are elucidated by laborious fieldwork. If true, vast regions could be roughly mapped using radar images and the resultant classes then defined through selected field mapping.

6. CONCLUSION

Radar imaging has proven its usefulness in a number of archaeological studies. For some applications, such as imaging through sand, it is the sensor of choice. Additionally, it is the only option that is satellite-borne, and has worldwide coverage that is commercially available. Other airborne or surface-based sensors are occasionally used for subsurface imaging, but these techniques are used once the area of interest is well defined, not for regional or wide-area survey. (see Cook, 1997, for a survey of the various geophysical techniques). For detailed subsurface mapping once the site is defined, ground-penetrating radar (GPR) has a significant following (Conyers and Goodman, 1997).

However, interpretation of radar images is not as straightforward as it is for optical images. First, the (often large) incidence angle of imaging radar must be taken into account when interpreting the signal reflected back to the sensor. Furthermore, an understanding of the phenomenologies affecting the return signal is necessary to interpret correctly a radar image. Scattering models have been developed to describe various mechanisms of interaction between radar waves and the target surface. To utilize fully radar images, digital image processing techniques are required. For example, every radar image is degraded by speckle. Suppression of this noise will help in visual detection of subtle features.

The advent of polarimetric multi-frequency radar data has called for more sophisticated processing, such as multi-band scatter modeling and roughness classification, making possible an increased capacity for discrimination and analysis. Some attributes of the target area, such as roughness, can be defined and quantified in relation to their interaction with varying wavelengths of incident radiation, allowing for detailed classification of pixels within the image. Individual researchers have developed site and application specific scattering models optimized for the scatter mechanisms dominant in their study environments.

Research questions should be tailored to best exploit the information content of the data, recognizing that radar backscatter is largely moderated by vegetation, moisture, and dielectric constant of the target material. Vegetation classifications are widely used for indirect detection of potential and actual archaeological sites. Topographic variations, possibly indicative of human activity, often can be discerned through slight moisture and vegetation differences evident in radar imagery yet not seen on the ground or in optical images. Radar imagery is also well suited for paleohydrologic mapping, useful in both geologic- and vegetation-based approaches to archaeological investigations. Location of ancient watercourses impacted prehistoric selection of sites for human habitation. Geologic mapping of alluvial fans can reveal climatic history of an area. Such mapping of large-scale environmental factors was previously possible only through extensive laborious fieldwork.

The relatively low resolution of imagery from space-borne sensors is more appropriate for assessment of regional features than for direct detection of archaeological sites and artifacts. However, newer airborne sensors like AIRSAR/TopSAR operate at lower altitudes, making available the higher resolution of data with smaller pixels and the deeper penetration of longer wavelength P-band data.

Radar imaging should be seen as an adjunct to, not a replacement for, optical imaging. Data from both sensor types, correctly interpreted, will provide a more complete understanding of the imaged area than either data type alone; they are synergistic. Advanced analysis techniques are still in their infancy, for archaeological as well as other applications. As ever more sophisticated sensors are launched, so grows the challenge to interpret the vastly complex information contained in mixed pixels comprised of returns of varying frequencies and polarizations from varying target types. Yet, with the rapid advancement of image processing techniques, imaging radar promises to be a rich resource and a powerful tool, a valuable addition to the contemporary archaeologist's armamentarium.

REFERENCES

Abdelsalam, M., Robinson, C., El-Baz, F., and Stern, R., 2000, Applications of Orbital Imaging Radar for Geologic Studies in Arid Regions: The Sharan Testimony. *Photogrammetric Engineering & Remote Sensing* 66 (6):717–726.

Adams, R., 1980, Swamps, Canals, and the Locations of Ancient Maya Cities. *Antiquity* LIV:206–214.

Adams, R., 1982, Ancient Maya Canals. *Archaeology* (November/December):28–35.

Adams, R., Brown, W., and Culbert, T., 1981, Radar Mapping, Archeology, and Ancient Maya Land Use. *Science* 213:1457–1463.

Andrews, R., 1929, *Ends of the Earth*. Garden City Publishing. New York.

Berlin, G., Tarabzouni, M., Al-Naser, A., Sheikho, K., and Larson, R., 1986, SIR-B Subsurface Imaging of a Sand-Buried Landscape: Al Labbah Plateau, Saudi Arabia. *IEEE Transactions on GeoScience and Remote Sensing* GE-24 (4):595–601.

Blom, R., Clapp, N., Zarins, J., and Hedges, G., 1997, Space Technology And The Discovery Of The Lost City Of Ubar. Paper read at IEEE Aerospace Conf. February 1–8.

Blom, R., Crippen, R., and Elachi, C., 1984, Detection of Subsurface Features in SEASAT Radar Images of Means Valley, Mojave Desert, California. *Geology* 12:346–349.

Boerner, Wolfgang-Martin, Mott, H., Lunebug, E., Livingstone, C., Brisco, B., Brown, R., and Paterson, J., 1998, Polarimetry in Radar Remote Sensing: Basic and Applied Concepts. In *Principles and Applications of Imaging Radar, Manual of Remote Sensing (Third Edition)*, Vol.2, edited by F. M. Henderson and A. J. Lewis, pp. 769–777. John Wiley & Sons, New York.

Cloude, S. R., and Pottier, E., 1996, A Review of Target Decomposition Theorms in Radar Polarimetry." *IEEE Transactions on Geoscience and Remote Sensing* GE-34 (2):498–518.

Cloude, S. R., and Pottier, E., 1997, An Entropy Based Classification Scheme for Land Applications of Polarimetric SAR." *IEEE Transactions on Geoscience and Remote Sensing* GE-35 (1):68–78.

Conyers, L. B., and Goodman, D. 1997. *Ground-Penetrating Radar: An Introduction for Archaeologists*. John Wiley & Sons, New York.

Cook, F.A., 1997, Applications of Geophysics in Gemstone Exploration. *Gems and Gemology*, Spring:4–23.

Crippen, R., 1988, The Dangers of Underestimating the Importance of Data Adjustments in Band Rationing. *International Journal of Remote Sensing* 9 (4):767–776.

Crippen, R., 1991, The Iterative Ratioing Method of Determining Atmospheric Corrections for Scenes with Rugged Terrain. *Unpublished*.

Dainty, J., 1980, An Introduction to Speckle. *Proceedings of the Society of Photo Optical Instrument Engineers* 243:243–301.

Dubois, P. C., van Zyl, J., and Engman, T., 1995. Measuring Moisture with Imaging Radars. *IEEE Transactions on Geoscience and Remote Sensing* 33(4):915–926.

Durden, S., Freeman, A., Klein, J., Oren, R., Vane, G., Zebker, H., and Zimmermann, R., 1991, Polarimetric Radar Measurements of a Tropical Rainforest. *Proceeding of the 3rd AIRSAR Workshop*, JPL Publication 91–30. pp. 223–229.

Elachi, C., Roth, L., and Schaber, G., 1984, Spaceborne Radar Subsurface Imaging in Hyperarid Regions. *IEEE Transactions on Geoscience and Remote Sensing* GE-22(4):383–387.

El-Baz, F., 1997, Space Age Archaeology. *Scientific American* 277(2):40–45.

El-Baz, F., 1998, Groundwater Concentration Beneath Sand Fields in the Western Desert of Egypt: Indications by Radar Images from Space. *Egyptian Journal of Remote Sensing & Space Sciences* 1:1–24.

Evans, D. L., Farr, T. G., Van Zyl, J. J., and Zebker, H., 1988, Radar Polarimetry: Analysis Tools and Applications. *IEEE Transactions on Geoscience and Remote Sensing* 26(6):774–789.

Evans, D. L., Stofan, E.R., Jones, T.D., and Godwin, L.M. 1994, Earth from Sky. *Scientific American* (December) 271(6):70–75.

Farr, T. G., and Chadwick, O. A., 1996, Geomorphic processes and remote sensing signatures of alluvial fans in the Kun Lun Mountains, China. *Journal of Geophysical Research* 101:23,091–23,100.

Freeman, A., and Durden, S. L., 1992, A Three-component scattering model to describe Polarimetric SAR data. *Radar Polarimetry*, pp. 213–225. SPIE-1748.

Freeman, A., and Durden, S. L., 1998, A Three-component scattering model for Polarimetric SAR data. *IEEE Transactions on Geoscience and Remote Sensing* 36:963–973.

Freeman, A., Hensley, S., and Moore, E., 1998, Radar imaging methodologies for archaeology: Angkor, Cambodia. Paper read at Conference on Remote Sensing in Archaeology, at Boston University, Department of Archaeology and Remote Sensing Center, 11–19 April.

Ford, J. P., Cimino, J. B., and Elachi, C., 1983, *Space Shuttle Columbia Views the World With Imaging Radar: the SAR-A Experiment.* JPL Publications 82–95, Pasadena, CA.

Ford, J. P., Blom, R. G., and Crisp, J. A., Elachi, C., Farr, T. G., Saunders, R. S., Theilig, E. E., Wall, S. D., and Yewell, S. B., 1989, *Spaceborne Radar Observations: A Guide for Magellan Radar-Image Analysis.* JPL Publications 89–41, Pasadena, CA.

Fung, A. K., and Ulaby, F. T., 1983, Matter-Energy Interaction in the Microwave Region. In *Manual of Remote Sensing* (Second Edition), edited by R. N. Colwell, Vol. 1, pp. 115–164. American Society of Photogrammetry.

Gillespie, A., 1980, Digital Techniques of Image Enhancement. In *Remote Sensing in Geology*, edited by B. Siegal and A. Gillespie. pp. 127–149. John Wiley & Sons, New York.

Hedin, S., 1940, *Wandering Lake.* E. H. Dutton and Co., New York.

Holcomb, D. W., 1992, Shuttle Imaging Radar and Archaeological Survey in China's Taklamakan Desert. *Journal of Field Archaeology* 19:129–138.

Holcomb, D. W., 1995, Archaeological Applications of SIR-C Imaging Radar. In Research Center for Silk Roadology, Nara, Japan,*Silk Roadology 1; Space Archaeology*(1):83–84.

Holcomb, D. W., 1996, Radar Archaeology: Space Age Tools Aid in Uncovering the Past. *Earth Observation Magazine* (September):22–26.

Holcomb, D. W., 1998a, Applications of Imaging Radar to Archaeological Research. In *Principles and Applications of Imaging Radar, Manual of Remote Sensing (Third Edition)*, Vol.2, edited by F. M. Henderson and A. J. Lewis, pp. 769–777. John Wiley & Sons, New York.

Holcomb, D. W., 1998b, Radar Archaeology in Arid Regions. Paper read at Conference on Remote Sensing in Archaeology, at Boston University Department of Archaeology and Remote Sensing Center, 11–19 April.

Holcomb, D. W. 1998c, Progress Report 1 Feb–31 April, 1998. NASA Contract No. NASW-96011.

Holcomb, D. W. 1999a, Progress Report 1 May–31 July, 1999. NASA Contract No. NASW-96011.

Holcomb, D. W., 1999b, Orthoradar Accuracy Evaluation. Leica Geosystems, Unpublished Report, available on request.

Holcomb, D. W., 2001, Imaging Radar and Archaeological Survey: An Example from the Gobi Desert of Southern Mongolia. *Journal of Field Archaeology* 28(1-2):131–141.

Lee, J. S., Jurkevich, I., Dewaele, P., Wambacq, P., and Oosterlinck, A., 1994, Speckle Filtering of Synthetic Aperture Radar Images: A Review. *Remote Sensing Reviews* 8:313–340.

Leica Geosystems, 2003, *ERDAS IMAGINE Field Guide.* 7th edn. Atlanta: Leica Geosystems GIS & Mapping, LLC.

Lillesand, T. M., and Kiefer, R. W., 2000, *Remote Sensing and Image Interpretation.* John Wiley & Sons, New York.

Madsen, S. N. and Zebker, H. A., 1998, Imaging Radar Interferometry. In *Principles and Applications of Imaging Radar, Manual of Remote Sensing (Third Edition)*, Vol.2, edited by F. M. Henderson and A. J. Lewis, pp. 359–380. John Wiley & Sons, New York.

McCauley, J. F., Breed, C. S., Schaber, G. G., McHugh, W. P., Issawi, B., Haynes, C. V., Grolier, M. J., and Kilani, A. E., 1986, Paleodrainages of the Eastern Sahara–The Radar Rivers Revisited. *IEEE Transactions on GeoScience and Remote Sensing* GE-24 (4):624–647.

McCauley, J. F., Schaber, G. G., Breed, C. S., Grolier, M. J., Haynes, C. V., Issawi, B., Elachi, C., and Blom, R., 1982, Subsurface Valleys and Geoarchaeology of the Eastern Sahara Revealed by Shuttle Radar. *Science* 218 (3):1004–1020.

McHugh, W. P., McCauley, J. F., Haynes, C. V., Breed, C. S., and Schaber, G. G., 1988, Paleorivers and Geoarchaeology in the Southern Egyptian Sahara. *Geoarchaeology* 3 (1):1–40.

McHugh, W. P., Schaber, G. G., Breed, C. S., and McCauley, J. F., 1989, Neolithic Adaptation and the Holocene Functioning of Tertiary Paleodrainages in Southern Egypt and Northern Sudan. *Antiquity* 63:320–336.

Moik, J. G., 1980. *Digital Processing of Remotely Sensed Images.* NASA SP-431.

Moore, E., 1997, The East Baray: Khmer Water Management at Angkor. *Journal of Southeast Asian Architecture* 1:91–98.

Moore, E., 2000, Angkor Water Management, Radar Imaging and the Emergence of Urban Centres. *Journal of Sophia Asian Studies* 18:39–51.

Moore, E. H., and Freeman, A., 1996, The application of microwave scattering mechanism to the study of early Ankorean water Management. Paper read at European Association of Southeast Asian Archaeologists, 6th International Conference, Leiden, p. 1–13. http://southport.jpl.nasa.gov/reports/finrpt/

Moore, E. H., and Freeman, A., 1998, Circular Sites at Angkor: A Radar Scattering Model. *Journal of the Siam Society* 85 (1&2):107–119.

Moore, E., Freeman, A., and Hensley, S., 1998, Beyond Angkor: Ancient Habitation in Northwest Cambodia. Paper read at Conference on Remote Sensing in Archaeology, Boston University Department of Archaeology and Remote Sensing Center, 11–19 April.

Moore, E., Freeman, A., and Hensley, S., 1999, Angkor AIRSAR: Water Control and Conservation at Angkor. Paper read at NASA/JPL PacRim Significant Results Workshop, Maui, Hawaii.

Moore, E., Freeman, A., and Hensley, S., 1998, The NASA/JPL AIRSAR Mosaic of Angkor: A tribute to Bernard P.Groslier. Unpublished Manuscript.

Moore, E., Freeman, T., and Hensley, S., 2006, Spaceborne and Airborne Radar at Angkor: Introducing New Technology to the Ancient Site. This volume.

Olsen, J. W., 1998, The Paleolithic Prehistory of the Gobi Desert, Mongolia. National Geographic Society Grant #6174-98.

O'Neill, T. O., 1993, New Sensors Eye the Rain Forest. *National Geographic Magazine* (September):118–123.

Owen, L. A., Windley, B. F., Cunningham, W. D., Badamgarav, J. and Dorjnamjaa, 1997. Quaternary Alluvial Fans in the Gobi of Southern Mongolia: Evidence for neotechtonics and Climate Change. *Journal of Quaternary Science* 12(3):239–252.

Pope, K. O., and Dahlin, B. H., 1989, Ancient Maya Wetland Agriculture: New Insights from Ecological and Remote Sensing Research. *Journal of Field Archaeology* 16:87–106.

Pope, K. O., and Dahlin, B. H., 1993, Radar Detection and Ecology of Ancient Maya Canal Systems–Reply to Adams et al. *Journal of Field Archaeology* 20:379–383.

Pope, K. O., Rey-Benayas, J. M., and Paris, J. F., 1994, Radar Remote Sensing of Forest and Wetland Ecosystems in the Central American Tropics. *Remote Sensing of the Environment* 48:205–219.

Robinson, C. A., 1998, Potential And Applications Of Radar Data In Egypt. *Egypt. Jour. Remote Sensing & Space Sciences* V.I:25–56.

Roth, L. E., and Elachi, C., 1975, Coherent Electromagnetic Losses by Scattering from Volume Inhomogeneities. *IEEE Transactions, Antennas and Propagation* 23:674–675.

Sabins, F. F., 1997, *Remote Sensing: Principles And Interpretation*. Freeman, New York.

Schaber, G. G., McCauley, J. F., Breed, C. S., and Olhoeft, G. R., 1986, Shuttle Imaging Radar: Physical Controls on Signal Penetration and Subsurface Scattering in the Eastern Sahara. *IEEE Transactions on Geoscience and Remote Sensing* GE-24(4):603–622.

Sever, T.L., 1998, Validating Prehistoric and Current Social Phenomena upon the Landscape of the Peten, Guatemala. In *People and Pixels*, edited by D. Liverman, E.F. Moran, R. R. Rindfuss, and P. C. Stern, pp. 145–163. National Academy Press, Washington, D.C.

Sever, T.L., and Irwin, D.E., 2003, Remote-sensing Investigation of the Ancient Maya in the Peten Rainforest of Northern Guatemala. *Ancient Mesoamerica* 14:113–122.

Siemens, A. H., and Puleston, D. E., 1972, Ridged Fields and Associated Features in Southern Campeche: New Perspectives on the Lowland Maya. *American Antiquity* 37:228–239.

Singhroy, V. and St. Jean, R., 1999, Effects of Relief on the Selection of RADARSAT Incidence Angle for Geological Applications. *Canadian Journal of Remote Sensing* 25(3):211–217.

Ulaby, F., Held, D., Dobson, M., McDonald, K., and Senior, T., 1987, Relating Polarization Phase Difference of SAR Signals to Scene Properties. *IEEE Transactions on Geoscience and Remote Sensing* GE-25(1):83–91.

Van Zyl, J., 1989, Unsupervised Classification of Scattering Behavior Using Radar Polarimetry Data. *IEEE Transactions on Geoscience and Remote Sensing* GE-27(1):36–45.

Verstappen, H. T., 1977, *Remote Sensing in Geomorphology*. Elsevier, Amsterdam.

Walker, A. S., Olsen, J. W., and Bagen, 1987, The Badain Jaran Desert: Remote Sensing Investigations. *The Geographical Journal* 153:205–210.

Wang, S. J., 1985, Comparative Archaeological and Geographic Research on Certain Historical Questions of Loulan Region [in Chinese]. *Western Frontier History Discussions* 2:269–286.

Wendorf, F., Close, A. E., and Schild, R., 1987, A Survey of the Egyptian Radar Channels: An Example of Applied Archaeology. *Journal of Field Archaeology* 14:43–63.

Williams, R. J., 1992, In Search of a Legend–The Lost City. *Point of Beginning, 1992 Data Collector Survey* 17 (6):10.

Chapter 2

Radar Images
and Geoarchaeology
of the Eastern Sahara

FAROUK EL-BAZ, CORDULA A. ROBINSON,
AND TURKI S.M. AL-SAUD

Abstract: The first Shuttle Imaging Radar (SIR-A) instrument was flown in earth orbit in
November 1981. Data were obtained pertaining to a flat, sand-covered region in
the eastern Sahara of North Africa. These data revealed courses of three channels
or dry river courses varying in width from 8 to 20 km. This revelation increased
interest in the geomorphology of desert regions and implications thereof to
the geoarchaeology of prehistoric environments, particularly in southwestern
Egypt.
            For radar waves to penetrate desert sands and reveal the underlying topog-
raphy, the surface material must be dry and fine grained. Moisture reflects radar
waves and interferes with their penetration ability, and the size of sand grains
has to be less than one-fifth of the wavelength of the radar waves. These two
conditions are satisfied in the eastern Sahara. More data obtained by SIR-C and
Radarsat confirm these findings.
            Identification of the dry river courses explains the prevalence of prehis-
toric wet conditions, which allowed the existence of plants, animals, and man
throughout the eastern Sahara. It also explains the underlying reasons for the
location of oases in this hyper-arid region. The eastern Sahara must have been
supplied by water through the ancient rivers, from occasional rainfall, even
after their courses were buried by wind-blown sand. The analysis of unveiled
river courses shows that they originate from highland massifs in the Sahara
or from the sub-Saharan belt of the Sahel. The latest time water flowed in
these courses was during the last pluvial, from 11,000 to 5,000 years ago.
Alternation of wet and dry episodes brought life and death, respectively, to

this region as indicated by human implements as well as flora and fauna of a savanna-like environment.

Much of the water that flowed in these rivers must have seeped into the substrate to be stored as groundwater. This would have occurred along fractures in the host rock or through the rock's primary porosity, especially in areas that were water rich such as beneath the numerous inland lakes that persisted during humid periods. Ample proof of this phenomenon is provided by the high productivity of the "Nubian" groundwater aquifer in the eastern Sahara. Thus, a detailed knowledge of the geomorphology of the desert allows us to understand better its archaeological record as well as its groundwater resources.

1. INTRODUCTION

The eastern Sahara of North Africa includes some of the driest regions on the earth. In this desert, precipitation is extremely variable and unpredictable; in some parts it rains only once in 20 to 50 years. This condition necessitates a complete dependence on groundwater resources for human consumption and agricultural activities in the region. However, numerous remains of plants and animals (particularly pollen and fragments of ostrich eggshells) and implements fashioned by human hands indicate the prevalence of wetter climates in the past. This paper relates the recent geologic history of the region to both the wind (aeolian) and water (fluvial) processes that have shaped the land.

The interplay between aeolian and fluvial processes in the eastern Sahara is revealed by the interpretation of satellite images. These include the regional views from the National Oceanic and Atmospheric Administration (NOAA) satellites and weather satellites such as Meteosat of the European Space Agency (ESA). Further details are provided by photographs obtained by the Apollo-Soyuz mission, images of the Landsat Multi-Spectral Scanner (MSS) and Thematic Mapper (TM), as well as radar data from NASA's Space-borne Imaging Radar (SIR) missions and Radarsat of the Canadian Space Agency. These data provide unique perspectives that allow the recognition of regional influences on wind action as well as the exposed, or sand-buried, effects of surface water action.

Dependence on groundwater for land reclamation in the oases depressions of this arid region has led to a decrease in groundwater levels; thus, known water resources have become increasingly scarce during the last three decades. The growth of populations and the attendant food and fiber requirements threaten to exacerbate the situation in the future. It follows that the location of additional groundwater resources in the region would ameliorate a growing problem. Therefore, much emphasis in research today is placed on indications of the potential for concentration of water in the subsurface that accumulated there during humid phases in the past. This would assist in the planning for groundwater exploration in the future.

2. DATA USED

Several types of satellite images were used in this study, including global weather satellite data, photographs obtained by astronauts from earth orbit, multi-spectral images, and radar data. Below are descriptions of these data, followed by a section on methods utilized for image processing and analysis.

2.1. Multi-spectral Images

NOAA satellite data were studied to establish the regional setting of the eastern Sahara in the context of the Great Sahara of North Africa (Figure 1). These data allowed the extent of the major surface processes, particularly sand movement by the wind, in the eastern Sahara to be perceived on a regional scale. The results were compared to, and were found to agree with, regional wind direction maps (Figure 2) based on the analysis of Meteosat data (Mainguet, 1992).

Mosaics of Landsat images (80 m resolution) were used to establish the setting of large accumulations of sand in the eastern Sahara. This was supplemented by analyses of color photographs obtained by the astronauts, particularly on the Apollo-Soyuz mission (Gifford et al., 1979).

In addition, high spatial resolution Landsat Thematic Mapper (TM) images (30 m resolution) allowed the interpretation of the distribution patterns

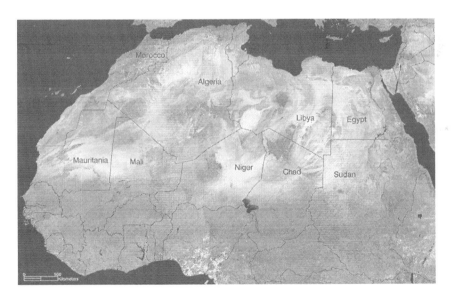

Figure 1. The Great Sahara and the Sahel of North Africa (AVHRR satellite image data from NOAA). Circular or elongate sandy zones (bright areas) represent the major sand seas in Egypt, Libya, Sudan, Chad, Niger, Algeria, Mali, and Mauritania.

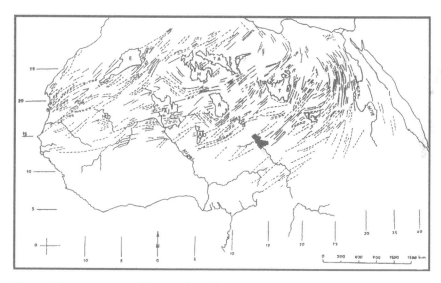

Figure 2. Linear patterns of dune paths in the Great Sahara as interpreted from Meteosat data. The Black area is Lake Chad (compare with Figure 1); lettered areas designate mountains (modified after Mainguet, 1992 and 1995).

of sand deposits and their relationships to topographic features (El-Baz, 1998). These data also provided a base map for correlation with radar data to establish the pattern of channels and the extent of their burial by aeolian sand.

2.2. Radar Data

Because of their utility and unique characteristics, radar data are discussed in more detail in this paper. Imaging radar instruments use active systems with side-looking geometry to produce non-intuitive images that require careful interpretation. For radar imaging, a coherent pulse of energy in the microwave region of the electromagnetic spectrum is issued to produce the beam (Elachi et al., 1982). The typical radar beam wavelengths used to study the earth are 3 cm for X-band, 6 cm for C-band, 10 cm for S-band, and 24 cm for L-band.

In this study, both SIR-C and Radarsat-1 data have been analyzed. SIR-C data were collected at two wavelengths (L-band and C-band) and multiple horizontal and vertical polarizations (HH, VV, HV, and VH). Color composites produced from these scenes have allowed detailed investigations of the imaged surface, because a greater number of wavelengths and polarizations provide more information about the physical characteristics of the terrain (e.g., Robinson et al., 2000).

Radarsat-1 data are collected at one wavelength (C-band) and one polarization (HH). These have the advantage over SIR-C radar in the continuity of data-collection scenes and images that have negligible interference patterns. Thus, a complete areal analysis of drainage patterns is possible using Radarsat-1 images and image mosaics.

Radar images are uniquely able to penetrate the desert sand cover to reveal the courses of ancient rivers and streams in the near surface, which reflect surface water flow during previous humid phases of climate. The principle of subsurface imaging was first alluded to, in theory, by Roth and Elachi (1975). This gained widespread scientific recognition and much interest when images by SIR-A were obtained in November 1981 (Figure 3).

Subsurface imaging occurs if the surface cover material is fine grained (relative to the radar wavelength) and physically homogeneous, which allow the radar signal to penetrate dry sand without significant attenuation (Roth and Elachi, 1975). The sand must be extremely dry, requiring a moisture content of less than 1% for maximum subsurface imaging (Hoekstra and Delaney, 1974). These criteria are satisfied within the sandy-plains environment of the Great Sahara.

Only when the radar wave penetrates the sand and encounters a surface that is rough enough to generate backscatter does subsurface imaging occur (Elachi et al., 1984). It happens preferentially at high look angles (>30°) as

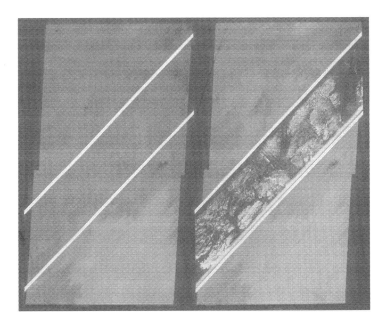

Figure 3. Landsat MSS image (left) with the coverage of the Space-borne Imaging Radar (SIR-A) image strip (right) superposed. The images cover part of the Darfur region of northwestern Sudan. The radar wave penetrated the sand cover to reveal the courses of three major dry rivers: the largest one in the upper right part of the strip is 20 km across. Source: Elachi and Granger, 1982.

this allows more refraction of the radar wave at the air-sand interface (Elachi and Granger, 1982). Conversely, features of low relief are enhanced by shallow look angles, including gently undulating dune faces. All images used in this study have high look angles (>30°) that favor subsurface penetration and imaging.

The depth of subsurface penetration decreases with an increased moisture content of the sand at the time of imaging. Similarly, the presence of an extensive soil and vegetation cover impedes sand penetration by the radar signal because of increased moisture and roughness effects. Experiments show that dry desert sand has a skin depth (the depth to which dry desert sand is attenuated to $1/e = 37\%$ the value at the surface; Blom et al., 1984) of five meters or more (Elachi and Granger, 1982). Penetration depths of this order have been confirmed by fieldwork in the southwestern part of the Western Desert of Egypt near the border with Sudan (McCauley et al., 1982; Schaber et al., 1997; El-Baz and Robinson, 1997; Robinson et. al., 1999; El-Baz et al., 2001).

Radar data served as a basis for the production of maps of subsurface palaeochannels. It was water from these channels that likely sustained prehistoric people in this part of the world during humid phases in the past. Therefore, radar data become indispensable tools in locating prehistoric human occupation sites in this and similar regions (El-Baz, 1991).

3. METHODS

Data processing began by georeferencing all images using PCI's Geomatica GCPworks software. Georeferencing is essential in order to: automatically mosaic the images; store them in a Geographical Information System (GIS) database; and enable the geolocation of satellite observations in the field. Multi-spectral and SIR-C images were georeferenced using ephemeris data, where available, or satellite header information. In the case of the latter, a first-order polynomial transform model was applied in the correction to minimize geometric distortion in the central parts of the scene, which lack control points (e.g., Robinson, 2002).

With respect to Radarsat-1 data, georeferencing proceeded by transferring Ground Control Points (GCP) and orbital layers contained in the original 16-bit image header files and using the information contained therein to apply the transformation. A second order model is applied for the transformation, which is preferred for orbital data, with the cubic resampling mode selected to avoid autocorrelation. For all scenes, a Universal Transverse Mercator (UTM) projection and the WGS84 datum were used where zones and rows were selected based on the scene center of each file.

The images were then digitally enhanced. Methods used include linear scaling of the 16-bit Radarsat-1 files to 8-bit images using the information contained in the image histogram. They also include applying Gaussian stretches, and in the case of SIR-C data, producing color composites. These straightforward procedures produce images with high contrast that are suitable for digitizing drainage features (e.g., Robinson et al., 2000; Robinson, 2002).

Landsat TM and MSS images were combined to produce false-color composites. For example, TM bands 7 $(2.08 - 2.35\,\mu m)$, $4(0.76 - 0.90\,\mu m)$, and 2 $(0.50 - 0.60\,\mu m)$ were combined to allow discrimination of soil-moisture anomalies, lithological variations, and to some extent, the mineralogical composition of sediments (El-Baz et al., 1979). Furthermore, these bands have low-correlation, thus they produce high-contrast images suitable for geological applications analysis (for example, Figure 4). Additional image enhancements included Gaussian, histogram equalization and contrast stretches, and edge enhancement routines with a 3×3 Laplacian filter.

In order to map the fluvial channels, a GIS spatial database was created using ESRI's ArcView and ArcInfo GIS software. Digital maps of fluvial features were generated in vector format, and mapping began by digitizing all wadi paths on individually processed scenes. Digitizing was performed using individual scenes, and not the mosaic, in order to preserve the full dynamic range of the images and to minimize the digitizing file size. Drainage was digitized to include all lines on each individually processed image by tracing

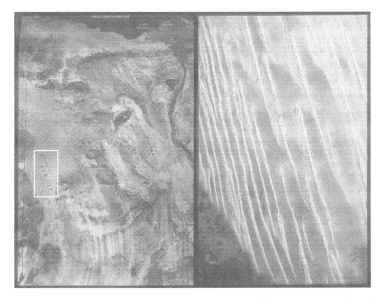

Figure 4. Clear distinction of the spectral reflectance of surface materials is shown by a mosaic of Landsat MSS images (left) of the Western Desert of Egypt (700 km wide). The box in the lower left center designates the area of the enlargement to the right (70 km wide), which shows linear sand ridges of the Great Sand Sea with wide inter-ridge corridors.

the center of a given channel. In the case of flood features, these were traced as polygonal areas. On completion, both the raster satellite images and the vector drainage layers were mosaicked independently to produce the final product. Mosaicking of the vector layers was performed using ArcInfo. The satellite images, on the other hand, were mosaicked using PCI's Geomatica Ortho-Engine software. The main objective while mosaicking was to minimize any loss in radiometric resolution by applying as few adjustments as possible. Several adjustments were needed to find the best combination of parameters. For example, the mosaic shown in Figure 5 was obtained using the following parameters:

- Applying a first-order normalization across images n21.22 e22.37 and n21.22 e21.34 only, leaving the other images unchecked;
- Color balancing in the area of overlap only; and
- Setting the cutline selection method to minimum relative distance.

Normalization was used to even out the maximum and minimum tone across an image to produce a more balanced result. First-order normalization does so by correcting the gradual change in brightness from one side of the image to the other. Color balancing applies tonal and contrast adjustments over the area where the images overlap to produce a more seamless mosaic. Finally, in automated mosaicking, cutlines constitute the areas where seams are least visible based on the radiometric values of the overlapping images. A minimum relative distance option is chosen in order to place the cutline in areas with the least amount of difference in gradient values between the images. Once the parameters are set, the program is left to run overnight to generate the mosaic (Figure 5).

Topographic data clearly show that sand accumulations in the eastern Sahara are restricted to depressions. The most recent Digital Elevation Model (DEM) was produced from data acquired by NASA's Shuttle Radar Topography Mission (SRTM). The elevation data were obtained in February, 2000, using interferometric techniques onboard the Space Shuttle Endeavor to generate a high-resolution digital topographic database of 80% of the earth. Figure 6 shows the 90-m resolution dataset covering the eastern Sahara. The model clearly depicts the depression that encloses the Selima Sand Sheet west of the Nile Valley and the deeper areas that encompass the Great Sand Sea in Egypt and neighboring sand seas in Libya. It also depicts the location of the mosaic in Figure 5.

4. AEOLIAN FEATURES

Surface winds in the eastern Sahara trend in an arcuate pattern that emanates from the direction of the Mediterranean Sea. The direction of the prevailing wind is clearly recorded by linear sand-dune orientations (Figures 2 and 4).

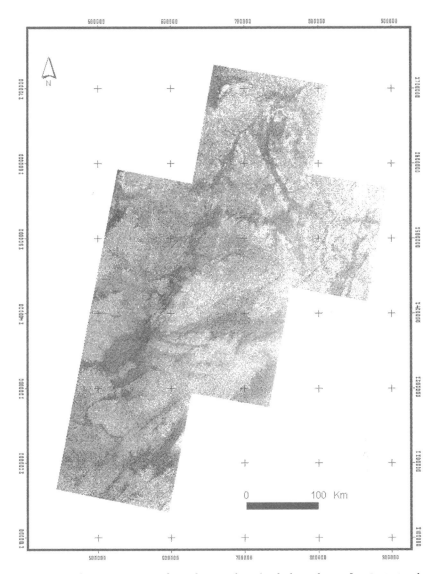

Figure 5. Radarsat-1 mosaic of southeast Libya (including the Kufra Oasis in the north) and northern Chad. It was produced by combining seven georeferenced images. The total file size is 2120 megabytes, with dimensions of 42,073 pixels by 52,824 lines, a size equivalent to an area of approximately 526 km by 660 km. Loss in radiometric resolution was minimized during mosaicking by applying as few adjustments as possible.

AVHRR data show that this pattern changes from southward in the northern part of the desert (Figure 4) to westward along the borders with the Sahel (Figure 2).

The arcuate wind pattern was first suggested for the eastern Sahara based on field mapping of sand dunes in the Western Desert of Egypt by Bagnold (1941). The pattern was observed in the rest of the Sahara from Meteosat data (Figure 2) by Mainguet (1992 and 1995). Sand-dune volumes in the Western Desert of Egypt prove the consistency of this regime at least during the past 5,000 years (El-Baz and Wolfe 1982; El-Baz et al., 2000). Furthermore, wind-produced erosional scars throughout the desert suggest that this regime was effective during much of the Pleistocene (Wendorf and Schild, 1980; Haynes, 1982 and 1985). The southward direction of the sand-carrying wind in the eastern Sahara may be locally affected by topographic prominences (Manent and El-Baz, 1980). Embabi (1982) measured the rate of motion of dune belts in the Kharga depression of the Western Desert of Egypt to vary from 20 m/yr for

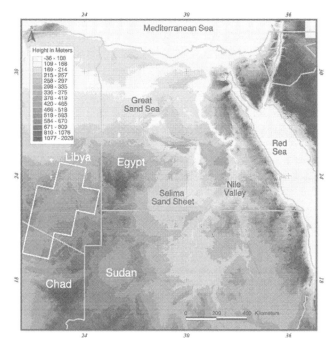

Figure 6. Digital Elevation Model (DEM) of the eastern Sahara (Egypt, and parts of Libya and Sudan) from the 90-m data of the Shuttle Radar Topography Mission (SRTM). The model clearly shows that areas of sand accumulation are in topographic depressions (compare with Figure 1). The region in Libya and Chad of the Radarsat mosaic (Figure 5) is shown in a box.

the large barchans to 100 m/yr for the smallest (1 m high) dunes. Seasonal winds from the south do occur, particularly in mid-spring (El-Baz, 2000). These are not significant transporters of sand, but they do cause dust storms. This research suggests that the Great Sahara is a region of sand export; an area of negative sediment balance (El-Baz et al., 2000). The transported sand accumulates in the Sahel belt south of the Sahara (Figures 1 and 2); an area of positive sediment balance (Mainguet, 1992).

Throughout the Sahara, as shown by DEM data and orbital photographs, sand accumulations occur within topographic depressions (for example, Figure 6). This must be explained in any theory regarding the origin of the sand and the evolution of dune forms in space and time.

Photographs obtained by the Apollo-Soyuz mission were combined with Landsat MSS images in order to map the dunes in the western desert of Egypt (El-Baz, 1977). The results show that in this desert, which covers an area of 681,000 km, over 23% of the total area (159,000 km) is covered by sand dunes (Gifford et al., 1979). The Great Sand Sea covers 72,000 km, where densely packed dunes are confined in a relatively low area in the north and linear forms prevail in the southern part with sand-free corridors (Figure 4, right). It was observed that the dune sand is composed mostly of well rounded quartz grains (El-Baz et al., 1979). The exposed rocks to the north of the sand seas are composed mostly of limestone; thus they could not have been the source of the vast amounts of quartz sand.

El-Baz (1998) proposed that the dune sand originated by fluvial erosion of sandstone rocks such as those of the "Nubian Sandstone" to the south of the dune fields of the eastern Sahara. Therefore, the rounding of the grains would have occurred in turbid water as the particulate matter was transported during humid phases in paleo-rivers and –streams. The sediment load must have been deposited in inland basins where the paleo-channels terminated (e.g., El-Baz and Robinson, 1997; Robinson et al., 2000). As the climate became drier and water evaporated, the particulate matter was exposed to the action of wind. The wind mobilized the sand and sculpted it into various dune forms (El-Baz, 1998 and 2000).

These observations therefore negate the conventional view of the origin of the sand by wind erosion and transportation from the north. The majority of the sand appears to have formed by fluvial erosion of sandstone rocks exposed in the southern part of the Sahara. Thus, it was transported from the south to the north in rivers and streams and was deposited into the topographically low areas where the sand resided.

When the wet conditions of climate changed to dry, the wind began to sculpt these sand deposits into the various dune forms. Repeated alternation of wet and dry periods throughout the Quaternary (1,800,000 million years), and probably much earlier, would result in a cleansing of the sand and causing a high quartz composition (Fairbridge, 2005; *personal communication*). Thus, the alternating climate episodes are responsible for both the location and composition of the material in the sand dunes and sand sheets seen today in the eastern Sahara.

The regional view afforded by observations from space allows the realization that the single most important characteristic of high concentrations of sand deposits is their location within topographic depressions (El-Baz, 1998). All the major sand fields of the Great Sahara occur in topographic basins (Figure 1); in some cases, sand emanates from these low areas as it is driven to higher ground downwind (El-Baz, 2000). In locations where the original fluvial deposits were thin, sand-free areas abound, such as the interdune corridors in the southern part of the Great Sand Sea (Figure 4, right).

5. FLUVIAL HISTORY

As stated above, SIR-C and Radarsat-1 data unveiled the location of numerous channels of former rivers and streams in the eastern Sahara. These channels abound in southwest Egypt, northwest Sudan, and southeast Libya.

In the southwestern part of the Western Desert of Egypt, SIR-C data depicted branching drainage channels (Seto et al., 1997; Robinson et al., 2000; El-Baz et al., 2000). Many of these channels emanate from the Gilf Kebir plateau, a prominent topographic high in southwestern Egypt. The southern Gilf is a nearly circular sandstone mass that is centered at approximately 23.5°N, 26°E, with a diameter of 320 kilometers. The plateau is bordered by numerous dry valleys (wadis) indicating that its edges were shaped by fluvial erosion (Maxwell, 1982).

Landsat TM images show that the wadis are truncated at the top, which suggests that an upper surface layer of softer sediments had been eroded by both fluvial and aeolian processes. The radar data enhance wadi definition, particularly in the plains. Three major drainage systems, Wadis El-Gharbi, Hamra, and El-Malik, start at the edge of the plateau and trend north toward the Great Sand Sea. Another, Wadi El-Rimal, starts northeast of the plateau (El-Baz et al., 2001) and also trends toward the Great Sand Sea region (Figures 7, top; and 8, upper part).

In northwestern Sudan there exists a vast flat plain over 600 km in diameter. It is known as the Great Selima Sand Sheet (Figure 6) after the Selima Oasis, which lies along its southeastern border in Sudan. This oasis is a prominent location along the Darb El-Arbain (the 40-day track) of camel caravans. Many drainage lines in the vicinity of the Great Selima Sand Sheet were revealed by SIR-C images with four major lines leading directly to it from the southwest (Figures 7 and 8, bottom). The northernmost drainage system trends due east and measures 150 kilometers in length. The longest wadi system is also very broad and is aligned in a NE-SW direction. Such broad channels usually develop under sheet flood conditions with plentiful surface water (El-Baz, 2000).

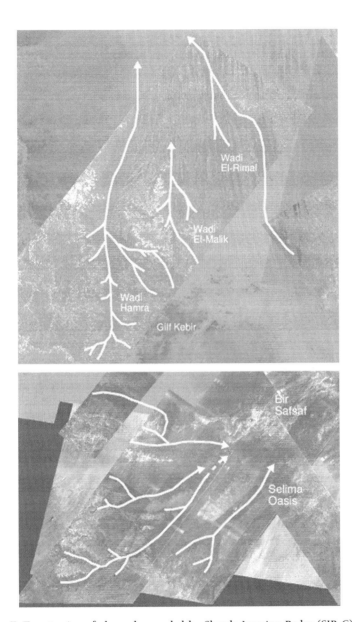

Figure 7. Top, tracing of channels revealed by Shuttle Imaging Radar (SIR-C) image strips in southwestern Egypt. Most of the sand-covered channels emanate from the southern Gilf Kebir plateau and trend northward to the central depression of the Great Sand Sea. Bottom, similarly unveiled channels leading to the Selima Sand Sheet along the border between southwestern Egypt and northwestern Sudan.

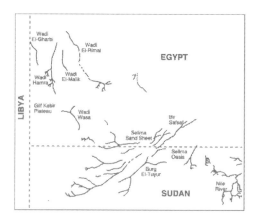

Figure 8. Sketch map of the radar-revealed courses of former rivers (braided at Bin Safsaf) in southwestern Egypt and northwestern Sudan, based on the tracings in Figure 7. Latitude 22°N demarks the border between Egypt and Sudan.

The high-resolution, high-precision SIR-C data show that some of these broad channels display braided streams in their floors, indicating that they are structurally confined. Field observations of trenches dug in May 1998 by a joint team of the Egyptian Geological Survey and Mineral Authority and the Desert Institute of Egypt indicate that moisture begins to appear at 25 centimeters depth in the sand cover of shallow channels in the Bir Safsaf region of southern Egypt. These observations suggest that moisture from occasional rainstorms is carried through, and retained by, the sand fill of the palaeo-channels (El-Baz et al., 2001).

In addition, Apollo-Soyuz and Landsat images of southeast Libya show that the Kufra Oasis region is the only inhabited area in this part of the eastern Sahara. The Oasis had been an important stop along the camel caravan route from Chad northward to the Mediterranean Sea. These images also show the pattern of circular irrigation farms northeast of the Kufra Oasis. These farms were developed starting in the 1960s by the Oxidental Oil Company as part of a concession to explore for oil. The farms are visible from space because of the contrast (in the visible, near-infra red, as well as radar data) between the vegetation and the surrounding sandy plain (El-Baz, 1977).

Both SIR-C and Radarsat-1 data reveal the courses of two sand-buried paleo-channels to the south and southwest of this region (Figure 9; see regional view in Figure 5). The longer and narrower western channel passes through the Kufra Oasis and appears to originate from the highlands to the southwest in the direction of the border with Chad. The wider eastern channel is oriented in a NW-SE direction and appears to originate from highlands southwest of the Gilf Kebir plateau of southwestern Egypt. Thus, it appears that the locations of both the Kufra Oasis and the circular irrigation farms are along the paths of these two former rivers (El-Baz, 2000).

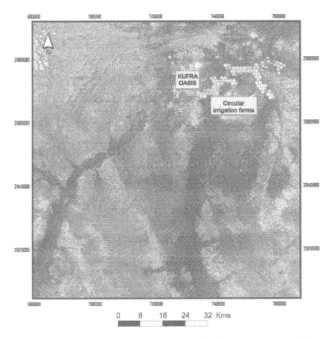

Figure 9. Two major river courses are seen here in Radarsat-1 data. The one on the left leads to the Kufra Oasis and trends southward toward the border with Chad (see Figure 1). The broader channel to the right leads to the region of circular irrigation forms northeast of the Kufra Oasis and appears to originate from highlands southwest of the Gilf Kebir plateau along the border with Egypt.

5.1. Implications for Human Occupation

The eastern Sahara is the largest of the driest regions on the earth (Henning and Flohn, 1977), where the received solar radiation is capable of evaporating 200 times the amount of received rainfall. At Siwa Oasis on the northern edge of the Great Sand Sea, it has barely rained in over four decades (El-Baz, 1979). This dearth of rain is symptomatic of the present-day conditions that are part of the latest of many dry cycles, which have alternated with much wetter climates during much of the Quaternary.

The sand-buried dry courses of rivers and streams are only one of the signs of the kinder climates. Along the northern border of the Selima Sand Sheet, the desert surface is strewn with objects fashioned by human hands. Places that appear to have been close to river channels or boundaries of lakes on the southeast edge of the Gilf Kebir plateau display core rocks from which hand-axes were chipped (Figure 10). Here and there, one encounters milling and grinding stones (for example, Figure 11, top), which indicate the prevalence of grain that was ground into flour for human consumption.

Figure 10. Top, faceted rock (composed of quartzite) from which hand-axes were chipped by prehistoric inhabitants, usually termed "core rock," near the Gilf Kebir plateau in southwest Egypt. Bottom, hand-axes collected in southern Egypt that were dated, by association, to be from 6,000 (second from left) and over 200,000 (last on right) years old.

Because humans roamed the land, we may infer that plants and animals must have existed too, and there is evidence for them. For example, ostriches appear to have existed in abundance, particularly along the northern boundary of the Selima Sand Sheet, where parts of the desert surface are strewn with fragments of ostrich eggshells (Figure 11, bottom) in numerous localities. These most likely denote areas where pre-historic humans stopped along the way and gathered to eat the eggs and discard their shells. In one location, fragments of eggshells were found, which were fashioned into perfectly round pieces with a hole in the middle to be strung into a necklace. A fragment from these eggshells was dated by ^{14}C to be about $7,800 \pm 20$ years old.

In addition, petroglyphs abound on walls of caves and escarpments in southwest Egypt. In one location, the petroglyphs denote a stratification of animal life (Figure 12). The oldest and most faint petroglyph depicts a giraffe. Next are depictions of an ostrich and a baboon. Superposed on these two are the most recent depictions of three grazing animals. This sequence suggests a succession of increasingly dry conditions from the presence of trees for an animal like the giraffe to roam the land, to a savanna-like environment

Figure 11. Top, grinding and milling stones abound in the Western Desert of Egypt, thereby indicating that prehistoric inhabitants harvested and ground grain for consumption. This pair was excavated along the bank of a dry river course southeast of the Gilf Kebir plateau. Bottom, fragments of ostrich eggshells along the northern border of the Selima Sand Sheet in southwestern Egypt. A sample from their suite was dated by ^{14}C method to be $7,800 \pm 20$years old, indicating that ostriches thrived in the region during the latest pluvial (11,000 to 5,000 years ago).

for ostriches and baboons to survive in abundance, to meager vegetation for grazing animals to exist before the drought set in causing dessication of the biota (El-Baz, 1988).

Numerous age-dating techniques have been applied to life remains and lake deposits in the eastern Sahara. These studies show that the present-day dryness began about 5,000 years ago. Before that the eastern Sahara experienced a wet climate dating back to approximately 11,000 years ago. This period is particularly represented by human habitation sites in the Nabta region in southwestern Egypt (Wendorf and Schild, 2001). This wet climate period is only one of numerous cycles that alternated with dry ones.

Archaeological evidence in the Western Desert of Egypt confirms earlier periods of greater effective moisture from remnants of playa or lake deposits (Wendorf and Schild, 1980; Haynes, 1982; Wendorf et al., 1987a, b; Maxwell and Haynes, 1989; Haynes et al., 1993; Szabo et al., 1995). An early Holocene pluvial cycle is well documented by geoarchaeological investigations

Figure 12. Petroglyphs on cave walls and escarpments in the Uweinat Mountain and the Gilf Kebir plateau in southwest Egypt depict numerous animal forms. In this case a giraffe (upper right) appears to be the oldest, most faint, depiction. An ostrich and a baboon are vaguely depicted in lower center, and three cows appear to have been added later in time, because they are superposed on the others. The superposition suggests a time sequence of abundant trees (for giraffes to exist), followed by a savanna-like environment (for ostriches and baboons to thrive), and ending with meager vegetation (for grazing animals to roam the land). Associated indications show that the area dried up and was deserted by most life from about 5,000 years ago. Only an occasional line of trees along a fault zone in the open desert or a meager human settlement near a spring in the Uweinat Mountain remained in modern times.

at Neolithic playa sites in Egypt (Wendorf and Schild, 1980; Pachur and Braun, 1980; Gabriel and Kroeplin, 1989; Haynes et al., 1989). Late Pleistocene lake deposits with associated early and middle Paleolithic archaeological sites are known from work in the Bir Tarfawi area of southwestern Egypt (Wendorf et al., 1987a) and similar associations occur in northwestern Sudan (Haynes, 1985; Haynes et al., 1989 and 1993).

Calcium carbonate deposits associated with some of the radar-revealed channels are believed to have been precipitated in the upper portions of the zone of saturation during pluvial episodes, when water tables were higher than they are today. Uranium-series analyses were applied to samples from southwest Egypt and northwest Sudan, including the Great Selima Sand Sheet localities (Szabo et al., 1995). Results indicate that five paleo-lake forming episodes occurred at about 320–250, 240–190, 155–120, 90–65, and 10–5 ka. Four of these five pluvial episodes may be correlated with major interglacial stages 9, 7, 5e, and one, the 90-65 ka episode, may be correlated with the interglacial substage 5c or 5a (Szabo et al., 1995). In addition, Wendorf et al. (1993) document radiocarbon alluviation from 20 to 12.5 ka as well as a wet period from 60 to 40ka.

Dating and field relationships suggest that the oldest lake and groundwater-deposited carbonates were much more extensive than those of the younger periods. In addition, carbonates of the late wet periods were geographically localized within depressions and buried channels (Szabo et al., 1989 and 1995). These results suggest a trend of decreasing wetness over time.

The prevalence of alternating wet and dry cycles in the eastern Sahara suggests the possibility of repeated migration of plants, animals, and man into and out of the land. As wet conditions set in, human populations spread throughout the land, utilizing the natural resources that accompany the presence of surface water in rivers, streams, and lakes. As a dry climate cycle prevailed, the people deserted the land in search of a stable water source. El-Baz (2001, 2003) suggested that the last migration of Saharan people about 5,000 years ago coincided with the initiation of civilization along the Nile River. He theorized that the migrants from the drying climate in the west possessed a culture of "desert wisdom." They mixed with those who, starting nearly 3,000 years earlier, had developed an advanced "river technology" along the banks of the Nile. El-Baz further theorized that the cross-fertilization of these two different cultures ignited the spark of ancient Egyptian civilization.

Similarly, earlier periods of dryness must have forced nomadic populations of North Africa to the south. These migrations may have resulted in their interaction with sedentary populations in the Sahel region of southern Sudan, Chad, Niger, Mali, and Mauritania (Figure 1). The mixing of these cultures may have invigorated the numerous ancient civilizations that spread throughout North Africa.

5.2. Implications for Groundwater

As the satellite image data have shown, much of the surface water during wet climates ended up in inland basins (El-Baz, 1998 and 2000). It follows that these basins would have stored much of the water in the underlying porous sandstone rocks and their fracture zones. Therefore, the potential of groundwater occurrence within the depressions is an important factor in the exploration for this increasingly scarce resource, throughout the eastern Sahara.

Concentration of groundwater in topographic basins has particularly been tested in Libya. Because of the encroachment of sea water into the coastal aquifers, Libya had to transport sweet groundwater from the southern part of the country to where the population is concentrated along the Mediterranean seacoast. Hundreds of wells were drilled for groundwater in five basins: Kufra, Sarir, Sirt, Hamra, and Murzuq (Salem, 1991). The water is transported in pipelines up to 4 m in diameter. When complete, the pipelines will measure 2,000 km in length and will be capable of delivering up to 2,000,000 m of water per day.

There are indications of a similar hydrogeologic setting in the Western Desert of Egypt. Groundwater exploration in the East Uweinat region of the northern part of the Great Selima Sand Sheet confirmed the presence of resources to support agriculture on 150,000 hectares for 100 years. Over 500 wells have been drilled to a depth of 350 m. In nearly all cases, the water rose in the wells to about 25 m below the surface. In 1989, this region was assigned to become a major land reclamation project by the private sector in Egypt and this practice continues to this day.

Furthermore, two wells drilled for oil exploration near the northern and eastern edges of the Great Sand Sea have proven the presence of vast amounts of water. These wells were drilled south of Siwa Oasis and southwest of Farafra Oasis, respectively, to the depth of 1,200 m The wells penetrated thick sandstone sequences that are saturated with groundwater. The water in these two wells fountains under artesian pressure up to 20 m into the air, indicating vast resources at depth (El-Baz, 1998 and 2000).

6. CONCLUSIONS

Analyses of color photographs as well as multi-spectral and radar images obtained from earth orbit relate the origin of the sand in the eastern Sahara of North Africa to fluvial erosion of the sandstone rocks that are exposed in the southern part of the desert. Interpretation of these data suggests the down-gradient transport of the sand grains to the north in the courses of ancient rivers, which subsequently deposited the sand within inland depressions. Much of the water that accumulated in these depressions during wet climate episodes would have seeped through the underlying rocks to be stored as groundwater.

As dry climates set in, the wind mobilized the sand and shaped it into various dune forms (such as in the Great Sand Sea) or sand sheets in flat plains (such as the Great Selima Sand Sheet). This hypothesis implies that sand was borne by water and sculpted by the wind. It follows that areas with large accumulations of sand host much groundwater beneath the surface. Because similar settings exist in other major deserts of the world, this hypothesis might apply to sandy arid lands worldwide.

This recent geologic history of the eastern Sahara has implications for human occupation in the region. During wet climates, surface water in the form of rivers and lakes allowed plants, animals, and man to thrive. Numerous archaeological remains of the latest wet episode, from 5,000 to 11,000 years ago, abound. These include hand axes, milling and grinding stones, as well as decorative items fashioned from ostrich eggshell fragments, and animal depictions in petroglyphs. The terminal date of this episode coincides with the initiation of civilization along the banks of the Nile. Similarly, previous ends to wet episodes might have caused mass migrations of Saharan people to dependable sources of water such as the Nile to the east and the rivers of the Sahel region to the south.

7. ACKNOWLEDGEMENTS

SIR-C data were obtained from NASA's Jet Propulsion Laboratory. Radarsat-1 data were supplied by the King Abdulaziz City for Science and Technology, Riyadh, Saudi Arabia. Satellite images were interpreted as part of the UNESCO-sponsored International Geological Correlations Program (IGCP) Project-391. Fieldwork was supported by the Arab League Educational, Cultural and Scientific Organization (ALECSO) in Tunis, Tunisia. Remote sensing software is provided by PCI Geomatics. The authors benefited greatly from a review of the manuscript by Fred Wendorf of Southern Methodist University, Dallas, Texas.

REFERENCES

Bagnold, R. A., 1941, *The Physics of Blown Sand and Desert Dunes.* Methuen and Co. Ltd., London.

Blom, R. G., Crippen, R. E. and Elachi, C. 1984, Detection of subsurface features in Seasat radar images of Means Valley, Mojave Desert, California. *Geology* 12:346–349.

Elachi, C., and Granger, J. 1982, Space-borne imaging radars probe "in depth." *IEEE Spectrum* 19:24–29.

Elachi, C., Roth, L. E., and Schaber, G. G. 1984, Space-borne radar subsurface imaging in hyperarid regions. *IEEE Transactions on Geosciences and Remote Sensing* GE-22:383–388.

Elachi, C., Brown, W. E., Cimino, J. B., Dixon, T., Evans, D. L., Ford, J. P., Saunders, R. S., Breed, C., Masursky, H., McCauley, J. F., Schaber, G., Dellwig, A., England, A., MacDonald, H., Martin-Kay, P., and Sabins, F., 1982, Shuttle imaging radar experiment. *Science* 218:996–1003.

El-Baz, F., 1977, *Astronaut Observations from the Apollo-Soyuz Mission.* pp. 98–100, Smithsonian Institution, Washington D.C.

El-Baz, F., 1979, Siwa resort of kings. *Aramco World Magazine,* 30 (4):30–35.

El-Baz, F., 1988, Origin and evolution of the desert. *Interdisciplinary Science Reviews* 13:331–347.

El-Baz, F., 1991, Remote sensing and archaeology. In *1991 Yearbook of Science and the Future,* pp. 144–161. Encyclopedia Britannica, Chicago, IL.

El-Baz, F., 1998, Sand accumulation and groundwater in the eastern Sahara. *Episodes* 21 (3):147–151.

El-Baz, F., 2000, Satellite observations of the interplay between wind and water processes in the Great Sahara. *Photogrammetric Engineering and Remote Sensing* 66 (6):777–782.

El-Baz, F., 2001, Gifts of the desert. *Archaeology* 54 (2):42–45.

El-Baz, F., 2003, Pyramids and the Sphinx. *Dahesh Voice Magazine* 9 (1 Summer 2003):13–18.

El-Baz, F., and Robinson, C. A., 1997, Paleo-channels revealed by SIR-C data in the Western Desert of Egypt: Implications to sand dune accumulations. In *Proceedings of the 12th International Conference on Applied Geologic Remote Sensing,* pp. 469–476. Environmental Research Institute of Michigan, Ann Arbor, MI.

El-Baz, F., and Wolfe, R. W., 1982, Wind patterns in the Western Desert. In *Desert Landforms of Southwest Egypt: A Basis for Comparison with Mars,* pp. 119–139. NASA CR-3611.

El-Baz, F., Robinson, C. A., Mainguet, M., Said, M., Nabih, M., Himida, I. H., and El-El-Etr, H. A., 2001, Distribution and morphology of palaeo-channels in southeastern Egypt and northwestern Sudan. *Palaeoecology of Africa* edited by Klaus Heine, 27:239–258.

El-Baz, F., Mainguet, M., and Robinson, C. A., 2000, Fluvio-aeolian dynamics in the north-eastern Sahara: Interrelation between fluvial and aeolian systems and implications to ground water. *Journal of Arid Environments* 44:173–183.

El-Baz, F., Slezak, M. H., and Maxwell, T. A., 1979, Preliminary analysis of color variations of sand deposits in the Western Desert of Egypt. In *Apollo-Soyuz Test Project Summary Science Report: Volume II: Earth Observations and Photography*, edited by F. El-Baz and D.M. Warner, pp. 37–262. NASA SP-412.

Embabi, N. S., 1982, Barchans of the Kharga depression. In *Desert Landforms of Southwest Egypt: A Basis for Comparison with Mars*, edited by F. El-Baz and T.A. Maxwell, pp. 141–155, NASA CR-3611.

Fairbridge, R., 2005, Personal communication to F. El-Baz.

Gabriel, B. and Kroeplin, S., 1989, Holocene lake deposits in northwest Sudan. In *Paleoecology of Africa and the Surrounding Islands*, edited by J.A. Coetzee and E.M. van Zinderen-Bakker, pp. 295–299. Balkema, Rotterdam.

Gifford, A. W., Warner, D. M., and El-Baz, F., 1979, Orbital observations of sand distribution in the Western Desert of Egypt. In *Apollo-Soyuz Test Project Summary Science Report, Volume II: Earth Observations and Photography*, edited by F. El-Baz and D.M. Warner, pp. 219–236. NASA SP-412.

Haynes Jr., C. V., 1982, Great Sand Sea and Selima Sand Sheet: Geochronology of desertification. *Science* 217:629–633.

Haynes Jr., C. V., 1985, Quaternary studies, Western Desert, Egypt and Sudan – 1979–1983 field seasons. *National Geographic Society Research Reports.* 16:269–341.

Haynes Jr., C. V., Eyles, C. H., Pavlish, L. A., Rotchie, J. C., and Rybak, M., 1989, Holocene paleoecology of the Eastern Sahara: Selima Oasis. *Quaternary Science Review.* 8:109–136.

Haynes Jr., C. V., Maxwell, T. A., Johnson, D. L., 1993, Stratigraphy, geochronology, and origin of the Selima Sand Sheet, eastern Sahara, Egypt and Sudan. In *Geoscientific Research in Northeast Africa*, edited by U. Thorweihe and H. Schandelmeier, pp. 621–626. Bekena, Rotterdam.

Henning, D., and Flohn, H., 1977, *Climate Aridity Index Map.* U.N. Conference on Desertification, UNEP, Nairobi, Kenya.

Hoekstra, P., and Delaney, A. 1974, Dielectric properties of soils at UHF and microwave frequencies. *Journal of Geophysical Research.* 79:1699–1708.

Mainguet, M. M., 1992, A global open wind action system: The Sahara and the Sahel. In *Geology of the Arab World*, Vol. II, edited by A. Sadek, pp. 33–42. Cairo University Press, Cairo, Egypt.

Mainguet, M. M., 1995. *L'homme et la Secheresse.* Collection Geographie, Mason, Paris. p. 335.

Manent, L. S., and El-Baz, F., 1980, Effects of topography on dune orientation in the Farafra region, Western Desert of Egypt, and implications to Mars. In: *Reports of Planetary Geology Program*, NASA Technical Memo 82385, pp. 298–300.

Maxwell, T. A., 1982, Erosional features of the Gilf Kebir and implications for the origin of Martian Canyonlands. In *Desert Landforms of Southwest Egypt. A Basis for Comparison with Mars.* edited by F. El-Baz and T.A. Maxwell, pp. 281–300. NASA CR-3611.

Maxwell, T. A., and Haynes, C. V., 1989, Large-scale, low-amplitude bedforms (chevrons) in the Selima Sand Sheet, Egypt. *Science* 243:1179–1182.

McCauley, J. F., Schaber, G. G., Breed, C. S., Grolier, M. J., Haynes, C. V., Issawi, B., Elachi, C., and Blom, R., 1982, Subsurface valleys and geoarchaeology of the Eastern Sahara revealed by Shuttle radar. *Science* 218:1004–1020.

Pachur, H. J., and Braun, G., 1980, The paleoclimate of the central Sahara, Libya, and the Libyan Desert. *Paleoecology of Africa.* edited by M. Sarentheim E. Siebold and P. Rognon, 12:351–363.

Robinson, C. A., 2002, Application of satellite radar data suggests that the Kharga Depression in Southwestern Egypt is a fracture rock aquifer. *International Journal of Remote Sensing* 23 (19):4101–4113.

Robinson, C. A., El-Baz, F., Ozdogan, M., Ledwith, M., Blanco, D., Oakley, S., and Inzana, J., 2000, Use of radar data to delineate palaeodrainage flow directions in the Selima Sand Sheet, Eastern Sahara. *Photogrammetric Engineering and Remote Sensing* 66 (6):745–753.

Robinson, C. A., El-Baz, F., and Singhroy, V., 1999, Subsurface imaging by Radarsat: comparison with Landsat TM data and implications to ground water in the Selima area, Northwestern Sudan. *Canadian Journal of Remote Sensing* 25 (3):268–277.

Roth, L. E., and Elachi, C., 1975, Coherent electromagnetic losses by scattering from volume inhomogeneities. *IEEE Transactions on Antennas and Propagation* 23:674–675.

Salem, O., 1991, The Great Man-Made River Project: A partial solution to Libya's future water supply. In *Planning for Groundwater Development in Arid and Semi-Arid Regions*, pp. 221–238. Research Institute for Ground Water, Cairo, Egypt.

Schaber, G. C., McCauley, J. F., and Breed, C. S., 1997, The use of multifrequency and polarimetric SIR-C/X-SAR data in geologic studies at Bir Safsaf, Egypt. *Remote Sensing of the Environment* 59: 337–363.

Seto, K. C., Robinson, C. A., and El-Baz, F., 1997, Digital image processing of Landsat and SIR-C data to emphasize drainage patterns in southwestern Egypt. In *Proceedings of the 12th International Conference on Applied Geologic Remote Sensing*, pp. 93–100. Environmental Research Institute of Michigan, Ann Arbor, MI.

Szabo, B. J., Haynes Jr., C. V., and Maxwell, T. A., 1995, Ages of Quaternary pluvial episodes determined by uranium-series and radiocarbon dating of lacustrine deposits of Eastern Sahara. *Paleogegraphy, Paleoclimatology, Paleoecology* 113:227–242.

Szabo, B. J., McHugh, W. P., Shaber, G. G., Haynes Jr., C. V., and Breed, C. S., 1989, Uranium-series dated authigenic carbonates and Acheulian sites in southern Egypt. *Science* 243:1053–1056.

Wendorf, F., and Schild, R., 1980, *Prehistory of the Eastern Sahara*. Academic Press, New York.

Wendorf, F., Close, A. E., and Schild, R., 1987a, Recent work on the Middle Paleolithic of the Eastern Sahara. *African Archaeological Review.* 5:49–63.

Wendorf, F., Close, A. E., and Schild, R., 1987b, A Survey of the Egyptian Radar Channels: An Example of Applied Archaeology. *Journal of Field Archaeology.* 14:43–63.

Wendorf, F., and Schild, R. and Associates, 2001, The Archaeology of the Nabta playa. In *Holocene Settlement of the Egyptian Sahara*, p. 707. Kluwer Academic, New York.

Wendorf, F., Schild, R., and Close, A. E., 1993, Summary and Conclusions In: *Egypt During the Last Interglacial*. pp. 552–573. Plenum Press, New York.

Chapter *3*

Southern Arabian Desert Trade Routes, Frankincense, Myrrh, and the Ubar Legend

RONALD G. BLOM, ROBERT CRIPPEN, CHARLES ELACHI,
NICHOLAS CLAPP, GEORGE R. HEDGES, AND JURIS ZARINS

Abstract: A location that likely inspired some elements of legendary accounts of the "lost city" of Ubar has been found at the edge of the Arabian Peninsula's Empty Quarter at the village of Ash Shisr in modern day Oman. The site consists of the remains of a central fortress surrounding a well. Artifacts from as far away as Persia, Rome, and Greece are found, indicating a long period of far-flung trade through this isolated desert location. More recent work in Oman and Yemen indicates this fortress is the easternmost remains of a series of desert caravansaries that supported incense trade. Legend was that Ubar perished in a sandstorm as divine punishment for wicked living. Actually, much of the fortress collapsed into the sinkhole that hosted the well, perhaps undermined by ground water withdrawal used to irrigate the surrounding oasis. Less fanciful interpretation of legendary and other accounts clearly indicates "Ubar" was actually a region— the "Land of the Iobaritae" identified by Ptolemy. Desert trade was probably abandoned because of three primary factors: frankincense diminished in importance with the conversion of the Roman Empire to Christianity, desert ground-water levels continued to fall and the oases dried up, and reliable sea transportation was developed. The archaeological site was located, and its importance recognized, by an unusual combination of historical research and application of space technology in support of traditional archaeology. The site was known earlier but its significance unappreciated, as it was never studied in adequate detail. The archaeological importance of the site is supported through regional context provided by carefully enhanced Landsat Thematic Mapper (TM) and other satellite imagery that shows a discontinuous network of trails that

converge at Shisr. Some of these trails are demonstrably old as they pass beneath
sand dunes 100 m tall. Thus the desert environment can preserve ancient evidence
of human occupation detectable in remote sensing data. Image analysis further
shows no evidence of major undocumented sites in this desert region (e.g. the
"Ubar" of legend). The interdisciplinary nature of this work demonstrates the
significant and still underutilized potential of using remote sensing and GIS
technology in support of traditional archaeology.

1. INTRODUCTION

While documenting the reintroduction of the Oryx to Oman, one of us (Clapp)
became enchanted with legendary accounts of a fanciful city in the Empty
Quarter of Arabia, now one of the most inhospitable places on earth. His
inquiries ultimately led to the formation of an interdisciplinary team to investigate. The team included the present authors, who contributed expertise
in archaeology (Zarins), geology and remote sensing (Blom, Crippen, and
Elachi), and history and campaign organization (Hedges and Clapp).

The fortress site at Ash Shisr would have existed to support trade in
frankincense and myrrh, which flourished from before 3,200 BCE to about
300 CE. Frankincense is the resin of *Boswellia sacra,* which grows principally in southern Arabia. Similarly, myrrh is from *Commiphora myrrha,*
which grows widely in the region. Precisely where principal sources of these
valuable commodities were, however, probably deliberately concealed at the
time. In the ancient world incense was widely used for religious ceremonies,
medicinal purposes, and generally to make life smell better. Frankincense
in particular was used in large quantities for Roman cremations until the
Emperor Constantine banned the practice in beginning the conversion of the
Empire to Christianity after 312 CE.

Trade routes across the desert had considerable economic appeal in that
they avoided the dangers of sea travel (pirates and storms), and the dangers
and expense of travel through more populated areas (thieves and taxes).
In ancient times, travel across such regions would have been aided by the
fact that aquifers would have been bank full after the last climatic optimum
which ended, $\sim$ 6,000 BCE in this region (Anton, 1984), and memories of
water sources in the desert would have been continuous. Since this region
is primarily a limestone terrain, near-surface water would have been very
irregularly distributed and restricted to regions where geological factors were
favorable (e.g. folds, fractures, geological contacts).

In the desert-trade scenario, outposts where frankincense and other cargo
were gathered and caravans were outfitted for the long trek across the desert
would have existed at reliable water sources. Inquiry into such varied sources as
the *Arabian Nights,* the *Koran,* and particularly the Greek geographer Ptolemy
in his *Geographia,* indicated the general region where an "Ubar" may have
existed. Ptolemy refers to the "land of the Iobaritae" located in the southern

Arabian Peninsula. Unfortunately, none of Ptolemy's original maps survives. Surviving are lists of coordinates, in Ptolemy's own latitude/longitude scheme, specifying many known places, and some unknown places, in the ancient world. In the "Land of the Iobaritae" Ptolemy also locates a specific place, "Omanum Emporium" (Ptolemy *Geo* VI:7,36; VII:12,22). In Ptolemy's time (ca. 150 CE), this would have been a functioning trading center and caravansary. Additionally, in Pliny's *Natural History* Book XII, there is reference to "octo mansiones" which in Bostock and Riley (1855) is translated as "at a distance of eight stations from this is the incense-bearing region". "Masiones" would more likely mean "fortress" or "castle," rather than a distance measure. So, while less than clear, "stations" may refer to eight caravan stops on a desert trade route. Zarins notes there are major spring sites in the desert of Oman and Yemen which are not yet fully explored (Zarins, 2001:141). Tkach (1897) summarizes earlier Arab mention of southern Arabian wealth from trade, including scholars Al-Tabari, Yaqut, Al-Thalabi, Al-Hamdani, and Al-Kisai. This evidence gave a logical basis to expect that desert outposts did in fact exist to outfit caravans. The difficulty was focusing the search into a manageable area. The ancient accounts were compelling in support of the notion that desert trade routes existed, and in that the legends were an amalgamation of exaggerated and fanciful fact and stories about a region that grew wealthy on incense trade, with a tiny kernel of underlying validity.

2. REFINING A SEARCH AREA

The most precise lead available at the outset of our efforts was the report by British explorer Bertram Thomas from his last expedition in the Empty Quarter in 1930 (Thomas, 1931, 1932). During this expedition, Thomas noted the presence of a track at 18 degrees 45 minutes north, by 52 degrees 30 minutes east, leading northward into the Empty Quarter in an area where no reasonable person would now go (Thomas, 1932: 161). His Bedouin guides revealed the legendary account of Ubar to Thomas, who marked the track on his map with remarkable precision. Along the projection of the track, he marked "probable location of the ancient city of Ubar," based on the Bedouin description of distance northward into the sands. Thomas had planned to return and follow the track, but was unable to do so. Thomas also reported the track to T. E. Lawrence (Lawrence of Arabia), who regarded Ubar as the "Atlantis of the Sands." Ubar was again forgotten until the early 1950s when two expeditions led by explorer and oil magnate Wendell Phillips managed to re-locate Thomas' road, but were unable to follow it, their vehicles defeated by the dunes. Following Bertram Thomas' track, and perhaps finding Ubar, required something beyond traditional approaches.

In 1982, one of us (Clapp) read of the results from the Jet Propulsion Laboratory's Shuttle Imaging Radar-A mission, which flew on the second flight of the Space Shuttle Columbia in 1981 (McCauley et al., 1982). Radar images from that mission covering the Sahara, in the region near the Egypt-Sudan border, revealed ancient, now abandoned river channels in what is at present one of the very driest places on Earth. Extreme dryness of the area had rendered the windblown sand cover, which blankets the old fluvial landscape with a variable thickness of sand up to several meters thick, transparent to the radar (McCauley et al., 1982). Thus the underlying landscape created under much wetter climatic conditions during the Tertiary was revealed on the radar images. During wetter periods in the Quaternary water occasionally flowed in the Sahara's older channels, creating a grassland environment intermittently favorable to animal and human occupation. With deterioration of the climate, humans and animals were forced to more favorable areas, most recently about 6,000 BCE. The Arabian Peninsula shares this climatic history. Clapp reasoned that because the legendary accounts indicated that Ubar was buried in sand, the technological edge provided by the ability of long-wavelength radar to see through 2 or more meters of dry sand cover, as demonstrated with SIR-A radar images and verified in the Western Desert of Egypt, might help locate the site. An overview of the sub-surface imaging potential of radar is outlined in Blom et al., 1984, and Elachi et al., 1984.

Clapp contacted JPL (Jet Propulsion Laboratory) and after nearly being dismissed as a crank, convinced us that his search was both serious and possible. Charles Elachi, Shuttle Imaging Radar Principal Investigator and now Director of the Jet Propulsion Laboratory, arranged to acquire radar images of the Thomas Road area as a target of opportunity during the flight of the Shuttle Imaging Radar-B in 1984. The resulting images showed intriguing features in a then largely uncharted area, but unfortunately no actionable sites were obvious. What was obvious was the need for complete regional image coverage not available from infrequent Shuttle flights providing limited geographic coverage.

3. REMOTE SENSING IMAGE DATA AND PROCESSING

In our search for potential archaeological sites in the desert, remote sensing data acquisition was targeted by the historical research outlined above. Remote sensing data covering a broad swath across the southern margin of the sands of the Empty Quarter from the Shuttle Imaging Radar, the Shuttle-based Large Format Camera, Landsat Thematic Mapper, and SPOT satellites were acquired, enhanced, and analyzed. Ultimately, we used every available type of remote sensing data having sufficient spatial resolution to detect something helpful. It is useful to review the key features of each in the sequence employed in the study, and review the most useful features of each. Since a goal of this paper is to motivate future applications of remote sensing in archaeology, brief non-exhaustive mention will be made of available new sensors with potential.

3.1. Shuttle Imaging Radar

As the name implies, this instrument is an imaging radar carried on board the Space Shuttle. The capability of this relatively long-wavelength (23 cm, or "L-Band") radar to image features in the immediate sub-surface under extremely arid conditions, such as in the Sahara, and the Empty Quarter of the Arabian peninsula, was in fact the original reason that JPL was called in to help with the search. The Shuttle Imaging Radar series flew only four times, SIR-A in 1981, SIR-B in 1984, and SIR-C twice in 1994 with a three wavelength system (in addition to L Band, C-Band or 5.6 cm, and X-Band or 3 cm wavelength), and also multipolarization capability. As we began our study, only limited coverage of the area of Bertram Thomas' road area in Oman was obtained during the 1984 flight of SIR-B aboard the Shuttle Challenger. The radar data did show evidence of a landscape created by wetter climatic conditions, with prominent fluvial channels carved in the desert similar to what had been seen in SIR-A images of the Sahara. However, no features were seen in the limited SIR-B radar image coverage to support the idea that a major caravansary such as Ubar might have existed nearby. Our plan was imaging of the area in the follow-on Shuttle Imaging Radar-C mission. However, with the Shuttle Challenger accident on its very next flight after SIR-B was carried, SIR-C was significantly delayed. So, although imaging radar, along with the potential for sub-surface imaging, was the original reason for JPL/NASA involvement in this project, lack of radar data at the time compelled us to broaden our data horizons. This in fact led to the realization that each data type has unique contributions to regional reconnaissance. Additional information on radar and the SIR series is available at http://southport.jpl.nasa.gov/.

3.2. Landsat Thematic Mapper

We have made extensive use of Landsat Thematic Mapper (TM) images for geologic studies in the southwestern US (Ford et al., 1990, Beratan et al., 1990). At the time Landsat satellites 4 and 5 were both operational with Thematic Mapper instruments onboard. These satellites are in 830 km high sun-synchronous polar orbits, with a morning overpass. Each pixel (picture element) in the image represents a 28.5 by 28.5 meter area on the ground, and records reflected light in six wavelength bands spanning 0.45 to 2.35 micrometers (plus one thermal channel which was not used because of poor ∼ 100 m spatial resolution). There are three channels in the visible (0.45–0.69 micrometers), and three in the reflective (i.e., non-thermal) infrared (0.76–2.35 micrometers). The reflected light at near-infrared wavelengths turned out to be most useful for geology and other applications in arid areas because many landscape features, such as rocks, soils, sand, and old desert tracks, are much more distinctive at these longer wavelengths. The desert tracks are very distinctive on the enhanced Landsat images, and can be traced over long distances using a single TM scene spanning over 30,000 square km. Some of the tracks seemed to narrow and vanish as they went into the dune areas, even though we knew we were looking at desert floor and not sand, as the desert floor is distinct from sand in the images. We reasoned that higher resolution was necessary and acquired SPOT image data to supplement the Landsat images (SPOT described next). It is important to note that properly enhanced images from the Landsat TM instrument were critical to the success of this work. It is also important to note for future work that the Landsat 7 TM instrument (launched in 1999) has a 15-meter panchromatic (black and white) band useful for resolution enhancement (http://landsat7.usgs.gov/). While the Thematic Mapper instrument on Landsat 7 is no longer functioning properly, a large archive of data exists for analysis (e.g., http://glcf.umiacs.umd.edu/data/landsat/ and others). Furthermore, the ASTER instrument (Advanced Spaceborne Thermal Emission Spectrometer http://asterweb.jpl.nasa.gov/) on board the EOS (Earth Observing System) Terra platform launched in 1999 has additional spectral channels and has significant, and so far largely unexploited, potential for archaeological studies.

3.3. SPOT

The SPOT (Systeme Probatoire d'Observation de la Terre) satellites were launched by and are operated by the French (http://www.spotimage.fr/html/_.php). Multiple SPOT satellites are in ~ 700 km high sun-synchronous orbits, with morning overpasses. A typical SPOT scene covers 60 km × 60 km, substantially less than the 180 km × 170 km of Landsat. SPOT satellite data used by us have two modes, a color mode with 20-m pixels, and a black and white mode with 10-m pixels. The color mode is less useful than the Landsat data because, although the spatial resolution is better, the spectral coverage does not extend to the longer near-infrared regions, from 1.2 to 2.4 micrometers, so useful for geological studies. We used the 10-m black-and-white images which, at the time, had the highest spatial resolution generally available to the scientific community from a satellite. We digitally combined the high-resolution 10-m SPOT images with Landsat Thematic Mapper images to achieve high spatial resolution with the spectral information from the Landsat images. Images resulting from this technique enabled us to detect and follow tracks not discernible on the Landsat images alone. In particular, detection of Bertram Thomas' road into the dune areas required the resolution of the SPOT images. This track is not detectable on Landsat 5 images alone.

3.4. Large Format Camera

For completeness it is important to mention the Large Format Camera (LFC), which was a Space Shuttle-borne photographic camera system, now largely of historical interest having flown only once and overtaken by available sub-meter satellite digital imaging. It was capable of 2–3 meter spatial resolution. Unlike the other data sets, LFC photographs are not digital. During its 1984 flight, some black-and-white photographs were acquired of portions of Oman. These were analyzed in conjunction with the satellite images for a more comprehensive view of possible trade routes. As with the other shuttle-borne data, the discontinuous strips and lack of regional coverage was a great disadvantage. Although superseded by modern high-resolution digital sources, the images from 1984 may provide information not otherwise available (http://edc.usgs.gov/products/aerial/space.html). It is also worth noting that a large archive of photographic data from the "Corona" intelligence satellites is available. While also not digital, and the best resolution is ∼2 m, the acquisition dates from 1960–1972 precede much modern development and so Corona photography may provide information not otherwise available. Information on Corona data, and links to searching the catalog is available from the US Geological Survey at (http://erg.usgs.gov/isb/pubs/factsheets/fs09096.html).

3.5. Image Processing

Digital images used in this project were enhanced with several standard and well known techniques. The methods include various band (or channel) combinations, ratios of various band combinations, utilizing the full dynamic range through contrast enhancements, spatial filtering, and edge enhancement. These procedures have been reviewed thoroughly elsewhere (Moik, 1980, Gillespie, 1980, and others), and are common to image enhancement software packages available today. In the case of this work, the major difference of our image-processing methods was principally in attention to detail as we were searching for subtle features. In particular, Landsat band-ratio composites and band composites sharpened with SPOT images were created from the image datasets for the areas where we wished to map the desert tracks. In making band-ratio composites, it is important to account precisely for atmospheric effects, which vary with wavelength, and for sensor gain settings for each band. Direct analytical solutions to achieve this result in the absence of surface calibration data are discussed in papers by Crippen (1987, 1988, 1989a). Also, the detectors in the Landsat TM scanner do not all have the same response, and banding on the images occurs under extreme enhancement (this is often especially noticeable on scenes which include some ocean). Simple along-line filtering algorithms do not work because single detector scan lines are converted into more than one line of image data during geometric rectification of the raw data. We developed a simple, non-fourier based technique that ameliorates this problem in images of natural terrain (Crippen, 1989b). The method does not work on highly structured scenes covering agricultural or urban areas, so appropriate application is required for this useful technique (such as the desert terrain of interest here).

Unusual techniques applied to the Landsat and SPOT datasets included "directed band ratioing" (Crippen et al., 1988), and "four component" processing (Crippen, 1989a). The latter methods proved very useful in detection and mapping of the desert tracks. The essence of the enhancement procedure is the following.

Four appropriately selected components of remotely sensed data can in fact be displayed and interpreted in a three-color composite image. An achromatic component, that is, the best enhanced black-and-white image, is selected for its depiction of topographic shading and band-correlated reflectance features. The black-and-white image is heavily spatially filtered to remove regional brightness variations not of interest when trying to detect small local features. This is accomplished via sequential high-pass filtering of the image, typically with values like 501×501 pixels, 51×51 pixels, and then 3×3 pixels. This procedure dramatically increases local image contrast. Three non-redundant chromatic components are selected for their depiction of band-variant reflectance information. Band ratios 3/1, 5/4, and 5/7 result in discrimination of primarily materials relatively rich in ferric iron (3/1), ferrous iron (5/4), and clays, carbonates, and vegetation (5/7). When computing these ratios one must precisely account for atmospheric effects and sensor gain and offset values or the resulting images will be unsatisfactory under extreme enhancement (Crippen, 1987, 1988). The 3/1 ratio is displayed in blue, 5/4 in green, and 5/7 in red, all multiplied by the achromatic image and scaled appropriately for display. In order to accomplish this, the coefficients of variation for the chromatic components are equalized, and the coefficients of variation for the achromatic component are adjusted so that all four components will be prominent in the composite. Mathematical convolution of the four components into three display channels (red, green, blue) is achieved by multiplying each successive chromatic component (blue, green, and red) by the achromatic (or black-and-white) component and then scaling for display. Little loss of information occurs during this mathematical convolution because the typical spatial properties of topographic features in the achromatic component differ sufficiently from those in the chromatic components to prevent perceptual confusion. In the resulting image, color is a function of the composition of the materials in the image in so far as they can be discriminated in Landsat TM data, while the topography and texture are displayed achromatically. The technique can produce highly informative and readily interpretable enhancements of Landsat Thematic Mapper (TM) or similar data sets. It can also be used for the merger of differing image datasets such as SPOT panchromatic and TM. Input data are enhanced high-resolution black-and-white image data from SPOT or the new Landsat 7 panchromatic channel, which is used to multiply the color channels. Here we used the method on both Landsat and SPOT/Landsat combinations. Crippen (1989a) provides a detailed explanation.

4. MAPPING ANCIENT AND MODERN TRACKS, GUIDING EXPEDITIONS

We acquired Landsat scenes, and as the data were enhanced by the methods covered above and analyzed many tracks began to emerge. We determined that even ancient tracks across the little-changing desert floor could be mapped using suitably enhanced image data. Some tracks were obviously old because they went directly under very large sand dunes. Analysis indicated that detection of the tracks would be improved with higher resolution than that provided by the 28.5-meter pixels of the TM sensor. We acquired and integrated panchromatic images from the French SPOT satellite, with 10-meter pixels in this mode. To achieve the best possible detection, the Landsat and SPOT image data were combined using the technique outlined above where the Landsat spectral information is used for image color, and the black and white SPOT data for improved image detail. Because of the report of a possible location for Ubar by Bertram Thomas, we initially targeted this area for this type of analysis. In the resulting images, we were able to follow and map the road Thomas' delineated in his 1932 map out into the Empty Quarter. Our enthusiasm and suspicion were both aroused by analysis of the images. We were excited by the appearance of an L-shaped feature in the enhanced SPOT/TM images next to the track more than 30 km into the dunes (Figure 1).

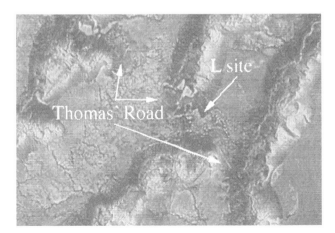

Figure 1. SPOT/Landsat TM four component merge of area 30 km north of road noted by Bertram Thomas. The road is nearly continuous from where Thomas' mapped it and here emerges from under a > 100 m tall sand dune at right center, demonstrating its antiquity. Road proceeds past an L shaped feature, which is actually a deflationary depression about 1 m deep, and continues northwest under dunes. Limestone desert floor is blue because of low reflectivity in band 7. The image is about 5 km wide; north is up.

This unnatural looking feature (in our fantasies), was located exactly where Thomas had noted "probable location of the Ancient City of Ubar" on his map (Thomas, 1932 map plate). However, the track to the "L" site seemed rather narrow to have been the highway to the center of desert frankincense transport, especially as compared to other tracks mapped elsewhere on the images. The "L" site seemed unlikely as a location for Ubar on several other grounds. It was well (about 30 km) out into the high dunes, had no obvious source of water, and was not in a position to control caravan traffic since this site could be bypassed. It is interesting to note that by using the enhanced satellite images, we had rediscovered Bertram Thomas' road while sitting at computer terminals, and could see the continuation of the track out into the desert where Thomas only dreamed of going.

Despite our misgivings about the "L" site, it was a prime target in the reconnaissance phase expedition in July of 1990. We were generously provided with a UH-1 helicopter and crew from the Royal Oman Air Force. Departing from Salalah, on the southern coast of Oman, and using the enhanced satellite images and a GPS receiver, we were able to fly a direct intercept to Thomas' road, and then fly the road directly to the "L" site. In one long morning's reconnaissance, we were able to cover the same distance that took Thomas 11 days by camel, and then follow his track into the dunes where he was never able to go. The track and the L are more distinctive on the image data than to the eye because of the fact that the limestone desert floor contrasts markedly with sand dunes at longer than visible wavelengths detected by the Landsat Thematic Mapper instrument. Had we not had both the images and the GPS navigation tool, we would have had great difficulty locating the site. As it turned out, the mysterious "L" appears to be a natural shallow depression formed by deflation of the surface around the sand dunes. Under wetter climatic conditions, a shallow pond would have existed here. Only a few minutes at the site were necessary for archaeologist Zarins to conclude that this was not the site of a mythic Ubar. While Bertram Thomas may have been disappointed, in fact it was a major Neolithic site, with many, many stone tools scattered about. This in itself was a remarkable find. Large dunes move slowly and in Neolithic times the path was likely clear, but the archaeological implication is that this was no longer a viable travel route by the Bronze Age. Figure 2a is a photograph from the helicopter showing Thomas' road emerging from under a large sand dune. Compare the expression of the desert floor and the dunes in the photograph to that in the satellite image in Figure 1. The limestone desert floor is distinctive in Figure 1 because limestone (and carbonates in general) has low reflectivity in band 7, compared to the other materials in the scene. The satellite image in Figure 1 is a band composite of Landsat band 1 (blue), band 4 (green), and band 7 (red), all multiplied by the SPOT satellite's 10-m panchromatic channel. Figure 2b shows Zarins examining artifacts at the L site; note the helicopter in the background for scale.

Figure 2a. Photograph from helicopter of Thomas' Road emerging from under a sand dune > 100 m tall. This is a southeastward looking view of the dune and road at the southern end of Thomas' road seen in Figure 1. Note lack of contrast between sand dunes and limestone desert floor in this visible image compared to satellite images.

Upon returning from the reconnaissance expedition, just before the first Gulf War, we realized three things. First, the Bertram Thomas' road results demonstrated that one could detect and map even very narrow and old tracks in the desert, which pre-dated major incense trade. This meant it was unlikely that a major trading center could escape detection if we acquired the right image data. Second, unlike the Sahara where we had used radar imagery, the sand cover was not continuous and therefore the limited radar datasets were not necessary. Very effective use of satellite images showing only surface features could be made. Third, the amount, ages, and diverse origin of

Figure 2b. Archaeologist Zarins at the "L" site. Helicopter at corner of L for scale. Located on the extension of Thomas' road 30 km into the dunes from where he mapped it. The L-shaped feature is a shallow natural depression with abundant Neolithic material surrounding. It was likely formed when the edges of dunes stalled for an extended period of time causing deflation.

archaeological material at other known, but poorly studied, sites investigated in the region strongly supported the notion that there was in fact significant desert trade. The goal then became to detect and document the infrastructure that would have supported this trade. The difficulty was in greatly expanding the search area. We acquired additional Landsat image datasets and began enhancing the data, and plotting additional tracks.

Analysis of the image data based on experience gained in our initial reconnaissance work enabled us to eliminate very large regions of the desert from consideration, and indicated precisely where to look in the field for significant sites. Continued analysis revealed that the major tracks followed well-defined routes to only a limited number of places. These sites were targeted for a subsequent ground-based reconnaissance expedition in fall of 1991. Most useful were the Landsat image data which revealed a network of tracks that converge at the modern village of Shisr (Figure 3). While some of these tracks are obviously reinforced by recent traffic, artifacts along the older interconnecting tracks indicate that these routes have been in use for a very long time. This continued use is entirely reasonable since Shisr is the key source of water in this area, and was in the past. It is instructive to compare the Landsat image in Figure 3 with a Shuttle Imaging Radar-C image in Figure 4. This image was acquired on April 13, 1994, after Shisr

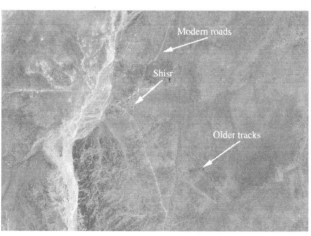

Figure 3. Sub-scene of Landsat Thematic Mapper image of the Ubar site (Path 160 Row 47, acquired 13 January 1986). Note the minor trails near the major modern gravel roads. The tracks show primarily because of increased quantity of sand and fine-grained material in the tracks as compared to the surroundings, resulting in higher reflectivity. Although used into modern times, the trails and their continuations to the west into Yemen are ancient, as evidenced by the archaeological materials along them. These trails form a regional network. North is at the top; scene is approximately 25 km in wide.

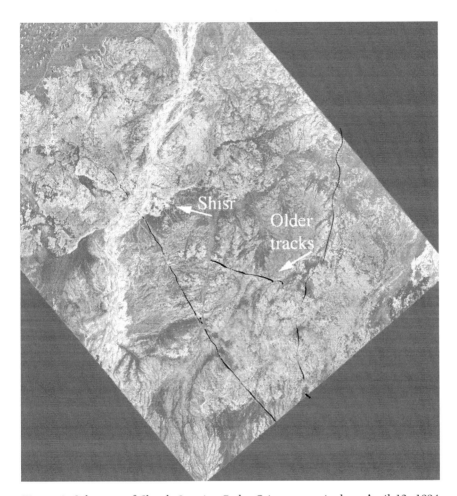

Figure 4. Sub-scene of Shuttle Imaging Radar-C image acquired on April 13, 1994 covering part of the area of Figure 3. Many tracks are visible in the radar image, and express themselves differently than in the Landsat image. The magenta color of the tracks results from increased sand suppressing the cross polarized return. It is interesting to note that the modern track largely obscured. Three SIR-C channels are shown, red represents L- band, HH (horizontal transmit and receive), blue is C- band HH, and L-band HV (horizontal transmit, vertical receive) is shown in green. The magenta areas are sand dunes and sand, bright reflectors at both L- and C-band for this imaging geometry. Green areas (L-HV) are rough limestone, a good depolarizer. Regional datasets of radar images would have provided indications of track convergence at the Shisr site.

was recognized as a key location, but many tracks are visible in the radar image as well. The image is constructed from three of the SIR-C channels. Red represents L- band, HH (horizontal transmit and receive), blue is C-band HH, and L-band HV (horizontal transmit, vertical receive) is shown in green. The prominent magenta colored areas are sandy areas and sand dunes, which are bright reflectors at both L- and C-band for this imaging geometry. Green areas (L-HV) are rough limestone which is good depolarizer giving the relatively large cross-polarized return. A regional dataset of similar radar images would have provided indications of track convergence at the Shisr site, just as the Landsat images. Hence, satellite radar images have considerable potential for archaeological applications as well. The Japanese L-Band multi-polarization L-Band imaging radar system launched in January 2006 may have very significant potential for archaeology (http://www.eorc.jaxa.jp/ALOS/). Shorter wavelength single polarization C-Band imaging radars of the Canadian Radarsat (http://gs.mdacorporation.com/) and the European ERS and ENVISAT (http://envisat.esa.int/) systems have utility as well, although the shorter wavelength and single polarization limits their ability for archaeological applications.

The Shisr site is a possible candidate for Ptolemy's "Omanum Emporium." From a commercial view, control of the frankincense trade would require additional fortress complexes at critical points such as waterholes along the trade routes. These would be the "octo mansiones" of Pliny. The site at Shisr is then the easternmost of a series of desert outposts leading westward into Yemen. Reconnaissance level work to search for and document continuation of these trade routes in Yemen by Zarins, assisted by Hedges, and Blom, has been underway as recently as 2000, but work in Yemen is proceeding slowly. Archaeological evidence gathered to date and presented in Zarins (2001)

Archaeological investigation of the ruins at Shisr indicates what likely inspired some aspects of the legendary accounts of a catastrophic end to Ubar. However, other sites in the region clearly contributed to the legendary accounts as well. Figure 5 shows a collection of archaeological material from the Shisr site. The key observation is that of the artifacts in Figure 5, only the circle and dot pottery is of local origin (Zarins, personal communication, and Zarins, 2001). Although no inscriptions have been found at the Shisr site, or elsewhere in southern Arabia, which prove the existence of an Ubar of legend, archaeological material recovered at Shisr in three seasons of excavation (winters of 93, 94, and 95) demonstrate that this site was clearly not an isolated desert watering hole (Zarins, 2001). Pot sherds, glass, coins, and fragments of an oil lamp, all testified to a long period of occupation and commerce with truly distant lands. Bronze and Iron Age material from Rome, India, and Persia spanned a long time period ending about 400 A.D. The cessation of activity at about this time is in broad agreement with legendary accounts, and also with the probable decline in the frankincense trade, and particularly in overland trade.

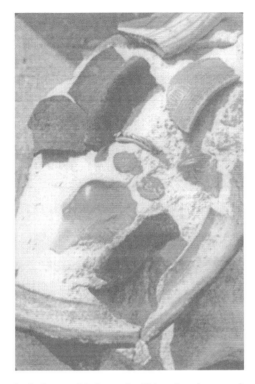

Figure 5. Archaeological material from the Shisr site. Among the potsherds, glass, coins, and other material, the ONLY local piece is the circle-and-dot pottery at the upper right. Zarins, 2001 provides complete analysis of artifacts from the region.

demonstrates that trade and travel across this region was common in the Bronze Age. As work in Yemen proceeds, details of an integrated picture of overland trade in southern Arabia will emerge.

The Shisr site makes geographic sense. This observation is important for future archaeological work using remote sensing images. Shisr has the only source of near-surface water for a very long distance, and clearly the water table was higher in the past. Indeed the modern village of Shisr is at the site because of water, now at a few meters depth. The site is also in an advantageous position with regard to frankincense source regions nearby on the north facing slopes of the Qara Mountains, and as an isolated, critical point from which to launch camel caravans across the desert. When evaluating potential archaeological targets in remote sensing data, then, geographic context becomes a significant factor.

The Shisr site is in accord with aspects of the legendary accounts if some considerable allowance for exaggeration is made. In the *Koran*, Ubar (called Irem in the *Koran*) is described as a many-towered city. Excavation revealed an eight-towered fortress which surrounded the water well (Zarins, 2001:101). Ubar supposedly perished in a calamity. While no definitive evidence has emerged on this matter, the Shisr fortress did in fact collapse into the sinkhole that hosted the well at about the time indicated by legend. Legendary allusions that this occurred in a sandstorm, or may perhaps have been triggered by a distant earthquake are plausible, but of course the sinkhole may have sponta- neously collapsed as they commonly do. In any event, the collapse would have rendered the water well useless until dug out, which may not have been possible if water supplies on hand at the time of the collapse were meager. Since incense trade was declining at this time as well, it is likely the site was merely abandoned.

5. SUMMARY AND IMPLICATIONS FOR FUTURE WORK

In summary, circumstantial evidence indicates the Shisr site is a likely source of some elements of the Ubar legend, but it is not the "Ubar" of legend. However, as Zarins (2001) documents, the incense trade was a regional activity in the land of the Iobaritae. Shisr was clearly a key place for the control of overland trade, but hardly the only important place. Analysis of the remote sensing data indicates there are no other archaeological sites of similar or greater economic importance in this region of the desert, and that it is only one of multiple desert outposts.

This investigation can serve as a case study for the application of remote sensing to archaeological investigations. The effective use of various types of remote sensing data show that reconnaissance, or Phase A-level archaeology can potentially benefit greatly when integrated into traditional studies in an interdisciplinary way. Remote sensing can be used to directly identify potential targets for field examination and evaluate the geographic context of sites, particularly when integrated into a Geographic Information System. Since the remote sensing portion of this work was conducted, new satellite systems sampling more wavelength regions have become available. Integrated GIS and image analysis tools have made the sorts of analyses conducted here routine. The key missing element is integration of a remote sensing component into appropriate archaeological investigations.

6. ACKNOWLEDGEMENTS

Critical to the success of this work was the overwhelming support of Omani businesses, and the Government of Oman. Expeditions were supported by the Sultanate of Oman, businesses in Oman, Europe, and the U.S. Additional support was provided by the Seaver Institute, the David and Lucille Packard Foundation, Magellan Corporation, nWave Entertainment, Andrew Solt, Miles Rosedale, and many others. R. B, R. C., and C. E. received partial support

from the Jet Propulsion Laboratory, managed /
Technology under contract with the National /
istration. In addition, British explorer Sir Ranu
the Omani government, and further support /
in the organization of the initial reconnaissar

REFERENCES

Anton, D., 1984, Aspects of geomorphologic evolution; paleosols and dunes in Saudi Arabia. .. Quaternary Period in Saudi Arabia Volume 2, edited by A. R. Jado and J. G. Zøtl, pp. 274–296.

Beratan, K.K., Blom, R.G., Nielson, J.G., and Crippen, R.E., 1990, Use of Landsat Thematic Mapper images in regional correlation of syn-tectonic strata, Colorado River extensional corridor, California and Arizona. Journal of Geophysical Research 95(B1): 615–624.

Blom, R.G., Crippen, R.E., and Elachi, C., 1984, Detection of subsurface features in Seasat radar images of Means Valley, Mojave Desert, California. Geology 12(6): 346–349.

Bostock, J., and Riley, H. (translators), 1855, The Natural History of Pliny. H.G. Bohn, London. http://www.perseus.tufts.edu/cgi-bin/ptext?lookup=Plin.+Nat.+toc

Crippen, R.E., 1987, The regression intersection method of adjusting image data for band ratioing. International Journal of Remote Sensing 8(2): 137–155.

Crippen, R.E., 1988, The dangers of underestimating the importance of data adjustments in band ratioing. International Journal of Remote Sensing 9(4): 767–776.

Crippen, Robert E., 1989a, Image display of four components of spectral data: Chapter 7 in Development of Remote Sensing Techniques for the Investigation of Neotectonic Activity, Eastern Transverse Ranges and Vicinity, Southern California. Ph.D. dissertation, University of California, Santa Barbara, University Microfilms # 90–09533.

Crippen, R.E., 1989b, A simple spatial filtering routine for the cosmetic removal of scan-line noise from Landsat TM P-tape data. Photogrammetric Engineering and Remote Sensing 55(3): 327–331.

Crippen, R.E., Blom, R.G., and Heyada, J.R., 1988, Directed band ratioing for the retention of perceptually independent topographic expression in chromaticity-enhanced imagery. International Journal of Remote Sensing 9(4): 749–765.

Elachi, C., Schaber, G., and Roth, L., 1984, Spaceborne Radar Subsurface Imaging in Hyperarid Regions. IEEE Transactions on Geoscience and Remote Sensing GE-22: 383–388.

Ford, J.P., Dokka, R.K., Crippen, R.E., and Blom, R.G., 1990, Faults in the Mojave Desert, California, as revealed on enhanced Landsat images. Science 248(4958): 1000–1003 and cover image.

Gillespie, A.R., 1980, Digital techniques of image enhancement. In Remote Sensing in Geology, edited by B.S. Siegal and A.R. Gillespie, pp. 139–226. John Wiley and Sons: New York.

McCauley, J.F., Schaber, G.G., Breed, C.S., Grolier, M.J., Haynes, C.V., Issawi, B., Elachi, C., and Blom, R., 1982, Subsurface valley and geoarcheology of the Eastern Sahara revealed by Shuttle radar. Science 218:1004–1019.

Moik, J. G., 1980, Digital Processing of Remotely Sensed Images. NASA SP–431.

Thomas, B., 1931, A camel journey across the Rub al Kahli. The Geographical Journal, LXXVIII, (3), 14 (intro): 210–242.

Thomas, B., 1932, Arabia Felix. Charles Scribner's Sons, New York.

Tkach, J., 1897, Iobaritai. In Pauly's Real-Encylopadie der Classischen Altertumswissenschaft, edited by G. Wissowa, pp. 1831–1837.

Zarins, J., 2001, The Land of Incense. Sultan Qaboos University Publications, Archaeology and Cultural Heritage Series V. 1.

Chapter *4*

The Use of Interferometric Synthetic Aperture Radar (InSAR) in Archaeological Investigations and Cultural Heritage Preservation

DIANE L. EVANS AND TOM G. FARR

Abstract: The availability of a near-global digital elevation model derived from the Shuttle Radar Topography Mission (SRTM) provides archeologists a new tool to complement other remote sensing data. SRTM data not only provide basemaps for geographic information systems and aid in geometric corrections, they also provide data for visualizations and animations that can be used in cultural resource management. SRTM data also provide a regional context in which sites can be studied. For example, slope and elevation information, combined with other factors used in predictive modeling such as availability of water and suitability for animal and crop domestication can be used to identify potential archaeological sites. The data can also help in the identification of possible migration pathways or trade routes along which settlements might occur. In addition, since topography played a role in the location of many religious sites (e.g., Borobudur in Central Java), it may be possible to recognize other topographically significant areas through a systematic analysis of the SRTM dataset. Topography has a clear influence on the location of floods and landslides as well as on soil development and erosion, and several studies have shown the value of topographic information in determining the timing and nature of wet and dry periods and the susceptibility of a region to environmental change.

This information, combined with the increased recognition of the relationship between environmental change and the histories of past civilizations, can further elucidate their evolution, migration, and in some cases extinction.

1. INTRODUCTION

Because the land surface is continually modified by climatic, tectonic, and human influences, understanding how landscapes change over time is important for both archaeology and cultural heritage preservation. Quantitative analysis of landscapes requires digital topographic data that have a uniform spatial resolution and vertical accuracy. Near-global digital elevation models (DEMs) that meet this requirement have recently become available using spaceborne interferometric synthetic aperture radar (InSAR). InSAR can also be used to measure centimeter-scale changes in topography for assessment of natural disasters and human-induced hazards. Understanding the rapidity and effect of such changes is critical in assessing their ecological and economic impact. It is also relevant for archaeological investigations in appreciating how similar events might have, for example, prompted shifts in the center of power of past cultures. This paper describes first the technical aspects of InSAR and then summarizes a number of potential applications for archaeology and cultural heritage preservation.

2. TOPOGRAPHY AND TOPOGRAPHIC CHANGE FROM INSAR

InSAR is a technique in which radar pulses are transmitted from a conventional SAR antenna and radar echoes are received either by two separate antennas or the same antenna on a subsequent overpass (e.g., Rosen et al., 2000). By coherently combining the signals from the two antennas, the interferometric phase difference between the received signals can be formed for each imaged point. The phase difference is related to the geometric path-length difference to the image point, which depends on the topography. Assuming the geometry of the antennas is known, the phase difference can be converted into an elevation for each image point.

During the European Space Agency's (ESA) European Remote Sensing Satellites (ERS-1 and ERS-2) Tandem Operations Phase in 1995, the satellites were maintained in the same orbital plane with ERS-2 following thirty minutes behind ERS-1. This procedure resulted in a swath on the ground acquired by ERS-2 one day later than by ERS-1, and approximately 73% of the global land mass being covered with baselines small enough for generation of DEMs (e.g., Rufino et al., 1998). In February 2000 NASA's Space Shuttle Endeavour carried the Shuttle Radar Topography Mission (SRTM), a dual antenna InSAR, which produced a DEM of the Earth's land surface (Figure 1) between about

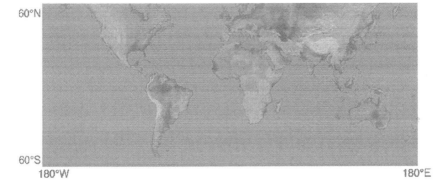

Figure 1. Image of the world generated with data from the Shuttle Radar Topography Mission (SRTM). The shade image was derived by computing topographic slope in the northwest-southeast direction, so that northwest slopes appear bright and southeast slopes appear dark. Color coding is directly related to topographic height, with green at the lower elevations to white at the highest elevations (NASA/JPL).

60° north and 56° south latitude with a 30-m pixel spacing and better than 10-m vertical accuracy (Farr and Kobrick, 2000). Topographic changes can also be observed using InSAR data. If the flight path and imaging geometries of all the SAR observations are identical, any interferometric phase difference is a result of changes over time such as variable propagation delay through the atmosphere, or surface motion in the direction of the radar line of sight. So-called "Repeat pass InSAR" was first discovered to detect centimeter-scale changes in the land surface by Gabriel et al. (1989). ERS-1, ERS-2, and ESA's follow-on mission ENVISAT, which was launched in March 2002, are all sources of repeat-pass InSAR data (e.g., Desnos et al., 2000). The Japanese Space Agency JERS-1 SAR, which flew from 1992 to 1998, provided InSAR data as will the Advanced Land Observation Satellite (ALOS), launched in January, 2006 (e.g., Furuya, 2004). Finally, The Canadian Space Agency's RADARSAT-1, which was launched in 1995, and future RADARSAT sensors will provide repeat-pass InSAR data (e.g., Bamler and Holzner, 2004).

3. APPLICATIONS OF INSAR

While the use of InSAR for archeological investigations and cultural heritage preservation have so far been limited owing to data access, the following examples demonstrate the varied applications of InSAR data for landscape analysis and hazard assessments. InSAR marks an important development in the use of SAR, given both its place within the rapidly developing field of interferometry and its increasingly proven ability to address significant ecological and archaeological issues.

3.1. Landscape Analysis

DEMs based on SRTM and ERS tandem data can be used as basemaps for geographic information systems and for geometric corrections, as well as for visualizations that can be used in viewshed analyses. Since these DEMs cover extensive areas, they provide a regional context in which archaeological sites can be studied.

Topography is the framework upon which subsequent natural and cultural forces interact to sculpt the cultural landscape as it is observed at any moment. Topography provides shelter from the elements, depredation by animals, and attack by other groups of humans. It is in that sense proto-architecture, and the placement and design of formal architecture is often influenced by it. Areas of high relief provide essential vantage points from which the movements of game and enemies can be observed from afar so that a timely response can be quickly organized. Topography creates varied environments that attract different assemblages of plants and animals. This diversity is attractive to human populations. As plant and animal diversity increases, the probability that food, and the materials needed for the production of tools and other useful items, occur within a relatively small area increases. Topography influences soil development, and so is a major factor in determining whether or not land is arable.

Topography, therefore, influences human settlement and land-use patterns on both the micro and macro scale, and is an important, if not essential, element in any model of settlement patterning. The use of radar-derived topographic data has recently been demonstrated by Blom and his colleagues, who utilized high-resolution DEMs derived from an airborne SAR (which works on the same principal as the SRTM system) to devise a predicative model used in the detection of archaeological sites based upon slope, slope inflection ("slope of slope"), distance to abrupt slope, aspect, and viewshed (Blom et al., 2003). Many of these sorts of analyses should be as valid on the macro scale. Truly accurate data for these kinds of analyses have become available now through the SRTM mission.

As one example of the influence of topography on a macro scale, it is generally accepted among archaeologists that river systems and drainages are the best predictors of culture-area boundaries. Rivers provide what is arguably the prime limiting factor in human population density, and the most reliable predictor of site location. Archaeological sites, and especially occupation sites, are never situated far away from water, and, generally, the larger and more reliable the source of water, the larger and more permanent the archaeological sites. As just one example of the influence of reliable water sources, the only human settlements that remained after the Maya collapse were those located adjacent to such sources, while the 95% human-occupation sites farther removed simply ceased to exist (Sever, 1998)

Topography also provides the natural paths for migration and diffusion of plant and animal populations, including human populations. One well known example of this is Wadi Araba, the Jordan Valley, which in reality is a continuation of the Great Rift of Africa. The presence of this topographic feature which ties Africa to the Middle East is likely one reason for the well known bounty of domesticable plants and animals along the Fertile Crescent. This plethora of domesticable species has been cited by many as one of the important reasons that agriculturally based complex societies arose here before anywhere else on earth (D. Comer, personal communication).

Analysis of DEMs can also be used to reconstruct past landscapes (e.g., Farr, 2004). For example, Figure 2 shows a regional view of the Borobudur area in Central Java where topography is thought to have played a major role in site location and evolution. According to Murwanto et al. (2004), collapse of the Merapi Volcano may have resulted in formation of an early Lake Borobudur where several large and many small Buddhist and Hindu temples were constructed between 732 and 900 A.D. Subsequent eruptions in which temples were destroyed are thought to have contributed to an abrupt shift of power and organized society to East Java in 928 A.D. (Newhall et al., 2000). While many hypotheses have been put forward to explain this shift, none fully explains the rapidity of the transfer of royal power to East Java (E. Moore, personal communication).

Figure 3 is a contour map of the area around Borobudur and Figure 4 reconstruction of what lake levels might have been in the past. Using Murwanto et al. (2004) as a guide, the reconstruction shows the extent of a lake in its late stages when a water surface at 250 m corresponds to their

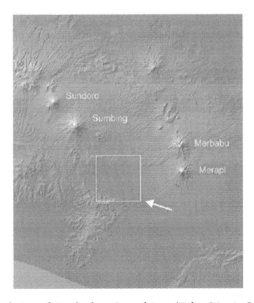

Figure 2. Regional view of Borobudur, Central Java (7 deg 36 min S; 110 deg 12 min E) with location of Figures 3 and 4 outlined.

Figure 3. Contour map of Borobudur area outlined in Figure 2. Heavy contours are 100 m. Borobudur sits on a small "hill" at tip of arrow just inside 250 m contour. Map is about 20 km across.

mapped lake deposits. The base of Borobudur is only about 15 m above this level with the monument sitting at the end of a peninsula in this reconstruction. This lake was the culmination of a series of lakes dating back to at least 20,000 years before present (ybp). Discrete lake stands were dated by Murwanto et al. (2004) at 20,000 ybp; 3,000 (calibrated radiocarbon years) B.C.; and 12th-14th century A.D. The sequence indicates the likelihood that a lake surrounded the site of Borobudur during its construction in the 9th century A.D.

3.2. Hazard Assessment

Several studies have shown the value of topographic information in determining the timing and nature of wet and dry periods and the susceptibility of

Figure 4. Map of Figure 3 with elevations below 250 m colored blue to simulate a lake of that level. Note Borobudur sits at end of a peninsula jutting into lake. Red contour is boundary of the 3,000 B.C. lake of Murwanto et al. (2004) and the orange contour is boundary of their lake of 12th–14th century A.D.

a region to environmental change (e.g., Farr and Chadwick, 1996). This information can elucidate the evolution, migration, and in some cases extinction of past civilizations. Topography has a significant influence on soil development and erosion. The ability to reconstruct the recent history of the land surface, as described earlier, plays an integral role not only in the interpretation of archaeological landscapes but in predicting the likelihood of future flood and landslide hazards (e.g., Tralli et al., 2005). By way of example, using the same technique as in the previous section, Figure 5 shows areas within 10 m above sea level to indicate areas vulnerable to flooding north of Phuket, Thailand. Comparison with satellite images before and after the Indian Ocean tsunami of December 26, 2004, show that this is a good indicator of where damage could be expected.

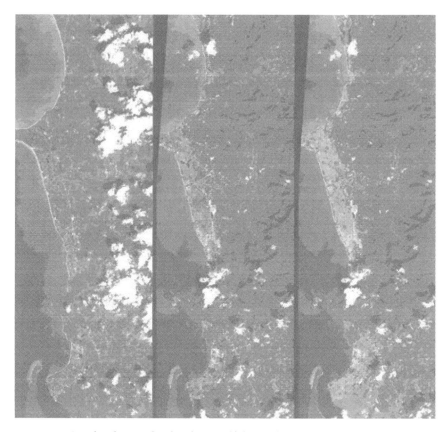

Figure 5. Simulated natural color ASTER (Advanced Spaceborne Thermal Emission and Reflection Radiometer) images showing a 27-kilometer-long stretch of coast 80 kilometers north of the Phuket airport on December 31, 2004 (middle), and also two years earlier (left). The changes along the coast from green to grey indicate where the vegetation was stripped away by the tsunami following the Sumatra earthquake of December 26, 2004 tsunami. The image on the right is a copy of the later ASTER scene with elevations within 10 meters of sea level derived from an SRTM elevation map highlighted in red. (NASA/JPL)

With the ability to detect cm-scale changes in topography, the potential also exists for monitoring ground deformation associated with natural or human-induced hazards such as subsidence, landslides, earthquakes, and volcanoes. Systematic acquisition of InSAR in combination with other techniques can also be used to develop mitigation strategies in response to these threats. For example, InSAR data have been used for monitoring landslides (e.g., Massonnet and Feigl, 1998; Kimura et al., 2000; Canuti et al., 2004), which are a significant risk in areas such as Machu Picchu World Heritage site in Peru (e.g., Sassa et al., 2003). In many cases, expanded development and human activities, such as modified slopes and deforestation, can increase the incidence of landslide disasters (e.g., CEOS DMSG Report, 2003), making this an increasingly important area of research.

Subsidence events and other ground instabilities can also become significant risk factors as infrastructure is developed in association with archaeological and cultural heritage sites. InSAR data have also been used to identify subsidence associated with mining operations (e.g., Carnec and Delacourt, 2000; Prakash et al., 2001; Raucoules et al., 2003a; Closson, 2003), underground tunneling (Tesauro et al., 2000; Le Mouelic et al., 2002), withdrawal of subsurface water and oil (e.g., Amelung et al., 1999; Fielding et al., 1999; Xu et al., 2001; Buckley et al., 2003; Hoffmann et al., 2003; Stancliffe and van der Kooij, 2001), ground settlement (Raucoules et al., 2003b; Ding et al., 2004), and soil compaction (Kim and Won, 2003). InSAR has recently been applied to monitoring deformation in Venice, Italy, which has subsided because of a combination of the unstable subsoils, consisting mainly of silt and clay, and pumping of water from the groundwater underlying the city (Tosi et al., 2002). Tosi et al. (2002) show changes in elevation from 1992 to 1996 (Figure 6) and conclude that the Venetian soil is quasi-stable over this period.

Such figures provide a baseline for future monitoring, an example applicable to a number of international and national cultural and natural heritage sites. At Angkor (Cambodia), for example, since the inscription of the site in the UNESCO World Heritage List in 1992, the expansion of the population and the tourism infrastructure has greatly accelerated the utilization of underground water sources. However, to date, InSAR data have not been employed in such studies or at other contemporary ancient cities of Southeast Asia such as Sukhothai, Ayutthaya, and Bagan (E. Moore, personal communication).

Other widespread uses of repeat-pass InSAR data have been in studies of volcanoes and earthquakes. In volcanic areas the pattern of deformation can be used to help determine the location, volume, and shape of a subsurface magma body and to anticipate the onset and course of an eruption (Dzurisin, 2000).

Figure 6. Changes in elevation (cm) from 1992-1996 In Venice, Italy, from Tosi et al. (2002)

For example, Lundgren et al. (2003) analyzed Etna deformation using InSAR data and were able to model changes in both the magmatic system and stability of its edifice. InSAR data were also used to determine the relationship between ground motions and composition of the hydrothermal fluids, and seismicity in the Phlegrean Fields, an active caldera structure, located on the periphery of Naples, Italy, which last erupted in the Monte Nuovo eruption of 1538 (Todesco et al., 2004).

InSAR is also a promising tool for earthquake-hazard assessment and mitigation planning, which is critical in countries where disasters have both direct and indirect impacts on economic development. The Dead Sea region is an example of an area with significant cultural heritage where disaster mitigation is important because of its tectonic setting and history of devastating earthquakes (e.g., Barakat and Daher, 2000). In this region Baer et al. (2002) measured rates of subsidence along the Dead Sea shores from InSAR and showed that in certain locations, subsidence appears to be structurally controlled by faults, seaward landslides, and salt domes, while Sarti et al (2003) used InSAR data to measure ground movement in playas within the southern Arava Valley that may correlate with pre-seismic and post-seismic events. In another example, Talebian et al. (2004) mapped co-seismic surface displacements and decorrelation effects from the M(w) 6.5 earthquake which devastated the town of Bam in southeast Iran on 26 December, 2003, using Envisat radar data and found that over 2 m of slip occurred at depth on a fault that had not previously been identified. With this study in place, areas of future potential damage in the area could potentially be identified. However, systematic InSAR observations would be required to test this possibility.

Because there is no dedicated InSAR satellite observing surface deformation, data are usually acquired after a seismic event and paired with pre-seismic data where they exist. This procedure poses problems because the InSAR technique requires specific baselines for separation between data pairs and can be limited by the presence of atmospheric water vapor during data acquisition in one or both images. In areas that are systematically covered with InSAR data, however, it has been possible to detect aseismic creep and to model build-up of stress related to potential seismic events (e.g., Rosen et al., 1998). The potential, therefore, exists to monitor areas where stress may be building in areas of archaeological and cultural heritage interest near faults such as Bam, which has recently been inscribed on UNESCO's World Heritage List, and Bagan, Myanmar, which suffered significant damage in an earthquake in 1975.

It is also possible to determine where stress is building up on specific fault segments in a fault zone. Bos et al. (2004) determined that the earthquake in Izmit, Turkey, on August 17, 1999, released most of the strain that had accumulated since the last main event on this stretch of the North Anatolian fault in 1719. Cakir et al. (2003) suggest that the distribution of heterogeneous slip and loading along the different fault segments may be important factors controlling the propagation of large earthquake ruptures along the North Anatolian Fault. The pattern of stress on this fault has particular implications for the seismic risk for the city of Istanbul and could be used to guide high-priority areas for risk-mitigation strategies such as building retrofitting.

4. CONCLUSIONS

The use of spaceborne data for archaeology and cultural resource management is an emerging field. In particular, the use of both polarimetric and interferometric SAR data has increased substantially over the last decade. With the availability of near-global DEMs generated from InSAR data it is now feasible to study landscape evolution for almost any site of interest. In addition, many studies in the recent literature point to the potential of developing mitigation strategies for archaeological and cultural heritage sites based on systematically acquired InSAR data. Topographic changes observed with InSAR can be used to identify areas vulnerable to landslides, ground subsidence, earthquakes, and volcanic eruptions, and inform mitigation strategies in response to these threats.

5. ACKNOWLEDGEMENTS

This work was done under contract to NASA. We acknowledge very helpful comments from Elizabeth Moore and Douglas Comer.

REFERENCES

Amelung, F., Galloway, D.L., Bell, J.W., Zebker, H.A., Laczniak, R.J., 1999, Sensing the ups and downs of Las Vegas: InSAR reveals structural control of land subsidence and aquifer-system deformation. *Geology* 27(6):483–486.

Baer, G., Schattner, U., Wachs, D., Sandwell, D., Wdowinski, S., Frydman, S., 2002, The lowest place on Earth is subsiding - An InSAR (interferometric synthetic aperture radar) perspective. *Geological Society of America Bulletin* 114(1):12–23.

Bamler, R., Holzner, J., 2004, ScanSAR interferometry for RADARSAT-2 and RADARSAT-3. *Canadian Journal of Remote Sensing* 30(3):437–447.

Barakat, S.Z., and Daher, R.F., 2000, The cultural heritage and nature of disasters in Jordan and Palestine. *CRM*, no. 6:39–43.

Blom, R.G., Comer, D., Yatsko, A, Holcomb, D., and Byrd, B.F., 2003, Progress on Application of Airborne Radar and GIS to San Clemente Island Archeology. *Geological Society of America Abstracts with Programs*, 2003.

Bos, A.G., Usai, S., and Spakman, W., 2004, A joint analysis of GPS motions and InSAR to infer the coseismic surface deformation of the Izmit, Turkey earthquake. *Geophysical Journal International* 158(3):849–863.

Buckley, S.M., Rosen, P.A., Hensley, S., and Tapley, B.D., 2003, Land subsidence in Houston, Texas, measured by radar interferometry and constrained by extensometers. *Journal of Geophysical Research-Solid Earth* 108(B11), Article No. 2542, November 26.

Canuti, P., Casagli, N., Ermini, L., Fanti, R., and Farina, P., 2000, Landslide activity as a geoindicator in Italy: significance and new perspectives from remote sensing. *Environmental Geology* 45(7):907–919.

Cakir, Z., de Chabalier, J.B., Armijo, R., Meyer, B., Barka, A., and Peltzer, G., 2003, Coseismic and early post-seismic slip associated with the 1999, Izmit earthquake (Turkey), from SAR interferometry and tectonic field observations. *Geophysical Journal International* 155(1):93–110.

Carnec, C., and Delacourt, C., 2000, Three years of mining subsidence monitored by SAR interferometry, near Gardanne, France. *Journal of Applied Geophysics* 43(1):43–54.

Closson, D., Karaki, N.A., Hansen, H., Derauw, D., Barbier, and C., Ozer, A., 2004, Space-borne radar interferometric mapping of precursory deformations of a dyke collapse, Dead Sea area, Jordan. *International Journal of Remote Sensing* 24(4):843–849.

CEOS (=Committee on Earth Observation Satellites), 2003. The Use of Earth Observing Satellites for Hazard Support: Assessments and Scenarios. Final Report of the CEOS Disaster Management Support Group, November.

Ding, X.L., Liu, G.X., Li. Z.W., Li, Z.L., Chen, Y.Q., 2004, Ground subsidence monitoring in Hong Kong with satellite SAR interferometry. *Photogrammetric Engineering and Remote Sensing* 70(10):1151–1156.

Desnos, Y.-L., et al., 2000. The ENVISAT Advanced Synthetic Aperture Radar System. *Proceedings IGARSS 2000*, vol. 1, pp. 1171–1173.

Dzurisin D, 2000, Volcano geodesy: challenges and opportunities for the 21st century. *Philosophical Transactions of the Royal Society of London Series A-Mathematical Physical And Engineering Sciences* 358(1770, May 15):1547–1566.

Farr, T.G., 2004, Topographic Signatures of Geomorphic Processes at Desert Piedmonts. *American Geophysics Union*, December.

Farr, T.G., and Chadwick, O.A., 1996, Geomorphic processes and remote sensing signatures of alluvial fans in the Kun Lun mountains, China. *Journal of Geophysical Research-Planets* 101(E10, October 25):23091–23100.

Farr, T.G., and Kobrick, M., 2000, Shuttle Radar Topography Mission Produces a Wealth of Data. *Eos, Transactions of the American Geophysics Union* 81(48, November 28):583, 585.

Fielding, E.J., Blom, R.G., and Goldstein, R.M., 1999, Rapid subsidence over oil fields measured by SAR interferometry. *Geophysical Research Letters* 25(17, September 1):3215–3218.

Furuya, M., 2004. Localized deformation at Miyakejima volcano based on JERS-1 radar interferometry: 1992–1998. *Geophysical Research Letters* 31(5), Article No. L05605, March 4.

Gabriel, A.K., Goldstein, R.M., and Zebker, H.A., 1989, Mapping small elevation changes over large areas - differential radar interferometry. *Journal of Geophysical Research-Solid Earth and Planets* 94 (B7, July 10): 9183–9191.

Hoffmann, J., Galloway, D.L., and Zebker, H.A., 2003, Inverse modeling of interbed storage parameters using land subsidence observations, Antelope Valley, California. *Water Resources Research* 39(2), Article No. 1031, February 13.

Kim, S.W., and Won, J.S., 2003, Measurements of soil compaction rate by using JERS-1 SAR and a prediction model. *IEEE Transactions on Geoscience and Remote Sensing* 41(11, Part 2, November):2683–2686.

Kimura, H., and Yamaguchi, Y., 2000, Detection of landslide areas using satellite radar interferometry. *Photogrammetric Engineering and Remote Sensing* 66(3, March):337–344.

Le Mouelic, S., Raucoules, D., Carnec, C., King, C., and Adragna, F., 2002, A ground uplift in the city of Paris (France) detected by satellite radar interferometry. *Geophysical Research Letters* 29(17), Article No. 1853, September 1.

Lundgren, P., Berardino, P., Coltelli, M., Fornaro, G., Lanari, R., Puglisi, G., Sansosti, E., and Tesauro, M., 2003, Coupled magma chamber inflation and sector collapse slip observed with synthetic aperture radar interferometry on Mt. Etna volcano. *Journal of Geophysical Research-Solid Earth* 108(B5), Article No. 2247, May 14.

Massonnet, D., and Feigl, K.L., 1998, Radar interferometry and its application to changes in the earth's surface. *Reviews of Geophysics* 36(4, November):441–500.

Murwanto, H., Gunnell, Y., Suharsono, S., Sutikno, S., and Lavigne, F., 2004, Borobudur monument (Java, Indonesia) stood by a natural lake: chronostratigraphic evidence and historical implications. *Holocene* 14 (3, May):459–463.

Newhall, C.G., Bronto, S., Alloway, B., Banks, N.G., Bahar, I., del Marmol, M.A., Hadisantono, R.D., Holcomb, R.T., McGeehin, J., Miksic, J.N., Rubin, M., Sayudi, S.D., Sukhyar, R., Andreastuti, S., Tilling, R.I., Torley, R., Trimble, D., and Wirakusumah, A.D., 2000, 10,000 Years of explosive eruptions of Merapi Volcano, Central Java: archaeological and modern implications. *Journal of Volcanology and Geothermal Research* 100(1-4): 9–50.

Prakash, A., Fielding, E.J., Gens, R., Van Genderen, J.L., and Evans, D.L, 2001, Data fusion for investigating land subsidence and coal fire hazards in a coal mining area. *International Journal of Remote Sensing* 22(6):921–932.

Raucoules, D., Maisons, C., Camec, C., Le Mouelic, S., King, C., and Hosford, S., 2003a, Monitoring of slow ground deformation by ERS radar interferometry on the Vauvert salt mine (France) - Comparison with ground-based measurement. *Remote Sensing of Environment* 88(4, December 30):468–478.

Raucoules, D., Le Mouelic, S., Carnec, C., Maisons, C., and King, C., 2003b, Urban subsidence in the city of Prato (Italy) monitored by satellite radar interferometry. *International Journal of Remote Sensing* 24(4, February 20):891–897.

Rosen, P., Hensley, S., Joughin, I., Li, F., Madsen, S., Rodriguez, E., Goldstein, R., 2000, Synthetic Aperture Radar Interferometry, *Proceedings of the IEEE* 88(3, March):333–382.

Rosen, P., Werner, C., Fielding, E., Hensley, S., Buckley, S., and Vincent, P., 1998, Aseismic creep along the San Andreas Fault northwest of Parkfield, CA measured by radar interferometry. *Geophysical Research Letters* 25(6, March 15):825–828.

Rufino, G., Moccia, A., and Esposito, S., 1998, DEM generation by means of ERS tandem data. *IEEE Transactions on Geoscience and Remote Sensing* 36(6, November):1905–1912.

Sarti, F., Arkin, Y., Chorowicz, J., Karnieli, A., and Cunha, T., 2003, Assessing pre- and post-deformation in the southern Arava Valley segment of the Dead Sea Transform, Israel by differential interferometry. *Remote Sensing of Environment* 86(2, July 30):141–149.

Sassa, K., Fukuoka, H., Shuzui, H., and Hoshino, M., 2003, Landslide Risk Evaluation in the Machu Picchu World Heritage, Cusco, Peru. *Nippon Koei Technical Forum* 11:45–64.

Sever, T.L., 1998, Validating Prehistoric and Current Social Phenomena upon the Landscape of the Petén, Guatemala. In*People and Pixels: Linking Remote Sensing and Social Science*, edited by Diana Liverman, pp. 145–163. National Academy Press, Washington, D.C.

Stancliffe, R.P.W., and van der Kooij, M.W.A., 2001, The use of satellite-based radar interferometry to monitor production activity at the Cold Lake heavy oil field, Alberta, Canada. *AAPG Bulletin* 85(5):781–793.

Talebian, M., Fielding, E.J., Funning, G.J., Ghorashi, M., Jackson, J., Nazari, H., Parsons, B., Priestley, K., Rosen, P.A., Walker, R., and Wright, T.J., 2004, The 2003 Bam (Iran) earthquake: Rupture of a blind strike-slip fault. *Geophysical Research Letters* 31(11), Article No. L11611, June 11.

Tesauro, M., Berardino, P., Lanari, R., Sansosti, E., Fornaro, G., and Franceschetti, G., 2000, Urban subsidence inside the city of Napoli (Italy) observed by satellite radar interferometry. Geophysical Research Letters 27(13, July 1):1961–1964.

Todesco, M., Rutqvist, J., Chiodini, G., Pruess, K., and Oldenburg, C.I.M., 2004, Modeling of recent volcanic episodes at Phlegrean Fields (Italy): geochemical variations and ground deformation. *Geothermics* 33(4):531–547.

Tosi, L., Carbognin, L., Teatini, P., Strozzi, T., and Wegmuller, U., 2002, Evidence of the present relative land stability of Venice, Italy, from land, sea, and space observations. *Geophysical Research Letters* 29(12), Article No. 1562, June 15.

Tralli, D. M., Blom, R.G., Zlotnicki, V., Donnellan, A., and Evans, D.L., n.d (in press), Satellite
     Remote Sensing of Earthquake, Volcano, Flood, Landslide and Coastal Inundation Hazards.
     ISPRS Special Issue: Applications of Remote Sensing to Disaster Management.
Xu, H.B., Dvorkin, J., Nur, A., 2001, Linking oil production to surface subsidence from satellite
     radar interferometry. *Geophysical Research Letters* 28(7, April 1):1307–1310.

Chapter 5

Detection and Identification of Archaeological Sites and Features Using Synthetic Aperture Radar (SAR) Data Collected from Airborne Platforms

DOUGLAS C. COMER AND RONALD G. BLOM

Abstract: Protocols for archaeological inventory of large areas using synthetic aperture radar (SAR) are presented here. They were developed during a 2002–2005 research project sponsored by a Department of Defense Strategic Environmental Research and Development Program Research Project (SERDP CS-1260) on San Clemente Island, California. Archaeological evidence has established that San Clemente Island has been occupied for almost 10,000 years. Some protocols can also be used with other aerial and satellite datasets, such as those acquired by multispectral sensors. Protocols rely upon algorithms to merge data from multiple flight lines, collection of spatially precise ground data with which to develop signatures, knowledge of site morphology, and elegant statistical treatments of sensing-device return values to automate the development of site signatures (in contrast to using the "trained eye" of remote sensing experts). The Introduction of the article presents the study rationale, SAR basics essential to this research, study-site description, and an overview of fieldwork done. The Results section details eight SAR protocols that were developed: data collection, including look angles, flight lines, and choice of band and polarization; data processing and image production, including orthorectification and the merging of data from multiple flight lines; image post-processing, including statistical techniques and iteration of images; corroborative use of multispectral datasets; spatial modeling; procedures for incorporation into and analysis with GIS; establishment of statistically based site signatures; and procedures for ground verification. In the Discussion, the authors extend some of their findings to archaeological research at San Clemente Island and beyond. The article concludes by highlighting some important management implications for the use of SAR as an archaeological evaluation tool for sizeable land areas. In environments not too dissimilar to those on San Clemente Island, these protocols can produce planning level inventories much more rapidly and inexpensively than field methods commonly used at present can.

1. INTRODUCTION

1.1. Project Background and Objectives

This chapter presents protocols for using airborne synthetic aperture radar (SAR) in the inventory of archaeological sites. The research was conducted by the NASA Jet Propulsion Laboratory at CalTech (JPL/NASA) and Cultural Site Research and Management, Inc. (CSRM) on San Clemente Island, California, with funding from the Department of Defense (DoD) Strategic Environmental Research and Development Program (SERDP) as project CS-1260. SAR data were collected by the JPL/NASA platform AIRSAR and the private sector platform GeoSAR.

The problem that the research was intended to address can be stated succinctly: *The presence of unidentified and unevaluated archaeological sites (and other cultural resources such as historic sites in ruins) on United States Department of Defense (DoD) land (and on lands administered by other federal agencies, e.g., the Department of Energy) poses the continual risk of costly delays in training, testing, and construction.* In particular, approximately 19 million acres of DoD land remain to be surveyed, with evaluation and mitigation required before disturbance. Numerous Phase I cultural resource surveys (Identification) and Phase II research projects (Evaluation) are conducted each year; in fact, at many areas managed by the military there has been an effort to accomplish a 100% inventory and evaluation. However, conventional inventory methods are very costly, at $30 to $35 per acre, and evaluation costs average more than $15,000 per site.

What is needed are ways to accurately and rapidly inventory wide areas using noninvasive, nondestructive technologies that do not involve artifact collections. SAR has for years held out the promise of a cost-effective way to find archaeological sites and thus alleviate these problems. SAR collects a series of radar echoes over a study area from a moving platform such as an airplane or a satellite, thereby creating a synthetically large antenna that produces high-resolution images of landscape features. Until the current project, however, protocols for the application of SAR to survey and aid in the evaluation of large areas had not been developed. The purpose of project SC-1260 was to develop such protocols.

In the remaining parts of the Introduction, we provide a brief background on the use of SAR and then describe the research test area and give an overview of the fieldwork that was completed. In the second major section we present, in some detail, the primary results of this study: eight protocols for applying SAR in archaeological evaluation. In the third section we discuss some implications of our findings for archaeological research at San Clemente Island and elsewhere. In the fourth section, our Conclusion, we consider practical management and research implications of our findings.

1.2. Basics of SAR

Most aerial and satellite remote sensing systems are passive. As enormously informative as passive sensing devices are, radar is an active sensing system uniquely capable of sensing physical, as opposed to chemical, aspects of the environment, and thus for detecting *physical* aspects of environmental change that are associated with human occupations.

SAR systems have been carried by both space and aerial platforms. Space-borne SAR systems have been used to map over 80% of the earth by means of the Shuttle Radar Topography Mission (SRTM). This application provides an example of how effective SAR is in providing high resolution over large areas from a platform at a high altitude.

SAR systems carried on airplanes deliver, as one might expect, both higher resolution and higher signal-to-noise ratios than do those carried on space platforms. Also, the ability to carry radar instrumentation along several flight lines results in multiple observation geometries, providing multiple image perspectives. This ability can be used to overcome a serious problem in the interpretation of radar images and data: shadowing. Because radar is an active sensing technology that illuminates the ground surface at an angle, radar "shadows" form on the far side of hills, ridges, and, in fact, anything that reflects radar waves.

To fill in the data gap that produces the shadow, the surface must be illuminated from another direction, or perhaps from even more than two directions. Multiple illuminations produce several images, each having areas of no data unless they are merged. Software developed during this research project does just that: it merges data collected during multiple flight lines, thereby mosaicking the imagery produced from the data.

The NASA/JPL Airborne Synthetic Aperture Radar system (AIRSAR), flown on a retrofitted DC-8 jet, was one of the platforms utilized in this research project. AIRSAR can provide multiband and multipolar SAR data from which high-resolution imagery can be produced. In addition, AIRSAR utilizes more radar bands and polarizations than any other air- or space-borne radar device known to the principal investigators. Therefore, AIRSAR was suited perfectly to this research, which was intended, in part, to identify those bands and polarizations that will be most useful to archaeological site inventory and evaluation. The other SAR platform that provided data to this research, GeoSAR, represents the migration of technology developed by NASA to the private sector. In the future, such platforms will supply data required for the application of the protocols in this report.

SAR returns are influenced largely by four phenomena, each of which is implicated in human alteration of the environment: topography, structure, surficial roughness, and dielectric property.

The first of these, *topography,* influences environment at a basic level, and environment patterns human settlement. Thus, topography influences settlement pattern in an elementary way. Prior to the advent of architecture, topographic irregularities such as ridges, hills, caves, and ledges offered shelter to human populations. They also formed diverse econiches for various species of plants and animals that, in turn, encouraged human occupation.

Topography is key to hydrology, and water is the most important human attractor. Looked at another way, water is arguably the most important limiting factor determining human settlement patterning. In turn, human occupation of the landscape produced topographic change on scales from micro to macro. Humanly designed structures can be thought of as a kind of topographic change. Often, humans worked to enhance landscape character-istics that provided suitable habitats to plants and animals, thereby gradually altering topography. In most areas, this alteration occurred long before formal agriculture was introduced.

Structure, as the term is used here, refers to formal angular geometric form. Such structure is found, for example, in architecture and vegetation. Some structural forms are more easily detected by radar than others. Radar is not well reflected by spherical objects. Rather, radar highlights angles and corners, and it is especially sensitive to angular constructions, either made by humans or occurring in nature.

Radar waves can be used in ways that detect structures of differing orientations. Radar waves can be polarized either vertically or horizontally. Some polarizations are reflected well by vertical structures and poorly by horizontal structures, while the reverse is true of other polarizations. Because AIRSAR uses three bands that can each be polarized in four different ways, the spectrum of returns from any given target (e.g., species of vegetation) can be used to formulate a signature based upon the physical structure of the target. *Surficial roughness* can be characterized similarly. Polarized shorter waves (C- and L-band in the case of AIRSAR) often provide the best results here. Whereas structure is measured mostly with polarization, roughness is measured mostly by wavelengths. As a result, it is possible under certain circumstances to determine what target characteristics have produced radar returns. This capability will be discussed in more detail later.

Radar is also returned differentially based upon *dielectric property* (measurable responses of a material to electric fields). Often, variations in dielectric property in the field are related to soil moisture, although other attributes, such as conductivity of material, can also affect return. It is sometimes possible to determine which returns have been produced by dielectric properties, as it is with returns from surficial roughness and structure. The SAR datasets used are listed in Table 1.

1.3. Research Test Area and Overview of Fieldwork

San Clemente Island lies approximately 60 miles off the coast of southern California (see Figure 1). The geomorphology of the island is best described as a series of marine terraces culminating in a high plateau. Vegetation is primarily grasses and small shrubs. The island was selected because it contains a wide variety of archaeological sites and a set of environmental zones that are representative of those found in the western United States (in which DoD and the Department of Energy, among other federal agencies, own large tracts of land for which they must provide environmental stewardship).

San Clemente Island was also selected because excellent baseline data exist for the island, including precise locations and characterizations of many archaeological sites, as well as environmental information, including accurate topographic, hydrological, soil, vegetation, and geomorphology characterizations. These baseline data were collected by Dr. Andy Yatsko, Cultural Resource Manager for Navy Region Southwest. Dr. Yatsko made numerous contributions to this research throughout its four-year duration, utilizing his comprehensive knowledge of the archaeology and the environment of the island. It should be said at the outset that a favorable outcome such as we have enjoyed in this research is probably not possible without the committed involvement of an expert on the archaeology of the survey area to be investigated, such as Dr. Yatsko provided.

AIRSAR collected data over San Clemente Island on 7 April 2002 utilizing different look angles and modes. This was done to establish the first of the protocols listed in the next section, that is, to determine the optimal manner in which SAR data should be collected to best discover archaeological sites. Four field sessions were conducted in 2003; five additional field sessions were conducted in 2004; and three in 2005. Ground control points (GCPs) were established to sub-meter accuracy using differential processing geographical positioning system (DGPS) equipment. GCPs were ultimately used in signature development, but initially used to determine the spatial precision of images produced from orthorectification protocols. GCPs were initially collected by CSRM at four survey plots (see Figure 2), each located in a separate environmental zone on San Clemente Island. Later, additional GCPs were obtained from the Navy Southwest Regional Office through Dr. Andy Yatsko, Cultural Resource Manager for that region. Finally, other GCPs were collected by CSRM, most of these in the Coastal Terrace Geomorphological Zone. This zone is a small band of land on the western side of San Clemente

Table 1. Radar bandson Polarizations used in this research

Instrument	Frequency band	Bandwidth (MHz)	Band Length (cm)	Single-look range resolution (m)	Polarizations	Interferometric	Pixel Size in This Study, After Orthorectification and Post-Processing
AIRSAR	P	20	68	7.5	HH, VV, HV,VH	No	5,5
AIRSAR	L	40,80	25	3.7,1.8	HH, VV, HV,VH	Yes	5,5
AIRSAR	C	40	5.7	3.7	HH, VV, HV,VH	Yes	5,5
GeoSAR	P	160 (max)	86	0.9	HH, HV	Yes	N/A
GeoSAR	X	160	3	0.9	VV	Yes	3,3 DEM 5,5 Image

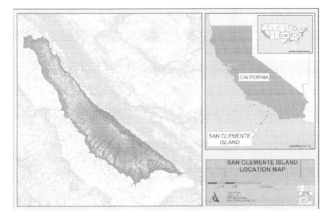

Figure 1. Test area: San Clemente Island.

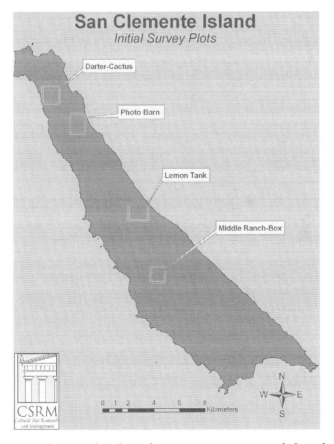

Figure 2. Initial survey plots, located in representative geomorphological zones.

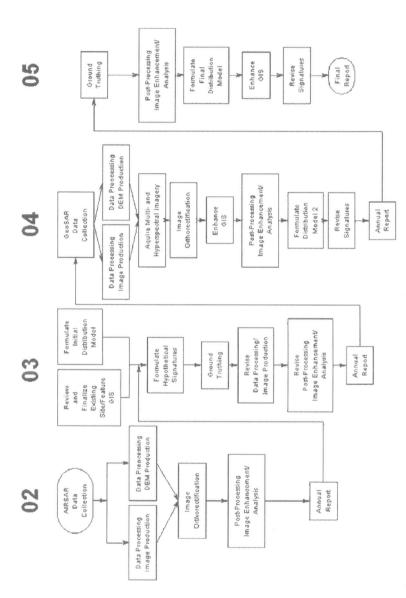

Figure 3. Summary representation of work flow over the four-year extent of the project.

Island, where soils and general environmental conditions are quite different than elsewhere on the island. In great part this difference is because of the salts introduced by the nearby ocean. In all, 100 sites were recorded to sub-meter accuracy in the Coastal Terrace Zone, while 733 sites were collected on the rest of the island. GCPs were collected not only for the center point of each site within each survey plot, but in many cases also for site features that might affect the radar return from each site. Therefore, in most cases, not only were highly accurate points recorded by means of DGPS units for these sites and features, but also each site and feature was delineated with a DGPS, forming a polygon that could be retrieved as a layer in the GIS for San Clemente Island. Once images were produced through processing at JPL, they were post-processed at Cultural Site Research and Management, Inc. (CSRM) in ways that facilitated development of signatures for archaeological sites.

The chart in Figure 3 provides a summary representation of the work flow over the four-year extent of the project. In presenting the eight SAR protocols that were developed in the research in the next section, we provide a more detailed examination of separate aspects of the project.

2. RESULTS: EIGHT SAR PROTOCOLS

2.1. Protocol 1: SAR Data Collection: Look Angle, Mode, and Optimal Flight Lines

2.1.1. Look Angles and Flight Lines

Optimal data collection is accomplished by means of *complementary look angles*. All SAR platforms illuminate the area to be examined by means of transmitting electromagnetic waves of microwave length. Only planes facing in the direction of the transmission are illuminated and subsequently sensed, with other surfaces remaining undetected and thus not subject to character-ization until illuminated from the proper angle. Complementary look angles are produced by executing flight lines from different sides of the area to be examined. Ideally, this would be done by scanning the same area from the north, south, east, and west. Depending upon topography, however, only two or three angles might be required. At San Clemente Island, a racetrack flight plan that provided three different look angles worked very well.

2.1.2. Choice of Band and Polarization

AIRSAR provides optimal SAR versatility by utilizing three different radar bands—C, L, and P—that can be polarized either vertically or horizon-tally when transmitted or received. This multipolar, multiband toolkit can be instructive about a great variety of features that might be found on a landscape because each band is scattered when it encounters features as small as one-third of the band's wavelength. For example, the C-band, ca. 5.7 cm in length (all band lengths given in this chapter are for AIRSAR and GeoSAR), can provide great detail about small structures that are often associated with landscape roughness, such as that produced by gravels or small rocks. Vegetation can also scatter 5-cm wavelengths, allowing differentiation, for example, of grasses from shrubs. The L-band, 25 cm in length, reacts noticeably with landscape features that produce variation in gross texture, such as large stones or small boulders, plants on the scale of tall grasses, medium to large shrubs, and trees. P-band, about 68 cm in length, interacts with larger features such as human-made structures, boulders, and trunks of very large trees. However, depending on its polarization, the megahertz at which it is transmitted, and the water content of material that it encounters, P-band can at times pass through the trunks of all but the largest trees. The available private-sector aerial SAR platforms, notably GeoSAR (which was designed by JPL), carry, at most, P- and X-band (X-band length is 3 cm). The GeoSAR P-band can be polarized in only two ways, as compared with AIRSAR's four, and X-band is polarized only VV (as seen in Table 1).

Because use of the P-band can interfere with television and citizen's band radio, only the AIRSAR C-band and GeoSAR X-band are routinely deployed and analyzed interferometrically to produce digital terrain models or digital elevation models (DTMs or DEMs). Because these are short wavelengths, the radar signal may not penetrate to ground surface. However, at sites such as San Clemente Island with sparse vegetation comprised of grasses and small shrubs, X-band and C-band radar penetration to the surface is likely, and longer wavelength radar waves may penetrate through the vegetation and into the ground surface (Henderson and Lewis, 1998:166).

The most informative radar-band polarization will depend upon the structural characteristics of the sites to be found and the general environment of the area to be examined. At San Clemente Island, for example, we see strong evidence that both C- and L-bands interact with the rock scatters associated with the presence of archaeological sites. They may also be reflected by grasses and brushes that are anomalous to surrounding vegetation. Soils created by human occupation often attract distinctive sorts of plants, which frequently grow with more vigor than do those surrounding them. These soils, in general, and by empirical observation on San Clemente Island, are richly organic, ashy, and contain rock, stones, shell, and artifacts. These soils tend to clump, creating interstices that fill with water. Interstices between material introduced by human occupation (e.g., rock, shell, artifacts) also fill with water. Water-soaked soils are highly conductive to electricity. The strongly conductive properties of midden soils on San Clemente Island reflect those radar bands that penetrate overlying vegetation layers. More specifically, P-band returns are of a magnitude that suggests the manner in which a flashlight in a dark room interacts with a mirror on the floor at an angle. While the cone of light that strikes its surface illuminates the mirror, most of the electromagnetic waves skip from the surface and off into space. The return is relatively weak. The phenomenology of the P-band in its interaction with the typical archaeological site on San Clemente Island follows this model. To the P-band, the site acts much like a mirror, as seen in Figure 4.

Signature development for archaeological sites, approached scientifically, demands some prior knowledge of site morphology. As is true in general, we must know what we are looking for before we can hope to find it. The more known about the physical and chemical attributes that comprise the site, the better. Therefore, background research that either draws from prior fieldwork or collects information about site morphology by means of preliminary fieldwork is essential to the protocols that are developed in this paper: the typical sorts of archaeological sites must be determined in order to provide the universe of values from which signatures are created. Similarly, the radar bands used should be selected on the basis of what is known prior to the investigation about the structure and physical characteristics of archaeological sites in the inventory area. As to mode of polarization, structures that are marginally detectable because of their spatial orientation by bands of a certain wavelength are generally more detectable in cross-polar mode. It is this versatility inherent in the AIRSAR platform that made it the best choice for the determination of the value of polarization to the current application. While the shorter wavelengths proved to be most valuable in directly detecting archaeological sites at San Clemente Island, this might not always be the case. In other areas, sites are to be found beneath soils or under forest canopy. For example, on a spinoff project, which used AIRSAR in Central America on mostly Mayan sites, initial findings are that masonry features located beneath even a tropical rain forest canopy can be detected by the unnotched P-band polarized vertically and transmitted at 40 MHz, higher than would have been done near developed areas (Comer et al., 2005).

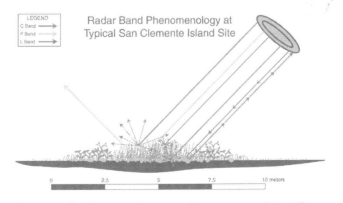

Figure 4. Schematic of radar wave phenomenology at a typical San Clemente archaeological site.

2.2. Protocol 2: SAR Data Processing and Image Production

2.2.1. Orthorectification

It became apparent early in this research that in order to exploit fully the capabilities of SAR to detect archaeological sites, images would have to be rigorously orthorectified. Previously, research utilizing radar imagery had been focused on very large features, often long and linear, or on areas that were generally homogenous in regard to broad taxonomic categories of interest, for example broad agricultural areas, geomorphologic zones, large ice sheets, and oceanic wave patterns (e.g., Moghaddam, 2001; Schmullius and Evans, 1997; Durden et al., 1989; Crippen and Blom, 2000; Gabriel et al., 2000). To find relatively small archaeological sites, a much greater degree of precision was necessary. The most common sites on San Clemente Island, the habitation sites, are on average 9.2 m in diameter. Because most radar imagery was made up of 5-m-square pixels, it was obvious that spatial accuracy would be imperative. (The statistical approach described later also made the use of relatively large pixels feasible.)

2.2.2. Merging Data from Multiple Flight Lines

It was necessary that radar imagery be available for all locations within the survey area. To accomplish merging of data from multiple flight lines, JPL radar engineers developed what they dubbed "Jurassic Proc" software (for the gigantic body of data used and the enormous processing power required to produce images) to merge data from two or more flight lines. Interferometric data merged in this way provided a digital elevation model of the terrain that is much more accurate than any produced before. The images orthorectified by means of the interferometric digital elevation model proved to be amazingly accurate. For all of the radar bands utilized by AIRSAR (P, L, and C), an accuracy on the order of 5 meters was obtained. Ground control points for this test were supplied by a chicken-wire square that had been utilized by San Clemente Island botanists to protect cultivated plants from foraging animals and an approximate figure-eight arrangement of barbed wire left over from Marine maneuvers in the area of the island's old airfield.

2.3. Protocol 3: SAR Image Analysis (Post-Processing)

2.3.1. Iteration of Image Analysis

Image analysis was conducted numerous times, as described below and as indicated by the diagram in Figure 3. As this diagram indicates, image analysis is a key to virtually all other signature development activities.

2.3.2. Quantitative and Replicable Analysis Protocols: Statistical Techniques of Value in Establishing Signatures for Archaeological Sites

Our initial attempts at image analysis (sometimes called "image post-processing") were intended to make archaeological sites more visible in imagery. The first field session in which ground-truthing was conducted revealed, for example, that archaeological sites in certain areas were extremely visible as bright locations in the black-and-white C-band imagery. Having been thus encouraged to examine imagery, we developed ways to amplify the bright returns that were associated with archaeological sites, not only in the C-band imagery, but also what we saw in the L-band imagery. We were, in fact, quite successful in enhancing imagery so that sites could be identified through examination by trained observers. As can be seen in Figure 5, when L-band imagery produced from the three polarized bands (HH, HV, and VV) was enhanced by means of pixel averaging, the great majority of sites were visible. The most effective protocol for pixel averaging was the Gamma-MAP filter for the three-by-three-pixel window (ERDAS , 1997). There are, however, great impracticalities in utilizing this "trained eye" approach. First of all, not all eyes are equally trained. Also, an image may reveal site locations more or less satisfactorily depending upon the peculiarities of the screen upon which it is displayed or the printer that is used to produce a hard-copy image. Brightness is a relative term and is unquantified in the "trained eye" approach.

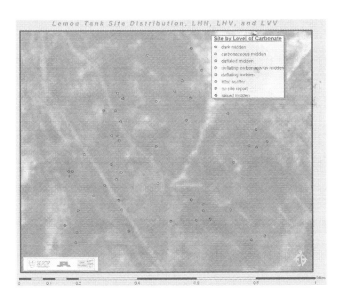

Figure 5. Locations of all known archaeological sites in Lemon Tank survey plot over speckle-reduced radar image.

As seen in Figure 6, sites appeared bright in some cases merely because returns from sites were brighter than an unusually dark background. The statistical protocol, nevertheless, established that although returns from archaeological sites were distinctive, they could be either brighter or darker than immediate surroundings. As we can see by looking at the histogram of values seen in Figure 6, returns collected from archaeological sites show a central tendency distinct from that shown by returns collected from the survey area with no archaeological sites. It is this statistical comparison that can be used to develop archaeological-site signatures, rather than the appearance of an image in a given area.

Statistically, it is a straightforward exercise (although one that has numerous steps) to determine whether or not returns from archaeological sites are different enough from returns from the rest of the survey area to conclude, within determinable probability parameters, that the values were taken from two different populations. To do this, we used a variation of a difference-in-means test. This test was carried out in two steps. The first step was to determine if, simply, there was enough difference between values obtained

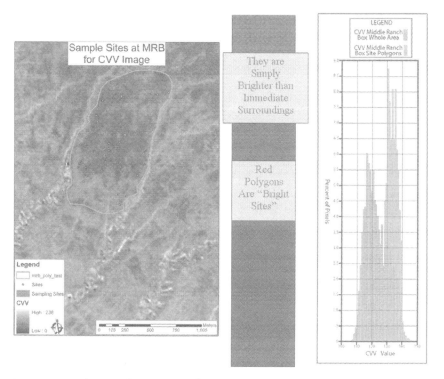

Figure 6. Distribution of CVV returns from site and non-site areas in the Middle Ranch Box survey plot.

from the pixels within established archaeological-site boundaries and values representative of all pixels not within archaeological-site boundaries to justify the conclusion that the two sets of samples actually represented two different populations. The second step was to determine *which pixel values* were most strongly associated with the set taken from archaeological sites. That is, which pixel values were statistically different enough to justify the assertion that they were obtained from a population distinctly different from the rest of the island? Both steps 1 and 2 utilized the following statistical protocol:

Our null hypothesis was that there is no difference between the population of values that lies within site boundaries and the population of values that lies outside site boundaries, that is:

$$H0 : u^1 = u^2$$

This was equivalent to testing the hypothesis that the mean of x1–x2 is 0. If the null hypothesis were upheld, of course, it would mean that pixel values associated with archaeological sites could not contribute to signatures for those sites. If, on the other hand, the null hypothesis were disproven, pixel values associated with sites could be used to develop signatures for those sites.

We tested this hypothesis with the formula : $\left(\sum_n x^1/n) - (\sum_n x^2/n\right)$

$< 1.96\sqrt{\sigma_1^2/n + \sigma_2^2/n}$, which is to say that the difference of the means of the site and randomly selected samples, or $\left(\sum_n x^1/n\right) - \left(\sum_n x^2/n\right)$, will be less than 1.96 standard deviations apart, with the standard deviation of the difference of means being determined by the elementary formula: $\sigma_{x1-x2} = \sqrt{\sigma_1^2/n + \sigma_2^2/n}$ (see, for example, Hoel, 1971:172).

2.3.3. Optimal Use of SAR Bands, Polarizations, and Combinations of These

Step-1 results indicated conclusively that, for all vectors[1] tested, values within and outside of site boundaries were drawn from different populations. Figure 7 illustrates some of these results. Note that the difference of means for LHV is at 14.89 standard deviations. A difference of 1.96 standard deviations would

---

[1] A *vector* here is a dataset that has been derived from the application of a certain type of remote sensing. In this research, vectors were generated by the use of specific bands and polarizations of SAR; the surface models that were developed from interferometric SAR (including measures of slope); and multispectral returns treated by means of a variety of algorithms (e.g., Tassled-Cap, NDVI).

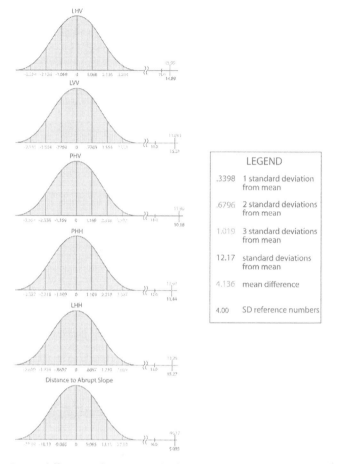

Figure 7. Step-1 difference-of-means results for several key vectors. Note that site and non-site differences of mans are separated by many standard deviations.

indicate a 95% probability that the two samples were from different populations. Three standard deviations would indicate a 99% probability that this was the case. The number of standard deviations associated with all of the radar datasets indicated virtual certainty. However, the LHH Step-1 test indicated the greatest number of standard deviations, by a small margin, and so this was selected as the most instructive L-band radar vector. This procedure was carried out for all radar bands and multispectral standard algorithms (e.g., NDVI and Tasseled Cap) available to us. The most instructive vectors were determined to be the SAR band polarizations CVV, LHH, PVV (from AIRSAR), and XVV (from GeoSAR); slope, derived from interferometeric analysis of X-band data (from GeoSAR), and the multispectral vector NDVI (the normalized difference vegetative index), calculated from IKONOS satellite data. Multi-spectral data from Landsat were also used in various protocols. As a check of this statistical protocol, it was applied to random samples obtained from areas outside of the known locations of archaeological sites. When the null hypothesis was tested for two random samples taken from areas outside of archaeological sites, it was confirmed. As shown in Table 2, results indicated that random samples are indeed drawn from a single population for every vector tested.

The Step-2 statistical analysis utilized the same formula, but it compared the means of all individual pixel values within archaeological site polygons with the means of all individual pixel values within an equal number of randomly selected polygons of equal size. These would be the same if the populations from which values were drawn (i.e., site and random) affected returns in the same way and to the same degree. That is, the null hypothesis in this case would be upheld. The pixels for which the null hypothesis was not upheld would be those that are associated with archaeological sites. In Figure 8, these pixels are highlighted in green. Pixel values for which the null hypothesis is proven are highlighted in red.

2.4. Protocol 4: Corroborative Use of Multispectral Datasets

Certain multispectral image data also correlate very strongly with the locations of archaeological sites. While radar image values were greatly influ-enced by vegetation structure and dielectric properties, multispectral imagery highlighted spectral attributes of vegetation, including those associated with vegetative health. NDVI (normalized difference vegetative index) values could be used to locate the greener and more vigorous vegetation that developed at San Clemente Island archaeological sites (and that typically develop at non-structural, and many structural, archaeological sites everywhere) by virtue of the enriched, organic soils produced by human occupation. Taken together, then, radar and multispectral imagery were used to sense the following differ-ences (as schematically depicted in Figures 4 and Figure 9):

2.4.1. Topographic Roughness Produced by the Cluster Distribution of Stone and Rock Associated with Archaeological Sites (Sensed by SAR)

This material was probably brought in both for use in erecting shelter and as tools. Given our knowledge of radar wave phenomenology, we can postulate that the L-band is strongly affected by surface roughness of this sort. L-band waves are roughly 25 cm in length. Features that are as small as one-fourth the length of the radar band affect radar waves. This is a dimension that fits with the sizes of stone and rock associated with archaeological sites.

Table 2. Statistical tests of two randomly selected sets of pixel values drawn from areas approximately the size of archaeological sites. The results indicate they are taken from the same universe

Vector	Rand Stdey 1	Rand Stdey 2	Standard Deviation	mean difference	Rand. Mean 1	Rand. Mean 2	Scaling
				Random vs. Random Mean Difference Test			
cvv	11.51011157	10.31388397	0.5708	−0.9228	127.0410479	126.118261	−1.6165261
lhh	17.33674375	18.02115265	0.9236	−1.8093	125.8912324	124.0818963	−1.958928
ndvi-N	18.01451571	19.02954193	0.9679	−1.6876	235.0855593	233.3980011	−1.7435894
pvv	0.032980673	0.038006617	0.0019	−0.0008	0.134966307	0.134131511	−0.4491392
slp	7.999462582	8.3211335	0.4263	−0.6238	12.30585348	11.68204118	−1.4631895
xvv	3.808589696	3.97097976	0.2032	−0.2657	−7.528561041	−7.794241331	−1.3073027

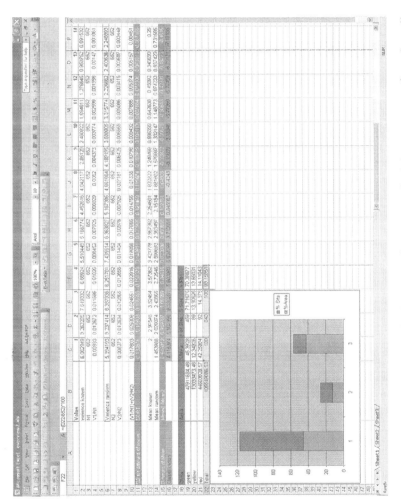

Figure 8. Step-2 calculations: pixel values more than two standard deviations apart from the null hypothesis mean.

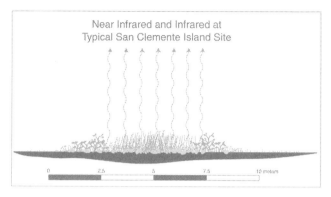

Figure 9. Schematic of NIR and IR radiation at a typical San Clemente Island archaeological site.

While there are many factors that influence radar backscatter, the Rayleigh scattering criterion is a guide to radar backscattering behavior as a result of surface roughness (Peake and Oliver, 1971; Schaber et al., 1976). Surfaces become rough enough to begin backscattering significant radar energy at approximately 1/4 of the imaging wavelength. Surfaces smoother than this will be dark in radar images, surfaces rougher than this will be increasingly bright. Other key factors include imaging geometry, dielectric constant (largely a function of moisture content), and surface slope.

2.4.2. Vegetative Structures Associated with Archaeological Sites (Sensed by SAR)

Vegetation on San Clemente Island is restricted to shrubs or grasses, in part because of scarce precipitation. Vegetation growing on an archaeological site is frequently of a different type than vegetation surrounding it because of soils enriched during human occupation of the site, as well as the presence of rock, shell, and other materials that alter soil moisture and chemical composition. Most archaeological sites on San Clemente Island seem to have grasses at their center. These are frequently either distinct from surrounding vegetation or substantially taller and thicker than any surrounding grasses. The grasses at the center of the site provide a thick biomass structured in a predominantly linear way. The stalks of grass provide a reflector to radar waves, which are scattered differently depending upon their length (short waves seem more readily reflected) and polarization (the strongest reflection results from VV polarization). L-band radar waves seem to be most sensitive to the density/distribution of vegetation on archaeological sites. One might speculate that this phenomenon is at least in part because of the presence of long and tall grasses at the center of many of these sites. Thicker vegetation mass, made up of many individual reflective planes, provides a more effective scattering or reflecting surface than does thinner.

2.4.3. Dielectric Property (Sensed by SAR)

Because soils associated with archaeological sites are richer in organic materials than surrounding soils, they are also moister. Soil particles at archaeological sites tend to clump, producing interstices in which water is trapped. These characteristics, and perhaps the carbon in the soils from campfires, affect dielectric property, rendering the soil very conductive to electricity. The results of soil-conductivity tests at archaeological sites at the old airfield on San Clemente Island established this fact quite well. These tests, conducted by Larry Conyers (2000), showed that all sites tested were enormously conductive. This conductivity would affect the propagation of radar waves much as a mirror affects propagation of light waves. Since radar waves are transmitted at an angle, longer waves would be reflected obliquely into space via specular reflection and not back to the radar platform. This is analogous to light from a flashlight striking a mirror obliquely. In both cases, the return from the area illuminated by the beam might be discernable, but would be at best weak.

2.4.4. Greenness and Vegetative Vigor
(Sensed by Multispectral Data)

As previously mentioned, archaeological soils are conducive to thick and vigorous vegetation. Such greenness and vigor are readily discernible by examination and analysis of NDVI images produced from multispectral data using a standard algorithm.

2.5. Protocol 5: Radar-derived Spatial Modeling to Detect Archaeological Sites and Features

This research produced statistically based site signatures, not site-distribution models. One aspect of spatial modeling, however, that is directly associated with the use of radar data is the surface model, called here the *digital elevation model* (DEM). DEMs were generated by interferometry. A C-band DEM was produced by data collected by the AIRSAR platform, and an X-band DEM by data collected by the GeoSAR platform. The horizontal accuracy of both the AIRSAR and the GeoSAR DEMs, as tested in field sessions at San Clemente Island, was generally five meters or better. Therefore, the merging of interferometric SAR data appears to have been highly beneficial to accuracy, in addition to eliminating radar shadows and resulting gaps in DEMs.

AIRSAR and GeoSAR DEMs were used to generate a slope model with degree of slope being calculated for every 5-meter pixel within the AIRSAR

DEM. The GeoSAR DEM that was developed using X-band data as opposed to C-band data provided great height elevation accuracy—about 0.5 m compared with about 1 m—in part because of the shorter length of the X-band radar. Also, because this X-band DEM was comprised of pixels of about 3 m, as opposed to 5 m for the C-band DEM, slope could be determined for every 3 m instead of every 5 m. This more accurate GeoSAR DEM was used to determine those slopes most associated with the presence of archaeological sites. Interferometric DEM data were analyzed utilizing the same Step 1 and Step 2 protocols described above. The GeoSAR DEM provided us with a third and very powerful vector for determining archaeological site signatures.

2.6. Protocol 6: Procedures for Incorporation into, and Analysis with, GIS

Once the vectors that were very strongly associated with archaeological sites were identified (see 2.3.3, Step 1), they were combined by means of the GIS. A circle with a radius of 15 m was drawn around the center point of each archaeological site. As noted above, center points had been established to less than 1-m accuracy. Although sites averaged 9.2 m in diameter, allowance was made for larger sites. Another consideration was that often the effects of site occupation on the landscape were attenuated rather than abruptly ending. The critical rationale for this protocol was that it was most essential to reliably capture areas that were within site boundaries; the protocol thus allowed for a certain degree of spatial error. While this procedure virtually ensured that some areas unrelated to the site would be included in the sample, the statistical nature of the image-analysis protocols developed from this research could admit of such dilution of the pivotal data. Therefore, values were harvested from each of the pixels within sampling polygons. For the control sample, that is the universe of values known to lie outside archaeological sites, other 15-m radius circles were established at randomly selected points outside of archaeological sites on the landscape. The statistical tests were conducted utilizing values associated with these two sets of circles.

2.7. Protocol 7: Establishment of Signatures Based upon Radar Returns Within Given Environmental and Cultural Zones

Subsequent adjustment of the GIS analytical protocol (Protocol 6) was based on iterative episodes of fieldwork and analysis. For example, when an initial test of the signature model was made, it was discovered signatures held up to field examination very well in most areas of San Clemente Island. A pronounced exception was the coastal terrace on the western side of the island. In this area, the signatures were much less reliable. Additional sites, therefore, were collected to serve as sampling or training areas within what became a separate test area, the Coastal Terrace Zone. Not only did this procedure produce valid site signatures for the Coastal Terrace, but results also improved for the rest of the island. With this in mind, we established three testing zones, based upon the three major geomorphological areas on the island: the Coastal Terrace Zone, The Marine Terrace Zone, and the Plateau Zone (see Figure 10).

Relevant environmental and cultural zones were determined to coincide with previously established geomorphological zones through the iterative process of data collection, field investigation and verification, and additional analysis based on the results of field investigations. This process highlighted data that did not fit comfortably within the existing framework of environmental and site categories, and it recommended reanalysis or reconfiguration of categories. Reconfiguration of categories could not have been accomplished without thorough knowledge of the archaeological record and cultural history of the survey area. In this case, we had at our disposal the leading expert in the archaeology of San Clemente Island, Dr. Yatsko, who had written his dissertation on that subject. We also conducted a thorough archival research of sites in excavation records, many of which were obtained at UCLA.

The *statistically based site signatures* that we developed by our statistical protocol resulted from quantitative differences in the sensing device returns from archaeological sites, as compared to returns from areas not within archaeological sites. We developed a way to establish archaeological-site signatures using return values from as few as 15 archaeological sites within a given environmental zone. A statistical-analysis protocol developed during this research (see Protocol 3) was carried out for each site and nonsite pixel value for each vector. Signatures were formed by pixel values that were

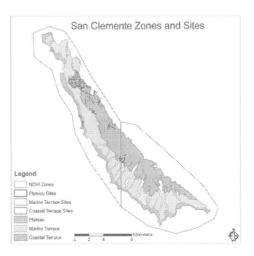

Figure 10. Three testing zones, based upon the three major geomorphological areas on the island: Coastal Terrace Zone, Marine Terrace Zone, and Plateau Zone.

two standard deviations away from the null hypothesis mean for pixel values taken from areas within two discreet sets of polygons, the first describing archaeological sites, the second around randomly selected areas not within known archaeological sites. When the null hypothesis is disproven, we can be 95% sure that these two sets of pixels were drawn from different populations. The simplest, and therefore most likely, explanation for the difference of these special sets is that characteristics of archaeological sites are affecting return values.

An example of signatures developed from a single vector is seen in Figure 11 (CVV Results). Here, by using a sample of 15 sites from each of the three major geomorphological areas, signatures were developed that covered 10.65% of the island, and contained 48% of the remaining 701 sites for which location was known to an accuracy of less than a meter. For the three zones that contributed to this overall result: Using CVV on the Coastal Terrace produced a one-vector signature that occurred over 12.70% of the Coastal Terrace yet caught 64.76% of sites. When used on the Plateau geomorphological zone, these figures were different. CVV single-vector signatures covered 15.15% of that area and included 49.51% of sites (see Table 3). For the Marine Terraces, these figures were 6.09% and 37.50%. Thus, the local

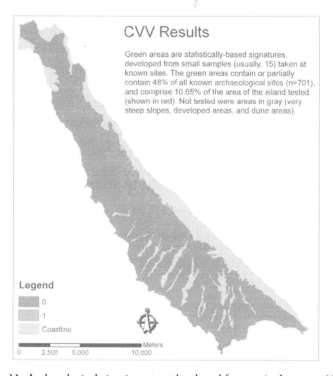

Figure 11. Archaeological-site signatures developed from a single vector (CVV).

Table 3. Key vectors. Note especially column 1 (% of survey area contained within signatures) and column 3 (% of sites falling within signatures)

	% of area	Plateau # of pixels	% of sites	# of sites
CVV	15.5	283812	49.51	204
LHH	3.86	72379	35.92	148
NDVI	35.06	657006	80.58	332
PVV	8.06	151065	44.66	184
SLOPE	5.19	97216	38.83	160
XVV	14.2	268986	42.72	176
SUM		1873762		412
	% of area	Coastal Terrace # of pixels	% of sites	# of sites
CVV	12.7	74029	64.76	68
LHH	2.65	15441	25.71	27
NDVI	4.78	27890	42.86	45
PVV	24.15	140839	80.95	85
SLOPE	2.89	16836	30.48	32
XVV	18.38	107165	64.76	68
SUM		583100		105
	% of area	Marine Terrace # of pixels	% of sites	# of sites
CVV	6.09	128370	37.5	69
LHH	3.05	64220	21.74	40
NDVI	17.95	378202	85.87	158
PVV	16.8	353954	50	92
SLOPE	2.86	60204	34.52	58
XVV	5.23	110318	41.85	77
SUM		2107341		184
	% of area	Whole Island # of pixels	% of sites	# of sites
CVV	10.65	486211	48.64	341
LHH	3.33	152040	30.67	215
NDVI	23.29	1063098	76.32	535
PVV	14.15	645858	51.5	361
SLOPE	3.82	174256	35.66	250
XVV	10.59	483469	45.79	321
SUM		4564203		701

environment played a strong role in how effective CVV was in developing signatures. Table 3 displays these figures for other important vectors, NDVI, for example, is 23.29% and 76.32%. When different vectors are combined, they produce signatures that are both more reliable, because they are based upon sensing of diverse attributes by the different vectors, and more useful because they pare down the areas that are most likely to contain archaeological sites.

Two important applications of this approach are seen in combinations of vectors that would be possible given different research scenarios. For example, if only the AIRSAR platform were available, the results at San Clemente Island

would be those appearing in Figure 12. Private-sector SAR aerial platforms
today carry only X- and P-band; however, P-band cannot regularly be used
in most of the United States because it interferes with commercial transmissions, including television and citizens-band radio. Commercial multispectral
satellite data are available for virtually the entire earth, however, and so could
be used as seen in Figure 13, to generate the NDVI vector.

It should be clear by now that, while some protocols apply only to
radar data (e.g., protocols for data collection and for data processing to
produce images from radar data), others were found to be useful for developing signatures from multispectral datasets, and there would seem to be
no reason why they could not be used to develop archaeological-site signatures from hyperspectral and other sorts of data. Notable among these are the

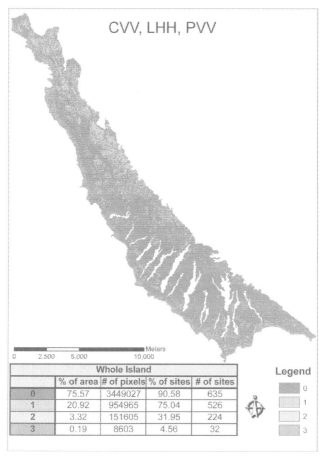

Whole Island				Legend
	% of area	# of pixels	% of sites	# of sites
0	75.57	3449027	90.58	635
1	20.92	954965	75.04	526
2	3.32	151605	31.95	224
3	0.19	8603	4.56	32

Figure 12. Signatures developed from private sector airborne SAR and private sector
multispectral satellite data.

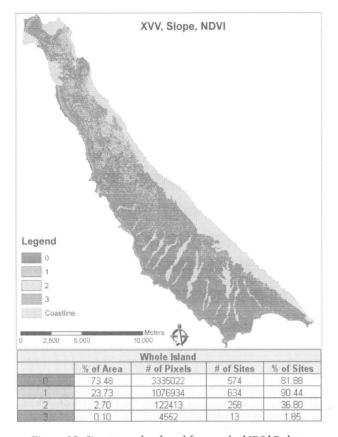

Whole Island				
	% of Area	# of Pixels	# of Sites	% of Sites
0	73.48	3335022	574	81.88
1	23.73	1076934	634	90.44
2	2.70	122413	258	36.80
3	0.10	4552	13	1.85

Figure 13. Signatures developed from only AIRSAR data.

statistical protocols developed for image analysis (using grid algebra instead of the standard functionalities of image-processing software) and protocols for ground truthing.

2.8. Protocol 8: Procedures for Ground Verification of Site Locations

Our protocol for ground truthing was to establish locations of a great number of archaeological sites to a high degree of precision and to acquire highly accurate information about other sites from other researchers. We ourselves collected precise locations for several hundred archaeological sites (particularly in the Coastal Terrace Zone and other areas toward the southern end of the island where this had not been done previously), and we had the good fortune to be presented with precise locations for hundreds more by Dr. Yatsko. This insured good representation for different environmental zones and provided latitude to alter zone boundaries as suggested by interim results without having to collect many more precise ground control points for other archaeological sites.

Having this store of locational information, we could simply randomly select sites within a certain environmental zone (or sites by type, for that matter). We utilized a statistical protocol that permitted the use of relatively small samples (we settled on n=15). After developing signatures based on samples, the veracity of the signatures could be tested against the locations of the sites not selected at random.

3. DISCUSSION: APPLICATIONS OF FINDINGS TO ARCHAEOLOGICAL RESEARCH

The research conducted under SERDP grant CS-1260 was intended to produce the protocols for utilizing airborne SAR to find archaeological sites that were presented in the beginning of the paper. These we have presented.

What also bears mentioning here are a number of findings derived from the datasets and statistical protocols that we developed that are relevant to current archaeological research interests. In the section below, vectors are regarded qualitatively, in terms of what site characteristics they can be used to determine. AIRSAR and GeoSAR data, as well as data collected by means of other remote sensing devices, contributed in many valuable ways to understanding site distribution at San Clemente Island. To begin with, they shed light on several primary determinants of site distribution that have been well established in all areas worldwide where archaeological-site models have been formulated. These include proximity to water, slope, soil drainage characteristics, and, in many cases, aspect. Using the first of these, proximity to water, as an example of how remote sensing data can contribute information to archaeological site modeling, we were able to identify the locations of standing water, using LANDSAT imagery, merely by assigning Band 7 to red, Band 4 to green, and Band 2 to blue. This was effective even in detecting very small "tanks," that is, depressions in ravines where water collects naturally.

The accurate DEMs from AIRSAR and GeoSAR data added another dimension to understanding site distribution by forming the basis for a prediction of flow accumulation. The flow-accumulation model shows where water would be likely to collect given the current topography. If we assume that the topography is not much changed over the approximately 10,000 years during which the island has been occupied, the model could also be used to show where water was likely to collect in the past. Precipitation in southern California is strongly bimodal: one mode is a long period of no precipitation during the late spring, summer, and early fall, and the other a short period in the winter characterized by rain events that are often intense. A flow-accumulation model of San Clemente Island is shown in Figure 14, which

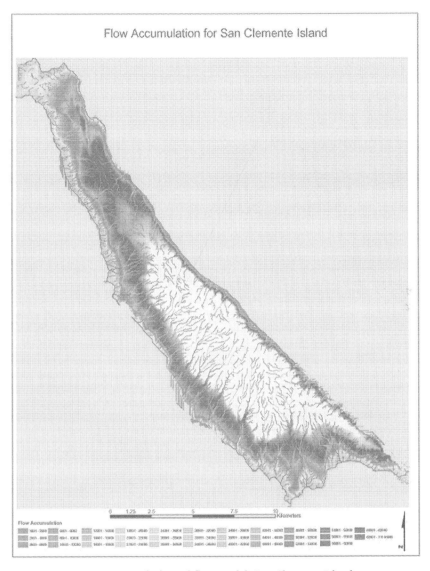

Figure 14. Hydrological flow model, San Clemente Island.

indicates where water is likely to accumulate during the winter season. Such accumulations might have influenced site selection during that time of year in ways that were beyond the scope of this research project, but might be examined in the future. In societies in which water-management systems have not been developed, sites must be within a few hundred meters at most from sources of water.

The precise AIRSAR and GeoSAR DEMs were also used to model slope. Virtually all archaeological sites occur at slopes of less than 15%. Exceptions, based upon field inspection, appear to be those instances in which erosion is exposing a buried site along the side of a ravine and where a site has been discovered in a cave or rock shelter located in the wall of a ravine or canyon. Furthermore, almost all occur at slopes of less than 8%.

Aspect is also often a strong factor in the determination of prehistoric site selection. Aspect suggests certain functions for the sites found in each of the survey plots. At the northernmost plot, Darter Cactus, 95% of sites are looking to the west (55% SW, 25% W, and 15% NW). These sites are positioned in a way that would provide a view toward the direction of prevailing winds. Wind would move objects of value in the direction of prehistoric occupants of the island (e.g., logs or dead sea animals). By keeping an eye out for materials that were pushed by the wind near the island, these might be collected. Sites also face the Cortez Banks and the channel between the Cortez Banks and San Clemente Island. The Banks provide an environment that has produced a large population of blue whales, and the channel is home to many large pods of dolphin and other sea mammals. Thus, the sites at Darter Cactus might have been lookout spots for pods of sea mammals (Figure 15). That sea mammals comprised a large percentage of the diet of prehistoric islanders is well documented by archaeological excavations on the island. Whales provided not only food, but also many materials that the prehistoric inhabitants of San Clemente Island found extremely valuable. The prime example of this is that

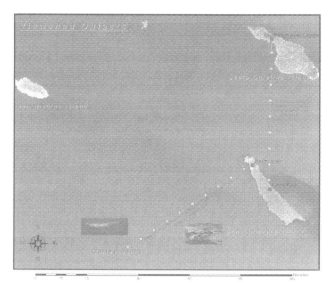

Figure 15. Viewshed from two key San Clemente Island locations to marine-resource locations and locations of cultural importance on Santa Catalina Island.

whale ribs were used as structural support for dwellings on the island, such as those that have been found in archaeological excavations at the Eel Point and Nursery sites. The Darter Cactus sites could also have served, of course, as locations from which surveillance was kept for valuable items that might wash ashore or come near enough to the island to be retrieved.

Just to the south of Darter Cactus, the Photo Barn survey plot (Figure 2) occupies a position on the spine of the island, that is, the highest plateau running north and south. The plateau itself gains elevation steadily as it moves to the south. The aspect analysis indicates that 29.41% of Photo Barn sites face to the west, while 60.8% of sites look toward the east. The sites with an eastern aspect are increasingly blocked as one moves farther east from the prevailing winds, which are northwesterly in summer and more northerly in winter, yet these sites are located near to places that have a view to the west. Views to the west would have been useful, for reasons detailed above in the discussion of the Darter Cactus sites.

The Lemon Tank survey plot (Figure 15) contains many sites that archaeological excavation indicates to have been ceremonial in nature. These ceremonies seem to have been associated with the Chinigchinich cult, which was a component of what is called in anthropology a revitalization movement. Such movements have been recorded among societies that have been subjected to severe stress. In this case, the social organization experiencing stress would have been the Gabrielino Indians, a group that occupied portions of what is now southern California, including the islands of San Clemente, Santa Catalina, and San Nicholas, at the time of contact with Europeans. Like many Indian cultures, the Gabrielinos lost numerous members, especially the very young and the old, to European diseases, and more generally to the disruptions precipitated by European contact. These disruptions included conflicts not just between Indians and Europeans, but also among Indian groups as they jockeyed for viable positions in a rapidly and erratically (from their standpoint) changing political and economic environment. Revitalization movements predictably promote group solidarity by appealing to the ancestors to return to earth and restore the order of the past. Ceremonies often involve supplications or other communications with ancestors, who are usually thought to occupy locations of geographical prominence, such as hilltops or mountaintops, or valleys or caves in some cases. Also, ceremonies were intended to draw all surviving members of the group together, under the protective umbrella of the ancestors.

We see at the Lemon Tank sites that the aspect is toward Santa Catalina Island. This orientation might be expected, as Gabrielinos occupied Santa Catalina Island. The locations of the highest point on Santa Catalina Island and of Two Harbors, a spot on the island where two well protected harbors on the east and west sides of the island almost converge at a narrow isthmus, and the location of the largest historically recorded village on the island, are almost due north of the Lemon Tank sites. The aspect of Lemon Tank sites

(24% due north, 20% to the northeast, 24% to the northwest, and in total almost 75% of sites facing generally northward) would suggest that these key points of cultural reference were incorporated into ceremonies held at Lemon Tank sites (see Figure 15). Almost all of the remaining sites, 20%, faced west.

At the Middle Ranch Box survey plot (see Figure 2), slightly over 60% of sites again face in a westerly direction (40.38% to the southwest, 19.23% to the west, and 1.92% to the northwest). The predominance of western aspect sites and the large percentage of sites facing the most likely location of sea mammals again suggest the importance of these resources. In this case, we also have a significant percentage of sites with a view to the south (17.31%) and southeast (17.31%), toward other locations on the sea that might provide resources.

In summary, the analysis of site distribution on San Clemente Island suggests that elements in addition to those that provide limiting factors at all archaeological sites, e.g., water and slope, might well have played an important role in determining site location. These elements can be used to guide archaeological research at San Clemente Island, and perhaps other Channel Islands, in the future. Thus, there are research applications of the protocols developed by means of this research, in addition to the application of this research to cultural resource management.

4. CONCLUSION: MANAGEMENT AND RESEARCH IMPLICATIONS

There are numerous management as well as research applications of our research results for San Clemente Island. Almost 37% of sites detected fell within the two vector signatures seen in (XVV, Slope, NDVI), whereas these comprised less than 3% of the area examined. This 2.7% of the land area could be set aside with a high degree of certainty that it contains the locations of many sites. This area could be reduced over time by on-ground survey of that small area of the total landmass in question. One-vector signatures encompass over 90% of the sites, and less than 24% of the island. Therefore, 75% of the island could be surveyed relatively quickly, because few sites would be found in those areas. Following this relatively inexpensive survey, most of the island could be cleared for all activities. These two survey efforts would, as well, constitute an additional empirically based test of the site signatures. It is also important from a management and financial standpoint that little material would be collected that would need to be treated by chemical and physical means for conservation, kept in special containers, and stored in environmentally controlled spaces. Finally, even in the absence of such surveys, developments and other ground-disturbing activities are enormously less likely to encounter archaeological sites, and thus trigger archaeological evaluation, if they are located in areas that do not contain site signatures.

The development of SAR protocols that grew out of our research also has important implications for archaeological-site evaluation beyond San Clemente Island. Our findings strongly suggest that SAR can be an effective tool for evaluation of other large land areas at a significantly lower cost than current methods that require expensive Phase II evaluation and artifact storage costs (see above section 1.1). Use of these protocols to avoid Phase II evaluation, which is time-consuming, expensive, and attended by the long-term and unbudgeted, but still very real cost of artifact storage, would be beneficial from both a financial and a preservation perspective.

5. ACKNOWLEDGEMENTS

We acknowledge with thanks the cooperation and assistance of Dr. Michael Abrams, NASA Jet Propulsion Laboratory/California Institute of Technology; Mark E. Brender Vice President Communications & Marketing, GeoEye; California State Archaeological Archives, University of California, Fullerton; Dr. Bruce Chapman, NASA Jet Propulsion Laboratory/California Institute of Technology; Dr. Elaine Chapin, NASA Jet Propulsion Laboratory/California Institute of Technology; John Cook, President, ASM, Associates; Dr. Robert Crippen, NASA Jet Propulsion Laboratory/California Institute of Technology; Earth Data Corporation; Dr. Robert Holst, SERDP Program; Katherine Kerr, SERDP Support Office (HydroGeoLogic); Dr. Mahta Moghaddam, Associate Professor, Department of Electrical Engineering and Computer Science, University of Michigan; National Archives and Records Administration (NARA), College Park, Maryland; Stacey Otte, Executive Director, Catalina Island Museum, Catalina Island Conservancy; Dr. Mark Raab, University of Missouri-Kansas City; Eileen Regan, SERDP Support Office (HydroGeoLogic), now Earthcare Associates; Dr. James Reis, Earth Data Corporation; San Diego County Archaeological Center; University of California at Los Angeles (UCLA) Archaeology Laboratory; Dr. Andy Yatsko, Cultural Resources Program Manager, Navy Region Southwest.

REFERENCES

Comer, D., Blom, R., Golden, C., Quilter, J., and Chapman, B., 2005, Inventory of Archaeological Sites Using Radar and Multispectral Data. Lecture presented at National Geographic Society, Washington, D.C.

Conyers, Lawrence, 2000, Field Report of Geophysical Investigations at the Old Airfield, San Clemente Island. Scripps Institution of Oceanography, La Jolla, CA.

Crippen, R. E., and Blom, R. G., 2000, Unveiling the Lithology of Vegetated Terrains in Remotely Sensed Imagery. *Photogrammetric Engineering and Remote Sensing* 67:935–943.

Durden, S., Van Zyl, J., and Zebker, H., 1989, Modeling and observation of radar polarimetric signature of forested areas. *IEEE Transactions, Geoscience and Remote Sensing* 27:290–301.

ERDAS Corporation, 1997, *ERDAS Imagine Field Guide*, Fourth Edition. ERDAS Corporation, Atlanta, GA.

Gabriel, A. G., Goldstein, R. M., and Blom, R. G., 2000, ERS Radar Interferometry: Absence of Recent Surface Deformation Near the Aswan Dam. *Journal of Engineering and Environmental Science* VII:205–210.

Henderson, F. and Lewis, A., 1998, Principles and Applications of Imaging Radar. *Manual of Remote Sensing*, Third Edition. John Wiley and Sons, Inc., New York.

Hoel, Paul G., 1971, *Elementary Statistics, Third Edition.* John Wiley and Sons, Inc., New York.

Moghaddam, M., 2001, Estimation of comprehensive forest variable sets from multiparameter SAR data over a large area with diverse species. *International Geoscience and Remote Sensing Symposium 2001*, Sydney, Australia.

Peake, W. H., and Oliver, T.L., 1971, The response of terrestrial surfaces at microwave frequencies. Technical Report No. AFAL-TR-70-301. Ohio State University Electroscience Lab., 2440-7, Columbus Ohio.

Schaber, G. G., Berlin, G.L., and Brown, W.E., 1976, Variations in surface roughness within Death Valley, California-geological evaluation of 25 cm wavelength radar images. *Geological Society of America Bulletin* 87:29–41.

Schmullius, C.C., and Evans, D.L., 1997, Synthetic aperture radar (SAR) frequency and polarization requirements for applications in ecology, geology, hydrology, and oceanography: a tabular status quo after SIR-C/X-SAR. *International Journal of Remote Sensing* 18:2713–2722.

Chapter 6

Putting Us on the Map: Remote Sensing Investigation of the Ancient Maya Landscape

WILLIAM SATURNO, THOMAS L. SEVER, DANIEL E. IRWIN,
BURGESS F. HOWELL, AND THOMAS G. GARRISON

Abstract: The Petén region of northern Guatemala contains some of the most significant
Maya archaeological sites in Latin America. It was in this region that the Maya
civilization began, flourished, and abruptly collapsed. Remote sensing technology
is helping to locate and map ancient Maya sites that are threatened today by accel-
erating deforestation and looting. Thematic Mapper, IKONOS, and QuickBird
satellite, and airborne STAR-3i and AIRSAR radar data, combined with Global
Positioning System (GPS) technology, are successfully detecting ancient Maya
features such as sites, roadways, canals, and water reservoirs. Satellite imagery is
also being used to map the bajos, which are seasonally flooded swamps that cover
over 40% of the land surface. The use of bajos for farming and settlement by the
ancient Maya has been a source of debate within the professional community
for many years. But the detection and verification of cultural features within
the bajo system within the last few years are providing conclusive evidence that
the ancient Maya had adapted well to wetland environments from the earliest
times and utilized them until the time of the Maya collapse. In the last two
years, we have discovered that there is a strong relationship between a tropical
forest vegetation signature in IKONOS satellite imagery and the location of
archaeological sites. We believe that the use of limestone and lime plasters in
ancient Maya construction affects the moisture, nutrition, and plant species of the
surface vegetation. We have mapped these vegetation signatures in the imagery
and verified through field survey that they are indicative of archaeological sites.
However, we have not yet determined the nature of the spectral signature in
the imagery and that is a focus of our ongoing research. Through the use of

remote sensing and GIS technology it is possible to identify unrecorded archaeological features in a dense tropical forest environment and monitor these cultural
features for their protection.

1. INTRODUCTION

Since Stephens and Catherwood (1841) first investigated the ancient Maya
of Central America in the late 1830s, researchers have been hindered by
the dense vegetation of the tropical forest. Ricketson and Kidder provide an
accurate description of the field conditions that archaeologists encountered in
Maya research, which is quoted below.

> Slow and laborious travel, a hot, humid climate, swarms of insects, and
> prevalence of tropical diseases have greatly retarded exploration of the
> Maya country. Even in such parts of it as can be reached the traveler is
> so buried in the "bush," so shut in and engulfed by the mere weight of
> vegetation that he can literally never see more than a few feet or yards
> and so is almost totally in the dark as to the topography of the regions he
> is examining. A rise of ground crossed by the trail may, for example, be
> an isolated hill or part of an extensive ridge; it may be the highest land
> in the vicinity or a saddle in an important divide. It is almost impossible
> to reach any point for extended views over the terrain. For this reason
> no proper knowledge has been gained of the lay of the mountains, hills,
> plains, swamps, and drainages; of their position, their interrelation, and
> the nature of their forest cover. Without detailed information as to such
> vitally significant environmental factors it is manifestly impossible to gain
> a true understanding of the people whose history they must have played
> so large a part in molding, (Ricketson and Kidder, 1930:178).

Seventy-four years later, conducting field research in the dense tropical forests
of the Petén in northern Guatemala is as difficult today as it was for Ricketson
and Kidder. However, airborne and satellite imagery allow us to locate natural
and cultural features upon the landscape and improve our ability to investigate ancient Maya settlement, subsistence, water management, and landscape
modification. The use of remote sensing technology is a major advancement
over the traditional methodologies that have been used in the region since the
1800s (Stephens, 1841).

NASA and the University of New Hampshire have created a Space Act
Agreement to collaborate for five years (beginning in 2004) to revolutionize
settlement survey in the Maya Lowlands. Several remote sensing and ground-
based approaches are being undertaken to understand better the relationship
between the spectral signatures seen in the imagery and the vegetation parameters (nutrition, moisture, species) on the ground. Data analysis is also being
conducted to detect archaeological features in various remote sensing data sets,
and field verification activities are scheduled over the course of the five-year
agreement.

2. STUDY AREA: PETÉN, GUATEMALA

The state of Petén in northern Guatemala covers 36,000 sq km and, as are other regions of Central America, is severely threatened by deforestation activities and agricultural practices (Figure 1). The wetlands and intact tropical forest of the Petén represent one of the world's richest areas of biological diversity. The ecosystem contains over 800 species of trees, 500 species of birds, and large populations of mammals including monkeys, jaguars, and tapirs. The Petén contains some of the most prehistorically significant Maya archaeological sites in Latin America with many more potential sites remaining to be discovered. A few indigenous Maya descendants still live in the Petén, although the population of the inhabitants is increasing rapidly in the wake of migration and settlement by the Q'eqchi Maya (Cahuec and Richards, 1994). In the 1970s, over 90 % of the forest remained. However, half of the forest has been cut since that time and if deforestation rates continue, the Petén forests

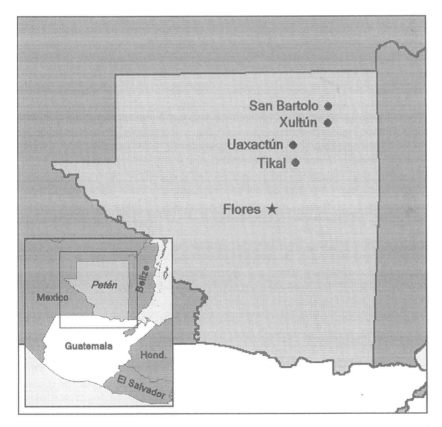

Figure 1. Regional map showing the location of San Bartolo and other archaeological sites in the Patén, northern Guatemala.

could disappear within 10 to 20 years, destroying both biological habitats and archaeological sites (Sever, 1998).

As part of the Maya Lowlands, the Petén is composed of karst topography with seasonal wetlands (*bajos*) comprising 40 % or more of the land surface. The porous limestone bedrock acts as a sponge. Instead of flowing into streams and rivers, rainwater flows down through sinkholes to underground drainage systems far below the surface. Obtaining water for household use during the dry season is a problem for most of the lowland inhabitants (Culbert, 1993). It is apparent that the ancient Maya were once able to manage water successfully in this area as evidenced by their intense population and large-scale construction of temples, houses, and other structures upon the landscape.

Water availability is one of the primary obstacles in conducting archaeological survey and excavation in the area today. During the 2004 field season at San Bartolo, for instance, 8,500 gallons of drinking water had to be hauled from 35 miles away and 200 gallons for bathing had to be transported 8 miles almost daily to the field camp.

3. THE SAN BARTOLO PROJECT

In March of 2001, while exploring in northeastern Guatemala, William Saturno stumbled into one of the most important archaeological discoveries in the Maya area in half a century. While following local guides to verify reports of new hieroglyphic monuments uncovered by looting in the area, his party happened upon a recently looted archaeological site, though not the one they had been seeking. It was in a looter's tunnel at this site that he discovered elaborate Maya murals (Figure 2) in a structure beneath a pyramid. This was the first major Maya mural discovery since 1946 when the now famous murals of Bonampak were first seen by outsiders. The murals at the site now called San Bartolo predate those at Bonampak by more than 800 years and, because of their beauty and their depiction of Maya creation mythology, have been called the Sistine Chapel of the Maya world (Kaufmann 2003:72-77). The current project is designed to place this remarkable work within its temporal and cultural context through understanding the nature of ancient settlement in the region. The fact that the ancient Maya constructed such an elaborate mural at San Bartolo in the Preclassic and that they intentionally buried it to make room for a more elaborate structure, suggests that other such works of art may exist at other locations in the region.

The area surrounding San Bartolo is basically unexplored and the use of remote sensing imagery greatly enhances research at both the local and regional level. The larger aim of the project at San Bartolo is to evaluate environmental, demographic, economic, and ritual factors in the development of complex society during the Preclassic, and how changes in those factors

Figure 2. The mural on the north wall, San Bartolo, revealing the corn god (center) receiving a gourd of water as he prepares to relay life-giving food and water to others while they ride atop the plumed serpent when it emerges from Flower Mountain, the home of Maya gods and ancestors. This mural along with the west wall mural (not depicted) represents one of the earlies records of Maya creation myths and predates the murals at Bonampak by centuries.

may have contributed to apparent Early Classic political reformulation in the region centered around Xultún (Figure 1) and by extension throughout the Maya Lowlands at the time.

The San Bartolo project combines high-resolution airborne and satellite imagery, advanced remote sensing analyses, large-scale regional survey, intensive and extensive excavation of residential and non-residential features, exploration of the paleo-environment, resource management, and site planning through the use of geographic information systems (GIS). This approach provides the capability to examine the transformations that took place during the Preclassic-Classic transition. The results of the project will contribute new models of organization, stability, and change in Preclassic Maya society and is expected to have an impact on the study of archaic states throughout the world. Understanding how the Maya civilization arose is an important component in understanding the processes that led to its collapse.

4. LANDSCAPE ARCHAEOLOGY

Our research incorporates a branch of research known as "landscape archaeology." Landscape archaeology is a branch of modern ecology that deals with the relationships between humans and their open and built-up landscapes. Remote sensing imagery provides a view from a different perspective and serves as a backdrop for archaeological landscape studies. This technique is particularly important in the remote and unexplored areas of the Maya Lowlands. Examining the sites and their settings from above leads to the realization that everything observable in an ancient landscape has meaning and interconnection. As a result, we can discover features previously unnoticed on the ground and their relationship to each other as revealed from the vantage point of space.

To assist us in our landscape-archaeology analysis, a GIS is being developed for the San Bartolo Project that incorporates multi-spectral remote sensing datasets, ancillary information, and the results of interdisciplinary field research. (A detailed description of this GIS will be discussed later in this chapter). Our ability to understand better the ancient Maya landscape provides us with a clearer comprehension of the complex evolution of the human factors for settlement patterns, subsistence strategies, use of resources, and the social, political, and economic relationships that connect the study of the past with the present. The San Bartolo project integrates information from various disciplines and research areas to address a wide range of research objectives. As a result, the landscapes of the ancient Maya and those of the inhabitants today can be virtually represented and the data can be preserved, shared, and applied.

5. LINDBERGH, KIDDER, AND VEGETATION SIGNATURES

Ever since O.G.S. Crawford (1928) demonstrated how vegetation and crop-marks could be used to locate archaeological sites, researchers around the world have looked for these clues to aid them in their archaeological survey. Unfortunately, tropical rainforests are one of the most difficult environments in which to conduct this type of research. One of the first attempts to do this for the Maya region occurred two years after Charles A. Lindbergh had successfully completed his famous trip across the Atlantic. After completing his mission to chart an air-mail route to Panama for Pan American Airways in 1929, Lindbergh explored the northeastern Yucatan Peninsula with the intention of locating emergency landing fields. It was during this trip that the aviator saw ancient structures protruding through the canopy, causing him to become intrigued by this ancient civilization. As he soared over the landscape, Lindbergh envisioned how the airplane could benefit archaeological research. Returning to the U.S., he was eventually referred to archaeologist Alfred V. Kidder at the Carnegie Institution, a leading authority on Pueblo and Maya research. After a successful collaboration in the American southwest during the summer of 1929, the two pioneers focused their attention in October of that year upon the Maya of Central America, an effort later known as the Lindbergh-Carnegie expedition.

For five days in early October of 1929, Lindbergh, Kidder, and archaeologist Oliver Ricketson, Jr., flew over the little-known or unexplored areas once inhabited by the ancient Maya in a twin-motor Sikorsky Amphibian aircraft (Kidder 1930; Ricketson and Kidder, 1930). During their journey, they flew for over twenty-five hours and covered over 2,000 miles. Kidder stated:

> And above all things, we wished to get an idea of what the Maya country really looks like, for in spite of the fact that archaeologists have for many

years been pushing their way into the region, they have been so buried
in the welter of forest, their outlook has been so stifled by mere weight
of vegetation, that it has been impossible to gain a comprehensive understanding of the real nature of this territory once occupied by America's
most brilliant native civilization ... Our problem was clear. We must cover
as much of the area as possible and learn as much as possible about it
(Kidder, 1930:194-195).

Kidder's writings document that he noticed distinct vegetation associated
with ancient sites. For instance, he states that "Particular attention was paid
to the lesser mounds. They were seen to be marked by a heightening and
bunching of the bush and a definitely darker green color of the foliage"
(Ricketson and Kidder, 1930:193). In another example he says: "One hundred
yards east were three lower pyramids, and the forest all about had a suspicious
lumpy appearance which seemed to indicate the presence of a number of
smaller mounds" (Ricketson and Kidder, 1930: 198-199). Kidder summarizes
his observations as follows.

The bird's-eye view attainable from the air renders easy the finding of
sources of water supply such as aguadas and cenotes, and near these
one may expect always to encounter ruins. With a little practice one
can learn to recognize the sort of terrain the Maya were accustomed to
pick for their temples and what varieties of trees flourish on the soil best
fitted for their system of agriculture. Thus much barren territory can be
disregarded and all available flying time be devoted to examination of the
areas most likely to have been former centers of population (Ricketson
and Kidder, 1930:205).

Unfortunately, despite his keen observations of the vegetation, Kidder did not
provide a systematic approach of how various vegetation types were associated
with archaeological sites. Although some black-and-white aerial photography
was acquired, the investigators were primarily conducting their research with
the naked-eye. In short, Kidder was onto something, but he did not follow up.
As a result, critics would look back and criticize, saying that the expedition
fell short in approach and method. As Deuel states:

That buried structures, not by themselves visible from above, could
conceivably produce striking effect to announce their presence and dimensions was just barely suggested by Kidder's acute botanical observations,
which were never properly worked out and applied. Furthermore, all
flights were geared to scouting by the human eye (Deuel, 1969:211).

Inspired by the Lindbergh-Carnegie expedition, one of the authors (Sever)
began looking for vegetation signatures in multispectral imagery as an archaeological indicator beginning in the mid-1980s. His attempt to correlate various
ground observations with spectral signatures in airborne and satellite imagery,
was unsuccessful. The datasets included 30-meter Landsat Thematic Mapper

satellite data, airborne 5-meter Thermal Infrared Multispectral Scanner (TIMS) data, and airborne color infrared (CIR) photography. Although analysis of the imagery did result in finding other archaeological indicators such as causeways, canals, and reservoirs (Sever, 1999, 1998; Sever and Irwin, 2003), the concept of vegetation as an archaeological indicator could not be established. Various supervised and unsupervised classification techniques were attempted, including trying to isolate sapote trees in the imagery because local gum tappers (*chicleros*) claimed the species was associated with archaeological sites. Another attempt was to classify stands of high-density ramon that Dennis Puleston had theorized was used in Maya subsistence (Puleston, 1982). In general, the 30-meter resolution of the Landsat satellite imagery precluded success since biodiversity is highest in tropical rainforests: boreal forests can contain one tree species per hectare while tropical forests can contain over 200 tree species per hectare.

Only recently has the ability to use vegetation as an indicator of archaeological sites been realized through the use of commercial high-resolution, one-meter pan-sharpened IKONOS data. Since its launch in September 1999, Space Imaging's IKONOS earth imaging satellite data have provided high-resolution imagery to the general public. The technical approach for using IKONOS data for extraction of vegetation signatures as indicators of archaeological sites will be discussed later in this chapter.

6. PREVIOUS REMOTE SENSING RESEARCH IN THE PETÉN REGION

Since 1987, the overall goal of NASA funded research in the Petén has been to use a combination of remote sensing, GIS, GPS technology, and climate analysis and modeling in order to understand the dynamics of how the Maya successfully interacted with their karst topography, the effects of natural and human-induced changes to that landscape, the consequences of these changes to water resources and climate change, and the modeling of these changes for application by current societies.

Specific NASA research objectives included the following.

• Correlation of settlement patterns with vegetational, hydrological, and geological features upon the karst topographic landscape.
• Isolation of topographic features in great detail where sites are likely to be located (*bajo* islands).
• Mapping of drainage patterns that may provide insight into human settlement, water management, and storage capability.
• Detection of linear features of an unknown origin that may be related to roadways or canal systems.

- Creation of detailed vegetation maps in order to provide insight into the relationship(s) between land cover, paleo-hydrology, *bajo* systems, and subsistence strategies.
- Use of global and regional climate models (Mesoscale Model 5 version 3, known as MM5; and the Regional Atmospheric Modeling System, known as RAMS) to evaluate the relative roles of land-use changes and natural climate variability in the rise and fall of Mayan civilization.

Although the dense tropical vegetation hinders our ability to conduct an archaeological survey, the multispectral capability of our remote sensing instrumentation allows us to detect unrecorded archaeological features and navigate to them using GPS technology. Through the use of Landsat TM, IKONOS, and high-resolution STAR-3i airborne Interferometric Synthetic Aperture Radar (IFSAR) imagery, we have been able to detect, map, and locate archaeological features such as ancient causeways (Figure 3), temples, reservoirs (Figure 4), and agricultural areas (Sever and Irwin, 2003; Sever, 1999, 1998; Miller et al., 1991). We have also mapped features which

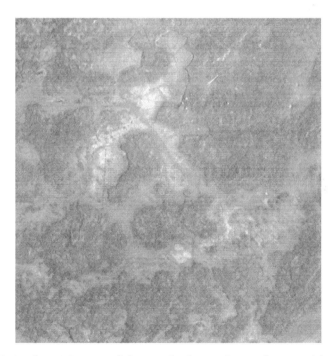

Figure 3. Landsat TM image of the Mirador basin, Guatemala, created by ratioing Band 4 with Band 5. The dark, straight lines represent Maya causeways and natural geological features. Ground survey and sometimes excavation are required to separate the cultural from the natural features. Throughout the rest of the image the bright areas are lowland swamps while the darker areas are higher elevations.

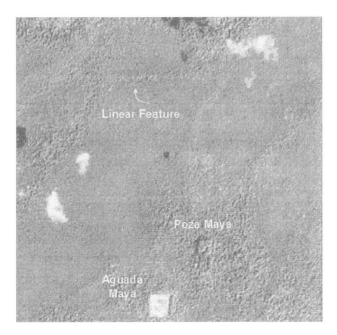

Figure 4. IKONOS satellite data revealing a linear feature, a reservoir (Aguada Maya), and the archaeological site of Poza Maya in the Bajo la Justa. Features such as these are often indistinguishable in Landsat TM data.

we have determined from ground verification are not of recent origin that need to be investigated through future survey and excavation (Figure 5).

STAR-3i is an X-band system that can be used day or night and has cloud penetration capability. Over 2,000 sq km of STAR-3i data were collected over the eastern Petén consisting of a 2.5-m resolution, orthorectified image (ORI) and a 10-meter resolution (3-meter vertical) Digital Elevation Model (DEM). Although the X-band data maps the vegetation canopy and not the ground surface, the data were still extremely useful at this resolution since the vegetation canopy generally mirrors the surface topography. The data allowed us to create three-dimensional images that improved our ability to visualize the terrain.

Using the STAR-3i imagery, we were able to demonstrate that almost every rise in elevation that was above the level of seasonal inundation in the *bajos* revealed evidence of prehistoric occupation (Figure 6). The *bajos* represent between 40 and 60 percent of the land surface in the area and over 100 of these *bajo* elevations have been visited (Sever and Irwin, 2003). Almost all of these contained evidence of an archaeological site and they ranged from isolated mounds to small groups of structures, larger formal plaza arrangements, the medium-size center of Poza Maya, and major centers such as Yaxha and Nakum. The sites date from the Late Preclassic through

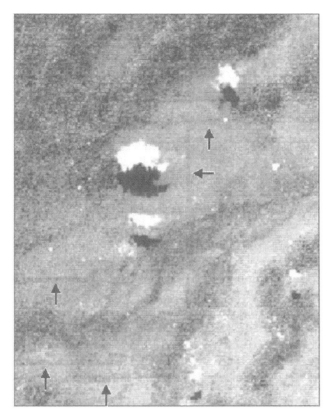

Figure 5. IKONOS satellite image revealing linear features in Bajo Santa Fe east of the archaeological site of Tikal. It is unknown whether these features represent ancient causeways or canals or are the result of landscape-modification activities. Initially, it was thought that these features might be the result of oil exploration activities or botanical transects from the last few decades. However, field research by Culbert, Fialko, and Sever indicated that these features do not appear to be of recent origin.

the Terminal Classic periods, and the occupation of these bajo sites seems to correlate with that of the urban centers (Sever and Irwin, 2003; Kunen et al., 2000).

The imagery was also successful in the investigation of the Holmul River in delineating land forms and vegetational associations (Figure 7), as well as dry and ancient water courses (Sever and Irwin, 2003; Culbert and Fialko, 2000). The imagery also documented the approaching deforestation that is altering the modern-day drainage patterns. The survey and excavations along the Holmul indicated that the river either dried up or became swampy between A.D. 750 and 850 (Culbert and Fialko, 2000), adding to growing speculation concerning the existence of a major drought during this period

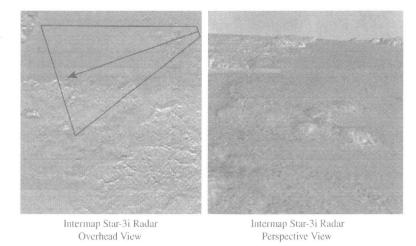

Intermap Star-3i Radar   Intermap Star-3i Radar
Overhead View   Perspective View

Figure 6. Visualization of STAR-3i DEM revealing bajo-island features. Research from the past five years indicates that archaeological sites and features are found on almost every bajo island. The DEM is generated with 10-meter postings from interferometric radar data to map detailed topographic information.

(Brenner et al., 2001; Hodell et al., 2001; Gill, 2000). In addition to our archaeological research, we have used satellite imagery to map current land cover as well as monitor land cover/land use changes (Figure 8) over the past twenty years (Sader et al., 1996; Sever, 1998). This information can be used by managers and decision-makers for the monitoring and protection of the natural and cultural features within the Petén. In short, without the use of remote sensing, archaeological and environmental projects of this scope would not be possible.

7. VEGETATION AS AN INDICATOR OF ARCHAEOLOGICAL SITES

Responding to a request from William Saturno, Daniel Irwin and Tom Sever provided satellite imagery to the San Bartolo Regional Archaeology Project for the spring 2003 field season. At the time, Sever and Irwin were working with Patrick T. Culbert and Vilma Fialko in an area adjacent to Saturno's study area. Various three-band false-color composites were created using both Landsat TM and IKONOS satellite imagery. They were delivered to Saturno to assist him in his survey of previously unexplored areas in the San Bartolo region. During the course of his field survey, Saturno noticed that a vegetation signature in the IKONOS imagery (Figure 9) indicated the locations of archaeological sites on the ground. The use of these images enabled him to identify with remarkable

Figure 7. IKONOS imagery used in the investigation of the Holmul River for delineating land forms and vegetational associations as well as dry and ancient water courses. Notice the approaching deforestation in the bottom of the image.

accuracy the presence and location of previously unknown archaeological sites beneath the forest canopy. The vegetation signature characterized to within a few meters of accuracy a variety of sites ranging from diminutive settlements of only a few small housemounds to once thriving cities. Saturno quickly recognized that the imagery had the potential completely to revolutionize archaeological survey in the tropics by dramatically reducing the time involved in systematically covering vast areas.

Saturno visited Sever and Irwin in the fall of 2003 at the NASA Marshall Space Flight Center in Huntsville, Alabama, to analyze the imagery in greater detail and develop a strategy for future field verification activities. Four scenes of IKONOS imagery were acquired in 2003 with less than 10 % cloud cover over the San Bartolo study area. The imagery was processed and analyzed using ERDAS Imagine (1999) software. With its high resolution and multispectral capability, IKONOS is a significant improvement over Landsat TM imagery for archaeological analyses. The IKONOS high-resolution satellite carries both panchromatic and multispectral sensors. IKONOS provides 1-m resolution

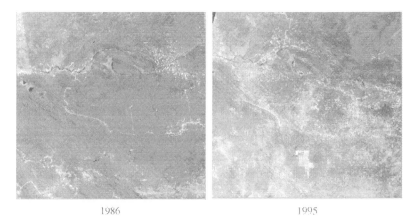

1986   1995

Figure 8. Change detection analysis using Landsat TM data showing the effects of deforestation between a) 1986 and b) 1995 in the northern Petén. The red color represents healthy standing forest, while the cyan color represents deforestation. The protection of the rainforest is synonymous with the protection of cultural features. Satellite imagery such as this provides managers and decision-makers with the capability to monitor the natural and cultural resources in the region.

panchromatic imagery and four 4-m resolution multispectral bands (visible and near infrared). The satellite has a polar, circular, sun-synchronous orbit at an altitude of 681 km, and both sensors have an at-nadir swath width of 11 km. Approximately 700 km^2 of IKONOS data were collected and analyzed over selected areas of the Petén. Features not apparent in the Landsat TM imagery are easily visible in the IKONOS data, as indicated in Figures 10 and 11.

8. TECHNICAL APPROACH

Spatial resolution is a measure of a sensor's ability to record detail across a scene. For Earth-sensing instruments, it is commonly defined by the sensor's instantaneous field of view (IFOV) or resultant image pixel size, and expressed as a linear measure at the target surface (1 m, 4 m, 30 m, etc.). A sensor's spatial resolution is governed by the optics of the system and the height of the sensor above the target.

Spectral resolution may be defined as the extent of the portion of the electromagnetic spectrum that can be recorded as a band of data by the sensor, and is expressed as a range of wavelengths. Spectral resolution increases as the width of the band decreases.

For a given IFOV, as the band width is narrowed, less energy is received by the sensor and, at some point, the amount of energy falls to a level insufficient for the system to register a clean, usable signal. Likewise, for a given band

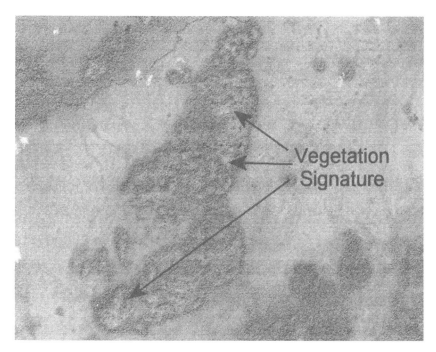

Figure 9. False-color composite of IKONOS imagery showing the vegetation signature (indicated as yellow) that is indicative of ancient Maya occupation. So far, field verification has been remarkably accurate in correlating the vegetation signature with the location of archaeological sites to within a few meters of accuracy. We believe that the decomposition of excavated limestone and lime plasters over the centuries affects the surface vegetation, which can be mapped in satellite imagery. As a result, a detailed analysis of forest composition can reveal the location and dimensions of ancient settlements that lie beneath it.

width, as the IFOV is decreased, less energy is received, and the same clean signal threshold is reached. Therefore, for any sensor, there is a practical limiting inverse relationship between spectral resolution and spatial resolution with the result that, within that system, narrow-band multispectral data are recorded at a coarser resolution than broad-band panchromatic data (Lillesand and Kiefer, 1994).

In order to enhance the spatial resolution of the multispectral IKONOS data used in this project, a data-fusion process known as pan sharpening was performed. In pan sharpening, one of several techniques is employed to generate new data sets that combine the high-frequency spatial domain data of the panchromatic band with the higher information content, but lower spatial-resolution data of the multispectral bands. There are several classes of pan-sharpening techniques, the most common of which are spatial domain models, spectral domain models, and algebraic models.

 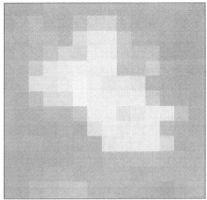

Figure 10. Comparison of high-resolution IKONOS imagery (left) and Landsat TM imagery (right) over Tikal National Park in Peté, Guatemala. The high-resolution capability of IKONOS data clearly reveals the temples, stelae, amd other structures from the surrounding tropical forest vegetation.

In a spatial domain model, high-frequency spatial domain information is isolated from the high-resolution image, then combined with the multispectral data (Schowengerdt, 1997). A simple example of this type is the High Pass Filter method in which the high-resolution image is subjected to a small filter kernel, which reduces lower frequency spatial information. The filtered high-resolution data is then added to the multispectral data, with the result divided by some factor, usually 2 (Chavez et al., 1991).

Spectral domain models transform the multispectral data from spectral space into a feature space where one of the resulting bands is correlated with the high-resolution data. The high-resolution band is then substituted for the correlated band and a reverse transform applied to translate the bands back to spectral space. An example of this method is the Principal Components Transformation (PCT) in which information content common across the multispectral bands is mapped to a single output data set called the Principal Components Image (PCI). The content of this image is typically highly correlated to brightness or intensity (Jensen, 1996; Chavez et al., 1991). The high-resolution panchromatic data are contrast-matched to the PCI, then substituted for the PCI in an inverse PCT operation which transforms the data back into spectral space. (Schott, 1997).

Algebraic models employ image-arithmetic operations on a pixel-by-pixel basis, incorporating the respective values of the higher resolution and lower resolution data sets into equations to produce the sharpened output. A common algebraic method is the Brovey Transform (BT). BT divides the value of one multispectral band by the sum of the others, then multiplies the resulting quotient by the value of the panchromatic band (ERDAS, 1999).

While the ultimate goal of pan sharpening is to gain resolution while maintaining fidelity to the underlying multispectral data values, the different techniques achieve this goal with varying degrees of success. We tested the Multiplicative Model and the Brovey Transform, both algebraic methods, and a spectral domain transform technique, the Principal Components Transformation. While all three methods provide similar resolution enhancements, slight color shifts resulting from the several processes caused the output data sets to be of more or less utility in identifying areas of interest in the forest. In particular, utilizing PCT, the anomalous canopy signature associated with sites of interest appeared significantly brightened when subjected to common contrast stretching. Additionally, the canopy anomalies in the PCT

Figure 11. Comparison of ground photograph (a), with 1-meter pan-sharpened IKONOS satellite image (b,c). These image demonstrate the capability of high-resolution data to detect small features from space, such as Maya stelae at Tikal National Park, Guatemala.

Figure 11 (Continued).

data displayed a conspicuous yellow tone not apparent in the other pan-sharpened datasets. This yellow signature was unlike any other region in the data, unique as an identifier and obvious to the viewer.

In the spring of 2004, Sever and Irwin, as well as a NASA television crew, joined Saturno's San Bartolo field-survey team to observe first-hand and document whether or not the spectral signatures were an indicator of archaeological sites. Exploring an area to the northwest of San Bartolo where Saturno had established a base camp from which to operate, numerous sites were visited over the course of several days. Using courses determined from the georeferenced satellite imagery, and navigating with the aid of a GPS receiver, access paths were chopped through the vegetation to the predicted

Figure 12 Pan-sharpened IKONOS data (bands 4,2,1) of the San Bartolo site, overlain by surveyed archaeological details. Note that the locations of architectural structures are highly correlated with the anomalous yellow spectral signature apparent in these images. Similar signatures surrounding the surveyed site probably indicate other areas of archaeological significance.

areas. The results of the survey were astounding. Once again, as indicated in the previous year, the signature was accurate in predicting archaeological site locations, their boundaries and dimensions, to within a few meters of accuracy. Whether or not this accuracy will be maintained in future surveys remains to be seen. Although we have not encountered false positives to date, we remain alert to their potential existence.

All four IKONOS scenes, although acquired at different dates, reveal vegetation signatures indicative of archaeological sites. This result demonstrates that the vegetation signature that we have extracted can be transferred to other IKONOS data sets. At this point, however, we have not determined if the vegetation signatures are seasonally dependent. What is intriguing, however, is that the location, size, and dimension of the signature in the imagery we have analyzed to date correlate precisely with archaeological sites on the ground. Although the vegetation signature can be observed in the true-color-composite image which uses bands 3, 2, and 1 (from the red, green, and blue visible portions of the spectrum, respectively) of the IKONOS data, the signature is accentuated in the false-color composite utilizing bands 4, 2, and 1, substituting the near-infrared band for the red. In our final imagery, this signature is a bright yellow color (Figures 12 and 13).

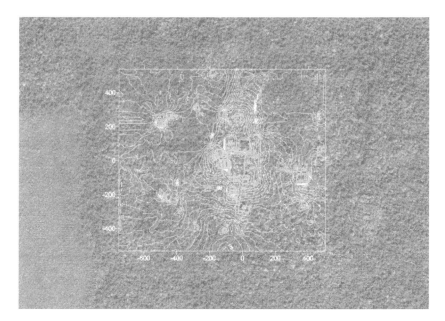

Figure 13 Pan-sharpened IKONOS data (bands 4,2,1) of the San Bartolo site with surveyed archaeological details and 1-meter topographic contours. Although not thoroughly discussed here, it is apparent that, as is the case with the distinct spectral signature, elevated topographic features are well correlated with the location of architectural features. Future investigations will explore the specific relationship between the vegetation signature, topography, and settlement patterns.

The Ancient Maya used limestone and lime plasters in the construction of their cities, towns and even their smallest hamlets. We believe that the decay of these structures over the centuries provides a unique microenvironment for the growth of vegetation as the levels of moisture and nutrition within the ruins vary substantially from those in the surrounding forest. These microenvironmental differences on the ground are likewise represented by compositional differences in the forest canopy both in the species present and in leaf color (representing moisture/nutritional stress) visible through the analysis of high-resolution satellite data. As a result, a detailed analysis of forest composition can reveal a detailed picture of the ancient settlements that lie beneath it. As noted previously, our results to date using these techniques have been very successful and we are refining and expanding our methodology in order to comprehend efficiently the details of ancient Maya settlement in the Lowlands.

Further fieldwork will be conducted in order to understand the true nature of the spectral response of the vegetation signature in the imagery. First, Sever and Irwin will join Saturno's research team to explore new locations in the San Bartolo region to document further that the vegetation signature in the imagery is indicative of archaeological sites. This activity will be followed by members of Dr. Barry Rock's research team from the University of New Hampshire who will visit selected areas to examine the nature and composition of the plant species that are creating this signature. Rock's research and publications have focused on the remote sensing of vegetation, specifically on basic and applied research dealing with biophysical properties (pigment concentrations, anatomical characteristics, and moisture conditions) of leaves and their influence on reflectance features that may be remotely detected. If it can be determined why the spectral properties of the vegetation are being recorded in the satellite imagery, it is possible that the analysis techniques used in the Petén could be applied to archaeological projects in other tropical forests of the world.

The analytical techniques used for the IKONOS imagery will be applied to recently acquired high-resolution QuickBird data to determine if vegetation signatures can be extracted that are similarly indicative of archaeological settlements. Currently, there is over 500 sq km of QuickBird scenes in our database that were acquired in the San Bartolo region. We will also analyze STAR-3i and AIRSAR radar data, both of which (Sever and Irwin, 2003; Freeman et al., 1998) have been used successfully for archaeological research, to determine their utility in the unexplored areas of the northeastern Petén.

In March, 2004, high-resolution AIRSAR data were collected throughout Central America. AIRSAR acquired polarimetric and/or interferometric data for thirty sites throughout eight countries including the entire northeastern Petén. The NASA/JPL AIRSAR system operates in the fully polarimetric mode at P-, L-, and C-band or in the interferometric mode in L- and C-band simultaneously. It is anticipated that the high-resolution imagery will record the height of the vegetation, the contour of the ground surface, and, potentially, the dimensions of archaeological structures in between. We will also merge the elevation data of the STAR-3i imagery (resolution of 2.5-m radar backscatter and 10-meter posting Digital Elevation Model) and that of the AIRSAR imagery (3-meter resolution) with the QuickBird and IKONOS data to provide detailed contour maps in order to better understand relationships between ancient settlement patterns with natural features on the landscape.

9. SAN BARTOLO GEOGRAPHIC INFORMATION SYSTEM (GIS)

The GIS for the San Bartolo project is under constant development. As of the 2005 field season, the primary hardware consists of a purpose-built, small footprint μATX computer, utilizing an AMD Athlon XP 3200+ processor, 2GB PC3200 RAM, 400GB SATA 150 hard disk, double-layer DVD±RW drive, and Apple flat-panel display. Because of the extreme operating environment, 4 additional cooling devices have been fitted. The entire system is designed to be luggable, stowed in a shock-resistant, waterproof, MIL-C-4150-J specification, polypropylene-resin transport case. As better, more suitable computer hardware becomes available, the system will be upgraded or replaced.

The San Bartolo GIS project is constructed within the ESRI ArcGIS framework. Current layers covering the site include the following data sets.

Raster

Satellite

• Landsat multispectral – complete mosaics of Guatemala, acquired 1990 (Thematic Mapper) and 2000 (Enhanced Thematic Mapper).
• IKONOS multispectral and panchromatic – approximately 250 km^2, acquired December, 2002.
• QuickBird multispecral and panchromatic – approximately 650 km^2, acquired March/April, 2003.

Airborne

• AIRSAR synthetic aperture radar – approximately 750 km^2, acquired March, 2004.
• STAR-3i interferometric synthetic aperture radar – approximately 6250 km^2, acquired July, 1999.

Topographic

• Shuttle Radar Topographic Mission – complete coverage of Central America at 60-m post spacing, acquired February 2000.

Vector (Arc coverages)

Surveyed on site

• Topographic contours for site
• Locations of major structures on site
• Physical description of major structures
• Excavations

From external sources

• Political boundaries
• Population centers
• Forestry concessions
• Cattle concessions
• Surface hydrography
• Transportation

10. CONCLUSIONS

The archaeological features in the Petén, Guatemala, are threatened today by expanding deforestation, human migration, and looting. A methodology that can predict the location of archaeological sites to safeguard against destruction would provide us with the ability to prioritize areas for survey and excavation and recover scientific information before it is lost. The imagery also allows us to monitor and protect these cultural resources for the future.

The dense, uninhabited tropical forest of the Petén presents a challenge for archaeologists working in the area. Traditional-based survey methodologies are time-consuming and costly and provide limited information at the regional level. However, through the use of remote sensing imagery we are revolutionizing archaeological survey and inventory for Maya Lowland research. Using IKONOS satellite imagery we have discovered a spectral vegetation signature that correlates with the location, boundaries, and dimensions of ancient Maya sites. To date, our ground verification of these areas has been extremely accurate. We plan to apply our data analysis techniques to other airborne and satellite imagery, and determine from the plant composition, exactly why we are seeing this vegetation signature. The IKONOS imagery is one example of how modern technology can aid in our search and improve upon traditional ground-based survey techniques. Space-based information, combined with ground-based measurements and predictive models, promises to increase our ability to understand the Maya past, apply that knowledge to the contemporary world, and preserve the cultural and natural heritage of the Petén for future generations.

11. ACKNOWLEDGEMENTS

The archaeological survey in the San Bartolo region was conducted under the San Bartolo Regional Archaeology Project through permission of the Guatemalan Institute of Anthropology and History. The authors wish to thank Joshua Kwoka, Anatolio Lopez, Miguelángel Rodríguez, Melvin Barrientos, Rocaél Méndez, and Robert Griffin for their tireless assistance on the ground in Guatemala. Saturno's participation in this research was made possible by grants from the National Geographic Society, The Peabody Museum of Archaeology and Ethnology at Harvard University, and the Reinhart Foundation. Research by Sever, Irwin, and Burgess Howell was funded by NASA.

REFERENCES

Brenner, Mark, Hodell, D.A., Curtis, J.H., Rosenmeier, M.F., Bindford, M.W, and Abbott, M.B., 2001, Abrupt Climate Change and Pre-Columbian Cultural Collapse. *Interhemispheric Climate Linkages*, edited by Vera Markgraf, Chapter 6. Academia Press, San Diego.

Cahuec, E. and Richards, J., 1994, La Variacion Sociolinguistica del Maya Q'eqchi.*Boletín Lingüístico 40.* Guatemala.

Chavez, P. S. Jr., Sides, S. C., and Anderson, J. A., 1991, Comparison of Three Different Methods to Merge Multiresolution and Multispectral Data: Landsat TM and SPOT Panchromatic. *Photogrammetric Engineering and Remote Sensing* 57(3):295–303.

Crawford, O.G.S., and Keiller, Alexander, 1928, *Wessex from the Air.* The Clarendon Press, Oxford.

Culbert, T.P., 1993, *Maya Civilization.* St. Remy Press and Smithsonian Institution, Washington, D.C.

Culbert, T.P., and Fialko, Vilma, 2000, Reconnaissance of the Course of the Holmul River and Bajo de Santa Fe in the Region of the Tikal National Park, Guatemala, 2000 Season. Report to NASA, Global Hydrology and Climate Center, Marshall Space Flight Center, Huntsville, AL.

Deuel, Leo, 1969, *Flights into Yesterday.* St. Martin's Press, New York.

ERDAS, 1999, *Field Guide, Version 8.5,* Fifth Edition. ERDAS Inc., Atlanta, GA.

Freeman, A., Hensley, S., and Moore, E., 1998, Radar imaging methodologies for archaeology: Angkor, Cambodia. Paper presented at Conference on Remote Sensing in Archaeology, Boston University Department of Archaeology and Center for Remote Sensing, 11–19 April 1998.

Gill, R. B., 2000, *The Great Maya Droughts: Water, Life, and Death.* University of New Mexico Press, Albuquerque.

Hodell, D.A., Brenner, M., Curtis, J.H., and Guilderson, T., 2001, Solar Forcing of Drought Frequency in the Maya Lowlands. *Science* 292 (18 May):1367–1370.

Jensen, J. R., 1996, *Introductory Digital Image Processing: A Remote Sensing Perspective,* 2nd edn. Prentice Hall, Upper Saddle River, NJ.

Kaufmann, Carol, 2003, Sistine Chapel of the Early Maya. *National Geographic Magazine* (December):72–77.

Kidder, A.V., 1930, Five Days over the Maya Country. *The Scientific Monthly* (March):193–205.

Kunen, J. L., Culbert, T. P., Fialko, V., McKee, B. M., and Grazioso, L., 2000, Bajo Communities: A Case Study from the Central Petén. *Culture and Agriculture* 22(3):15–31.

Lillesand, T. M., and Kiefer, R. W., 1994, *Remote Sensing and Image Interpretation,* 3rd ed. John Wiley & Sons, Hoboken, NJ.

Miller, Frank, Sever, T., and Lee, D., 1991, Applications of Ecological Concepts and Remote Sensing Technologies in Archeological Site Reconnaissance. In *Applications of Space-Age Technology in Anthropology,* edited by Clifford Behrens and Thomas Sever. NASA, Stennis Space Center, MS.

Puleston, Dennis E., 1982, *The Role of Ramon in Maya Subsistence.* Academic Press, New York.

Ricketson, O. Jr., and Kidder, A.V., 1930, An Archaeological Reconnaissance by Air in Central America. *The Geographic Review* XX(2):177–206.

Sader, S.A., Sever, T., and Smoot, J.C., 1996, Time Series Tropical Forest Change Detection: A Visual and Quantitative Approach. International Symposium on Optical Science, Engineering and Instrumentation, Denver, CO. *SPIE,* Volume 2818-2-12.

Schott, J.R., 1997, *Remote Sensing: The Image Chain Approach.* Oxford University Press, Oxford.

Schowengerdt, R. A., 1980 Reconstruction of Multispatial, Multispectral Image Data Using Spatial Frequency Content. *Photogrammetric Engineering and Remote Sensing* 46(10):1325-1334.

Sever, Thomas L., 1998, Validating Prehistoric and Current Social Phenomena upon the Landscape of the Petén, Guatemala. In *People and Pixels: Linking Remote Sensing and Social Science,* edited by Diana Liverman, pp. 145–163. National Academy Press, Washington, D.C.

Sever, Thomas L., 1999, The Ancient Maya Landscape from Space. In Thirteen Ways of Looking at a Tropical Forest, edited by James D. Nations. Published by Conservation International, Washington, D.C.

Sever, Thomas L., and Irwin, D., 2003, Landscape Archeology: Remote Sensing Investigation of the Ancient Maya in the Petén Rainforest of Northern Guatemala. In *Ancient Mesoamerica* 14:113–122.

Stephens, John L., 1841. *Incidents of Travel in Central America, Chiapas and Yucatan.* Harper and Brothers, New York.

Chapter 7

Creating and Perpetuating Social Memory Across the Ancient Costa Rican Landscape

PAYSON SHEETS AND THOMAS L. SEVER

Abstract: For most of the time that Native Americans lived in ancient Costa Rica, we think
their travel across the landscape was task-oriented and thus sufficiently randomized
to leave no detectable trace. That changed about 500 B.C. in the Arenal area
when people separated their cemeteries from their villages, and travel between
them was ritually mediated into travel precisely along the same path, in single
file, in as straight a line as possible. The inadvertent erosion over centuries of use
resulted in paths entrenching to 2 m or more deep. We believe the cultural standard
developed over the centuries was that the preferred way of entering a special place
was by an entrenched path. Thus, people approaching a special place would have a
highly restricted view of their surroundings, but upon entering the special place, it
would dramatically open up to view. People created and perpetuated social memory
across their landscapes with generation after generation of use. The construction of
meaning developed as the paths entrenched, ultimately embedding that meaning
deep in people's belief systems as their pathways were embedding themselves into
the landscape. Although chiefdoms never developed in the Arenal area, a series of
chiefdoms did develop east of the area at about A.D. 1,000. To satisfy the emergent
chief's needs for monumentality, we suggest here that chiefs chose the "proper
entrenched entryway" as exemplified in the Arenal area, for elaboration. On the
rocky slopes of the volcanoes monumental entryways were built of stone, and on
the fine-grained alluvial plains they were of earthen construction. For instance,
the radiating entrenched roads entering Cutris are many meters long, as wide as
50 kilometers, and many meters deep.

161

1. INTRODUCTION

The Proyecto Prehistorico Arenal has been conducting research in northwestern Costa Rica since the mid 1980s, documenting Native American occupation from about 12,000 years ago to the Spanish Conquest. Most of the research objectives, methods, and results are presented in Sheets and McKee (1994), with earlier research results presented in Sheets and Mueller (1984). The previous publications dealing with remote sensing in the Arenal area include McKee and Sever (1994), McKee et al. (1994), Sheets (1994a, b), Sheets and Mueller (1984), Sheets and Sever (1988, 1991), and Sever (1990). Publications on relevant archaeology include Bradley (1984, 1994), Bradley et al. (1984), Chenault (1984), Hoopes and Chenault (1994a, b), and Mueller (1992, 1994).

Evidence of Paleoindian and Archaic occupations has been found, with sedentism achieved by 2,000 B.C. Four phases of sedentary occupation have been defined, with societies in all phases remaining egalitarian, or slightly pushing the boundaries of egalitarian organization. Compared to other areas of Middle America, population densities remained slight, agriculture minimal, and social stability remarkable over long periods of time. The avoidance of complex society, competition, and warfare contributed to social stability.

In the project's second year a collaboration was established with Tom Sever of NASA for remote sensing. If remote sensing for archaeology could be successful in the Arenal area tropical rainforest with minimal human impact on the ancient environment and multiple layers of volcanic ash draping the landscape, then it was believed that it should be applicable to other areas. Many kinds of analog and digital remote sensing imagery have been used in the research area. The most important, and wholly unanticipated, remote sensing success was detecting ancient footpaths as linear anomalies. The reason we have not been able to detect ancient paths for most millennia of native occupation, we believe, is because people had no need to walk along precisely the same line for decades. However, ancient footpaths are detectable for almost two millennia, beginning about 500 B.C. When people separated cemeteries from their villages, they followed ritually prescribed routes single-file to connect these two special places. The sustained use of these paths had unanticipated erosional and entrenching effects that, we believe, led to their becoming the favored entryways into special places. Later chiefdoms "writ large" these entrenched paths into monumental and very impressive entryways into their central places.

2. RESEARCH BACKGROUND

2.1. Arenal Research Project

In the early 1980s, Payson Sheets decided to investigate the relationships among egalitarian peoples, explosive volcanism, and tropical rainforest environments in the Arenal area of Costa Rica (Figure 1). He received funding

Figure 1. Map of the Arenal research area in Northwestern Costa Rica. The eastward Silencio phase path might lead from the cemetery to the village sites on the shore of Lake Arenal. The westward Silencio phase path does lead to the Tovar source of stone for tomb construction, just west of Tilaran. The earlier Arenal phase path leads from the G-156 village on the lake shore westward to the cluster of cemeteries in the Castrillo and Mandela area.

from the National Science Foundation and National Geographic Society for a few seasons of fieldwork, beginning in 1984. The only form of remotely sensed imagery used during the first field season was commercially available black-and-white aerial photography from the local Instituto Geografico, in the form of 9" × 9" contact prints and enlargements. The first season was successful in establishing general phases of occupation (see next section), and in documenting settlement patterns during the past four or five millennia of sedentary occupation (Sheets and McKee, 1994).

Independent from, but concurrent with the first field season at Arenal, Tom Sever at NASA and John Yellen at the National Science Foundation were looking for a regionally oriented research project that might benefit from remote sensing enrichment. When that was offered to Sheets, it took less than a nanosecond to accept, and the project research was permanently transformed. Remote sensing for archaeological objectives was seen as challenging in the Arenal area, because population densities were minimal compared to other areas of Middle America, human impacts on the environment were slight, the tropical rainforest or heavy pasture grasses obscured the ground surface, and multiple layers of volcanic ash buried the ancient ground surfaces.

Arenal volcano erupted explosively some 10 times during the past 4,000 years (Melson, 1994), deeply burying many archaeological features, and making remote sensing for evanescent features difficult in many areas. However, the volcanic ash (tephra) layers provided invaluable time-stratigraphic horizon markers for dating footpaths that have virtually no associated artifacts.

The radiocarbon-dated tephra layers provided a means of dating ancient footpaths or other phenomena forming linear anomalies, after the anomaly was detected in the analog or digital remote sensing imagery (see below). Excavations and stratigraphic interpretations could identify the paleosol that was contemporary with the initiation of path use. The ending of path use is dated by the tephra layer that uniformly covered the path, with no evidence of the erosion that resulted from continued use (see McKee et al., 1994, for details of methodology, assumptions, path formation, and preservation, and other key issues). Thus we can date ancient paths to a particular archaeological phase, or in the case of 19th-century ox-cart roads or more recent features, the stratigraphy clearly distinguishes Precolumbian linear features from historical and recent ones. The criteria we developed for dating the various linear features, from ancient to recent, are presented in McKee et al. (1994).

It is important to point out that, in all the years of excavations into confirmed ancient paths, we have yet to find a single instance of deliberate construction activity. The actual path surface upon which people walked is rarely wider than a half meter, but the erosional effects on sloping terrain routinely affect 2–4 m on each side of the path, resulting in a linear erosional feature wide enough to be detected quite readily in medium-resolution imagery.

2.1.1. Phases of Human Occupation

People have lived in the Arenal area for approximately 12,000 years, and for ten of those 12 millennia they left no traces of human movement across the landscape that we have been able to detect. A Clovis-style projectile point found on the south shore of Lake Arenal is clear evidence of Paleoindian occupation about 12,000 years ago, and occasional Archaic projectile points and campsites were found in the area, dating to 5,000 bp and earlier (Sheets, 1994). The first phase with semi-sedentary to sedentary villages is the Tronadora phase, beginning possibly by 3,000 B.C., but certainly by 2,000 B.C. (Sheets and McKee, 1994; Hoopes, 1994c) and ends at 500 B.C. Burials during the Tronadora phase were adjacent to the house, in small pits that presumably were for secondary burials, along with ceramic vessels as grave goods. The very moist, acidic soils do not preserve ancient human bones or teeth in this area, with the sole exception of the Silencio phase (see below), when stone boxes were built as tombs that protected the skeletons. All kinds and scales of remote sensing imagery have been examined to try to find footpaths dating to any time during these ten millennia, and none have been found. We suspect that human travel across the landscape was task-oriented, in that when someone needed to collect some food, hunt, obtain water, or get some stone or firewood, they simply went there, got it, and returned. Task-orienting travel has a randomizing effect, leaving no long-lasting or permanent impact on the environment that we can detect. Surely this is fortunate, that our environment does not become permanently trampled by such travel. Our concern in this chapter is not to elucidate the nature of travel prior to the

Arenal phase, as we do not have direct evidence of it. Rather, our focus is the new structured kind of travel that emerged in the Arenal phase, its impact on the environment and its ideational implications, and finally how it may have become the seed for later monumentality in more complex societies.

In contrast to travel in preceding phases, a different kind of travel evidently developed during the Arenal phase, 500 B.C. to A.D. 600, when people evidently separated their villages from their cemeteries. We cannot claim that all Arenal-phase villages were separated from their cemeteries, but separation was common and routinized travel began to have repercussions that we can still detect today. A sense of place, and paths as sentient, may have been similar to what Snead (2002) found in the U.S. Southwest, and with the Yurok in northern California. For the Arenal research area, population densities were at their greatest during this phase, but when compared to other areas of Mesoamerica or the Andes, population was still sparse. Cemeteries show at least some indications of differentiation (Hoopes and Chenault, 1994a), and it is possible that society was pushing the boundary of egalitarianness. Cemeteries were built up of individual tombs made by hauling in large numbers of subrounded river rocks, with impressive remains of post-interment ritual, feasting, and breaking of ceramic vessels and occasionally elaborate metates (Hoopes and Chenault, 1994a). The new kind of travel linked cemeteries to villages, and people followed the same precise path generation after generation. The path itself was as straight as possible, denying topographic irregularities, and even going up and over a hill rather than around it in one case. The straight-as-possible walking along the same path created path entrenchments that were commonly 2 or 3 meters deep, and in some places much deeper (Figure 2). What we are suggesting here is that the inadvertent path entrenchment among these egalitarian Arenal-area societies began to create a cultural standard. We suggest that the "proper" way to travel between "special places," and to enter a "special place" such as a cemetery, was in an entrenched path, seeing little of the surrounding countryside while in transit. But when one walked into the cemetery, the path opens up into a dramatic broad vista. In a sense it is like the birthing process, or a rite of passage from one domain to another. We assume both villages and cemeteries were considered to be special places, and we believe what must have developed as people first separated villages from cemeteries is the concept that travel between them was a sacred act. Thus people created and perpetuated social memory by performing repetitive ritual travel that embedded itself deeper in the landscape through the generations.

The case of an Arenal-phase path going up and over a hill rather than around it is important here, as it informs us that people went out of their way to have an entrenched path, rather than following along the "path of least resistance" along the floodplain of the Rio Piedra. The associated Arenal-phase village (G-180) and cemetery (G-184) are 1.1 km apart, and the hill is just outside the village. As the path reached the top of the hill, it was no longer

Figure 2. Color infrared airphoto taken in March 1985, rendered in black-and-white, covering 1.8 km horizontal distance. The Arenal phase footpath extends from Trench 21 down to Trench 28, where it was stratigraphically confirmed in both places. The path is double at Trench 21, but unites into one near the landslide scar in the center. An oxcart road from the late 19th and early 20th century is visible at the bottom.

incised because of the lack of erosional potential. People built two small stone features, which could be called platforms, of subrounded river cobbles on either side of the path. We found no evidence of burials there, but there were some broken pottery vessels, presumably indicating ritual activities, and it is tempting to envision rituals with mourners/celebrants passing single file between people standing on the stone platforms, supporting or somehow commemorating their passage.

Many other cases of entrenched pathways entering special places have been found during Proyecto Prehistorico Arenal research. The path that enters the Silencio cemetery (G-150) from the west is sunken, as are the three that enter that cemetery from the east. The four paths that arrive at the spring that supplied cemetery celebrants with water are sunken, as are the two that enter the spring from the east. The Arenal-phase path is sunken as it enters the village site (G-156) on the lakeshore. The two parallel Arenal-phase paths are sunken as they enter the Poma Cemetery (G-725, publications in preparation) and other cemeteries in the Poma area. And there are other apparent cases outside our research area, as exemplified by the Mendez site in the Rio Naranjo-Bijagua area 29 km north northwest of Tilaran (Norr, 1982). Sheets has detected a linear anomaly that appears to be an ancient entrenched path entering the Mendez cemetery from the north, and has inspected it on the ground, where it also appears to be an ancient eroded footpath. It has yet to be confirmed by excavations and analyses. Other sites such as Rivas (Quilter, 2004) have never been examined in aerial photography or other remote sensing imagery for evidence of possible footpaths (Jeffrey Quilter, personal communication, 2004), although Quilter is eager to have it done.

The physical phenomena of path formation and preservation are reasonably well understood (McKee et al., 1994). Compaction is the initial result of people walking along precisely the same path, which creates a linear depression. What we have found is that on slopes greater than 5°, and especially on slopes greater than 10°, erosion begins a progressive lowering, or entrenchment, of that path. The eastern end of the research area receives over 6,000 mm of mean annual precipitation, and the central area about 3,000, so there is sufficient moisture to do erosion on slopes where people were creating lines of susceptibility. The paths themselves were narrow, generally about 0.50 m wide, indicating single-file use was the actual and proper mode of transit. As the path itself became entrenched, the soil on both sides of the path was affected by the erosion. Because the tephra layers and soils are minimally consolidated in the Arenal area, the "angle of repose" is a widened "V" shape, generally sloping upward 25° to 30° from horizontal on both sides of the path (see McKee et al., 1994). In many places the modern land surface still retains the indentation of the ancient path, partially or largely filled in by later tephra deposits but still visible. In other places the entrenched path is entirely filled in by later tephra deposits, and it can only be detected by remote sensing sensitive to subtle vegetation differences, noticeable, in this case, because plant roots growing in the ancient path are in a sufficiently different matrix than plant roots on either side of the path. In this case color infrared aerial photography is the most sensitive remote sensing technique. When the paths traverse flat areas, they entirely disappear in all remote sensing imagery and from all indications in the field. Because flat areas are small and rare, and confirmed paths from different phases are generally far from each other, we usually can extrapolate from confirmed segments on both sides of the flat area, and continue to follow the same path. When we do extrapolations across flat areas we check stratigraphic indications especially carefully for synchronicity. Flat areas rarely extend for more than a few meters to tens of meters.

One challenge has been distinguishing ancient footpaths from historical-period and more recent features. Dating and stratigraphic interpretation procedures help distinguish linear anomalies such as ox-cart roads when lower elevations of the Arenal area were under sugarcane cultivation (McKee et al., 1994). Sugarcane cultivation, introduced from the lowlands to the west in the late 1800s, extended as high as 500–600 m and was cultivated as late as the 1950s. Ox carts were used as a common transport mode, and after an ox-cart road has been abandoned for a few decades, it can look much like an ancient footpath. A difference is the footpaths were routed as straight as possible, while ox-cart roads curved in response to topography.

Some aspects of funerary practices changed during the Silencio phase (Figure 3), A.D. 600–1300. Tomb construction turned to using the naturally flat-fracturing andesitic stone slabs called "*lajas*" and elongated "headstones" called "*mojones*." Sheets has observed the elongated "*mojones*" still in situ in looted graves in the Silencio cemetery. But what did not change was the ritually prescribed travel along straight-as-possible paths between villages and cemeteries. In the case of the Silencio cemetery, perched atop the continental divide, on the highest ridge in the area, the paths also led to a nearby spring used for drinking and cooking to support the impressive amount of feasting that took place (Figure 4). Three paths leading from the cemetery down to the spring can be seen in Figure 4, and each continues its arc under the rainforest canopy. One of the three splits into two just before entering the rainforest. All four come together at the spring. The other path from the cemetery, that heads westward, goes straight down a ridge and then makes an obtuse-angle bend to travel straight downhill. In that angle bend was a *laja* and headstone repository (Chenault, 1984; Hoopes and Chenault, 1994b), with stones placed here for intended future use. The path down the steep hill to the stream had eroded down to the pre-Arenal eruption clay-laden slippery "Aguacate formation" (McKee et al., 1994:147–149), and therefore they created a parallel path. Both paths emerge on the other side of the stream to climb the steep

Figure 3. IKONOS satellite image from March 2001, draped over digital terrain model, with Silencio-phase path indicated by dots. Horizontal distance covered in foreground is 4.8 km. The path heads to the Tovar source of stone for cemetery construction just off the image to the left.

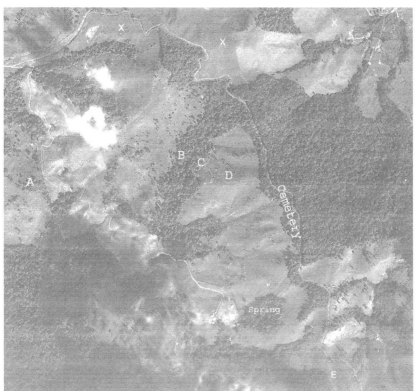

Figure 4. Black-and-white airphoto taken in 1971; horizontal distance is 2.1 km. The continental divide runs along the 2-track jeep road from the northern edge, past the Silencio "Cemetery" and past the letter "E." The earlier Arenal-phase path is at the top, marked by the series of "Xs." Three Silencio-phase paths lead south from the cemetery to the spring, and then continue to the hill marked "E" where they split into two parallel paths, and then reunite. The westward path exits the cemetery and makes an obtuse angle bend at "D" where there is a laja and mojon repository. The path splits into two and heads steeply downhill to "C" and crosses a stream, to re-emerge at "B" and head uphill to another stone repository. The path continues westward at "A" and continues downslope past the town of Tilaran to the Tovar stone source of lajas and mojones.

slope, and, after another obtuse-angle bend, head straight atop a broad sloping ridge toward the Pacific. Inside that second obtuse-angle bend is another *laja* and headstone repository. Because these repositories were both on the west side of the cemetery, we generated the working assumption that the source of *lajas* and headstones was to the west, along this path. That hypothesis was confirmed by geological field and laboratory research that chemically, petrographically, and morphologically confirmed the principal source at Cerro

Tovar, just southwest of Tilaran. The straight-line distance from source to cemetery is 7 km, a long distance to carry stones that weigh on average about 25 kilos.

Thinking anthropologically, we realized that a cemetery needs a steady supply of dead bodies to stay in operation, and of mourners to do construction and post-interment rituals. The location, however, of the village, or villages, that buried their dead in this cemetery is yet unknown. It is a good possibility that it lies at the terminus of the eastward path that has been traced and confirmed for two kilometers to the east-southeast of the cemetery and no farther. Because of the trajectory of the eastward path, we believe the two most likely candidates are the sites designated G-177 and G-176 on the lakeshore (Figure 1). Both are villages that date to the same phase. There are other villages of the same phase to the north of these, which are less likely and thus not on the map. We have spent many months examining images and walking this area and doing test-pitting, during the years of 1987, 1991, 2002, 2003, and 2004, but as of yet to no avail. Certainly, the huge earthquake of 1973, which resulted in many landslides, destroyed many segments of the eastward path. Also, the road that runs down to Rio Chiquito is on top of what appears to be the most likely ridge where the ancient path may have run. If ancient and modern routes in fact share the same ridgetop, the road would have destroyed much of the path. We have trouble believing that landslides and the road completely destroyed the ancient path. Therefore, we do not understand why we have been unable to find any vestiges of the path. We do not know what concatenation of factors is causing us such frustration. And, of course, one or more villages on the western side of the cemetery, past the Tovar stone source, could have been participating in burials. We have not found a continuation of the Silencio-phase path past the Tovar source, but that does not mean it does not exist. And the fact that the decorated ceramics at the Silencio cemetery exhibit much stronger similarities to lowland communities on the Pacific side (Hoopes, 1994a) than the Caribbean side may indicate that most cemetery users came from the west. The relative amount of foot traffic on the westward and the eastward paths was approximately equal, based on the amount of erosion on comparable slopes. So it is clear to us that what we know about the Silencio-phase path network is greatly outweighed by what we do not know.

Following the peak population in this research area during the Arenal phase, the Silencio phase witnessed a decline in population to less than half of what it was, based on the numbers and sizes of sites occupied. The sites found by survey should be representative of population densities, and our excavations were in sites from all phases. We consider the survey to be the most reliable source of data for exploring demographic trends. A less reliable, but we believe not worthless, indirect proxy for demographic trends could be the relative numbers of ceramics identified of the various phases. If we assume that the total phase-diagnostic ceramics collected by the project are at least roughly proportional to ancient populations, the decline may have been even more dramatic, as in the entire research area Hoopes (1994a:205) classified 7,106 sherds to the the Arenal phase and 1,879 sherds to the Silencio phase, a decline of about 75%. This decline is in spite of the fact that more excavations were done in Silencio-phase sites than Arenal-phase sites.

The decline in sizes and numbers of sites continued in the succeeding Tilaran phase (A.D. 1300–1500), to about half of those in the Silencio phase. The ceramics show a similar decline, to only 819 (Hoopes, 1994a:205), or 44% of the previous total. All of the digital and analog remote sensing imagery has been searched for evidence of paths associated with Tilaran-phase sites, and not a single one has been found. It is likely that the ritually prescribed movement that characterized the previous two phases was eliminated during this phase. The Tilaran phase was a time of culture change, as ceramics broke from the previous painted tradition and adopted plastic decoration characteristic of peoples to the southeast. This is either a dramatic change in cultural orientation by diminishing but ongoing local residents, or it is the result of an immigration of new peoples and elimination of the former residents. Although this is a very small sample, we found only one village with burials dating to this period (Bradley, 1994), and it is important to note the lack of separation between the two. The burials were in plain pits with ceramic vessels and no stone tomb construction or any post-interment feasting or other ritual evidence. One possible other Tilaran-phase habitation site was mentioned by Hoopes (1984). Other publications exploring demographic trends in the Arenal area are by Mueller (1994) and Hoopes (1994b).

Relatively standard archaeological procedures of survey and excavation, along with volcanology, radiocarbon dating, and stratigraphy, resulted in the above-mentioned phases. It is within this chronological-spatial framework that remote sensing has made significant contributions. Below we review the kinds of remote sensing and their variable contributions to the discovery of archaeological features.

3. REMOTE SENSING INSTRUMENTATION AND IMAGERY USED TO DETECT ANCIENT FOOTPATHS

3.1. Analog Imagery

3.1.1. Color and Color Infrared

The color infrared (CIR) aerial photography is presented first here, because it was in that imagery that Sever first detected a linear anomaly that began the footpath research. Sever arranged for a NASA research aircraft overflight that took low-level high-resolution infrared photography in March 1985 (the drier season) that included the area of the Silencio cemetery and the pasture to the west. In that pasture he detected a faint but slightly redder line that indicated somewhat healthier pasture grasses, and excavations of three trenches the next day confirmed the footpath atop the ridge and that it divided into two paths down the steeper slope. Other overflights, at elevations between 1,000 and 30,000 feet above terrain, have proven very useful in detecting anomalies that turn out to be ancient footpaths, particularly when they are flown in March or April, the drier months when plants can experience slight moisture deficits.

Color aerial photography was taken during the same low-level overflight, and is of some utility in detecting linear anomalies. As it is less sensitive to variation in photosynthetic exuberance than the CIR, it has been used much less for our purposes. All of the low-elevation aerial photography is of exceptionally fine resolution, so we can detect features a few centimeters in size. For instance, we can resolve children's hopscotch chalk marks on sidewalks in the town of Tilaran, and individual rocks in Arenal-phase cemeteries that are exposed on the surface, i.e., not buried by volcanic ash or vegetation. We gave landowners copies of color and CIR aerial photography of their houses and/or their ranches, which they greatly appreciated. We did have one interesting case where a man was excited to identify his house in Tilaran in the CIR imagery, but then he noted "this is Jose Luis's pickup truck in front of my house. When was this taken?"

3.1.2. Black-and-white Aerial Photography

After noting how clearly the ancient paths can be detected in the CIR photography, we went back to the very inexpensive black-and-white 9"× 9"contact prints made from negatives from professional large-format cameras, which are available from the Costa Rican Instituto Geografico, and we found that we could readily detect the paths. In the areas where the paths still leave an indentation today, the black-and-white airphotos were almost as useful to us in tracing their routes as the color infrared airphotos. However, in areas where there is no present surface indentation, the color infrared airphotos were the only ones sensitive to the buried paths, because the root matrix of vegetation was different along the path from that away from the path.

When we identified a section of a 9"× 9" contact print that was of interest, we had 3"× 3" sections of the negatives enlarged to 1 m × 1 m, so we could resolve features less than a meter in diameter, and the footpaths became highly visible. In addition to the modest cost of only a few dollars, another advantage of the black-and-white photos is that large-format cameras have been flown over Costa Rica about once a decade since the 1940s, providing time-sequential coverage. Sun angle has proven to be a crucial factor in path detection, and the variation inherent in so many overflights highlights the microtopographic features on various slopes in ways a single overflight cannot. Often a path anomaly is not visible with one sun angle but highly visible with another. Another advantage of this historical aerial photography is that it helps resolve some big problems. For instance, a large dam built in the late 1970s more than doubled the surface area of Lake Arenal, flooding many areas of interest. Fortunately, pre-1980 airphotos record the area prior to that impoundment. And the huge earthquake of 1973 dislodged massive landslides that destroyed much of the Silencio-phase path to the east of the cemetery, but we have found portions of it in pre-1973 airphotos. We encourage high-tech remote sensing enthusiasts **not** to neglect this low-tech, inexpensive, easily available, and effective resource.

3.2. Digital Imagery

3.2.1. *Thermal Infrared Multispectral Scanner (TIMS)*

The Thermal Infrared Multispectral Scanner (TIMS) was flown in the NASA research aircraft in March 1985 at 6,250 feet, resulting in a 5-meter ground-resolution pixel over the research area. CIR photography (discussed above) was acquired simultaneously at 60% overlap for stereoscopic viewing. The six bands of the TIMS range from 8.2 to 12.2 micrometers. These bandwidths were successful in detecting linear anomalies by very subtle temperature differences, and the anomalies were later confirmed as Silencio-phase footpaths to the northwest and the southeast of the cemetery (Figure 5). The thermal response of a land surface is dependent upon complex interactions between many physical and vegetational factors. In agricultural areas and grasslands, the amount of vegetation cover, its composition, and soil moisture all influence the thermal response. Differences in forest structure present a complex range of surfaces such as the type of forest canopy, depth, and architecture. In addition, ground slope can become an important factor in the flow of radiant energy fluxes. As a result of these various factors the TIMS was able to detect various pathway locations based on thermal inertia differences.

3.2.2. *Radar*

Radar imagery was obtained at the same time TIMS data were acquired (Figure 6). Of all the imagery used for Arenal project research, radar was the farthest from the visible portion of the electromagnetic spectrum. Perhaps not surprisingly, it was also the most challenging to interpret. This L-band, 24-cm, 1.225GHYz radar system was flown aboard a NASA aircraft and gathered microwave data with four polarizations at 30- and 10-meter resolutions. The data acquisition resulted in a plethora of linear anomalies being detected in uncut tropical rainforest, in secondary regrowth, and in pastures. Following ground truthing many linear anomalies turned out to be historical and recent features, but others were not clear even after ground inspection. Only a few were ancient footpaths, verified by excavations and stratigraphic interpretations. Several filtering techniques were utilized to attempt to reduce the noise in the data, but they were unsuccessful. Part of the reason was

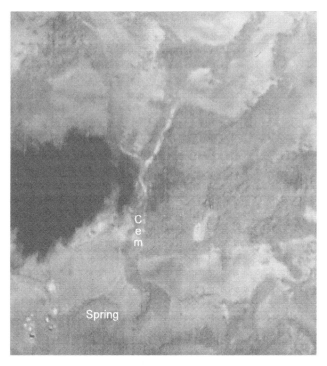

Figure 5. Thermal Infrared Multispectral Scanner image obtained in March 1985 of the Silencio Cemetery and the Spring. Horizontal distance shown is approximately 1.3 km. The three ancient Silencio-phase paths linking the cemetery to the spring can be seen, because of very subtle differences in thermal inertia. The large dark area northwest of the cemetery is the shadow of a cloud.

that prior to the time of data acquisition the primary antenna had been damaged, and a substitute antenna was used, generating uncalibrated data. Consequently it was difficult to reduce the data for meaningful archaeological interpretation.

3.2.3. Landsat

Landsat Thematic Mapper satellite imagery was obtained by NASA of the northwestern portion of Costa Rica in 1985, and it proved useful for documenting the dramatic moisture differences in the research area. That ranges from the wet tropical forest on the Caribbean side in the east to the quite dry conditions and sparse vegetation cover on the Pacific side in the west. Seasonality is highly developed in the west and almost nonexistent in the east. The areas affected by the explosive eruption in 1968 and the subsequent lava flows from the Arenal volcano that are continuing even to today (2004)

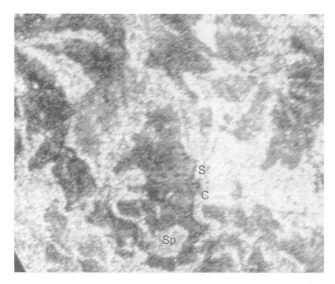

Figure 6. Radar image acquired in March 1985 including the Silencio Cemetery ("S C") and the spring ("Sp"), both in heavy tropical-rainforest vegetation. Pastures are darker. Note numerous linear anomalies, none of which turned out to be an ancient footpath. Some anomalies are fence lines while others are roads and unknown phenomena. Approximate horizontal distance is 2.5 km.

are clearly recorded, as well as the continuing gas emissions from the crater. The 30-m pixel size is not suitable for detecting the footpaths as anomalies, or the habitation sites. The only archaeological features visible in LANDSAT imagery are the large and looted Arenal-phase cemeteries. As they had been discovered by looters during the past century, and are well known to people living in the area, there is no reason to use LANDSAT to rediscover them.

3.2.4. Ikonos

Satellite technology has improved vastly in recent years, and four scenes of IKONOS satellite imagery were obtained recently, when cloud cover was less than 5%. IKONOS provides 1-m resolution panchromatic imagery and four multispectral bands (visible and near infrared) at 4-m resolution. The satellite has a polar, circular, sun-synchronous 681-km high orbit, and both sensors have an at-nadir swath width of 11-km. Pan-sharpening techniques were used to "fuse" the 1-m resolution high-resolution panchromatic with the 4-m multispectral imagery. The resultant high-resolution color imagery was sufficient to detect the linear anomalies created by ancient footpath use during both the Arenal and Silencio phases. Of the several available pan-sharpening techniques, our datasets were most successful when we used Brovey and Principal Component transformations.

Various band combinations, contrast enhancements, and filtering techniques were applied to the data to extract linear and curvilinear patterns. Since the data were georeferenced, features in the imagery could be located within 2-m accuracy on the ground using a GPS unit. In addition to ancient footpath segments, the data revealed erosional drainages, the location of historical-period fence lines, and a few century-old ox-cart roads, all of which were investigated and identified through ground survey, excavations, and stratigraphic analyses.

3.3. Image Analysis

The evolution of image analysis software, laptop computers, and GPS technology has revolutionized our ability to conduct field research. Originally, in the mid 1980s, we processed the data on mainframe computers in our offices and labs in the U.S., and made hard-copy outputs that we took to the field in Costa Rica. In recent years our analysis capability has become more proactive, efficient, and quick. Now we take our laptop computers to the field, which allows us to review imagery in real time, both at our field lab and literally in the field as we are walking the terrain. This ability allows us to quickly resolve questions that we incur in the field as well as determine the areas to be surveyed the next day. In addition, the image analysis software has become more user-friendly. Some of the software packages we used include ERDAS Imagine (1999), Remote View (2003), and Skyline Terra Explorer (2003).

These software packages allowed us to review both historical and recent imagery and provided the capability to monitor the land-cover/land-use changes that are occurring in the study area, threatening to destroy, or actually destroying the ancient footpaths. The Skyline software allows us to analyze the data in a virtual mode. For instance, the georeferenced remote sensing imagery was merged with topographic information and the locations of known footpath segments, cemetery areas, and excavated trenches were labeled and overlaid onto the data. In this way we were able to investigate the landscape as if we were flying over the study area in a helicopter with complete control over altitude, direction, and speed, and land at any spot with a 360-degree view of the horizon. We were also able to measure both the true and linear distance between features. Through these virtual methods, we were able to get a better understanding of the relationships between natural and cultural features upon the landscape and to become familiar with an area before beginning our ground survey. As a result of the advancements in computer hardware and software technology, all of the information associated with the footpath study can be stored, preserved, and easily shared with others.

For our objectives, to discover linear anomalies in remote sensing imagery that could be investigated and verified as ancient footpaths, the most effective means was visual inspection of color infrared aerial photographs, followed by black-and-white and true-color aerial photographs. In the digital domain, the new IKONOS satellite imagery was found to be dramatically superior to any of the other imagery. TIMS came in a distant second, with LANDSAT and radar imagery being the least useful.

4. IMPLICATIONS: INADVERTENT PATH EROSION TO MONUMENTALITY

The earliest erosional paths that we have detected connecting special places in the Arenal area date to about 500 B.C., and the phenomenon continued for almost 2 millennia, to about A.D. 1300. Societies in that area remained relatively simple, never developing into chiefdoms. In contrast, to the east of the Arenal area a series of chiefdoms developed toward the end of that time span, at about A.D. 1000. (These observations do not presume a unilinear evolutionary model, as societies can and in fact do increase and decrease in complexity, in response to a wide range of internal and external factors.) We assume that the emergent chiefs to the east of Arenal were in need of monumentality to impress people. We define monumentality as controlling labor to construct large facilities that were beyond the practical domain in nature and scale, that demonstrate centralized authority in the process of construction and in maintenance (Trigger, 1990). They needed something that, after it was built, would impress their followers as well as visitors from other polities. We suggest chiefs selected as the core of their concept of monumentality what had developed in the Arenal area centuries before, that the favored or proper way to enter a special place is by a sunken, long, straight entryway. We are not suggesting a direct and unique or exclusive relationship between inadvertent early erosional paths around Arenal and the later constructed sunken entryways. Rather, we are suggesting the kind of non-constructed sunken entryway that developed at Arenal and other areas of ancient Costa Rica became a valued cultural ideal that was later adopted by rulers seeking monumentality. Of course, a simple sunken path that developed inadvertently would not satisfy chiefly needs for monumentality, so not surprisingly chiefs, seizing that cultural ideal, constructed massive sunken entryways. In lowland areas where the gradient was slight and the alluvial sediments were fine-grained, they used earthen construction. On the sloping flanks of the volcanoes the most abundant building material is stone, and so the causeways and the facings of buildings were built of stone.

The clearest example we can offer of constructed earthen monumentality on the Atlantic plain is that of Cutris, located 52 km east of Arenal volcano. The Cutris site is a challenge to deal with in this chapter, as it has few substantive publications. Quesada (1980) was the first to mention Cutris in print, followed by Guerrero et al. (1988). Much of what we present here derives from our work with black-and-white airphotos from the Instituto

Geografico, and from a personal site visit by Sheets in 2003. Some information is gleaned from a recently written manuscript that, if/when it is accepted for publication, will be the first substantial article on the site (Vazquez et al., n.d.) Two radiating deep roadways can be seen on the map and in the airphoto (Figures 7 and 8) heading over 4 km to the northeast and the northwest from the site. Vazquez et al. (n.d.) have been able to trace the roadways farther, from at least 6.7 to as much as 9.4 km from Cutris. Each of the four roadways ends at a community smaller than Cutris, and likely dependant upon it. As Vazquez et al. (n.d.) note, this can be considered a localized interaction sphere (*"una esfera de interaccion inmediata"*). As the roadways approach Cutris center, in their last kilometer, they widen considerably, to almost the width of a football field from berm top to berm top. The original constructed depth below the surrounding terrain is unknown, but from present-day surface inspection it appears that the long roads were excavated to a depth of at least 4 m. One

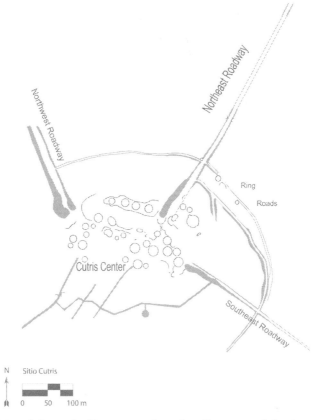

Figure 7. Map of Cutris chiefdom site, with roads radiating out of the site center. Two ring roads run around the site and connect the radii.

Figure 8. Airphoto of Cutris, site center at bottom, and radiating roads. The roads were excavated deeply into the alluvium, and broaden to about the width of a football field as they enter the site center. The Northwest and Southwest roads run for almost 4 km from the site center. The ring roads are barely visible. Someone entering the site from one of the roads would see little of the surrounding countryside for a long distance, before entering and having the view of the entire site center opening up.

wonders how water runoff was controlled, as the area receives over 3,000 mm of mean precipitation annually. And, at regular intervals of slightly more than 300 m, on top of the berms people constructed stone platforms, at least as the entryways get close to the site center (Guerrero personal communication 2003). We suggest these could be more formal versions of the small stone features/platforms on either side of the narrow path on top of the hill at Rio Piedra, about a millennium earlier. People standing on the platforms could consecrate the people in procession below, and those people below almost certainly had abandoned the single-file transit employed in the Arenal area. In addition to the long straight roadways, Vazquez et al. (n.d.) encountered what we call "ring roads" that curve and connect the main roadways. The "ring roads" are not as deep or wide as the main roadways. Vazquez et al (n.d.) also report zigzag sunken roadways that connect two of the main roadways. One can only imagine the complex processions that approached and entered the site center from the four villages, and the intercommunications on the ring and zigzag roads.

And Cutris is not the only chiefdom site east of Arenal with sunken straight entryways. The Fortuna site, 18 km east of Arenal volcano, has broad sunken roadways that are of the same scale, as they enter the site center, as those at Cutris. Unfortunately the Fortuna area has been more heavily plowed and affected by modern activities, and the roadways are not as well preserved. Stone and Balser (1965) conducted limited excavations at Fortuna.

In the rocky foothills of the volcanoes, chiefdom centers such as Guayabo de Turrialba (Aguilar, 1972; Fonseca, 1981; Murillo, 2002) built long *calzadas*, at least some of which were built of stone, to connect other centers about 4 km away. And they built impressive lowered entryways of *calzadas* and staircases to enter the site center, passing between twin large stone platforms (Figure 9). The large stone platforms have been interpreted in military-defensive ways, and considered to have functioned as "guard towers." Based on the above-mentioned discoveries, we believe that the interpretations of these features as defensive are no longer tenable. Entryways, whether *calzadas* or sunken earthen roads, that radiate out from site centers for many kilometers, to connect with other settlements, make no sense as defensive features. Roadways that connect central places with outlying settlements were constructed to

Figure 9. Entrance to Guayabo up a large stairway, past the two "Guard Towers," and then to the site center. The two towers more likely functioned in a ceremonial manner. Guayabo is in a more mountainous and rocky environment than Cutris, and subrounded river rock was used for construction. The sense of a sunken entryway is preserved.

facilitate contact rather than to hinder it, and we believe researchers should be looking into the kinds of contact that were being facilitated. The possibilities run the gamut from religious through social, economic, and others. These long-constructed features make a lot more sense as chiefs' demonstrating their mentality of monumentality, of the need to stage impressive processions and displays, the need to facilitate a variety of contacts with surrounding communities, and the need to impress their subjects and visiting dignitaries.

5. CONCLUSIONS

Native peoples have lived in what is now Costa Rica for some 12,000 years. For about 10,000 years of that time span their travel across the landscape left no trace that we have been able to detect. We believe that travel was task-specific, and thus sufficiently randomized across the countryside not to leave permanent marks. We see that result as fortunate. The environment of the tropical rainforest is sufficiently resilient to erase the footsteps of ancient inhabitants for 83% of the time they occupied Costa Rica. The cognitive structure of how they experienced their landscape changed at about 500 B.C. in the Arenal area when people separated villages from cemeteries, and began formal ritual movement between them. As they traveled the straight-as-possible route single-file across the countryside between these special places, their footprints initiated path compaction. On slopes, during heavy rainstorms, the linear compaction eroded downward. That erosion progressively entrenched paths downward, and the unused sides also followed downward. In many places the path and its sides became entrenched to a few meters below the surrounding landscape (Figure 10). We suggest here that a new ideal way of entering a cemetery developed in the last centuries of the first millennium B.C. That was to travel in a deep linear path with limited visibility, which changed dramatically as one arrived at the cemetery and it opened up to view. Thus, an unanticipated result of prescribed procession became the ideal, with no effort of construction. People created more than they initially intended by repeated passage from village to cemetery and back, literally embedding that passage into the landscape, and embedding it into social memory.

About A.D. 1000 more complex societies developed to the east of the Arenal area, and chiefs developed the mentality of monumentality. They "writ large" the proper straight entrenched entryways of simpler earlier times, in some cases with huge earthen constructions and in other cases with massive stone features. These long entryways often ran for a few kilometers, with platforms flanking them. What began a simple standard of travel between special places in earlier times, became the way elites constructed monumentality. We suggest this Costa Rican case is not unique. Rather,

Figure 10. From humble beginnings we see the seeds of monumental entryways. Ramon and his horse riding in the vestige of the Silencio-phase path, on the continental divide, southeast of the spring. While in use, the path would have been more than 3 meters below the surrounding ground surface. The four volcanic ash layers from Arenal volcano that were deposited across the landscape after the path ceased to be used have smoothed out the topography and filled-in the path depression somewhat.

elites seeking to institute monumentality are going to be more successful if they can incorporate a deeply held belief about what is proper, desired, and needed, that has centuries of authenticity. The incorporation of the simple into the monumental must have occurred many times in world prehistory.

6. ACKNOWLEDGEMENTS

It is with great gratitude that we acknowledge the support for this project provided by John Yellen and the National Science Foundation. The National Geographic Society has provided key funding at important times. NASA has been very supportive of the project research goals by assisting with personnel, equipment, software, and imagery at various times since the 1980s. We appreciate the assistance of personnel in the Museo Nacional de Costa Rica, and the Comision Nacional de Arqueologia. In particular, Ricardo Vazquez, Juan Vicente Guerrero, and Maritza Gutierrez have been especially helpful to us. During many years dedicated crews of students from the US and Costa Rican universities have done excellent fieldwork as well as technical analyses. In particular I thank Brian McKee, John Hoopes, John Bradley, Mark Chenault, and Marilynn Mueller. The local landowners in the Arenal-Tilaran have been extraordinarily gracious in allowing us to walk their properties and occasionally to excavate trenches to verify (or not) anomalies as ancient paths. We appreciate the patience of our wives, Candy Sever and Fran Sheets, for our extended absences as we chased the occasionally evanescent ancient path across the gorgeous Costa Rican landscape, and puzzled about their functions and their meanings.

REFERENCES

Aguilar Piedra, Carlos. H., 1972, *Guayabo de Turrialba; arqueología de un sitio indígena prehispánico*. Editorial Costa Rica, San José.

Bradley, John E., 1984, The Silencio Funerary Sites. *Vínculos* 10:187–192.

Bradley, John E., 1994, The Peraza Site Burials. In *Archaeology, Volcanism, and Remote Sensing in the Arenal Region, Costa Rica*, edited by P. Sheets and B. McKee, p. 121. University of Texas Press, Austin.

Bradley, John E.,1994, *The Silencio Site: An Early to Middle Polychrome Period Cemetery in the Arenal Region Barchaeology, Volcanism, and Remote Sensing in the Arenal Region, Costa Rica*. University of Texas Press, Austin.

Bradley, John E., Hoopes, John W., and Sheets, Payson D., 1984, Lake Site Testing Program. *Vínculos* 10:75–92.

Chenault, Mark, 1984, Test Excavations at Neblina and Las Piedras. *Vínculos* 10: 115–120.

ERDAS, 1999, *Field Guide, Version 8.5*, 5th edition. ERDAS Inc., Atlanta, GA.

Fonseca, Oscar, 1981, Guayabo de Turrialba and its Significance. In *Between Continents, Between Seas*, edited by Suzanne Abel-Vidor, pp. 104–111. Abrams, New York.

Hoopes, John W., 1984, Prehistoric Habitation Sites in the Río Santa Rosa Drainage. *Vínculos* 10(1–2):121–128.

Hoopes, John W., 1994a, Ceramic Analysis and Culture History in the Arenal Region. In *Archaeology, Volcanism, and Remote Sensing in the Arenal Region, Costa Rica*, edited by P. D. Sheets and B. McKee, pp. 158–210. University of Texas Press, Austin.

Hoopes, John W., 1994b, La arqueología de Guanacaste oriental. *Vínculos* 18–19(1–2):69–90.

Hoopes, John W., 1994c, The Tronadora Complex: Early Formative ceramics in northwestern Costa Rica. *Latin American Antiquity* 5(1): 3–30.

Hoopes, John W. and Chenault, Mark, 1994a, Excavations at Sitio Bolívar: A Late Formative village in the Arenal basin. In *Archaeology, Volcanism, and Remote Sensing in the Arenal Region, Costa Rica*, edited by P. D. Sheets and B. McKee, pp. 87–105. University of Texas Press, Austin.

Hoopes, John W., and Chenault, Mark, 1994b, Proyecto Prehistórico Arenal Excavations in the Santa Rosa River Valley. In *Archaeology, Volcanism, and Remote Sensing in the Arenal Region, Costa Rica*, edited by P. D. Sheets and B. McKee, pp. 122–134. University of Texas Press, Austin.

McKee, Brian R., and Sever, Thomas L., 1994, Remote Sensing in the Arenal Region In *Archaeology, Volcanism, and Remote Sensing in the Arenal Region, Costa Rica*, edited by P. D. Sheets and B. McKee, pp. 135–141. University of Texas Press, Austin.

McKee, Brian, Sever, Tom, and Sheets, Payson, 1994, Prehistoric Footpaths in Costa Rica: Remote Sensing and Field Verification. In *Archaeology, Volcanism, and Remote Sensing in the Arenal Region, Costa Rica*, edited by P. Sheets and B. McKee, pp. 142–157. University of Texas Press, Austin.

Melson, William, 1994, The Eruption of 1968 and Tephra Stratigraphy of Arenal Volcano. In *Archaeology, Volcanism, and Remote Sensing in the Arenal Region, Costa Rica*, edited by P. Sheets and B. McKee, pp. 24–47. University of Texas Press, Austin.

Mueller, Marilynn, 1992, *Prehistoric Adaptation to the Arenal Region, Northwestern Costa Rica*. Ph.D. dissertation, University of Colorado. University Microfilms, Ann Arbor.

Mueller, Marilynn, 1994, Archaeological survey in the Arenal Basin. In *Archaeology, Volcanism, and Remote Sensing in the Arenal Region, Costa Rica*, edited by P. D. Sheets and B. McKee, pp. 48–72. University of Texas Press, Austin.

Murillo Herrera, Mauricio, 2002, *Análisis crítico de las investigaciones en el sitio Guayabo (UCR-43), de Turrialba y las repercusiones sociales con relación al manejo de sus recursos culturales* Documento inédito. Trabajo Final de Graduación para optar por el grado de Licenciado en Antropología con énfasis en Arquelogía. Escuela de Antropología y Sociología, Universidad de Costa Rica, San Pedro.

Norr, Lynette, 1982, Archaeological site survey and burial mound excavations in the Rio Naranjo-Bijagua valley. In *Prehistoric Settlement Patterns in Costa Rica*, edited by Fred Lange and Lynette Norr. *Journal of the Steward Anthropological Society* 14: 135–156.

Quilter, Jeffrey, 2004, Cobble Circles and Standing Stones: Archaeology at the Rivas Site, Costa Rica. University of Iowa Press.

RemoteView, 2003, *User's Guide, Professional*. Sensor Systems, Inc., Sterling, VA.

Sever, Thomas, 1990, *Remote Sensing Applications in Archeological Research: Tracing Prehistoric Human Impact on the Environment*. Ph.D. dissertation, University of Colorado. University Microfilms, Ann Arbor.

Sheets, Payson, 1992, The Pervasive Pejorative in Intermediate Area Studies. In *Wealth and Hierarchy in the Intermediate Area: A Symposium at Dumbarton Oaks, 10th and 11th October 1987*, edited by F. W. Lange, pp. 15–42. Dumbarton Oaks, Washington, D.C.

Sheets, Payson, 1994a, Summary and Conclusions. In *Archaeology, Volcanism, and Remote Sensing in the Arenal Region, Costa Rica*, edited by P. Sheets and B. McKee, pp. 312–325. University of Texas Press, Austin.

Sheets, Payson, 1994b, Percepción remota y exploraciones geofísicas aplicadas a la arqueología en areas volcánicas activas de Costa Rica y El Salvador. *Vínculos* 18(1–2): 31–54.

Sheets, Payson and McKee, Brian, eds.1994, *Archaeology, Volcanism, and Remote Sensing in the Arenal region, Costa Rica*. University of Texas Press, Austin.

Sheets, Payson and Mueller, Marilynn, 1984, Investigaciones arqueológicas en la Cordillera de Tilarán, Costa Rica, 1984. *Vínculos* 10 (1–2).

Sheets, Payson and Sever, Thomas, 1988, High Tech Wizardry. *Archaeology* 41: 28–35.

Sheets, Payson and Sever, Thomas, 1991, Prehistoric Footpaths in Costa Rica: Transportation and Communication in a Tropical Rainforest. In *Ancient Road Networks and Settlement Hierarchies in the New World*, edited by C. D. Trombold, pp. 53–65. Cambridge University Press, Cambridge.

Skyline, 2003, *Terra Explorer User's Manual*. Skyline Software Systems, Inc. Woburn, MA.

Snead, James, 2002, Ancestral Pueblo Trails and the Cultural Landscape of the Pajarito Plateau, New Mexico. *Antiquity* 76: 756–765.

Stone, Doris, and Balser, Carlos, 1965, Incised slate disks from the Atlantic Watershed of Costa Rica. *American Antiquity* 17(3):264–265.

Trigger, Bruce, 1990, Monumental Architecture: a thermodynamic explanation of symbolic behaviour. *World Archaeology* 22(2):119–131.

Vázquez Leiva, Ricardo, Guerrero Miranda, Juan Vicente, Y sánchez Herrera, Julio César, N.d., Cutris: Descripcion, Crnologia, Y Afiliacion De Un Centro Arquitectonico Con Caminos Monumentales En La Llanura De San Carlos, Costa Rica. Unpublished manuscript, to be Submitted to Vinculos, Museo Nacional, San Jose, Costa rica.

Chapter 8

Spaceborne and Airborne Radar at Angkor: Introducing New Technology to the Ancient Site

ELIZABETH MOORE, TONY FREEMAN, AND SCOTT HENSLEY

Abstract: Defining an archaeological context for the temples of Angkor has been a major challenge in the interpretation of the ancient site. The authors' collaborative research addressed this issue using data acquired with the NASA Jet Propulsion Laboratory's Spaceborne Imaging Radar-C/X-band Synthetic Aperture Radar (SIR-C/X-SAR) in 1994 and the Airborne Synthetic Aperture Radar (AIRSAR) in 1996. The first archaeological use of combined SIR-C and AIRSAR datasets in Southeast Asia, and the ensuing perceptions identified new strategies of innovative research that have fundamentally changed our understanding of the breadth and complexity of the ancient Khmer achievement. The results summarized in this article are both technical and archaeological, ranging from the analysis of polarimetric and interferometric data to the transition from prehistoric habitation to the Hindu-Buddhist cities of Angkor in the 9–13th centuries A.D.

1. THE ARCHAEOLOGICAL LANDSCAPE

1.1. Radar Imaging and the Ancient Landscape

The first radar imaging data of Angkor, a Cambodian city of the 9th to 13th centuries A.D., was acquired in 1994 by the NASA Jet Propulsion Laboratory Space Imaging Radar (SIR-C/X-SAR) (Figure 1). In 1996, the first Airborne Synthetic Aperture Radar (AIRSAR) mission followed with three flight runs over Angkor. The data from these two pioneering acquisitions were used collaboratively by the authors to define an archaeological context for the

Figure 1. SIR-C 1994 image of Angkor. Across the central portion of Angkor, the linear arrangement of rectilinear water reservoirs and temple enclosures mark the city of Angkor in the 9th–13th centuries A.D. The radar imagery highlights the location of the city between the Kulen massif (light green area in the upper right corner) and the large seasonal lake, the Tonle Sap (black area, lower left). The northeast to southwest incline of the floodplain is reflected in the pattern of rivers flowing from the Kulen uplands across the floodplain, with water-management features designed to control excess waster in the rainy season and supply rice fields in times of drought. The distribution of small circular pre-Angkor sites can be seen along the edge of the floodplain (upper left).

period of transition in which the ancient city emerged. While the Hindu and Buddhist royal temples are often described as the apex of Khmer culture and innovation, our study focused on the equally, if not more radical, pre-Angkorean transformation of the landscape.

In our view, the curvilinear patterns of manmade moats and dikes associated with earlier village-based cultures figuratively and literally laid the foundations that enabled the emergence of the urban capital. In short, the origins of the city lie in earlier pre-Angkorean traditions of land alteration to venerate and control water. The expression of Khmer genius did not begin with the building of Hindu and Buddhist temples in the mid-first millennium A.D. The temples of Angkor provided rich articulation of their worldview, but the conceptual roots of the ancient Khmer cognitive model are in the land and its waters.

Imaging radar proved to be an ideal tool to differentiate manmade curvilinear elements in the flat landscape of Angkor, at times long stretches of rice fields and at others a denser canopy (Figures 2a, b). The SIR-C/X-SAR and AIRSAR data offered a uniquely extensive and quantified

Figure 2. Ground vegetation at Angkor. The landscape of Angkor varies between a) the typical rice field with scattered taller vegetation and b) denser canopy regions like this area in the Angkor Thom complex. In a), an ancestral "spirit" boat sits amidst the rice fields south of Angkor.

The generally flat terrain of the Angkor floodplain often makes detection of manmade elements difficult at ground level. The ability of radar data to generate images reflecting moisture and vegetation as well as elevation differences provided multiple views of anomalies linked to both the prehistoric and historical periods of Angkor.

source of information about subtle manmade alterations to the land. These alterations were made many centuries ago, but can be detected today in the patterns produced by interaction of the microwave pulses transmitted from the radar instrumentation with features in the target area. As this statement suggests, without the proper historical and cultural context to frame appropriate questions, radar imaging data alone cannot generate useful archaeological results. Testing the premise that changes in ancient social and religious belief systems were reflected in changes in settlement form, our studies thus fit within the wider archaeological field of investigations not only of where, how, and when, but also why ancient civilizations modified and controlled their physical environment (Hensley et al., 2001:197).

1.2. The Ancient City of Angkor

The word "Angkor" is derived from an ancient Sanskrit word meaning "city" and is used to refer to a succession of cities between the 9th and 13th centuries A.D. on the Siem the Siem Reap floodplain of northern Cambodia. This geographical feature upon which Angkor was built lies between the Kulen massif to the northeast and the Tonle Sap ('Great Lake') to the south (Figures 3a, b). Earlier cities at Angkor such as Hariharalaya (9th century A.D.) and Yasodharapura (10th century A.D.) also took their names from Sanskrit sources. While these contributed the priestly and dynastic nomenclature of Angkor, the location of the temples, massive reservoirs, and extensive dike-road systems of these cities was determined not by reference to the Hindu and Buddhist deities to whom they are epigraphically and iconographically dedicated, but to the landscape upon which they were built. The builders of these cities masterfully exploited Indic iconography and texts. However, the essence of the outstanding beauty and complexity seen in Angkor's temples and water-management structures lies in pre-existing patterns of ancestral veneration linked in particular to veneration of tutelary guardians and mastery of the life-giving waters of the land.

At its height of power, the kingdom of Angkor commanded the allegiance of a vast distribution of smaller cities and villages spread across the borders of present-day Cambodia and Thailand, as well as part of Vietnam and Laos. Every time a temple was built, water-control structures were built. Tanks were dug and dikes were raised. These structures controlled flooding during the monsoon months from May to October and conserved water during the dry months from November to April. In addition to the annual precipitation, the Khmer at Angkor benefited from a network of rivers flowing across the Siem Reap floodplain. South of the ancient city and the present town of Siem Reap is the Tonle Sap, one of the natural wonders of Southeast Asia. The lake doubles in size during the six months of rain and then acts as a relief valve for the overburdened drainage of the Mekong River into the South China

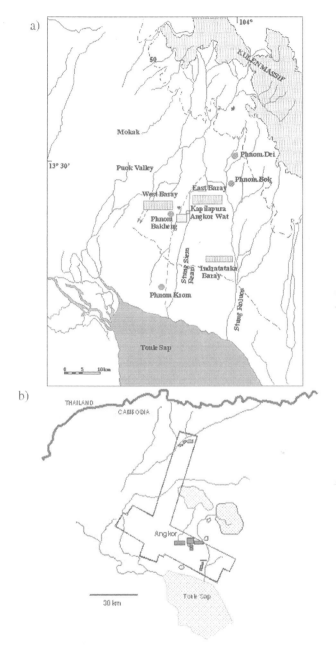

Figure 3. Maps of Angkor floodplain rivers and AIRSAR mosaic. a) The area of the Angkor floodplain imaged by the 1996 AIRSAR data acquisition was b) an L-shaped mosaic that cut across the central region of Angkor and extended northwest almost to the Thailand border.

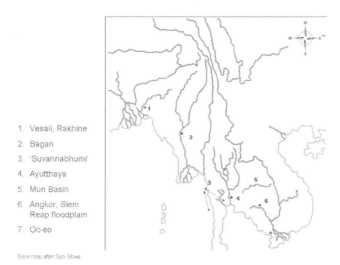

1. Vesali, Rakhine
2. Bagan
3. 'Suvannabhumi'
4. Ayutthaya
5. Mun Basin
6. Angkor, Siem
   Reap floodplain
7. Oc-eo

Base map after San Shwe

Figure 4. Map of Southeast Asia showing major sites discussed in the text. 1. Vesali, Rakhine coast of Myanmar. 2. Bagan, Myanmar. 3. "Suvannabhumi," Myanmar. 4. Ayutthaya, Thailand. 5. Mun Basin, Thailand. 6. Angkor, Siem Reap floodplain, Cambodia. 7. Oc-eo (Funan) in Vietnam. Base map after San Shwe, 2002.

Sea. This seasonal inundation brings an abundance of fish, the main source of protein in the local diet. The receding floodwaters are captured by dikes north of the lakeshore, supplying a reliable means to begin the year's wet rice cultivation. Much of the Angkor floodplain is given over to rice grown in bunded fields, which are interspersed with open savannah forest (Figures 5a, b). There are few landmarks in the flat terrain, its open expanse marked by only four mountains, or "phnom" and the shadow of the Kulen Massif on the horizon to the northeast (Figure 5c).

1.3. Regional Comparison

The questions and interpretations discussed in this paper address ways that vestiges of the past reflect pre-Angkorean spiritual preferences and perception of place. As described, the Angkor site is rich, endowed with fertile alluvial soils and benefiting not only from the annual doubling of the Tonle Sap at the end of the rainy season, but also from subsurface aquifers and the river-borne waters flowing down from the Kulen. Despite the many natural advantages of the floodplain, or perhaps because of their knowledge of these, the Khmer altered the terrain at every opportunity.

When compared to other contemporary powers such as the Bamar capital of Bagan (9th—13th centuries A.D. Thai capitals of Sukhothai and Ayutthaya (13th to 17 centuries A.D.), Angkor is unrivalled in the potential of its natural

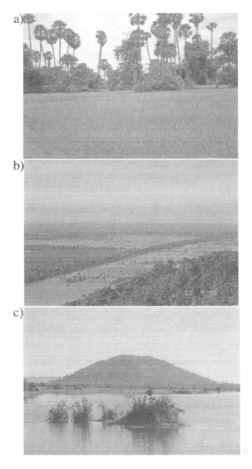

Figure 5. Ground views of vegetation and topography at Angkor. a) Much of the terrain around Angkor is a mix of tree and open areas. Bunded rice fields often flank b) manmade water-control features. Only a few higher features, such as c) Phnom Bok, are located to the northeast.

setting (Figure 4)Like Angkor, these mainland Southeast Asian countries experience long months of alternating aridity and monsoon wind and rain. Unlike Angkor, however, none of these other capitals was built within such an extraordinary ecological context. Bagan is sited within an arid zone, rice having come from areas to the east and south. Sukhothai and Ayutthaya rest within the fertile central plain formed by the Chao Phraya River. But the floodplain where Angkor is found is uniquely active, continually drawing in water and fertility from the lands all around it, and then releasing these to sustain the ancient population.

The Khmer understanding of the need to balance resource exploitation with veneration of the powers inherent in those resources saw earlier expression in the circular earthworks and moats constructed around small remnant mounds. These offered the chance to define spiritually purified spaces, defensively secure and agriculturally enhanced. These words may seem a complex theoretical knowledge with which to credit the builders of simple circular earthworks. However, the repeated reconfiguration and rebuilding of the terrain at Angkor was anything but simple. As briefly illustrated below, these traits are equally applicable to early walled sites in other areas of mainland Southeast Asia.

Variations of circular earthwork-moat constructions range from the lateritic sites of "Suvannabhumi" in Lower Myanmar (Burma) to the Khmer areas of the Mun River Basin in Lower Northeast Thailand. The Mun Basin lies less than 200 km north of Angkor (Figure 4). Like the Siem Reap flood-plain, the Mun basin has a uniform relief and an extended dry season. River and aeolian action over time created terrace remnants, which, combined with occupational debris, form slightly raised mounds favoured for habitation. These mounds, generally circular in shape, are surrounded in periods of flood by water, which was then retained around the perimeter during the dry season. Many of these naturally moated sites are sited along rivers offering year-round water supply (Moore, 1989). Over time, these features were enhanced with the digging and construction of additional earthworks and moats. This elabo-ration of the moated form corresponded to an expansion of settlement onto the terrace zones, which was linked possibly to utilization of local ore sources for the manufacture of iron in the late first millennium B.C. and the early first millennium A.D. (Moore, 1988; Higham, 2003).

In addition to these technological links, there are traces of similar beliefs connected to the land and its waters shared by the Khmer peoples of northern Cambodia and those of Lower Northeast Thailand. For example, traditional accounts from the moated site of Ban Takhong, Buriram Province, link the digging of the moat to the spirit of the land in the guise of a serpent or naga (Sorajet, 1985). In the course of our radar studies, all the circular sites surveyed yielded stone tools and had spirit posts planted in the center of the mound. This practice, discussed further later in this paper, is unique to this area of Cambodia, with its significance akin to the beliefs expressed in Northeast Thailand.

Given these similarities, the remains of the more than two hundred moated sites located in the Mun Basin provided a morphological and percep-tional model for our investigation of circular sites in the region of Angkor. The Angkor sites included sixty-six identified on the basis of maps and various types of images for the area south of the large rectangular tank of Nôkôr Phéas (13.36°N 103.43°E). The 1996 AIRSAR mission provided further coverage in the most promising region northwest of Angkor, the northern "Mokak" run beginning at Nôkôr Phéas and ending near the circular mound of Prey Phdau (14.08°N 103.48°E) close to the Cambodia-Thailand border (Moore, 1994, 2000). A close relationship between the circular sites at Angkor and those of Northeast Thailand also reinforces an understanding of the historical period of Angkor as a city of the north with minimal antecedents in the cultures of southern Cambodia.

In a wider context, the dispersal of enclosed sites on the mainland of Southeast Asia roughly parallels areas where silver coins have been found, from Vesali in the west to Oc-eo (Funan) on the east (Figure 4). This distribution, from the Rakhine coast of western Myanmar to the southern part of Vietnam, has been interpreted as evidence of early trade routes where centers of production shifted over time (San Shwe, 2002). Like silver coins, the enclosed sites gradually disappear in political terms by around the 8th century A.D. with the rise of more hierarchical kingdoms (Wicks, 1992:139). The laterite sites of Myanmar's Mon State occupy a key area within this early network. As with the Khmer sites, these are remnant mounds surrounded by multiple earthworks and moats in an area of lateritic soils. Several of the Mon State sites have multiple walls, their curvilinear forms mirroring the local topography. Traditional accounts of the Mon State also include serpents, with one of those responsible for the bringing of Buddhist teachings being descended from the union of a hermit and a female serpent or *nagini* (Moore and San Win 2007). However, rather than being in an inland area as Northeast Thailand and Angkor are, the Mon State sites are on the coast.

Thus the inhabitants of the Mon State sites probably benefited from both maritime and overland exchange networks, including trade with a further group of enclosed sites associated with the Pyu of Upper Myanmar. The Northeast Thai sites also had a network of relations, although overland routes, with both Mon-Khmer sites in Central Thailand and those of northern Cambodia. The interaction was not exclusively centered on trade, however, and notably along the peninsular coast of Myanmar, the early use of fortified enclosure seems to correspond with increasing incorporation of a variety of Buddhist and Brahmanic norms.

With some exceptions, such as Thaton in the Mon State and Phimai in Nakhon Ratchasima Province, the walled sites of Lower Myanmar and Northeast Thailand were not transformed into regional capitals. This pattern does appear to have been the case at Angkor. Here, habitation remained constant, with a succession of cities spanning more than four hundred years. This continued construction has virtually extinguished the traces of earlier land transformation within the central region of Angkor. Only one earlier site in the heart of the urban area is included in this paper, the small circular mound of Kapilapura located northeast of the 12th-century temple of Angkor Wat. The mound has not been excavated, although its form suggests it was part of the pre-Angkorean circular-site culture, and temples on the mound are associated with the 10th-century city of Yasodharapura. The Kapilapura mound encapsulates important questions about earlier occupation of central Angkor. These are discussed near the end of this paper, preceded by descriptions of the radar imaging data acquisitions and our use of the data to interpret the circular sites northwest of Angkor.

2. RADAR DATA ACQUISITION

2.1. 1994 Spaceborne Imaging Radar (SIR-C/X-SAR)

The first radar imaging data of Angkor was acquired on September 30, 1994, during the 15th orbit of the space shuttle Endeavour by the Spaceborne Imaging Radar-C/X-band Synthetic Aperture Radar (SIR-C/X-SAR). As radar data may be unfamiliar to many readers in the archaeological community a brief overview of radar image characteristics is provided here (See Henderson (1998) for a more detailed discussion of imaging radars).

Radar is an active sensor, sending out pulses from an antenna and digitally recording the backscatter or energy reflected back (Jet Propulsion Laboratory, 1997; Freeman et al., 1998). Imaging data from the various radar sensors come in a variety of resolutions and "flavors" governed by the characteristics of the radar used to acquire the data. The primary radar characteristics that determine the appropriateness of a given set of radar data to a particular application are the wavelength (or equivalently, frequency), polarization and bandwidth. The radar wavelength refers to the distance between amplitude peaks of the transmitted electromagnetic signal. Wavelength is inversely proportional to the wavelength with proportionality constant, the speed of light. Wavelength, or frequency of operation, is often designated by a letter code, for example, a C-band radar refers to a radar operating with a wavelength of approximately 5 cm.

A radar signal impinging on an area of the terrain will in general scatter energy in all directions. The strength of the reflected signal is a function of factors such as the dielectric constant of the soil and the structure of the surface and local surface slope. The amount of energy reflected from the ground back to the radar, called backscatter, has a wavelength dependency since radar waves interact most strongly with objects of similar scale or larger than the wavelength (similar scale means objects greater than about 1/10 a wavelength). Dependence of radar backscatter on the structural arrangement of objects in the scene is characterized by the surface roughness, that is, the variation of the height of objects in a pixel. It is roughness relative to the radar wavelength that in part determines how bright or dark an image pixel appears in a radar image. Smooth surfaces act like mirrors and reflect energy away from the radar and thus appear dark in a radar image whereas rougher surfaces reflect more energy back to the radar and hence appear brighter. Surfaces that appear smooth at one wavelength may appear rough at another, but in general the rougher the surface, the greater the backscatter (Hensley et al., 2001:151).

The other major factor affecting the amount of backscatter is the dielectric constant, an electrical property of a material determined by its chemical composition, which determines the amount of radar energy reflected and absorbed from an object. Of particular importance for the application of radar to Angkor is the fact that the dielectric constant of soil changes depending on the amount of water present. Soils with the same roughness but differing amount of soil moisture have different brightness values in a radar image.

Radar bandwidth determines the resolution of the radar, i.e., the smallest distance between two objects for which their images can be resolved. The higher the resolution the smaller the features that can be identified. Polarimetric radars exploit the fact that the electric and magnetic fields that comprise an electromagnetic wave have an alignment and regularity of these components in the plane perpendicular to the direction of propagation. The amount of backscatter energy returned to the radar depends on the relative orientation of the electric field of the radar wave and of the objects being imaged. This orientation dependency of the returned signal provides additional information that can be exploited when analyzing radar data.

Although there is an infinite family of polarization states, these can all be mathematically synthesized if an appropriate pair of transmit and receive polarizations are used by the radar. The most common pair of polarization states used in polarimetric radars is the so-called horizontal and vertical polarizations denoted by H and V respectively. A radar image obtained from a "B"-band radar with a transmitted polarization of T and received polarization of R is designated as BTR. For example, data from a C-band radar transmitting H polarization and receiving V polarization is designated as CHV data.

The SIR-C/X-SAR radar collected data at L- (23 cm wavelength), C- (5.6 cm wavelength) and X- (3 cm wavelength) bands. The data are polarimetric at L- and C-bands and were posted at a spacing of 25 m. This resolution is roughly equivalent to the ground-projected-resolution after multi-looking or spatial averaging of the raw processed-imagery data to improve interpretability by reducing speckle, a type of image noise that is present in SAR imagery.

Quite unusually in relation to normal planning procedures, Angkor was added to the acquisition roster after the Endeavour was launched, thanks to the efforts of the World Monuments Fund (WMF) in contacting NASA's Jet Propulsion Laboratory. Co-ordinates for Angkor were supplied by the Royal Angkor Foundation (RAF, Budapest), and a joint WMF-RAF workshop was convened in Princeton in early 1995. From the huge amount of data acquired and processed in the 1994 mission, one image has rapidly become the "classic" view of the site of Angkor (Image P-45156, Jet Propulsion Laboratory, 2004) (Figure 1).

On the classic SIR-C/X-SAR image, the cradle-like position of Angkor between the Kulen massif and the Tonle Sap or 'Great Lake' is well displayed. The image, covering some 55 kilometers by 85 kilometers, is centred on 13.43 degrees north latitude and 103.9 degrees east longitude (Jet Propulsion

Laboratory, 2004). Data inserted into each channel were processed repeatedly to enhance the depiction of the terrain and to make full use of the polarimetric capabilities of the SIR-C instrumentation. To produce the Red-Green-Blue (RGB) image, data acquired from C-band pulses horizontally transmitted and vertically received (CHV) were displayed in the blue channel. L-Band was used in both the green and red channels, horizontally sent and vertically received pulses (LHV) displayed in green and horizontally sent and received (LHH) in the red channel.

As there is little or no return from smooth water surfaces, the Tonle Sap, remains of the large rectangular reservoirs or baray, smaller rectangular tanks, and rectilinear moats appear as black on the image. The roughness of the Kulen massif means that the green of the HV data dominates this area on the upper right of the image. Also well displayed are the rivers coming off the Kulen, some going south towards the ancient city and others stretching westwards towards the present town of Puok and the northward-bound distribution of circular moated sites such as Phum Reul and Lovea (13.28°N 103.40°E). The brightness of the broad U-shaped water-management structure midway between Reul and Lovea adds to the picture of methods of curvilinear water management in the Puok area. (Figures 6, 7.)

The Phum Reul-Lovea U-shaped structure also displays well in a scattering model developed by Freeman and used by the authors to detect curvilinear patterns around circular mounds in the Angkor region. This model classifies polarimetric radar observations in relation to three scattering mechanisms or three types of behaviour of the radar signal when it hits the terrain (Freeman and Durden, 1992; Norikane and Freeman, 1993). The three scattering mechanisms are double-bounce scatter from a pair of orthogonal surfaces with differing dielectric constants, volume or canopy scatter from a cloud of randomly oriented dipoles, and Bragg, odd, or surface scatter from a moderately rough surface. Freeman had previously applied the three-component model in various terrains including an arid semi-desert site in Wyoming, a boreal forest site in Alaska, and a land-classification study of a tropical rain forest area in Belize. However, prior to its application at Angkor, it had not been used at an archaeological site or within Southeast Asia (Moore and Freeman,1998).

Our use of the model to study curvilinear features at eleven circular sites northwest of Angkor yielded results similar to the earlier land-classification study in Belize. At Angkor, classification of land cover is essential in detecting settlement preferences and land alterations of both the prehistoric and later Angkorean phases of the site. Decomposition of the radar signatures into different scattering mechanisms, for instance, allowed raised dikes and mounds to be distinguished from zones prone to flooding. This procedure proved to be an effective tool to detect curvilinear patterns—remnants of earlier moats and dikes—at the circular sites to the northwest of Angkor. Repeated ground survey of the zone around Phum Reul (13.33°N 103.42°E),

Figure 6. SIR-C scattering model of features west of Angkor. The area west and north of Angkor has a number of curvilinear water-management features, classified in this image into three scattering mechanisms and displayed using data from the L-band data of the 1994 SIR-C acquisition. Near the lower left corner is the prehistoric circular site of Lovea encircled by several manmade earthworks and moats. A U-shaped water-management structure to the right of Lovea lies midway along a linear pattern of scatter linking it to a second prehistoric circular site, Phum Reul. The rounded form of the floodplain fills the upper right area of the image and highlights the location of the prehistoric features just off the edge in areas with greater moisture. Because of the lack of scatter off more inundated areas such as these, the areas around the two prehistoric mounds and the U-shaped feature are black.

for example, confirmed the modelling results. This mound, some 250 m in diameter, is on the northeast tip of an eroded upland area. It appears to have been only partially enclosed by moats and earthworks, specifically on the downslope southwest portion of the perimeter, where vestiges of two earthworks remain.

2.2. 1996 Airborne Synthetic Aperture Radar (AIRSAR)

In 1996, the Airborne Synthetic Aperture Radar (AIRSAR) flew three 10 km × 60 km flight lines over the Angkor region (Jet Propulsion Laboratory, 1998). Two runs were over central Angkor. The third, called the "Mokak run" after

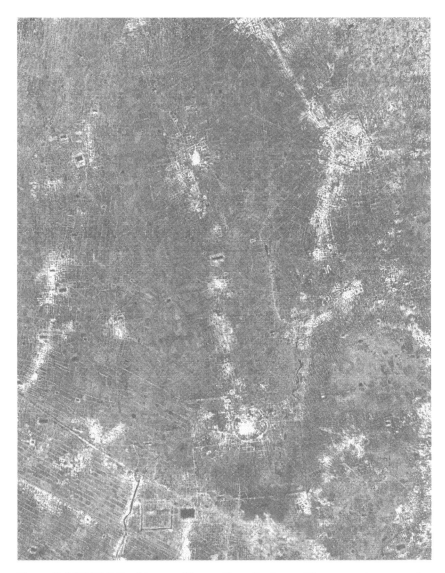

Figure 7. AIRSAR P-Band image of Lovea-Phum Reul area. The contrasting vegetation patterns associated with archaeological features west and north of Angkor are seen in this P-Band image generated from the 1996 AIRSAR acquisition. The U-shaped feature between the circular mounds of Phum Reul and Lovea is bright. While Phum Reul and Lovea are circular, later Angkor-period rectangular structures can be seen to the left of Lovea and south of the road running diagonally across the lower left portion of the image. The continued use of the area south of the road is also apparent in the series of lines paralleling the road that mark the bunded margins of rice fields. (Detail in Figure 11).

a circular site in this area, extended to the northwest toward the Thai border. As mentioned earlier, the northwest flight line was included to study the pattern of circular and rectilinear alteration to the terrain in hinterland areas outside Angkor. For this portion of the data, inaccessibility meant that maps were less accurate or, for archaeological features, non-existent (Figure 8). The 1996 AIRSAR data was collected at three wavelengths: C- (5.6 cm), L- (24 cm) and P-Band (68 cm). The NASA Jet Propulsion Laboratory's TOPSAR (Topographic Synthetic Aperture) system uses two spatially separated antennas in the plane perpendicular to the flight direction that are flush mounted on the JPL/NASA DC-8 aircraft. This augmented the conventional synthetic aperture radar (SAR) system enabling the simultaneous

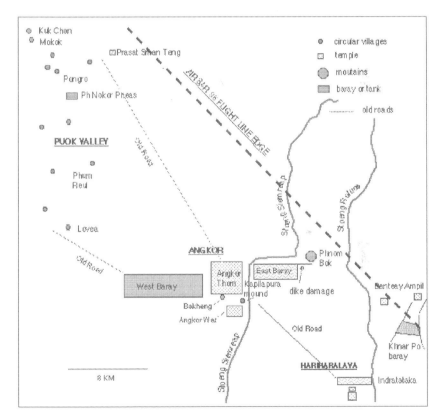

Figure 8. Sketch map of Mokak area northwest of Angkor. The Mokak leg of the 1996 AIRSAR acquisition includes a distribution of circular mounds such as Lovea, Phum Reul, Pongro, and Kuk Chan. The rectangular tank of Nokor Pheas and the previously unexplored temple of Sman Teng are also part of this area, one poorly documented in earlier archaeological literature.

extraction of topographic information (Hensley et al., 2001:153). Using the Angkor data acquisition, Hensley generated a 1500-square-kilometer mosaic including land close to the Cambodia-Thailand border to areas far to the east of Angkor.

Processing of the radar data proceeds in two basic steps. Radar data collected onboard an aircraft or spacecraft needs extensive signal processing before a recognizable image is obtained. This signal processing is first done for each of the flight paths (or strips) and includes the simultaneous generation of an elevation map for interferometric radars. These individual strip maps are then mosaicked to form a seamless image and elevation map. The C-band image and elevation data were posted at 5-m resolution and the polarimetric data were posted at 10-m resolution. Interpretation of the resulting imagery used an image-display program developed at JPL (by Scott Shaffer and available from JPL) that allows the user to display the data in a variety of formats designed to aid in the analysis of the data

The features of the display software that were most often employed in the analysis of the data were grey-scale images of the radar backscatter data where it was possible to scale the brightness values to enhance desired features, elevation data displayed as color contours with varying height scales (i.e., the amount of elevation change between colors was adjustable), a combined image where the color contours are overlain on the radar backscatter, and finally shaded relief images in which the elevation data were mathematically illuminated from a given direction and converted to a grey-scale image. This facility to transform the image data into various visual representations was central to the analysis conducted on the radar data and was the primary means of doing the analysis.

Another program developed by JPL/NASA, Sigma0, was particularly designed to display the raw radar-data files (in .stk format). This is publicly available on the JPL/NASA SIR-C Educational CD-ROM (SIR-CED CD-ROM). The authors also used various commercial software packages including Photoshop and other imaging programs.

One site covered about two-thirds of the way northward in the Mokak run, is the small (180 m in diameter) mound of Phum Pongro (13.37°N 103.43°E), where the P-Band data generated useful images of the circular features at this site. Here, the ratio between the P-Band HH and VV backscatter has helped to distinguish between what appear to be old moats from drier earthworks. When the HHVV ratio was placed in a three-image composite, along with PHH and PVV, curvilinear vestiges of eroded earthworks were well detailed. The vestiges on the southwest can be seen on the CVV amplitude and height images. However the PHH and to a lesser extent the PVV brings up additional segments on the north and northwest. These are further enhanced when put into a color composite, with the pink of the HH return in the red channel contrasting the blue of the HHVV ratio in the blue channel (Moore et al., 1999, 1998) (Figure 9).

Figure 9. Image of Phum Pongro area generated using P-Band ratios. Three circular mounds form a triangle just south of a rectangular water tank on this image generated from P-Band data acquired during the 1996 AIRSAR acquisition. The mound on the lower right is Phum Pongro. When the ratio of different wavelength combinations was placed in red, green, and blue channels, the resulting image contrasting the degree of soil moisture highlighted curvilinear vegetation patterns not visible on optical images. In several cases, such as Phum Pongro, vestiges of multiple circular earthworks and moats are apparent, linking this more remote part of the distribution of circular mounds to the more accessible prehistoric circular mounds to the south.

2.3. Radar Imaging and Archaeological Features

As noted, the 1994 SIR-C and 1996 AIRSAR data acquisitions were the first use of imaging radar at Angkor and also the first archaeological application of imaging radar within a non-arid environment. Imaging radar, including AIRSAR, was previously used archaeologically in the Peten of Guatemala. Previous studies, such as that of linear features associated with the Great Wall of China and the "lost city" of Ubar in Oman, or climatic changes affecting habitation such as the Saf Saf Oasis near the Egypt/Sudan border had demonstrated the utility of imaging radar in archaeology. In the Saf Saf case, archaeologists had recorded an unusual number of stone hand axes within an area of current desert. Using data from the earlier SIR-A and SIR-B, and then the SIR-C acquisitions, patterns of former braided rivers channels were detected in desiccated subsurface water channels lying beneath the wind-blown sands covering the land today (Jet Propulsion Laboratory, 2003).

Angkor offered a unique opportunity to test the applicability of imaging radar at an archaeological site with virtually the opposite characteristics from

Saf Saf: temples located in a forested area of high soil moisture within a tropical monsoon region. However, despite different types of vegetation patterns, in both cases radar was used to assess prehistoric interaction of man with the environment. By far the greatest potential of radar imaging at Angkor is in the discrimination of surface variations resulting from the sensitivity of radar imaging to moisture and vegetation (Moore and Freeman, 1998).

Equally important, our comparison of curvilinear and rectilinear forms made good use of both the polarimetric and interferometric capabilities of the radar imaging data. For example, it was the height data that clarified the shape of the Kapilapura mound. In contrast, as described above, in the Mokak area some 25 km northwest of Angkor, the earthworks around Phum Pongro were best displayed in P-band (68 cm) images.

2.4. Other Sources of Remote Sensing

Many types of remote sensing and maps provided useful comparison to our use of the radar imaging data. British cover from the World War II Williams-Hunt Collection (SOAS, University of London) and post-war French aerial photographs from the 1950s proved particularly valuable, given the post-World War II development of the region. SPOT imagery often highlighted features in a complementary way to the radar imaging data, and Finnish aerial cover from the early 1990s provided a more recent view of the terrain. Cartographic sources included American and Vietnamese 1:50,000 maps from the 1970s and 1980s. These were invaluable, although they become increasingly poor in the Mokak portion of the study area, from Nôkôr Phéas (13.36°N 103.43°E) to Prey Phdau (14.08°N 103.48°E) near the border between Cambodia and Thailand.

The notes and maps of Aymonier (1901) and Lajonquière (1911) were another source of valuable information. Often these sources contradicted each other, however, and lacked the accuracy of the radar imaging data. For instance, several French scholars, including Trouvé and Marchal in the 1930s and Stern in the 1960s made note of the Kapilapura mound in the area north of Angkor Wat, but none gave its accurate location as balanced on the corner of the outer moat (Pottier, 1993; Stern, 1965). Northwest of Angkor, in trying to locate earlier references to our identification of the 11th-century-A.D. rectangular enclosure of what is locally known as the "Sman Teng" (a type of rice) temple, we discovered that the works of Lajonquière dealing with this part of Angkor showed that in the majority of cases he had to rely on villager reports of a temple, and had not been able to access the area himself (Lajonquière, 1911). Sman Teng (13.37°N 103.46°E) is oriented to the northeast rather than the east following the local hydrology—useful knowledge for understanding both the rectilinear and curvilinear features in this area.

3. CIRCULAR SITES AT ANGKOR

3.1. The Mokak Run of the AIRSAR Cover

As explained above, imaging of the northern reaches arose from a hypothesis that Khmer curvilinear water management preceded the construction of rectilinear forms. In this context, curvilinear features are generally moats and earthworks associated with circular mounds while rectilinear features include temples, tanks, and dikes attributed to the Hindu-Buddhist period of the 9th–13th centuries A.D. This is a heuristic model not a rough substitute for a chronological classification arising from excavation of a range of the circular sites, many of which have proved inaccessible to date. Indeed, even in the case of the rectilinear features at Angkor, dating is commonly based on a single inscription. Such dating cannot accommodate the repeated repair and reconstruction of temples, let alone likely occupation and veneration practices in a prehistoric context.

Previous study of the circular sites at Angkor divided the distribution into three geographical zones: the Roluos River east of Angkor, sites of the central Angkor area of Siem Reap, and the Puok valley northwest of the site (Moore, 1998). While the more southerly circular sites such as Phum Reul (13.38°N) and Lovea (13.28°N) were included in the SIR-C/X-SAR data acquisition, the circular-site study prompted the addition of the 10-km by 60-km northern, or Mokak run to the 1996 AIRSAR mission. The sites of this Puok-Mokak northwest area have been the focus of our further study. Not all of the sixteen distinct sites in this group have been accessible on the ground, with the most northern pair checked jointly by the authors in 1998 being the mound (200 m in diameter) of Kuk Chan (Sala Kum Ta Saom, 13.39°N 103.41°E) and the mound (150 m in diameter) located 500 m east of Phum Mokak (13.39°N 103.42°E) (Figures 10a, b).

The Mokak run of the 1500-sq-km AIRSAR mosaic rises in elevation from 20 to 222 m, its northern end along the Stung Srêng midway between Samrong and Anlong Veng (14.07°N 103.48°E). This northernmost zone has no temples recorded on maps or in inventories, but there are both rectilinear and curvilinear features. Thus the Mokak run provided unique cover of a little-known region of northern Cambodia. Many vestiges of circular earthworks remain in this area. In the central Angkor area, however, as illustrated with the Kapilapura case below, rectilinear constructions have obscured interpretation of the circular features.

3.2. Circular Sites and Curvilinear Patterns of Water Management

The circular mounds average 150–250 m in diameter and rise 4–5 m above the surrounding terrain. The most well known is Lovea (13.28°N 103.40°E), located some 10 km to the northwest of central Angkor. The Lovea mound

Figure 10. Two views of the moat at the village of Kuk Chan, northwest Cambodia. Ground-truthing of the 1996 AIRSAR Mokak run by the authors extended northwest of Angkor to Kuk Chan, where the moat encircles much of the perimeter of the small village. Today, as was probably the case in prehistoric times, the water-management features have greatly enhanced the local ecology. They are used for plant cultivation and also attract a range of domestic animals and a range of wildlife from birds to small mammals.

is encircled by two or three earthworks and moats, each some 15–20 m wide. The circular mound is north of a rectangular reservoir (2 km × 2 km) called Kuk Chan and an Angkor-period temple, Banteay Sras, with a 600-m by 600-m temple enclosure surrounded by a 100-m-wide moat. Lovea, as with all the circular mounds, has not been excavated: stone tools, bronze, and bones have been either found or reported from the site. It is on the edge of the old alluvial fan coming off the Kulen, with an altitude of 18 m. Lovea was thus more at risk of flooding from the Tonle Sap than the circular sites to the north where the altitude is 20–28 m (Moore, 2000, 1989; Malleret 1959) (Figure 11).

Additional mounds are distributed along the edges of the alluvial fan coming off the western portion of the Kulen massif. Like Lovea, some of these, such as Kuk Chan and Phum Mokak (13.38–39°N 103.41°E) and Phum Pongro, have remnants of circular earthworks around the habitation mound (Figure 12). The lack of archaeological investigation in this region is highlighted by noting that Aymonier made the most useful reference to Kuk

Figure 11. P-Band image of Lovea and Banteay Sras, west of Angkor (see also Figure 7). The circular earthwork and moat, as well as the elevated central mound of Lovea, appear as bright patterns on this image generated from P-Band data acquired during the 1996 AIRSAR acquisition. South (left) of Lovea is the rectangular enclosure of the later temple of Banteay Sras. Both are aligned to optimize water control on the slight northeast to southwest slope of the Angkor floodplain.

Chan ninety-seven years before our 1998 ground check. Aymonier referred to a pond 500 m long cited on a sandstone boundary pillar at the site inscribed with two words translated as the "pond of leeches" (1901:378). Our survey showed this to be a remnant of the circular moat around the Kuk Chan village

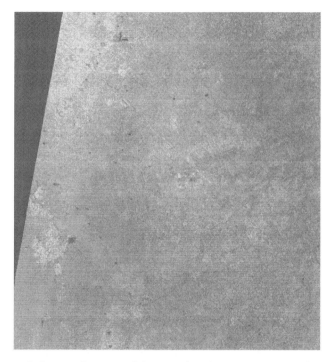

Figure 12. Mokak or northern run of the 1996 AIRSAR acquisition over Angkor. Color intervals have been draped over a height and brightness image so that the slightly elevated Kuk Chan and Phum Mokak mounds appear as blue in the surrounding pink of the lower rice fields.

mound. As noted earlier, vestigial earthworks around a site farther to the north called Phum Pongro showed up well when displayed with P-Band data. This is attributed to sparse woody vegetation, perhaps of the trunks of trees (Figure 11).

3.3. Stone Tools and Spirit Posts

As noted earlier, two somewhat unexpected results of our study were that all the circular sites had stone tools and centralized spirit posts. In the course of ground verification, a pattern emerged: all the circular villages had *phraa phum* posts, the "navel" in the center of the village. These were generally made of wood, and often enclosed within a shrine. At Phum Reul, the wooden posts had been replaced with concrete ones, painted with several bright blue horizontal rings (Figure 13a). In most cases, there were five posts under a meter in height, with four points formed around the center (Figure 13b).

This tradition is separate from the placement of a *néak tâ* or ancestral shrine on the northeast of a village, and to date appears to be distinct from other areas

Figure 13. Spirit posts at circular mounds northwest of Angkor. Near the center of the circular mounds northwest of Angkor are five 'navel' posts or *phraa phum*. The posts provided an unforeseen correlation of site form and animist veneration. a) Concrete posts at Phum Reul. a) Most sites had wooden posts, as shown in this photo.

of Cambodia. While the use of the post and the configuration of five has clear links to the pillar of the stupa and the Hindu cosmology elaboration centered on Mt. Meru, caution should be given in understanding the Khmer *phraa phum* post as merely a simplified version of these (Ang Chouléan, 1997). There is in addition a great difference in the post planted in and marking the earth itself and the funerary-pavilion aspect of the stupa pointing skyward. The *phraa phum* appear to represent pre-anthropomorphic animist notions of the earth. There is also the possibility that the posts may, at their time of origin, have also been associated with burial. In this context, it is worth recalling that the Indian model probably followed the same evolution (Ang Chouléan, 1997).

In addition to posts, villagers at all the sites northwest of Angkor surveyed to date had stone tools, gathered during field cultivation, and kept for their magical properties. During a site survey in 1992 by Moore, fifteen stone tools were recovered. Sherds were also found, including thick white and medium-orange pottery, and from Lovea, thin black sherds. The grey-color stone tools were grouped into four types: elongated rounded points with triangular section, bifacially chipped rounded pebble-type tools, bifacial and possibly unifacial pointed adze/axes, and rounded adze/axes. The tools ranged in length from 3.9 cm to 10.6 cm, with most being about 8 cm long and 4–5 cm wide (Moore, 1993).

Stone tools were also recovered at a mound on the northeast edge of the Nokor Phéas tank during a 1998 ground check. Most of the finds had been brought up by ploughing, reinforcing the hypothesis that the prehistoric habitation layer, if it exists in this area, will be found under a thick layer of sedimentary deposits. As in other villages, such as Phum Reul, the tool was referred to as "*krop runteas*" or "debris from lightning." Villagers reported that leaving the tool in water gave the water medicinal properties. At a house in the village, a woman showed us another similar tool recovered recently by her son in the fields nearby. The tool was kept in the rafters of her house, although she cited no specific reason for keeping it high up rather than in a basket or on the ground.

3.4. The Kapilapura Mound and Angkor in the 10th Century A.D.

Although continued construction at Angkor has obscured evidence of prehistoric habitation, there has been brief but persistent mention of prehistoric finds in exploration of the site. Chance discoveries of the 19th century first documented lithic industries with bronze and ceramic finds, at Samrong Sen and Anlong Pdau near the southeast highwater bank of the Tonle Sap (APSARA, 1996). In the Angkor area, prehistoric finds have been made under the temple sites of Ak Yum, Chau Say Tevoda, and at Trapeang Phong south of Hariharâlaya. Most of these finds were made by Groslier, who also unearthed evidence of earlier habitation during reconstruction of Baksei Chamkrong at the foot of Phnom Bakheng, at the center of Angkor thought to have been, in the mid-10 century A.D., the capital city Yasodharapura (Moore, 1993).

As these finds suggest, interpreting any possible prehistoric feature at Angkor is complicated by the successive constructions found at most temples. This proved to be the case in assessing the small circular mound of Kapilapura located on the northeast edge of the mid-12th-century temple of Angkor Wat. For Kapilapura, some explanation of the extent of hydraulic intervention in this part of Angkor is necessary. Before assessing the mound, therefore, two features north of Kapilapura are described: the Stung ("river") Siem Reap, the principal water course flowing through Angkor, and the "Old Khmer Road", a circa 15 kilometer dike running from the 9th-century capital of Harihâlaya to its 10-century successor, Yasodharapura.

The move of the capital in the 10th century A.D. has never been satisfactorily explained, with most accounts linked to locational limitations. Harihâlaya was close to the Tonle Sap, which may have limited acreage for rice cultivation, or have meant the site was too inundated to control. In contrast, however, the location of Yasodharapura is not discussed primarily in relation to water, but in relation to the state temple erected on the summit (99 m high) of the Phnom ("hill") Bakheng, in the center of the new city.

The founding of the new capital also prompted the construction of the East Baray (7120 m × 1700 m), its four corners marked by dedicatory inscriptions in Sanskrit (Moore, 1997:94). The East Baray is oriented east to west in order to make use of the northeast to southwest land slope. Its construction involved a major change to the flow of water through the heart of city, and the area where the Kapilapura mound is located. The construction of the East Baray effectively controlled land within both the Stung Siem Reap riverbed and that of the Stung Rolûos, which had supplied water to the earlier city at Harihâlaya.

In addition, the East Baray is sited just south of a column of ponds that are particularly visible on the classic SIR-C image of Angkor. Images generated to compare the C-band and L-band return gave evidence of water between the ponds, suggesting that these may be former streams and their alignment in fact parallels that of present-day streams. These remnant watercourses, quantifiable in relation to the radar return and visible as slightly darkened patches, are not apparent on aerial photographs and maps, where the ponds appear as densely distributed but separate bodies of water. Ground verification in this area confirmed the interpretation formed from the radar imaging data.

The Siem Reap River flowing through central Angkor is one canalized many centuries ago, visible today in a barrage at the village of Bam Penh Reach north of the East Baray. Groslier, following excavations at the temple of Thommanom and study of aerial photographs, raised the question of old river beds in the Angkor area. For instance, he speculated that the Royal Terraces, now in the middle of the late 12th-century city of Angkor Thom, may have been on the banks of an earlier basin or canal. More recent work by Gaucher around the Phimeanakas temple within Angkor Thom, has also yielded a curvilinear pattern of river sediments, which forms part of his ongoing research at the site (Moore, 1997:94).

The magnitude of the 10th-century East Baray construction, however, does not mean that its operation and effect can be understood in an equally singular fashion (Moore, 1997:97). A notable example of such multiplicity may be seen in a linear dike connecting the reservoir at Hariharâlaya with Angkor. The dike, called the "Old Khmer Road" in our radar studies, begins on the northwest corner of the Indratataka Baray (3200 m × 750 m).

The dike runs for some 15 km across rice fields and past temples into the central area of Angkor. Whether, suggested, this massive dike ended at the foot of the 10-century temple on the top of Phnom Bakheng or continued east to the Phimeanakas temple within Angkor Thom, it would have functioned both as a road and as a water-control structure in the 10th-century system of water management at Angkor (Groslier, 1979; Moore, 1997; Moore et al., 1999).

As with all water-management structures at Angkor, the East Baray exploited natural resources and irrevocably altered the terrain. The additional detail provided by the radar data clarified not only how the East Baray was sited, but also how the landscape changed over time. This clarification in turn influenced our interpretation of the earlier hydrology of the region and the possibility of its prehistoric habitation. In particular, it offered a means to access the layout of Angkor not as it is seen today, and not couched in this intricate framework of kings, capitals, roads, and temples, but in the context of the spatial arrangement of apparently manmade and natural features at the site. Our study of Kapilapura illustrates the way that the radar perspective allowed us to explore the question of occupational continuity in the heart of the ancient capital.

3.5. The Kapilapura Mound and Angkor Wat

The temple of Angkor Wat was built on the downslope side of the Kapilapura mound, a relationship that shows well on the radar topography (Figure 14). Our investigation of Kapilapura, in fact, began because of this visibility, noted by Freeman, who at that time had never visited Angkor. As mentioned earlier, reference to Kapilapura was included in some earlier French maps, but it was not accurately sited. The mound is small but elevated, ca. 175 m in diameter and rising at least 5 m above the surrounding terrain. What our subsequent investigations revealed was that the architects of Angkor Wat seemingly knew and respected the Kapilapura mound. This knowledge is noteworthy, for the clarity of its form is not visible at ground level or on optical imagery. While known in the 12th century, to reveal this today has required the sophistication of spaceborne and airborne radar imaging technology (Figures 14a, b).

Ground checking of Kapilapura resulted in our recording more temples than the three noted by Finot (1925), with architectural fragments such as a lintel and colonette provisionally attributed to early Angkorean styles. Different phases of habitation are also suggested by a stretch of open land and remnants of an outer earthwork curving around the mound on the north and northwest. This terrain profile reinforced Kapilapura's similarity to other circular mounds with remnant moats and earthworks both northwest of Angkor and in Northeast Thailand (Moore, 1992).

Of the four Sanskrit inscriptions recorded by Ayomnier (1904) and Finot 1925, only one bore a date of A.D. 968, with the others dated on rather

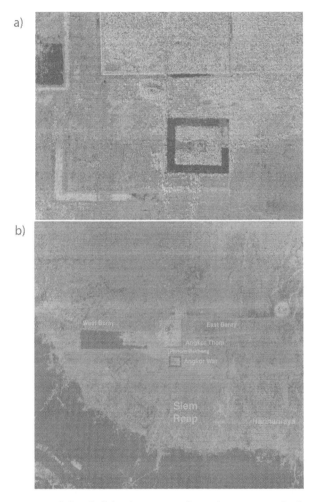

Figure 14. Mosaic and detail of Angkor Wat and Kapilapura mound. The wide moat surrounding the 12th-century temple of Angkor Wat is clearly seen on both b) the image of the central Angkor zone and a) the detail of the temple site. On the northeast corner of the Angkor Wat moat is the small circular mound of Kapilapura.

vague palaeographic grounds to the 14th century. One temple was oriented to the east with an inscription dedicated to Vishnu, with another dedication being made to Shiva by a guru of the king. The latter was inscribed on a large stone stele, and commemorates the digging of several ponds and reservoirs, the installation of a statue of a deity on an island at a sacred site along a lake or river, and further donations to the East Baray. However, if the palaeographic dating is accepted, this is a period when the East Baray was rendered non-functional by silt deposition and only the 8,000-m-long West Baray is thought to have been in operation.

The references in the epigraphic fragments, the east-facing temple dedicated to Vishnu at the same locale as the west-oriented state temple of Angkor Wat, and other epigraphy citing Shiva, recall not only the importance of place and continued veneration of a deity at a particular spot, but the associations of the site with several Hindu deities. Inscriptions elsewhere state that a new name may be given to a site and also that there could be multiple religious foundations at one point corresponding to local deities (Ang Chouléan, 1990:142). As argued above, Brahmanic veneration stemmed from pre-Indic ritual associated with locale. At Kapilapura this hypothesis is reinforced by the possibly pre-Angkorean stone pieces and further by the morphological links to much earlier sites.

The water-management structures reinforced man's demarcation of the inhabited zone and his efforts to control the "other," the forested areas. Concepts of man-village, the settled human and forested domains, are as significant a part of interpreting terrain alteration as are the more visible hydraulic works. Without village establishment, there is no need for communally organised water management to meet ritual and agricultural needs (Ang Chouléan, 1986, 1990). The clearest examples of these needs are seen in the circular sites northwest of Angkor described above. As noted several times, similar circular walled sites are abundant in Northeast Thailand where their development has been linked to the inception of localized iron production somewhere between the late first millennium B.C. and the early first millennium A.D. (Moore, 1988, 1992; Higham, 2003).

4. CONCLUSION

Earlier studies of Angkor often suggested that the pattern of manmade intervention was quite different during the late prehistoric and historical periods. Radar imaging data have contributed a new breadth and type of data to use in interpreting the relationship between human habitats and place in general, and utilisation of water resources in particular. The man-landscape relationship at Angkor cannot be reduced to a simple series of interventions marked by different dynasties, religious affiliation, and monument locale. It involved deeper ties to land and place expressed in the ancient traces demarcating areas of collective human habitation. Certainly these were transformed by Hindu and Buddhist norms, but despite the great variations in the form and extent of alterations, the water-management principles applied by the Khmer remained consistent: the detection of slight land slope combined with knowledge of localized hydraulic cycles.

Intimately known patterns to guide construction included above-ground water sources, pooling and flooding during the May-September rainy season, river and stream flow, and residual soil moisture indicating subsurface water during the dry months of October-March. The keen Khmer understanding of factors such as these is significant in understanding their ancient water-management expertise. It reflects not only great knowledge of the landscape but also a sense of time tied to the cycle of monsoon winds and rain. It demonstrates understanding of place and human agency linked to water sources and land contours, from lows where water might accumulate or be stored to high spots safe from inundation and perhaps attack.

Use of the localized Khmer principles of water-management spanned the transition from the earlier village-based cultures of the moated sites to the urban capitals of the 9th–13th centuries A.D. In both phases, these ensured the spiritual continuity and everyday efficacy needed to sustain the growing demands of royal and agricultural prosperity. It is these Khmer impulses to activate and empower tutelary and other natural forces through terrain alteration that motivated the case studies of spaceborne and airborne radar at Angkor summarized here.

As described above, differing fractions of the energy in the electromagnetic waves transmitted by the radar pulses are reflected back to the spaceborne or airborne instrumentation. The return is dependent on many factors, such as the angle of the antennae, the wavelength and polarisation of the pulse, the orientation of objects within the imaged scene, the electrical property on the surface of the observed features, and the polarization of the backscatter of the pulse returned to the radar instrument (Jet Propulsion Laboratory, 1997; Freeman et al., 1998). Analysis of this information was used by the authors to frame a manmade and natural context within which the 9th–13th century Hindu-Buddhist cities of Angkor emerged. Thus interpreting how the Khmer, both before and during the historical period of the ancient site, chose to structure the archaeological landscape seen at Angkor today formed the focus of our use of the SIR-C and AIRSAR data. This paper describes two aspects of this research centered on the investigation of circular sites: one in the center of Angkor and the other a distribution of sites northwest of Angkor.

The first of these, the Kapilapura mound, is strikingly close geographically to Phnom Bakheng, the 99-meter "mountain" at the center of the 10th-century capital of Yasodharapura at Angkor. The one dated inscription from Kapilapura is likewise close chronologically to the foundation of Yasodharapura as the new center of the kingdom, marking its shift from the earlier center of Hariharālaya to the southeast. If indeed there were two 'pura'— Kapilapura and Yasodharapura—so close in time and place, we clearly need to re-think the definition of city in a Khmer context. Rather than one supreme ruler and city, there may have been several rulers in a short period of time, their power repeatedly rising and falling. Previous scholars have made this observation in relation to genealogical succession, kingship, and regional control, but not specifically in relation to the foundation of the city that we call "Angkor." The idea that the kings of Angkor represent a single bloodline has long been dismissed, but the cities at Angkor are still too often viewed as a succession of constructions linked to specific rulers or dignitaries.

Such perceptions ignore changes brought by destruction and alteration of water-management features, and largely ignore the relationship of architectural foundations to competition between different ruling factions. Nor do they link power to place, or understand constructions as settlement markers of that place. If, instead, one works within precepts of continuity, place, and competition, the processes that created and maintained the Khmer village are fundamentally the same as those that created and maintained the cities of Angkor.

In studying the circular sites northwest of Angkor, the use of the three-component model described above allowed us to fit scattering mechanisms to polarimetric radar observations. From the results, a simple interpretation was generated, relevant to our ongoing investigation into the curvilinear water management features. This was possible because of the sensitivity of the polarimetric synthetic aperture radar not only to moisture, but also to all other contributors to the biomass. Many factors make up the present vegetation and hydrology of Angkor, ranging from the siting of temples and reservoirs to the earlier formation of the circular mounds. Despite many years of excellent scholarship, there is no generally accepted geographical or hydrological history of the city. It is commonly agreed that the archaeological landscape was created through a close understanding of elements such as vegetation and moisture. In the development of that landscape, the critical social transition from village to city was to a great degree reflected and perhaps catalyzed by land alterations exploiting water resources.

The basis and the evolution of the technological and conceptual dimensions of this transition have both been part of our radar studies. Thus the present interpretation is in part qualitative, for it concerns the ancient Khmer cognition of their land. We can only piece together an impression of past perceptions from the location of remains and the surviving artifacts of the material culture. The expectation is that in this way we glean some understanding of this culture's spiritual beliefs, in this case beliefs intertwined with the land and its waters. The resulting descriptions are necessarily fragmentary attempts to interpret the remaining traces of these within a coherent worldview.

The aspects summarized in this article, as the above words indicate, are variously theoretical, technical, and archaeological. They use the radar data to interpret the transition from prehistoric habitation to the 9–13th century A.D. Hindu and Buddhist temples of Angkor. As suggested by the temporal span and suggested by the temporal span and varied ritual practices, defining an archaeological context for the temples of Angkor has been a major challenge in the interpretation of the ancient site. The authors' collaborative research addressed this issue using data acquired with the NASA Jet Propulsion Laboratory's Spaceborne Imaging Radar-C/X-band Synthetic Aperture Radar (SIR-C/X-SAR) in 1994 and the Airborne Synthetic Aperture

Radar (AIRSAR) in 1996. The first archaeological use of radar imaging in Southeast Asia, our Angkor studies prompted the interpretation described here, one rooted in the data's quantification but one also growing from the radar's glimpse of time in the structure of the land, a living presence so well understood by the ancient Khmer.

5. ACKNOWLEDGEMENT

The work of the last two authors was performed at the Jet Propulsion Laboratory, California Institute of Technology, through an agreement with the National Aeronautics and Space Administration.

REFERENCES

Ang Chouléan, 1997, Becoming Ancestors or Reaching the Divine World; Representations of Death in Cambodia. Paper delivered at the National Gallery, Art Center for Advanced Study in the Visual Arts. Smithsonian Institution, Washington, D.C., September 27.

Ang Chouléan, 1990, La Communauté Rurale Khmère du point de Vue du Sacré. *Journal Asiatique* 278 (1–2):135–154.

Ang Chouléan, 1986, *Les êtres surnaturels dans la religion populaire khmère*. Cedoreck, Paris.

APSARA (Authority for the Protection and Management of Angkor and the Region of Siem Reap), 1996, *Angkor, a manual for the Past, Present and Future*. UNESCO-United Nations Development Program-Swedish International Development Agency, Phnom Penh.

Aymonier, Etienne, 1904, *Le Groupe d'Angkor et L'Histoire, III*. E. Leroux, Paris.

Finot, Louis, 1925, Inscriptions d'Angkor 8. Kapilapura, Notes d'épigraphie. *Bulletin de l'École Française d'Extreme Orient* (1902–15):365–369.

Freeman, A., and Durden, S., 1992, A three-component scattering model to describe the polarimetric SAR Data. *SPIE 1748, Radar Polarimetry*, pp. 213–224.

Freeman, A., Hensley, S., and Moore, E., 1998, Radar imaging methodologies for archaeology: Angkor, Cambodia. Paper presented at Conference on Remote Sensing in Archaeology, Boston University Department of Archaeology and Center for Remote Sensing, 11–19 April 1998.

Groslier, B., 1979, La Cité Hydraulique Angkorienne: exploitation ou surexploitation du sol? *Bulletin de l'École Française d'Extreme Orient* 66:161–202.

Henderson, Floyd M., 1998, *Principles and Applications of Imaging Radar*, Vol 2. In *Manual of Remote Sensing*. John Wiley and Sons, New York.

Hensley, S., Munjy, R., and Rosen, P., 2001, Interferometric Synthetic Aperture Radar (IFSAR). Chapter 6 in *Digital Elevation Model Technologies and Applications: The DEM Users Manual*. ASPRS.

Higham, C., 2003, *Early Cultures of Mainland Southeast Asia*. River Books, Bangkok.

Jet Propulsion Laboratory, 1997, What is imaging Radar? By Tony Freeman. http://southport.jpl.nasa.gov/.

Jet Propulsion Laboratory, 1998, AIRSAR images and description Angkor, Cambodia. http://www.jpl.nasa.gov/releases/98/angkor98.html.

Jet Propulsion Laboratory, 2003, Images and description of Saf Saf Oasis.http://southport.jpl.nasa.gov/cdrom/sirced03/cdrom/ROADMAP/PICSROOM/AFRICA/SAFSAF.HTM.

Jet Propulsion Laboratory, 2004, SIR–C image and description of Angkor, Cambodia. http://www.jpl.nasa.gov/radar/sircxsar/angkor.html.

Lajonquiére, E. Lunet de, 1911, *Inventaire Descriptif des Monuments du Cambodge* (3). Leroux, Paris.

Malleret, L., 1959, Ouvrages Circulaires en Terre dans l'Indochine Meridionale. *Bulletin de l'École Française d'Extreme Orient* 49:409–453.

Moore, E, 1988, *Moated Sites in Early NorthEast Thailand*. British Archaeological Reports (BAR), International Series no. 400. Oxford.

Moore, E., 1989, Water management in Early Cambodia: Evidence from Aerial Photography. *The Geographical Journal* 155 (2):204–214.

Moore, E., 1992, Water-enclosed Sites: links between Ban Takhong, Northeast Thailand, and Cambodia. In *The Gift of Water, Water Management, Cosmology and the State in South East Asia*, edited by J. Rigg, pp. 26–46. SOAS, University of London.

Moore, E., 1993, Ancient Habitation on the Angkor Plain. Unpublished report submitted to UNESCO, Zoning and Environmental Management Project (ZEMP).

Moore, E., 1997, The East Baray: Khmer water management at Angkor. *Journal of Southeast Asian Architecture* 1:91–98.

Moore, E., 1998, The prehistoric habitation of Angkor. In *Southeast Asian Archaeology 1994, Proceedings of the 5thInternational Conference of the European Association of Southeast Asian Archaeologists, Paris 24–28 October 1994*, Vol. I, pp. 27–36. University of Hull, Centre for Southeast Asian Studies.

Moore, E., 2000, Angkor Water Management, Radar Imaging, and the Emergence of Urban Centres in Northern Cambodia. *The Journal of Sophia Asian Studies* 18:39–51.

Moore, E., and Freeman, A., 1998, Circular sites at Angkor: a radar scattering model. *The Journal of the Siam Society, Bangkok* 85 (Parts 1 & 2, 1997, 1998):107–119.

Moore, E., Freeman, A., and Hensley, S., 1999, *Angkor AIRSAR, water control and conservation at Angkor. NASA/JPL PacRim Significant Results Workshop*. Maui High Performance Computing Center, Hawaii.

Moore, E., Freeman, A., and Hensley, S., 1998, Beyond Angkor: Ancient Habitation in Northwest Cambodia. Paper presented at Conference on Remote Sensing in Archaeology, Boston University Department of Archaeology and Center for Remote Sensing, 11–19 April 1998.

Moore, E., and San Win., 2007, The Gold Coast: *Suvannabhumi?* Lower Myanmar Walled Sites of the First Millennium A.D. *Asian Perspectives* 46(1):202–232.

Moore, E., Stott, P., and Suriyavudh, S., 1995, *Ancient cities of Thailand*. River Books, Bangkok; Thames & Hudson, London (1996).

Norikane, L., and Freeman, A., 1993, *User's Guide to MacSigma*. Jet Propulsion Laboratory, Pasadena.

Pottier, C., ed., 1993, *Documents Topographiques de la Conservation des Monuments d'Angkor*. L'École Française d'Extreme Orient, Paris.

San Shwe, 2002, The Culture of Vishnu Old City. Master of Research thesis, University of Yangon.

Sorajet Worakamwichai, 1985, *Survey of Ancient settlements in Thailand* (in Thai). Buriram Cultural Center, Buriram.

Stern, P., 1965, *Les Monuments Khmers du Style du Bàyon et Jayavarman VII*. Presses Universitaires de France, Paris.

Wicks, R., 1992, *Money, markets, and trade in early Southeast Asia. The Development of Indigenous Monetary Systems to A.D. 1400*. Southeast Asia Program, Ithaca.

Section II

Aerial Photography and Fractals

Chapter 9

Remote Sensing, Fractals, and Cultural Landscapes: An Ethnographic Prolegomenon Using U2 Imagery

Ezra B. W. Zubrow

Abstract: This paper is a preliminary discourse on how settlements fit into their landscapes. One is able to show the degree of similarity between the organizing principles of landscapes and the organizing principles of settlements. It is the first step in showing how landscapes and settlements are impacted by both natural and cultural forces. The paper measures landscapes surrounding settlements with fractals and compares these measurements to similar fractal measurements of the settlements themselves. The data are U2 remote imagery of ethnographic Rio Grande Pueblos. The landscapes surrounding Santa Ana, Pojaque, and Laguna are significantly more similar to their respective towns than the landscapes of Cochiti, Acoma, and Isleta. The technique is extendable to prehistoric landscapes and settlements.

I'm truly sorry man's dominion,
Has broken nature's social union,
An' justifies that ill opinion,
Which makes thee startle
At me, thy poor, earth-born companion,
An fellow-mortal! ... Robert Burns[1]

---

[1] From Robert Burns "To a mouse: On turning her up in her nest with the plough, November 1785."

Take the only tree that's left
And stuff it up the hole
In your culture ... .Leonard Cohen[2]

1. INTRODUCTION

There are many factors that create a landscape and many that have an impact on one's view of the landscape. On one hand there are natural forces and limits such as the amount of solar radiation and incident moisture. The unique lithosphere at a given location is frequently determinative of the unique biosphere. Interactions between the biotic and abiotic parts of the environment are complemented by the gradients determined by diversity, population size, competition, predation, and prey as well as the range of species. On the other hand, there are cultural forces. Economics, resource exploitation, environmental perception, cultural values, types of agriculture, pastoralism, and habitat protection all play roles.

One's view of the landscape is dependent upon what society one comes from as well as what status and role one has in the society. The reflexive view of the hunter, the city merchant, and the farmer is not the same. The theme of humanity's place in nature varies widely and is an important factor in the resultant landscape. The ideas of a "deity-designed landscape," a "humanity-designed landscape," a "landscape with human place," and a "landscape limiting humanity" result in very different perceptions of the same landscape. The conquest of the environment was altered drastically by the industrialization of the West in the 19th century and spread throughout the world. The change in viewing the landscape is reflected in the "contractual balance" of Burns and the "alienation" by Cohen in the quotes above.

There are orders of culture. It may be subtle or it may be apparent. These orders not only separate the "human" from the "non-human," but the developed from the undeveloped. Indeed, over the several million years of hominid culture it would appear that the amount of cultural organization is increasing and becoming more patterned and it is doing so at an ever increasing rate. These patterns not only contrast temporally but contrast culturally. The "orders" of one culture may be differentiated from another. One needs only to step into a village to tell the difference between a French Village, such as Minervois in the Languedoc, and an English village, such as Stalham in Norfolk. Within a society these cultural patterns frequently are consistent from one domain to another. Whether one looks at the architecture, listens to the music, or eats the food, there are patterns that are culturally specific that cross domains. The French food, language, architecture, church design, and settlement pattern of Minervois are of a type and they are different from Stalham.

---

[2] From Leonard Cohen "The Future."

In order to measure these patterned organizations that cross domains one needs a technique that is not only robust but which can be applied to numerous kinds of phenomena. One of these techniques is to use fractals. Benoit Mandelbrot (Mandelbrot, 1977) published the term "fractal" in 1975 to characterize geometrical objects that have a high degree of "self-similarity"[3] and a fractional dimension that is "less than" or "exceeds" the normal, or "topological," dimension ("D") for that type of object. Fractals are not just mathematical toys but have applied practical uses. They may be used to describe many highly irregular real-world objects. Examples include clouds (Barnsley, 1988), mountains (Turcotte, 1997), materials (Family, 1995), rivers (Rodríguez-Iturbe, and Rinaldo, 2001) and cities (Batty and Longley, 1994; Mapelli, 2001). Fractal techniques have also been employed in image compression (Barnsley and Hurd, 1993), as well as in a variety of scientific disciplines (Mandelbrot, 1983; Bunde et al., 2002).

Fractals are applied to archaeology. The earliest application was in 1985 by the author who applied it to a variety of prehistoric architectures and tool types (Zubrow, 1985). Soon after it was applied specifically to lithic tool analysis of projectile point shapes by Kennedy and Lin (1988). An informal team of archaeologists, Brown, Witschey, and Liebovitch, have investigated lithics, ceramics, and architecture using fractals (Brown, 2001; Brown and Witschey, 2003; Witschey and Brown, 2003, and Brown et al., 2005). Survey methodology and its results were explored by Cavanagh and Laxton (1994) and architectural patterns at Teotihuacan were examined by Oleschko et al. (2000). In the last few years, there have been a variety of conference sessions on the use of fractals in archaeology such as the one at the 2003 Society of American Archaeology Meetings (Witschey and Brown, 2003; Branting and Zubrow, 2003).

2. GOALS

The goal of this paper is to use remote sensing to provide an ethnographic prolegomenon to show how settlements fit or do not fit into their landscape. One should be able to show the degree of similarity between the organizing principles of the landscape and the organizing principles of the culture by measuring the landscape with fractals and comparing these measurements to similar fractal measurements for settlements in the landscape. Of course, there is continuity among nature, landscape, and settlement. The landscape surrounding the settlement is affected by both natural and cultural forces, as is the settlement. Today, there is no "pure nature" nor is there any "pure culture."

---

[3] Self-similarity means that sub-pieces within the object appear as scaled down, and perhaps translated and rotated, versions of larger pieces of the original object.

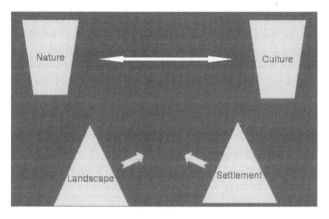

Figure 1. The Nature/Culture Continuum.

Instead, there are differing proportions of the "nature" versus "culture," which form a continuum between the settlement and the landscape (see Figure 1). A similar continuum existed throughout the human past.

This study is an ethnographic prolegomenon in the sense that if the techniques used are able to give a measure of how well ethnographic settlements fit into their landscape, they may also be applicable to prehistoric settlements and prehistoric landscapes.

3. SUBJECT

The Southwestern region of the United States includes western Texas, most of New Mexico, Arizona, eastern California, western Colorado, and eastern Utah (Kneese et al., 1981). There are some nine distinct ecosystems creating a land of extremes from the deserts to the mountains. The ecosystems are the Alpine Tundra, Subalpine – Montane Grasslands and Meadows; Subalpine Forests; Mixed-Conifer Forests; Ponderosa Pine Forests; Madrean Forests; Pinyon-Juniper Woodlands and Juniper Savanna; Encinal Oak Woodlands; Desert Shrublands and Semidesert Grassland; and Riparian Ecosystems. Of these the data for this paper focus geographically upon settlements and landscapes in eastern Arizona, western and central New Mexico, and the Pinyon Juniper (United States Forest Service, Southwestern Region, 1995), Desert Shrublands, and the Riparian ecosystems (Kneese et al., 1981; Ford, 1987; Sowell, 2001).

The landscape is striking and well known around the world. Even if one has not visited the Southwest, a clear image exists from the sophisticated artistic images of the photographs of Anselm Adams[4] and the paintings of

---

[4] For example "Moonrise, Hernandez, New Mexico," 1941.

Georgia O'Keeffe[5] to less sophisticated cartoons featuring the roadrunner and coyote. The landscape features mountains, mesas, canyons, and deserts. The area is arid or semiarid and, depending on the setting and elevation, may have temperate winters and summers, periods of bitter cold, or intervals of intense heat.

The organizing principles of the ecosystems are ecosystem type distinguished by altitude, and impact distinguished by human interaction (Zubrow, 1974; Zubrow, 1975). Air cools with higher altitude with a dry lapse rate of 1 degree C per 100 meters while moisture increases with a moist lapse rate of 0.6 degree per 100 meters. Thus, there is more precipitation and less evaporation with higher altitude. The result is more available moisture with higher altitudes and the resulting ecosystem zonation by elevation. If one drives up Mt. Lemmon some 2800 meters, one moves from the upland Sonoran desert through each ecosystem as one climbs, until one reaches the subalpine top. It is similar to driving up a piece of layer cake. Similarly, the population increase in Arizona and New Mexico has been phenomenal. Between 1990 and 2000, New Mexico's population increased 20.1% while Arizona's increased 40.0%. In 2003 New Mexico had a population of 1,874,614 while Arizona had 5,580,811(Census 2003).

The creation of the "Southwestern Desert" is partially the slow result of cultural impacts. The plant communities are being changed from grasslands to shrubland. The changes are caused by water availability, overgrazing, and fire suppression that are a consequence of cultural interventions. The result is a landscape of "shreds and patches" that favor the desert shrubs (creosote bush or mesquite) and bare ground. The change in some areas is simply a geographic redistribution, a zero-sum game, where the amounts of biomass and nutrients remain the same. However, in other areas there is a net loss, in which large bare patches are left. The shrubs form islands of fertility in an otherwise plantless landscape. Plant growth follows a pulse pattern in which pulses of plant production follow heavy-rainfall events (Kieft, 1988). Given these processes, the islands are not easily altered.

In these areas of the Colorado Plateau, Arizona, and New Mexico there is a complex ethnographic tradition including the well known Apache, Hopi, Navajo, Papago, Yavasupai, and the Rio Grande and Western Pueblos (Jorgensen, 1980).[6] Among the pueblo groups, there are cultural commonalities including similar environments as well as similar techno-logical and economic adaptations. Yet, they contrast in language, community organization, and religious ceremonialism. Many of these settlements are continuations of settlements begun more than a thousand years ago. These

---

[5] For example "View from my studio," 1930.
[6] Modern descendants of the Anasazi include the Nambe Tewa, Santa Clara Tewa, San Juan Tewa, San Ildefonso, Isleta, Taos, Jemez, Sia Keres, Sanda Ana Keres, Cochiti Keres, Santo Domingo Keres, Acoma, Hopi, and Zuni Native Americans.

towns evolved from collections of pithouses to multi-room and multi-building pueblos with their characteristic adobe structures. They were built and rebuilt since the times of the Anasazi (A.D. 750 Pueblo I to A.D. 1600 Pueblo V), Mogollon (from A.D. 550 Georgetown phase to A.D. 1300 Tularosa phase) and Hohokam (from A.D. 500 Colonial Phase to A.D. 1450 Classic Phase) with the same design elements. They are especially well adapted to the arid landscape of the desert climate. The adaptive success of these designs resulted in their being incorporated in the design being used for Spanish colonial architecture in the sixteenth and seventeenth centuries and are present today in the southwestern architecture of the Four Corner states[7] as well as California.

There is a clear spatial order based upon directions, verticality, and concentricity (Frazier, 1986). Robert Lister (Lister and Lister, 1981)[8] argued for a directional system based upon the "equinoxes" and an "up and down". Edward Dozier (Dozier, 1970)[9] suggested that the concentric system was not an "egocentric" but a "village oriented" one in which the village (safety) is at the spatial and cognitive center and malevolence (evil) is at the periphery. The temporal order is based upon when things emerge from the earth as well as the annual cycle (Ortiz, 1972).

4. DEFINING FRACTALS AND THEIR MANAGEMENT

Fractals are patterns that have non-integer dimensions. While points have zero dimension, lines one dimension, planes two dimensions and solids three dimensions, fractals may have dimensions such as 1.3 or 2.39.

Given fractals fractional dimensions it is not surprising that fractal is from the Latin "fractus" and is related to the English verb to fracture and suggests breaking (Oxford University Press, 1992). When one speaks of a fractal, one generally speaks of a pattern that one may divide infinitely into parts that are indistinguishable or statistically similar to the whole. In this common sense, since a fractal is a set that is self-similar, fractals are repetitive in shape, but not in size. No matter how much you magnify the elementary patterns of a fractal, it will always look the same. In other words, fractals are created recursively to have repetitive details visible at any arbritray scale. Thus, the part-to-whole relationships and the part-to-part relationships may be shown by such formal transformations as translations and rotations or reductions and zooming.

The fractal or Hausdorff dimension is a topological dimension calculated as the ratio logN/log(1/R) where N is the number of segments and R the length of each segment. Finally, fractals are defined through their self-similarity and for fractals there is no dominating scale.[10]

---

[7] Arizona, Colorado, New Mexico, and Utah.
[8] Robert Lister was the author's supervisor during the 1967 Wetherill Mesa excavations.
[9] Edward Dozier was the author's major advisor on his book on the Rio Grande Pueblos.
[10] These characteristics of fractals are (aside from dimensionality) remarkably difficult to define mathematically. The problems are: there is no precise meaning of "too irregular"; there are multiple ways for an object to be self-similar; and not every fractal has a recursive definition.

Real phenomena approximating fractals easily are found in nature. They include clouds (Bunde et al., 2002), mountains (Turcotte, 1997; Blenkinsop et al., 2000) and river networks (Rodríguez-Iturbe and Rinaldo, 2001; Mandelbrot, 2002). Surprisingly, they are less commonly studied in the cultural domain. There are notable exceptions by Blakeslee (Blakeslee, 2002) as well as by Eglash's (1998, 1999) studies of African settlement architecture and indigenous design. In addition, they have been found in systems of cities (Batty and Longley, 1985; Batty and Longley, 1986) and markets (Bunde et al., 2002; Mandelbrot and Hudson, 2004).

Despite their ubiquity, the importance of fractals was not recognized until the last few decades. Examples include Koch's curve, Cantor's dust, and Menger's sponge as well as Mandelbrot set, Lyapunov fractal, Sierpinski carpet, and the Peano curve.

Fractals are divided into deterministic or stochastic groups depending upon the influence of chance. There are three broad categories of fractals based upon the rule structure. They are as follows.

1. Iterated function systems with fixed geometric replacement rules (e.g., Sierpinski Carpet).
2. Escape time fractals defined by a recurrence relations at each point in a space (e.g., Lyapunov fractals).
3. Random fractals generated by stochastic rules and processes (e.g., Lévy flights).

How does one measure a fractal? Given that there is no single definition of fractal dimension, it is not surprising that there are a variety of methods. Most rely upon two factors: demonstrating the data follows an exponential function, and showing the distribution must show scale-invariant characteristics for pattern and complexity. One analyzes the variable of interest of the dataset using a length scale and a power law. The slope is estimated by fitting a line using the method of least squares. The results are plotted in log-log space, and, if the set is fractal, they should follow a straight line.

In a given image it is conceivable that there may be sets that have different fractal dimensions in different ranges of scales.[11] These are sometimes called multifractals and are subject to the same kinds of analysis as the simpler unidimensional fractal. These variations in fractal dimension may correspond to different physical or cultural processes acting at different scales in different areas. If so, it will be useful information for the archaeologist. Namely, the northern part of the site is different from the central site or that the northern landscape differs from the northern part of the site in terms of organizing principles or rules, complexity, pattern design, or pattern size.

---

[11] They should not be confused with length scales where the fractal dimension of the set changes.

In this study, the Box Dimension Estimation Method (TruSoft International Inc., 2004)is used to determine the fractal values. A square mesh of various sizes d is laid over the part of the image containing the object of interest. The number of mesh boxes N(d) that contains part of the image is counted. The fractal (box) dimension Db is given by the slope of the linear portion of a log(N(d)) vs log(1/d) graph. This is from the equation.

$$Log(N(d)) = Db \; log(1/d)$$

Or

$$Db = Log(N(d))/log(1/d)$$

where N(d) is the number of boxes of linear size d necessary to cover a dataset of points distributed in a two-dimensional plane. Note that d is equivalent to 1/R as defined earlier. For objects that are Euclidean, equation (1) defines their dimension. One needs a number of boxes proportional to 1/d to cover a set of points lying on a smooth line, proportional to $1/d^2$ to cover a set of points evenly distributed on a plane, $1/d^3$ for three dimensions, and so on. This dimension is sometime called grid dimension because, for mathematical convenience, the boxes are usually part of a grid.[12]

5. DATA: AERIAL PHOTOGRAPHS OF PUEBLOS AND THEIR IMMEDIATE ENVIRONMENTS

In 1967 and 1968, the Strategic Air Command flew a set of training missions over the Colorado Plateau and Rio Grande Valley using the U2 for the photographic platform. It is a single-seat, single-engine reconnaissance plane capable of flying above 70,000 feet (21.3 km) above sea level at more than 475 miles per hour (Mach 0.58, 730 km/hr) with a range beyond 6,000 miles (9,660 km). The aircraft carried a variety of sensors capable of atmospheric chemistry and dynamics for the detection of open-air atomic-bomb testing, as well as being able to take remote imagery of the earth's surface for troop movements. The U2 had cameras that could use wet-film photo, electro-optic, or radar imagery. One camera used a gyro-stabilizing framing system with a 66" (167.6 cm) focal length, producing a 9" (22.9 cm) by 18" (45.7 cm) contact negative. Another was a high-resolution panoramic camera with a 24" (60.9 cm) focal length where the lens moved across film producing a 30" (76.2 cm) by 5" (12.7 cm) negative. In both cases film had to be processed in laboratories after the aircraft landed.

---

[12] Ideally, one wants to define a box dimension where boxes are located and oriented to minimize the number of boxes needed to cover the set. This is a difficult mathematical problem. The above description is based upon the description provided by TruSoft with their software.

Examples of the images scanned from the 9" by 18" contact negatives showing particular pueblo settlements and landscapes follow.

6. METHODOLOGY

The methodology is the following. First, six pueblos were chosen to analyze. They were Acoma (Figures 2 and 3), Cochiti, Isleta (Figures 4 and 5), Laguna, Pojaque, and Santa Ana. A fractal measurement was made over each village. Then, four fractal measurements were made upon four landscape samples around the settlement (Figure 6). One is measuring all natural and cultural phenomena visible in the aerial photograph. The organizing principles of the patterning to and the patterns themselves may vary depending upon the particular geological, landscape, and cultural phenomena at any location. In general, one would expect the organization of the settlement to be the most different from the surrounding countryside.

On the other hand, if the settlement and the landscape are organized similarly the fractal Db value should be similar. In addition, the relationship between pattern width and pattern height should also be similar.

7. RESULTS

The results show that there are two groups of pueblos. Figure 7 is the distribution of the fractal values D(b) for each town and their related landscapes. The first point on each graph is the fractal value for the town and the next four points the fractal values for each landscape. The order of the points

Figure 2. Acoma Settlement.

Figure 3. Acoma Countryside.

corresponds to landscapes from the north, west, east, and south. The second point is the northern landscape, the third point the western, the fourth point the eastern, and fifth point the southern. Several patterns are apparent. First, and perhaps most important, is that the pueblos fall into two groups. Santa Ana, Pojaque and Laguna have landscapes more similar to their respective towns while Cochiti, Acoma, and Isleta have greater differences. This visual difference is confirmed by the means and standard deviations.[13] Second, most of the pueblos and most of the landscapes, that is 22 of 30 values, have generally similar D(b) values ranging between 1.5 and 2.0. Acoma is unique with all of its D(b) values below 1.2. Third, all towns with the exception of Acoma and Laguna have generally similar curve shapes. Santa Ana and

---

[13] The mean Db values and standard deviations for the six pueblos organized by increasing standard deviations follow:

Laguna	1.8588	0.0328
Santa Ana	1.8288	0.0787
Pojaque	1.814	0.125
Isleta	1.7494	0.1554
Cochiti	1.3942	0.1984
Acoma	0.858	0.307

Figure 4. Isleta Settlement.

Cochiti have almost identical curve shapes: i.e., middle value town, higher north, lower west, lowest south, and middle value east.

Different cultures use different sizes as part of their cultural rules. This is reflected in the size of the recursive unit that makes up the fractal pattern; e.g., the size of the "hexagonal crystal", "rhomboid crystal" or whatever element that makes up the units of the fractal design at different scales. Figure 8 is the relative size of the recursive pattern unit that makes up the fractal patterns for each pueblo and each landscape.

The results are variable. Consider first the relationship of the landscapes to each other. For Santa Ana all four landscapes (north, west, south, and east) are organized around a similar size pattern element. This is also true but to a slightly less degree for Cochiti and Acoma. However, for Pojaque, Laguna, and

Figure 5. Isleta Countryside.

Isleta, the landscape pattern elements have considerable more size variation. In the case of Pojaque there are three distinct values. The north is one value, then the west and south are equal, and finally, the east is very different and large). In the case of Laguna, the one outlier, the east, is much smaller than the element size of the rest of the landscapes while for Isleta the northern landscape is the outlier and is much larger than the others.

If one compares the pattern areas sizes of the pueblo to the landscapes Laguna, Cochiti, and Acoma are approximately of similar size to their landscapes. In the case of Santa Ana, Pojaque, and Isleta the pueblo pattern area sizes are more similar to some landscapes than others. If one disaggregates the pattern areas of both the pueblos and the landscapes into pattern area height and width as in Figure 9, there are only a few outliers that have very (very small pattern area height and pattern area width. They are from Santa Ana, Laguna and Isleta.

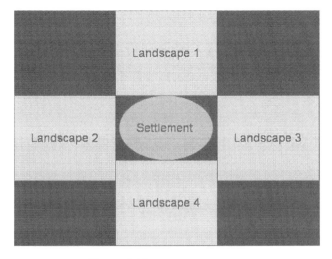

Figure 6. Measurement strategy.

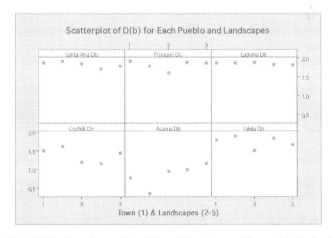

Figure 7. The D(b) values for each Pueblo and Landscape. The pueblo is the first point and each of the four landscapes follows.

8. CONCLUSIONS

This paper applies fractals to remote sensing to show how settlements fit or do not fit into their landscape. Fractals allow one to show the degree of similarity between the organizing principles of the landscape and the organizing principles of the culture. It is the first step in showing how settlements and landscapes are impacted by both natural and cultural forces.

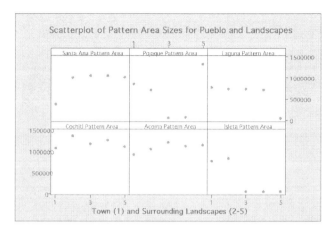

Figure 8. Scatterplot of Pattern-Area Sizes for Pueblo and Landscapes.

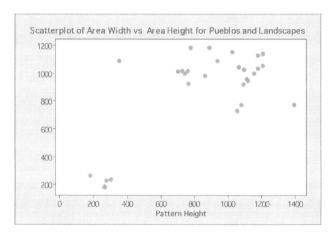

Figure 9. Scatterplot of Disaggregated Pattern Area by Height and Width for Pueblo and Landscapes.

The box dimension estimation method of calculating fractals was used on selected U2 imagery of the Rio Grande and Western Pueblos. Santa Ana, Pojaque and Laguna have landscapes more similar to the town while Cochiti, Acoma, and Isleta are not reflected in the landscape. Perhaps, it would be more appropriate to say that the landscapes of Cochiti, Acoma, and Isleta are not reflected in the settlement pattern of the towns. For most of the pueblos the western and the southern landscapes are most different—i.e. there is the greatest deviation from the settlement value. In short, the pueblos are oriented to their northern and eastern landscapes. If one adds the information derived from the pattern area, one narrows the conclusion. In general the element size of the northern landscape is most similar to the settlement. The conclusion is the pueblos are most clearly oriented towards the northern landscape.

This is a prolegomenon- a preliminary discourse. There are several important issues to consider. Although it works for the ethnographic present will it work for the prehistoric past? Clearly, if one had both a completely excavated prehistoric settlement and a completely excavated prehistoric landscape, the technique would be appropriate. Similarly, if one had a completely excavated prehistoric settlement and the prehistoric landscape was unchanged from the ethnographic landscape, the procedure would produce accurate results. If one could calculate a backwards trend in the fractal values from the present through the ethnohistoric and the historic to the prehistoric for the settlement and the landscapes, one might be able to compare the past projected values. Finally, there might be other ways to do this. However, these issues are the topic of another paper. Ultimately, the study of ideational issues and cultural rules is not limited to interpretation as suggested by the post processual archaeology. As this paper shows, using remote sensing and fractals one may actually measure the cultural and natural rules and their degree of similarity in differing domains.

BIBLIOGRAPHY

Barnsley, M. F., 1988, *Fractals Everywhere*. Academic Press, Boston.
Barnsley, M. F., and Hurd, L. P., 1993, *Fractal Image Compression*. Wellesley, Mass.
Batty, M., and Longley, P., 1985, *The Fractal Simulation of Urban Structure*. University of Wales Institute of Science and Technology, Dept. of Town Planning, Cardiff.
Batty, M., and Longley, P., 1986, *Fractal-based Description of Urban Form*. University of Wales Institute of Science and Technology, Dept. of Town Planning, Cardiff.
Batty, M., and Longley, P., 1994, *Fractal Cities: a Geometry of Form and Function*. Academic Press, London and San Diego.
Blakeslee, Donald J., 2002, Fractal Archaeology: Intra-generational Cycles and the Matter of Scale, an Example from the Central Plains. In *The Archaeology of Tribal Societies*, edited by William Parkinson, pp. 173–199. Monographs in World Prehistory. Ann Arbor, Michigan.
Blenkinsop, T. G., Kruhl, J. H., Kupkova, M., and International Symposium on Fractals and Dynamic Systems in Geosciences, 2000, *Fractals and Dynamic Systems in Geoscience*. Basel; Boston, Birkhäuser Verlag, Basel and Boston.
Branting, S., and Zubrow, E., 2003, Fractalizing Culture. Paper presented at Society of American Archaeology, 68th Annual Meeting, April 9–13, 2003, Milwaukee, Wisconsin.
Brown, C. T., 2001, The fractal dimensions of lithic reduction. *Journal of Archaeological Science* 28(6):619–631.
Brown. C. T., and Witschey, W. R. T., 2003, The Fractal Geometry of Ancient Maya Settlement. *Journal of Archaeological Science* 30:1619–1632.
Brown, C. T., Witschey, W. R. T., and Liebovitch, L.S., 2005, The Broken Past: Fractals in Archaeology. *Journal of Archaeological Method and Theory* 12(1):37–72.
Bunde, A., Kropp, J., and Schellnhuber, H.-J., 2002. *The Science of Disasters: climate disruptions, heart attacks, and market crashes*. Springer, Berlin and New York.
Burns, R., 2000, *Complete Poems and Songs of Robert Burns*. Leonard Books, Granton, or http://www.poetry-online.org/burns_to_a_mouse.htm

Cavanagh, W. G., and Laxton, R. R., 1994, The Fractal Dimension, Rank-Size, and the Inter-pretation of Archaeological Survey Data. In *Methods in the Mountains: Proceedings of UISPP Commission IV Meeting, Mount Victoria*, edited by I. Johnson, pp. 61–64. Sydney University Archaeological Methods Series Vol. 2.

Census, U.S., 2003, *US Census Bureau, State & County QuickFacts*. United States Government, Washington, D.C.

Cohen, Leonard, 2005, The Future. http:// www.a3lyrics.com /lyrics/leonardcohen/thefuture.html.

Dozier, E.P., 1970, *The Pueblo Indians of North America*. Holt, New York.

Eglash, R., 1999, *African Fractals: Modern Computing and Indigenous Design*. Rutgers University Press. New Brunswick, NJ.

Eglash, R.,1998, Fractals in African Settlement Architecture. *Complexity* 4(2, nov-dec):21–29.

Family, F., 1995, *Fractal Aspects of Materials: Symposium held November 28-December 1, 1994, Boston, Massachusetts, U.S.A.* Materials Research Society, Pittsburgh, PA.

Ford, R. I., 1987, *The Prehistoric American Southwest: a Source Book: history, chronology, ecology, and technology*. Garland Publ., New York.

Jorgensen, J. G., 1980, *Western Indians : comparative environments, languages, and cultures of 172 western American Indian tribes*. W. H. Freeman, San Francisco.

Kieft, T.L., White, C.S., Loftin, S.R., Aguilar, R., Craig, J.A., and Skaar, D.A., 1998, Temporal dynamics in soil carbon and nitrogen resources at a grassland-shrubland ecotone. *Ecology* 79(2):671–683

Kennedy, S. K., and Lin, W., 1988, A fractal technique for the classification of projectile point shapes. *Geoarchaeology* 3(4):297–301.

Kneese, A. V., and Brown, F. L., 1981, *The Southwest under Stress: national resource development issues in a regional setting*. Published for Resources for the Future by the Johns Hopkins University Press, Baltimore.

Lister, R. H., and Lister, F. C., 1981, *Chaco Canyon: Archaeology and Archaeologists*. University of New Mexico Press, Albuquerque.

Mandelbrot, B. B., 1977, *Fractals: Form, Chance, and Dimension*. W. H. Freeman, San Francisco.

Mandelbrot, B. B., 1983, *The Fractal Geometry of Nature*. W. H. Freeman, San Francisco.

Mandelbrot, B. B., 2002, *Gaussian Self-affinity and Fractals: Globality, the Earth, 1/f noise and R/S*. Springer, New York.

Mandelbrot, B. B., and Hudson, R. L., 2004, *The (Mis)Behavior of Markets: a Fractal View of Risk, Ruin, and Reward*. Basic Books, New York.

Mapelli, E. G., 2001, *Urban Environments*. Wiley-Academy, Chichester.

Oleschko, K., Brambila, R., Brambila, F., Parrot, J., and Lopez, P., 2000. Fractal analysis of Teotihuacan, Mexico. *Journal of Archaeological Science* 27(11):1007–1016.

Ortiz, A., 1972, *New Perspectives on the Pueblos*. University of New Mexico Press, Albuquerque.

Oxford University Press, 1992, *The Oxford English Dictionary Online*. Oxford University Press, Oxford and New York.

Rodríguez-Iturbe, I., and Rinaldo, A., 2001, *Fractal River Basins : Chance and Self-Organization*. Cambridge University Press, Cambridge.

Sowell, J., 2001, *Desert Ecology: An Introduction to Life in the Arid Southwest*. University of Utah Press, Salt Lake City.

TruSoft International Inc., 2004, *Benoit Fractal Analysis System*. TruSoft International Inc., St Petersberg, FL.

Turcotte, D. L., 1997, *Fractals and Chaos in Geology and Geophysics*. Cambridge University Press, Cambridge and New York.

United States Forest Service, Southwestern Region, 1995, *Restoring a degraded environment: understanding and caring for Piñon-Juniper woodlands = La reparación de un medio ambiente desgastado: la comprensión y el cuidado de los bosques del piñón y el junipero*. U.S. Dept. of Agriculture, Albuquerque.

Witschey, W. R. T., and Brown, C. T., 2003, Fractal Fragmentation of Archaeological Ceramics. Paper presented at the symposium "Fractals in Archaeology" at the 68th *Annual Meeting of the Society for American Archaeology*, April 9–13, Milwaukee, Wisconsin.

Zubrow, E. B. W., 1974. *Population, Contact, and Climate in the New Mexican Pueblos*. University of Arizona Press, Tucson.

Zubrow, E. B.W., 1985, Fractals, cultural behavior, and prehistory. *American Archeology* 5(1):63–77.

Zubrow, E. B. W., 1975. *Prehistoric Carrying Capacity, a Model*. Cummings Pub. Co., Menlo Park, CA.

Section **III**

Geographic Information Systems

Chapter 10

Understanding Archaeological Landscapes: Steps Towards an Improved Integration of Survey Methods in the Reconstruction of Subsurface Sites in South Tuscany

STEFANO CAMPANA AND RICCARDO FRANCOVICH

Abstract: The Department of Archaeology at Siena has been engaged for several decades in the testing of new methodologies, new approaches and new instruments for construction of the archaeological record. In relation to landscape archaeology and in particular with the South Tuscan landscapes the low level of visibility and heavy clay soils have directed us towards those techniques of remote sensing that leave a wide choice to the archaeologist in the periods for carrying out data capture. In particular we have begun to work on a systematic program of aerial survey, on Ikonos-2 and QuickBird-2 satellite imagery, and on micro-digital terrain modelling using digital photogrammetry. On the ground our infra-site analysis has been improved by applying extensive magnetic survey, recently integrated with GPR survey; other gains have come from the systematic use of differential GPS and PDA devices. Along with the development of new technologies we have continued the study of historical aerial photographs and the use of field-walking survey, both of which still constitute, in our opinion, undeniably valuable sources for the archaeological study of ancient landscapes. The results that we have obtained are encouraging and show clearly the need to use integrated sources. Source-integration now represents the prime focus of our research. In an area like that of South Tuscany without this approach we foresee little possibility of obtaining results that will have a real effect on our understanding of the development of the landscape across time.

1. INTRODUCTION

Graeme Barker, in a well known paper on landscape archaeology (Barker, 1986), identifies as one of the main differences between British and Italian studies on archaeological landscapes the separation in Italy between settlement and environmental patterns. His thesis maintains that the close relationship between ideas about ancient society and the environmental background in Great Britain is mainly a result of the early development of their aerial survey. The aerial photographs widely used by British archaeologists strongly emphasize the background context within which archaeological sites belong.

In the past few years, there have been changes that have at last allowed Italian archaeologists to photograph the landscape from the air and so to have at least a theoretical possibility of reconnecting the diachronic development of the environmental background to observed settlement patterns (Musson et al., 2005). Almost at the same time the availability of new technologies—especially spatial technologies such as GIS, remote sensing, and high-resolution satellite imagery (HRSI)—have provided innovative possibilities for the location, integrated analysis, management and monitoring of archaeological sites (Campana and Forte, 2001).

Prior to 1999, the Department of Medieval Archaeology at the University of Siena based its work in archaeological cartography on three methods of investigation: systematic field-walking in sample areas aimed at representing 20–30% of the whole landscape; the analysis of historical vertical air photographs through stereo viewing and ground-truthing in the field; and detailed surveying aimed at providing high-quality understanding of particular monuments or archaeological areas (Francovich and Valenti, 2001).

In this paper we will discuss in particular our experience with the progressive introduction of new methodologies and the problems of integrating different survey techniques in the archaeological mapping of South Tuscan landscapes, specifically in the administrative areas of Grosseto and Siena.

2. SOUTH TUSCAN LANDSCAPES

The need to test new instruments and new approaches to surveying derives from a certain dissatisfaction with the results obtained through previous methods. Our past work has allowed us to identify a large number of new sites and to collect new data about known sites. Notwithstanding these results, we still feel that we have not answered our questions about understanding the complexity that characterizes ancient landscapes, ancient settlement patterns, and their reciprocal relationships. In particular some specific chronological periods, such as the Early Middle Ages, or some specific historical questions, such as the change in the location of settlement from Roman villa to hillfort, remain particularly hard to confront (Francovich and Hodges, 2003). We focus in this paper on two main problems, the first largely qualitative, the second quantitative.

- In our previous strategy there was too large a difference between the nature of the information obtained from surface collection compared with that derived from stratigraphical excavation (which is by its nature too slow and too expensive). We clearly needed to develop our capacity to recover more detailed information without recourse to large excavation.
- The requirement to work on the basis of limited sample areas, combined with the opportunity to study our landscapes from the air only through vertical air photographs, represented a strategic shortcoming that resulted in a considerable loss of otherwise detectable sites.

It was clear to us that there was a need to improve our study of ancient landscapes in both of these respects. We therefore turned our interest first to remote sensing techniques, while remaining aware of the limitations that this methodology will encounter in a countryside like that of Tuscany and in particular of the Province of Siena and the northeast of Grosseto territory.

Fifty percent of Tuscany is covered in forests and other areas characterized by a low level of visibility, whether from the ground or from the air (Figure 1, Figure 2a–b). The remaining landscape consists in great part of agricultural cultivation on heavy clay soils that are known to constitute a particularly unfavourable surface for most remote sensing techniques (Figure 2c). A second limitation introduced by heavy clay soils is that the number of years

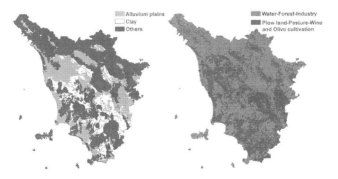

Figure 1. Geological map and land-use map of Tuscany.

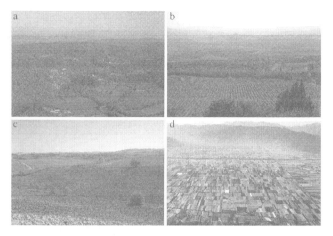

Figure 2. a) Hilly countryside covered by forest; a type representative of many stretches of Tuscany. b) Hilly countryside with vineyard and forest. c) Agricultural cultivation on heavy clay soils. d) Landscape pattern of reclaimed land.

when the meteorological conditions are likely to produce good archaeological traces is even smaller than on soils above substrata such as gravel or sand (Jones and Evans, 1975).

Areas with a higher level of visibility consist mainly of the alluvial plains of substantial rivers, in particular the Arno, Cecina, Ombrone, Serchio, Chiana, and Orcia (Figure 2d). In some of these areas, however, other problems arise from the great thickness of the alluvium and from the impact of modern industrial and residential development (Agnoletti, 2002). All these circumstances, as is already well known, have a direct influence on the results of research based on the use of the methods and instruments of remote and proximal sensing (Wilson, 2000; Clark, 1997).

This situation has directed us towards an integrated and interrelated use of those research techniques that leave a wide choice to the archaeologist in the periods for carrying out data capture, and in particular towards the study of parts of the electromagnetic spectrum not visible to the human eye (Doneus et al., 2002; Donoghue, 2001):

- Exploratory aerial survey and oblique air photography
- Multispectral high-resolution satellite imagery (HSRI)
- Digital photogrammetry
- dGPS survey
- Geophysical prospecting (gradiometer and GPR)

Along with the integrated development of these new approaches to the study of past landscapes we have of course continued our study of historical aerial photographs and the use of field-walking survey, both of which remain undeniably valuable sources for the archaeological study of settlement patterns (Guaitoli, 2003).

Riccardo Francovich

3. REMOTE SENSING AND DIGITAL RECONSTRUCTION OF ANCIENT LANDSCAPES

Our Department manages regional and sub-regional landscape projects. Whatever the scale, the study area is always based on the local administrative units, which in Tuscany range in size between 40 and 450 sq km with an average of 150 sq km (Francovich and Valenti, 2001). In order to obtain a total coverage of areas of these dimensions at relatively low cost, we began a project using multispectral imagery captured with the IKONOS satellite.

The results have been encouraging, within the limits of the geometric and spectral resolution of the data. At the end of the first phase of this work we have recorded 84 features, of which 39 were new sites, in two sample areas of about 470 sq km in total. In fourteen cases where anomalies had been identified previously through vertical or oblique air photography it was possible to add new data to the existing information. We should perhaps note two peculiarities of IKONOS-2 imagery (Campana, 2002a).

Firstly, through IKONOS-2 we can recognise many features that were visible in early vertical air photographs but which are no longer identifiable in those taken in recent years. This situation perhaps derives from the inappropriate time of year in which the later photographs were taken, or alternatively from the higher sensitivity and computer enhancement capabilities of the IKONOS-2 data. If confirmed, however, this trend will indicate HRSI as an important tool for monitoring and exploring the archaeological heritage (Figure 3).

Secondly, we believe that most of the results obtained from analysis of the IKONOS-2 imagery depend heavily on the multispectral properties of the sensor. Above all the near-infrared represents the most powerful band. This band is particularly sensitive to plant health and can often detect water stress in vegetation before it can be seen by the naked eye (Donoghue, 2001)

The following example highlights this possibility. The site in this case is in the district of Montalcino (Siena), in a zone in which, according to documentary sources, there stood from the middle of the VIIth century the medieval monastery of San Pietro ad Asso. Before our research the monastery was identified with the farm of San Pietro, which contains the still visible remains of a Romanesque building: a small church tower and several pieces of decorative stonework. All of the documentary sources, from the second

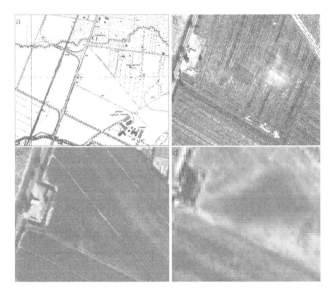

Figure 3. a) Technical map of the area of podere San Giorgio. b) Aerial photograph, 1954. c) Aerial photograph, 1996. d) IKONOS satellite imagery band 4 (near
infrared).

half of the XIIIth century onwards, refer to a village that must have been
present in the same area (Campana, 2004). The identification of the monastery
was determined through historical photographs analyzed by stereo viewing.
This analysis permitted the recognition of an abnormal hill form only 200 m
from the abandoned farm of San Pietro. The ground survey produced clear
evidence for shaping of the hill's profile, with terracing along the slopes as
well a as round the crest of the hill. In particular, on the topmost part on the
northern side there is a considerable spread of walling, perhaps attributable
to fortification works.

Analysis of the same area, carried out on panchromatic aerial photographs
from 1954, 1976, 1994 and 1996, revealed numerous features, while oblique
photographs taken during exploratory aerial survey in 2001 confirmed the
presence of many anomalies. A magnetometer survey of the hilltop provided us
with the possibility of mapping other archaeological features. The IKONOS-2
imagery revealed further evidence of linear anomalies in farmland adjacent
to the site, notably rectangular features and an anomaly on the top of a hill
about a kilometer from the monastery.

Some particular anomalies look very interesting. These lie in the area
of the farm of San Pietro where the data from surface surveying incline us
to identify the documented late medieval village of San Pietro. In this area
the vertical air photographs do not show anything. It is very interesting
to see that only the near-infrared band of IKONOS-2 imagery permits the

identification of two square features that ground survey allowed us to associate with buildings of the medieval village. Bands blue, green, and red in this case are entirely unproductive.

Despite these promising early results the true potential of this type of imagery is still not fully clear and needs to be further evaluated to test its effectiveness under a broad range of environmental conditions. A real limitation of the IKONOS products turned out to be the great difficulty of achieving with precision the desired capture time. The images ordered for the last week of May or the first week of June were not actually captured until the middle of July, a month which in our latitude corresponds to a very poor period of the year for the recording of archaeological traces. In spring 2002 we started testing three samples of Quickbird-2 imagery, two for the province of Siena and one near the coast in the province of Grosseto, covering a total area of about 200 km². On the basis of our experience with IKONOS-2, we focused our attention on two main problems: geometric resolution and best capture time.

Even though it has been possible to distinguish some small features through IKONOS-2 and to identify a first range of detectable site size, we feel it necessary to stress that there is still a risk of misinterpretation. When we captured Quickbird-2 imagery we acquired both the multispectral and the panchromatic data. Pan-sharpening of the four multispectral bands using the 0.7 m panchromatic image was then carried out to improve the spatial resolution. In this context it should be noted that a pixel of IKONOS-2 multi-spectral imagery corresponds to 32.65 pixels of Quickbird-2 pan-sharpened data (Figure 4). Our first impression, looking at the Quickbird-2 imagery, is that most features of the landscape can be easily and unambiguously recognized in this more recent source of data.

In relationship to the second problem, the IKONOS-2 imagery was captured in July, though we would have preferred the end of May. The QuickBird-2 imagery was captured after a delay of "only" 15 days from our preferred time, though this was probably enough to result in some significant loss of sites. There were two extenuating circumstances. Firstly, we did not consider the possibility of submitting a priority order (at a 50% increase in price), which would have given image capture within a maximum of five days from the specified date. The second was a typical problem of satellite imagery—though one not so significant in the Mediterranean region—that of poor weather conditions.

Our study of the Quickbird-2 imagery is still in progress and we do not yet feel able to present a fully considered report. However, our impression at this stage is that many of the limitations that we found in using IKONOS-2 imagery will be overcome with Quickbird-2 and that with the priority option of QuickBird it will at last be possible to achieve the right capture time for archaeological needs (Campana 2002b).

As we anticipated, the analysis of satellite imagery does not entirely remove uncertainty from the study of ancient landscapes and in particular

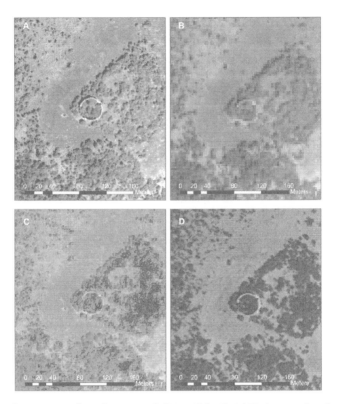

Figure 4. Comparison of resolution capabilities of the QuickBird sensor in relationship with the castles of Moscona. a) Panchromatic satellite imagery with a resolution of 0.70 m. b) Color composite 4-3-2 of multispectral satellite imagery with a resolution of 2.80 m. c) Pan-sharpened satellite imagery with a resolution of 0.70 m. d) Aerial photograph with a resolution of 1 m.

of complex territories like Tuscany. For instance, many of the archaeological discoveries that we made during field survey or in examining vertical air photographs are not visible on the satellite imagery.

For these reasons we started a program of aerial survey averaging 35–45 hours of flight per year, focused on the end of May and the beginning of June (Campana, 2001). The use of exploratory aerial survey in Italy has only become possible in recent years following legislative changes, but in many other countries of northern Europe it is a method with a long tradition of application. In ideal conditions this technique offers, also in South Tuscany, an extraordinary contribution to the search for new sites and for the continuous monitoring of the cultural heritage (Figure 5).

Even from our limited experience with the technique, we can point to the flexibility and rapidity of the aerial survey to assist in the discovery of

Figure 5. a) Settlement (Roman villa). b) Road system. c) Square and round prehistoric enclosures. d) Medieval field systems.

archaeological traces. These new capabilities allow us to be in a position to observe landscapes and document archaeological information at the most appropriate time for each individual year and each geographical area. In the air the archaeologist is free to choose conditions of lighting that range from soon after dawn to almost sunset (Musson et al., 2005).

The detail of the acquired information is remarkable and, despite the strong distortion of oblique images, the spatial information can be corrected using algorithms developed by Professor Irwin Scollar (Figure 6) and can thus be mapped and integrated without difficulty into our archaeological GIS (Herzog and Scollar, 1989; Scollar, 1998; Doneus, 2001). Currently at the University of Siena we have collected an archive of about 10,000 oblique air photographs, recording just under 1,000 archaeological sites of widely varying types.

We wish to emphasize here our view that oblique air photographs and satellite imagery are not in conflict with one another. Apart from the obviously varying degree of detail, the satellite images provide a total, continuous and objective view of the whole of the land surface within the chosen survey area. By contrast, every oblique aerial survey is dependent on the environmental conditions of the moment and is influenced by the experience of the individual archaeologist in choosing which parts of the landscape to document.

Along with the development of new methodologies, we are continuing our study of historical aerial photographs, which have an importance, which we think it unnecessary to emphasize here (Piccarreta and Ceraudo, 2000; Guaitoli, 2003). We are thinking particularly in this regard of the photographs of the Institute of Military Survey from the 1930s to the 1950s, along with the regional coverage of the 1970s. Our work has been concentrated

Figure 6. a, b, c) Example of mapping related to a single circular enclosure.
d, e, f) Complex site with a Roman villa and multiple chronological phases.

on the mapping of anomalies recorded in the last 20 years of aerial
activity and on repeated analysis of the images using traditional stereo
viewing, digital image processing techniques and more recently a digital
photogrammetric workstation. The availability of this last instrument has
proved extremely useful for the precise mapping of information from vertical
air photographs, as well as for the next research step, the collection
of data useful for infra-site analysis through the medium of detailed

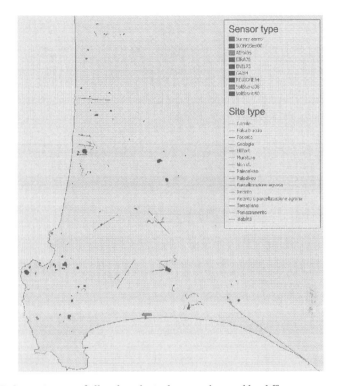

Figure 7. Synoptic map of all archaeological traces observed by different sensors around the city of Piombino (LI).

micro-topographic maps and photogrammetric digital elevation terrain models (Wheatley and Gillings, 2002).

The mapping of all the information recovered from satellite imagery and from vertical and oblique air photographs represents the main result of this strategy (Doneus et al., 2002). This operation allows us to create detailed georeferenced layers of sub-soil features that, as we will see in the following section, greatly enrich the data at our disposal and consequently our reconstruction of the archaeological record (Figure 7).

4. FIELDWORK RESEARCH METHODS AND RELATED PROBLEMS

The activities in the field are aimed at the systematic investigation of sample areas (Campana, 2004) and at the verification of the remotely sensed evidence. The main methodologies involved in the process are field-walking and geophysical survey.

Through field-walking survey there have so far been detected in the provinces of Siena and Grosseto about 9,000 sites (Francovich and Valenti, 2001; Carandini et al., 2002). This research method constitutes therefore an extremely important source for the archaeological study of settlement patterns, but the collected information often turns out to be incomplete, confused, and difficult to interpret because of post-depositional processes in the field (Boismier, 1991).

We should, for instance, take account of the progressive degeneration of many of the surface finds resulting from more than half a century of intensive ploughing, and vineyard and olive cultivation. Many years ago Tim Potter wrote about this subject in his *The Changing Landscape of South Etruria*, underlining how, by the beginning of the 1970s, the ideal moment for this kind of study had already passed (Potter, 1979).

Our experience in South Tuscan landscapes shows that progress in the development of interpretation methods for surface evidence has given us the possibility in the last 25 years to carry out successful programs of field-walking with an acceptable degree of uncertainly (Fentress, 2002). In recent surveys, however, there seems to have been a progressive change in the relationship between surface and sub-surface archaeology. We have realized that it is more and more rare to achieve a satisfactory interpretation of surface remains, and in some cases the process of collection within carefully predefined grids is no longer sufficient. This situation could arise, as already shown by authors in other study areas, from the total destruction of the once-present archaeological stratigraphy. In addition, there are at least two other scenarios which could produce on the surface a similar absence of finds.

Firstly there is the possibility that over the last fifty years the plough has worked almost always at the same depth, so that most of the surface finds have been completely destroyed. Beneath this "usual depth" of ploughing, however, it may be possible to identify other archaeological remains not yet damaged by ploughing. A second hypothesis could be related to the trend towards a reduction in the depth of ploughing as promoted, from the 1990s onwards, by the European Community with the objective of a better conservation of soil fertility (Agenda, 1996). In both cases the presence of surface remains would be very poor. The key issue for us is represented by the difference between the complete destruction of archaeological stratigraphy and problems related to the depth of ploughing. If the archaeological record has been totally destroyed the archaeologist has no chance of making worthwhile observations. The second option, associated with the depth of ploughing, reduces the applicability of field-walking methodology, directing research instead towards test excavation or remote and proximal sensing for the analysis of the sub-surface remains.

The discussion of these phenomena help to make clear a recurrent problem we encounter in the field. Visible traces detected through the analysis of remote sensed data do not always correspond with the presence of dense or well-defined scatters of archaeological material in the field. Considering that the minimum scale at which we operate is the local administrative area, with an average area of 150 sq km, and that the situations just described occur on average between ten and twenty times per administrative district, we cannot hope to address the situation through systematic test excavation. Moreover, we have to consider the bureaucratic difficulty of asking for permission for each excavation from the Italian Office of Heritage Conservation. In the last two years we have focused our attention on this topic in order to overcome or at least to reduce the consequences of this problem. The first attempt took the form of experiments with magnetic survey of a kind suited to our particular requirements.

In addition to the well known diagnostic characteristics of magnetic survey methods (Piro, 2001), this technique satisfies one of our fundamental needs: the capacity to cover large areas in a limited time (Powlesland, 2001). In a field survey carried out in Val d'Orcia, we progressively tested a system of acquisition that allowed us to cover one hectare per day at a resolution of 60 cm along traverses each set 1 meter apart.

So far we have acquired only 10 hectares of data, but the general trend of the results seems to confirm that the degree of detail, although not very high, is sufficient to show with a good approximation the position of the main features, depending on the characteristics of the material to which the magnetometer is reacting. This pattern of acquisition will allow us to contemplate the future acquisition of approximately 20–40 hectares per year, an area perfectly compatible with our research requirements.

In several cases we were able to improve the resolution by means of a sampling interval of 25 cm along traverses set only 50 cm apart. We may take as an example the case of Pieve di Pava in the community of San Giovanni d'Asso (Province of Siena), where we acquired 2 hectares of data at a resolution of 1 m between traverses and then reduced the resolution to 50 cm in order to make a comparison (Figure 8). The results from the closer sampling interval undoubtedly show an enrichment of the data and an improvement in definition of the shapes of features. At this stage this closer resolution looks to be the best choice for medieval sites generally characterized in our region by the absence of building materials with a high magnetic susceptibility.

The site of Pieve di Pava represents for us an important case study, in particular because in July and August 2004 we undertook an archaeological excavation at this site. The excavation represents for us the first chance to verify and compare the gradiometric data with the observed stratigraphy. It was for this reason that we intensified our geophysical work on the site by testing a wide range of different parameters of the gradiometer and by trying other instruments such as GPR.

In conclusion, in the immediate future the challenge will be on the one hand to enlarge the range of geophysical instruments systematically available

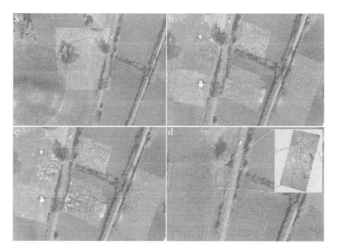

Figure 8. Integrated mapping process of the archaeological record. a) The red areas delimit surface findings collected during field-walking survey. b and c) Examples of increasing the resolution of magnetic survey. d) Comparison between magnetic anomalies and archaeological excavation.

and on the other side to make use more and more often of the practice of small test excavations on different sites so as to improve our experience. As has already been established, this last approach is essential to reconcile, at least to some extent, the different kinds of information given by geophysical survey and excavation evidence, allowing us to achieve a better overall understanding of our data.

5. BRIDGING REMOTE SENSING, INFRA-SITE ANALYSIS, AND ARTIFACT SCATTERS THROUGH MOBILE INFORMATION TECHNOLOGY

After discussing our work in the field we should emphasize that from the second half of the 1990s, when in Siena as in the rest Italy we began to use Geographic Information Systems for the management of archaeological data, we have felt a progressive disjunction between work in the laboratory and work in the field. While the availability of advanced technologies has been rapidly growing, activities in the field have continued to make use of instruments and methodologies developed in the 1970s. The risks arising from this situation are many. Firstly, there are problems inherent in the collection of data that lack the accuracy required by GIS systems, or which rely on a different kind of rational logic. Secondly, there is the problem that the large amount of data produced or available through GIS systems in the laboratory is available in the field only as hard copy, without the possibility of direct interaction or real-time data integration and interrogation.

The processes of field-walking survey represent a good example of this first problem. It has to be recognised that an important influence on the value of the data that we collect is the methodological process that we apply in the field and the possibility of georeferencing monuments or artifact scatters with an acceptable degree of accuracy. A first real improvement, as is well known, came with the introduction of GPS (Global Positioning System) measurement. Our own first experience in the use of GPS was strictly related to the mapping of artifact scatters and archaeological structures during field survey. We soon realized that GPS technology could be applied in the field for numerous tasks, such as the mapping of special finds, photo-locations, Quick Time Virtual Reality, movie sequences, shovel tests, tracking the actual path and distance between researchers during field-walking and monitoring the movement of artifact scatters, etc. Moreover, the GPS device was also extremely useful for activities such as aerial survey, for recording aerial tracks and the positions of the sites photographed from the air. GPS equipment has also been used to accelerate the recording of the geographic coordinates of grid intersections in magnetic survey, and finally for the collection of ground control points for photogrammetry etc.

Furthermore, our GPS not only permits the mapping of points, lines and areas, but also allows us to navigate on site on the basis of geographic coordinates exported from our laboratory-based GIS systems. Navigation with the support of a GPS represents an extremely useful tool in many archaeological applications. For instance, in relationship with anomalies a methodological problem that we encountered before the introduction of GPS technology was the problem of accurately identifying on the ground features previously identified on one kind of imagery or another. This difficulty was particularly acute in situations of limited visibility (wooded areas, vineyards, olive groves, etc.) or in the absence of identifiable artifacts or structures. The main limitation of the traditional GPS device for navigation is related to the interface and the non-availability of raster data as background to the locational information. Nevertheless, GPS represented the first device that was able to reduce the disjunction between the real world and its digital representation in our GIS systems. Data collection with the GPS allowed us to move from an approximate representation of reality (our archaeological GIS and digital data in general) to the real world and vice versa, making it easier to link and integrate the two environments with one another: reality and its GIS abstraction. The main restriction of the handheld GPS is that it remains only a GPS device. By this I mean that, with the exception of stored waypoints, we have no capacity to exchange data with the desktop PC and to take that data into the field with us.

This situation has now changed, however, with the advent of the PDA (Personal Digital Assistant) computer for use in fieldwork and other archaeological activities. Our first experience of the PDA computer came during work on the archaeological map of Siena (Ryan et al., 1999; Ryan and Van Leusen, 2002; Craig, 2000). We started by using a Compaq iPAQ on which we had installed the corresponding versions of the software used in the laboratory: ArcPad 6 as graphic interface with the geographic data and FileMaker Mobile as Database. In order to be able to work with satisfactory precision we used as our GPS the Trimble Pathfinder Pocket. This was the only device we knew of that was able to perform, with the appropriate GPS software, the differential correction of data either in post-processing or in real time.

On this topic, it is important to stress that, despite the cessation of selective availability, without access to a reference station it is still impossible to obtain with a survey device accuracy better that 5 or 3 meters. When using a GPS device such as the Trimble Geoexplorer 2 or 3 it was sufficient to correct the data in post-processing, but now the full exploitation of the possibilities offered by the PDA requires real-time correction. The installation in our own department of archaeology of a GPS base station (Trimble 5700) has given us at least two significant advantages. Firstly we can now plan the sampling interval for measurements in direct relationship to our needs, and secondly we have been able to connect to the GPS reference station a radio-modem that through telephone links allows the exchange in real time of the data collected by the rover device. The best solution available at the moment for the connection between the PDA and the mobile telephone is a Bluetooth link. Even though we are totally convinced about the iPAC system as the information vehicle, the fragility, bulkiness, and physical complexity of the instrument greatly limits its systematic use in the field (Figure 9).

Recently we have changed our choice of device and are now working with a Trimble GEO XT. Although this still has some hardware limitations it has proved itself to be totally satisfactory, with important increases in flexibility and efficiency and with an accuracy as close as 30 cm after real-time differential correction (now available in real time). This device has allowed us to load into the PDA all the GIS data that we require and therefore to take into the field a range of information that was formerly accessible only in the laboratory. Specifically, the main applications, and the resulting improvements to research in the field can be summarized as follows:

- real-time access to a large selection of the information stored in the database and GIS;
- the capacity to integrate topographical, thematic and historical maps, aerial photographs, satellite imagery, and geophysical data;
- quick, easy, and accurate navigation to any target whose coordinates are known through the interaction of a GPS point with the raster and vector data in the background;

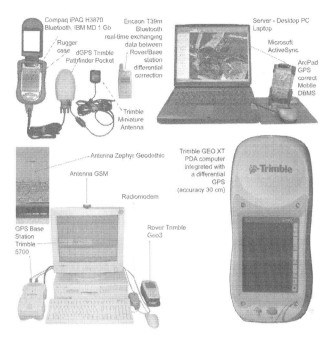

Figure 9. PDA, mobile GIS/GPS and DBMS solution tested during our landscape projects.

- the opportunity to make direct and accurate real-time comparisons between past survey data and the actual situation observed in the field, for instance in monitoring changes of land use or the movement of artifact scatters etc.;
- the capacity to compile the documentation directly in digital format in the field, giving significant savings of time through the use of software for the transfer to the server of data collected in the field (a future development could be the availability of an "always on" connection for data exchange with the cartographic server, though as yet the GPS connection has proved too unstable for this to be a practical proposition in Italy);
- the availability of a device that "connects" real landscapes (the material world as observed at a precise moment in time) with the digital representation of the past and present countryside.

But what we would really like to emphasize is that the technological merging between PDA and GPS devices goes far beyond the level of increased fieldwork efficiency, in at last making possible the systematic application of strategies and methodologies developed in the past but applied only rarely up till now because of the excessive amount of time involved in their use (Orton, 2000). One may think here, for instance, of some practices tested in the early eighties for field survey and the collection of artifacts, such as the georeferencing of every single object or the collection of artifacts within predefined grids etc.

In recent years the availability of GPS equipment has allowed us to consider entirely new ways of collecting surface material in the field, and the advent of the PDA device further extends these possibilities. To explain these developments more clearly we may take as an example a comparison between a traditional GPS device and the GEO XT PDA with regard to the creation of a survey grid.

In the case of the simple GPS instrument topographical recording has certainly become quicker in comparison with Total Station survey, but there still remains the need for physical construction of the survey grid, generally involving posts fixed into the ground at the corners of each cell. This is time-consuming and requires the researcher to carry all the tools and materials necessary for the physical construction of the grid. In our experience most sites involve an average of between 150 and 250 squares, of five metres each. To accelerate the topographical work we have adopted the practice of measuring *half* of the corners through the acquisition of 120 GPS measurements per corner, at a sampling interval of one second. This method took an average of between 1 and 1.5 hours for the building of the grid and between 2.5 and 4 hours for its georeferencing. We consider this to be a satisfactory outcome, bearing in mind that only one person was involved in building the grid, allowing other workers to concentrate on the collection of artifacts (Figure 10).

During our most recent survey, in the countryside of Montalcino in the province of Siena, the systematic use in the field of the PDA device has convinced us that it is possible to develop an entirely new solution for

Figure 10. Collection of artifacts organized into "physical" predefined grids georeferenced through the differential GPS.

the collection of artifacts within a predefined grid. In the laboratory GIS we generated vector grids in shapefile format of the areas to be surveyed, using three different sample intervals of 5, 10 and 20 m. Every cell of the "virtual" grid was allocated an identifier composed of the acronym of the local administrative unit, the reference number of the sample area, and finally a sequential number within that sample area. After transformation of the projection system from local grid to UTM, we loaded the data into the PDA. This allowed us to work in the field without the need to build a real grid. The merging between the digital grid visualised in the PDA and the real-time position of the DGPS allowed us to move directly to surface collection, without the need to carry with us all of the tools and other equipment needed for the construction of a physical grid (Figure 11).

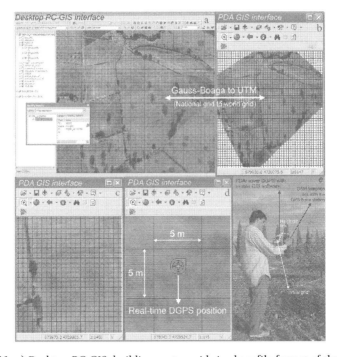

Figure 11. a) Desktop PC GIS: building vector grids in shapefile format of the areas to be surveyed, using three different sample intervals of 5, 10 and 20 m and generation of an identifier for every cell; b) PDA mobile GIS: coordinate-system transformation (Gauss-Boaga to UTM). c) PDA mobile GIS: detail of the grid overlaid on a rectified oblique aerial photograph. d) PDA mobile GIS: visualization of the merging between the digital grid and the real-time position of the DGPS. e) PDA mobile GIS at work: grid collection without the need to build a physical grid, thereby moving directly to surface collection.

The contribution to field-walking survey of this technological solution has not just been in the reduction of working time or the avoidance of conflict with land-owners, but, more importantly, in the capacity each time we find a new site to choose the best strategy for collecting artifacts at this particular location. Moreover, in field-walking survey we often need to replicate collection in different years or at different seasons of the year. The PDA allows us to go back into the field and repeat surface collection in precisely the same cells, without the effort of rebuilding a physical grid in exactly the same position.

For field-walking survey this technique represents only one of many new possibilities that a mobile GIS system can offer to archaeological research. Similarly significant changes can come from this new technology in the practice of aerial survey with the introduction of Tablet PC joined with Bluetooth GPS, which allow to have at our disposal in the air more flexible, rich, and interactive maps. Or again, in the ground-truthing of remotely sensed data with geophysical instruments such as GPR, the archaeologist could navigate, almost as we showed for the field-walking (with the difference in this case of using a topographic GPS), alongside predefined "virtual" grids built specifically for the prospecting.

In general we believe that giving more attention to the process of data collection, and in particular to the contribution that new technology can make in the process of fieldwork, is one of the best ways to achieve a real improvement in the acquisition of new data for our GIS systems and consequently in the type and quality of the analyses that we can then carry out.

In our experience the integration of PDA and GPS devices represents for archaeologists an extremely powerful combination, capable of transforming the practice, quality, and power of our work in the field in much the same way as happened more than 15 years ago with the arrival of desktop personal computers in the laboratory. Moreover, this new development goes a long way towards restoring the link between active work in the field and management and analysis of heritage data in the laboratory.

6. CONCLUSIONS

Much effort has been exerted at the University of Siena to improve archaeological research through the application of various remote sensing techniques in the laboratory and integrated with coherent methodological and technological developments into the fieldwork. A common risk in archaeological research on the use of technology in the study of cultural heritage is the obsessive pursuit of the latest technological device or software. In this short review of our work and our experience we have tried to show that the progressive integration of survey techniques directly responds to the need to answer specific historical and archaeological questions or to face specific methodological problems. This approach has allowed us to define a variety of research strategies for a better understanding of the past. In this context we must stress that we are not seeking to define standardized methodologies for site detection and data collection. We are aware of the risk of forcing data into predetermined descriptive frameworks unrelated to any particular historical question or concern (Boismier, 1991). On the other hand, we have needed to improve the range of methodologies and techniques at our disposal in order to apply each time different ways of meeting differing archaeological problems and different environmental backgrounds. Our challenge concerns the application of appropriate techniques and instruments using different parameters and related combinations to answer differing archaeological questions. We maintain that these efforts are inevitably conditioned by the physical context in which we operate and are closely conditioned by the kind of historical and archaeological questions that we wish to answer in our investigation of the landscape.

On the basis of our present experience we are firmly convinced of the advantages of using integrated sources and technologies. Even in favourable areas such as the Tavoliere in southern Italy, or in lowland Britain or parts of central and eastern Europe such as Hungary and Poland, it has been demonstrated that there are significant advantages in the use of integrated techniques (Powlesland, 2001; Grosman, 2000; Doneus et al., 2002; Gojda, 2002). This strategy becomes even more obvious in less favourable contexts such as Tuscany, and particularly in the Province of Siena and the hill-country of Grosseto (Campana, 2003).

Without the integrated use of multi-sensor approaches and the critical application of both traditional and new methods, we can rarely hope to achieve results that will have a real impact on the search for a better understanding of the development through time of regional settlement patterns.

Stefano Campana

7. ACKNOWLEDGEMENTS

We would like first to thank the editors of the present volume for inviting us to participate in their discussions. The authors are indebted to Professor Salvatore Piro (ITABC – CNR) and to Professor Dario Albarello for their invaluable comments and criticisms on geophysical survey techniques and data processing. Chris Musson has also helped by reading the translation and contributing related comments. Many researchers and students have collaborated and are still collaborating, in the Siena and Grosseto archaeological map projects. Special thanks are owed to the team of the Landscapes Archaeology and Remote Sensing Laboratories of the University of Siena at Grosseto: Dr. Cristina Felici, Dr. Emanuele Vaccaro, Dr. Anna Caprasecca, Francesco Pericci, and Maria Corsi.

REFERENCES

Agenda, 1996, La politica agricola comune. In *Commissone Europea, Agenda 2000*. Bruxelles.

Agnoletti, M., ed., 2002, *Il paesaggio agro-forestale toscano. Strumenti per l'analisi, la gestione e la conservazione*. Florence.

Barker, G., 1986, L'archeologia del paesaggio italiano: nuovi orientamenti e recenti esperienze. *Archeologia Medievale* XIII:7–27.

Boismier, W.A., 1991, The role of research design in surface collection: an example from Broom Hill, Braishfield, Hampshire. In *Interpreting Artefact Scatters. Contributions to Ploughzone Archaeology*, edited by A.J. Schofield, pp. 11–25. Oxford

Campana, S., 2001, *Carta Archeologica della Provincia di Siena, Vol. 5. Murlo*. Siena.

Campana, S., 2002a, IKONOS-2 multispectral satellite imagery in the study of archaeological landscapes: an integrated multi-sensor approach in combination with "traditional" methods. In *The Digital Heritage of Archaeology, CAA02 Computer Applications and Quantitative Methods in Archaeology*, edited by M. Doerr and A. Sarris, pp. 219–225. Athens.

Campana, S., 2002b, High resolution satellite imagery: a new source of information to the archaeological study of Italian landscapes? Case study of Tuscany. In *Space Applications for Heritage Conservation, Proceedings of the EURISI Conference (Strasbourg, 5–8 November 2002)*, edited on CD-ROM.

Campana, S., 2003, Remote Sensing, GIS, GPS e tecniche tradizionali. Percorsi integrati per lo studio dei paesaggi archeologici: Murlo-Montalcino e bassa Val di Cornia. Ph.D. thesis, University of Siena.

Campana, S., 2004, Ricognizione archeologica nel territorio comunale di Montalcino: campagne 1999–2001. Progetto Carta Archeologica della Provincia di Siena. In *Ilcinesia. Nuove ricerche per la storia di Montalcino e del suo territorio (Montalcino, 19 Maggio 2001)*, edited by A. Cortonesi, pp. 37–63. Roma

Campana, S., and Forte, M., eds., 2001, *Remote Sensing in Archaeology. InXI International School in Archaeology, Certosa di Pontignano, 6–11 December 1999, Florence*. Firenze.

Carandini, A., Cambi, F., Celuzza, M., and Fentress, E., eds., 2002, *Paesaggi d'Etruria. Valle dell'Albegna, Valle d'Oro, Valle del Chiarore, Valle del Tafone*. Rome.

Craig, N., 2000, Real-time GIS construction and digital data recording of the Jiskairomuoko excavation, Perù. *SAA Bulletin* 18 (1):24–28.

Clark, A.J., 1997, *Seeing Beneath the Soil. Prospecting Methods in Archaeology*. London.

Doneus, M., 2001, Precision mapping and interpretation of oblique aerial photographs. In *Archaeological Prospection* 8:13–27.

Doneus, M., Doneus, N., and Neubauer, W., 2002, Integrated archaeological interpretation of combined prospection data, Zwingendorf (Austria). In *Aerial Archaeology. Developing Future Practice*, edited by R. Bewley and W. Ryczkowski, pp. 149–165. IOS Press, Amsterdam.

Doneus, M., Eder-Hinterleitner, A., and Neubauer, W., eds., 2001, *Archaeological Prospection. Proceedings of 4th International Conference on Archaeological Prospection*. Vienna.

Donoghue, D., 2001, Multispectral Remote Sensing for Archaeology. In *Remote Sensing in Archaeology*, edited by S. Campana, and M. Forte, pp. 181–192. Firenze.

Fentress, E., 2002, Criteri tipologici e cronologici. In *Paesaggi d'Etruria. Valle dell'Albegna, Valle d'Oro, Valle del Chiarore, Valle del Tafone*, edited by A. Carandini, F. Cambi, M. Celuzza, and E. Fentress. Roma.

Francovich, R., and Valenti, M., 2001, Cartografia archeologica, indagini sul campo ed informatizzazione. Il contributo senese alla conoscenza e alla gestione della risorsa culturale del territorio. In *La carta archeologica. Fra ricerca e pianificazione territoriale, Florence, 6–7 May 1999*, edited by R. Francovich, A. Pellicanó, and M. Pasquinucci, pp. 83–116. Florence.

Francovich, R., and Hodges, R., 2003, *Villa to Village. The Transformation of the Roman Countryside in Italy, c. 400–1000*. London.

Gojda, M., 2002, Aerial Archaeology in Bohemia at the Turn of the Twenty First Century: Integration of Landscapes Studies and Non-destructive Archaeology. In *Aerial Archaeology. Developing Future Practice*, edited by R. Bewley, and W. Ryczkowski, pp. 68–75. IOS Press, Amsterdam

Grosman, D., 2000. Two examples of using combined prospecting techniques. In *Non-destructive techniques applied to landscape archaeology*, edited by M. Pasquinucci and F. Trémont, pp. 245–255. Oxford.

Guaitoli, M., ed., 2003, *Lo sguardo di Icaro. Le collezioni dell'Aerofototeca Nazionale per la conoscenza del territorio*. Rome.

Herbich, T., 2003, *Archaeological Prospection. Proceedings of 5ᵗʰ International Conference on Archaeological Prospection*. Warsaw.

Herzog, I., and Scollar, I., 1989, The Mathematics of Geometric Correction of Oblique Archaeological Air Photos. In *Into The Sun. Essays in Air Photography in Archaeology. In Honour of Derick Riley*, edited by D. Kennedy. Sheffield.

Jones, R.J.A., and Evans, R., 1975, Soil and crop marks in the recognition of archaeological sites by air photography. In *Aerial reconnaissance for archaeology*, edited by D.R. Wilson, pp. 1–11. CBA Research Report 12. London,

Lillesand, T.M., and Kiefer, R.W., 1994, *Remote Sensing and Image Interpretation*. New York.

Musson, C., Palmer, R., and Campana, S., 2005, In *Volo nel Passato Aerofotografia e cartografia archeologica*. Florence.

Orton, C., 2000, *Sampling in Archaeology*. Cambridge University Press.

Piccarreta, F. and Ceraudo, G., 2000, *Manuale di aereofotografia archeologica. Metodologia, tecniche, applicazioni*. Bari.

Piro, S., 2001, Integrazione di metodi geofisici ad alta risoluzione per l'indagine di siti archeologici. In *Remote Sensing in Archaeology*, edited by S. Campana and M. Forte, pp. 273–296. Firenze.

Potter, T.W., 1979, *The Changing Landscape of South Etruria*. London.

Powlesland, D., 2001, The Heslerton Parish Project. An Integrated multi-sensor approach to the archaeological study of Eastern Yorkshire, England. In *Remote Sensing in Archaeology*, edited by S. Campana and M. Forte, pp. 233–255. Firenze.

Ryan, N., Pascoe, J., and Morse, D., 1999, FieldNote: extending a GIS into the field. In *New Techniques for Old Times, CAA98 Computer Applications and Quantitative Methods in Archaeology*, edited by J.A. Barceló, I. Briz, and A. Vila, pp. 127–132. BAR International Series S757, Oxford.

Ryan, N. and Van Leusen, M., 2002, Educating the Digital Fieldwork Assistant. In *Pushing the Envelope, CAA01 Computer Applications and Quantitative Methods in Archaeology*, edited by G. Burenhult and J. Arvidsson, pp. 401–416. BAR International Series 1016, Oxford.

Scollar, I., 1998, AirPhoto – a WinNT/Win95 program for geometric processing of archaeological air photos. *AARGnews* 16:37–38.

Wheatley, D. and Gillings M., 2002, *Spatial Technology and Archaeology. The Archaeological Application of GIS*. London.

Wilson, D.R., 2000, *Air Photo Interpretation for archaeologists*. Tempus, Stroud.

Chapter 11

Remote Sensing and GIS Analysis of a Maya City and Its Landscape: Holmul, Guatemala

FRANCISCO ESTRADA-BELLI AND MAGALY KOCH

Abstract: Holmul is an ancient Maya city located in the dense tropical forest area of the Maya Bio-sphere reserve in northeastern Guatemala. The city is surrounded by subsidiary centers forming a yet to be understood complex political landscape. The current study of this largely unexplored region focuses on obtaining an understanding of landforms, ecozones, and their relationship to human settlement through the use of multi-source satellite images validated by archaeological foot survey. First, Tropical forest vegetation and landforms are identified by classifications of a combined set of imagery including Ikonos, Landsat ETM+, and airborne IFSAR sensors. From these data we derive a geomorphological map of the region which is used to identify the relationship of settlement with land features using a GIS modeling engine. The final products of these analyses are a better understanding of the environmental resources available to Maya settlement in the Holmul region as well as a predictive model for identifying new unexplored settlements.

1. INTRODUCTION

It is not a coincidence that in the northern Petén region of Guatemala are some of the largest and most complex cities built by the ancient Maya and the largest expanses of unexplored land in the entire Maya region. The lack

of large-scale archaeological surveys in this region is the result largely of
the difficulties of reconnaissance caused by the thick tropical forest and by
the lack of infrastructure and modern settlement. This condition in many
ways is unique to this part of Guatemala. This region today represents the
largest continuous tropical forest in Central America, which corresponds to a
conservation area known as the Maya Biosphere Reserve (MBR). The MBR was
created in 1990 by the Guatemalan government for the protection of precious
forestry, wildlife, and archaeological resources because an alarming rate of
deforestation was brought to the attention of the Guatemalan government by
a NASA team using satellite imagery (Garrett, 1989; Sever, 1998). Under the
provisions of the Maya Biosphere Reserve statute no new settlement or agricultural development is allowed in the core of this area; instead, sustainable uses
of its resources are fostered by several governmental and non-governmental
agencies (USAID, 1996). The remoteness of the area from most settlement and
paved roads is one of the many factors discouraging widespread mapping of
archaeological remains.

On the ground, the thick high-canopy tropical forest and hilly terrain
present a challenge for navigation and field survey. From the air, they make
traditional use of aerial photography and site-discovery completely ineffective.
The terrain is a hilly karstic landscape interspersed with wide expanses of
wetlands. Elevations range from 100 m to 300 m above sea level. The forest
of this area of Petén varies from scrub vegetation in the swampy areas to
20-m-high hardwood forest in the uplands (Ford, 1986; Lundell, 1937). As a
result, in the Maya Lowlands, large-scale surveys have been carried out mostly
in deforested lands areas such as the Belize River Valley (Willey et al., 1965)
or in the southeastern Petén district (Laporte et al., 2004). Examples
of systematic regional surveys such as the Three-Rivers Regional Survey
(Adams, 1995, 1999) spanning the northeast corner of Guatemala and
northwest of Belize are notable exceptions to this general pattern. Aside from
these rare examples, archaeological mapping in the forest has progressed in
a patchwork of site-centered maps, with narrow transects radiating out of
site centers, such as at Tikal (Puleston, 1983), or inter-site transects such as
Ford's (1986) Tikal-Yaxha "brecha".

In 2000, we initiated our investigation of the ancient city of Holmul, in
northeastern Petén with a regional perspective, in order to understand the
relationships of the settlement with its land base and near neighbors (Estrada-
Belli, 2000, 2002). We looked into a variety of remote sensing techniques
to overcome the challenges of a deeply forested and uncharted territory.
The pioneering work of Adams et al. (1981) with SEASAT radar mapping
of wetland features and Sever's search for causeways and wetland features
(Sever and Irwin, 2003) with a variety of new sensors were ideal precedents
to follow. We believed that integrating imagery of different types of sensors,
multispectral and microwave, at various spatial resolutions with field-surveyed
maps in a GIS database would help us discover and analyze ancient settlement

and landscape features in ways otherwise unattainable. Most importantly, satellite imagery combined with GPS navigation ensured that our crews could always find their way through the insidious terrain in spite of non-existent maps and confused guides. While site-discovery and prediction still remain difficult to implement with remotely sensed data in this environment, in general, the rate of success of these remote sensing and GIS methods have exceeded our expectations.

2. BACKGROUND HISTORY OF THE STUDY REGION

The ancient site of Holmul is known as the first Lowland Maya city to have been excavated by a proper scientific expedition. This historic excavation was directed by Harvard's Raymond Merwin in 1911. His work illustrated for the first time the developmental sequence of ceramic and architectural development in the Lowlands from the end of the Preclassic to the end of the Classic periods (ca. A.D. 150–900), with spectacular burial offerings and temple buildings (Merwin and Vaillant, 1932). Unfortunately, Merwin's incomplete report on Holmul because of his untimely death, the site's remoteness, and its lack of carved inscriptions discouraged further thorough study of the site for the following 90 years. The ceremonial core of Holmul remained unmapped and any sites around it unknown. In the 1950s, a mule-trail survey by Bullard (1960) came in the vicinity of Holmul and mapped a variety of residential groups. Then, in the 1980s Ian Graham explored this area in a search for inscribed stelae and discovered the Preclassic site of Cival, 7 km to the north of Holmul, and the site of La Sufricaya 1 km southwest of Holmul (Ian Graham, personal communication, 1992).

In 2000, the Holmul Archaeological Project began its explorations of the region within a 10 km radius of the ceremonial center. A preliminary tape-and-compass map of the site was produced in the first season and was further corrected in subsequent years with an electronic total station instrument (Estrada-Belli, 2000, 2002, 2004). We have surveyed the residential areas with traditional 250-m-wide transects extending 4.5 km to the west, 2.5 km to the east, and 2 km to the north (Estrada-Belli, 2002; Gardella, 2004), and we discovered and mapped a number of minor ceremonial centers within an hour-walk distance of Holmul center (Estrada-Belli, 2002; Estrada-Belli et al., 2004). These were K'o, Riverona, and T'ot, which were known to the local *chicle* (latex gum) collectors. In preparation of the first field season, we digitized topographic maps of the Guatemalan Instituto Geografico Militar (IGM) at 1:50,000 scale and created GIS layers for elevation, rivers, and roads. Subsequently, we generated a Digital Elevation Model (DEM) from the 20-m interval contour lines from the same topographic maps to aid in the study of the environment, navigation, and site discovery.

3. GIS LANDSCAPE ANALYSIS AND SITE DISCOVERY

In 2002, we discovered two ceremonial centers thanks to a prediction based on a GIS analysis of least-resistance pathways across the Holmul landscape. Recognizing that all known minor centers formed a ring-like pattern at regular distances around Holmul (roughly a 1-hour walk), we explored the possibility that these were located along major routes leading to the main center. Using the DEM, we created a *least-cost path* model with the GRASS-GIS program (Estrada-Belli et al., 2004). The first step was to develop an attrition surface for each direction of movement into the Holmul site, combining aspect and slope as the main factors determining the cost of movement across the land. Secondly, we added non-isotropic distances from Holmul, thus modeling the cost of movement according to attrition features in every direction rather than uniform ring-like distance zones.

In a previous study of the Maya settlement of the city of La Milpa, Belize, using viewshed maps we found that most settlement and especially elite settlement tended to be in view of the ceremonial center's temple pyramids (Estrada-Belli et al., n.d.). In addition, studies of formal pathways in the New and Old World show that, typically, these tend to be tied to civic or sacred locations on the landscape by visual "anchors" along the path. In these cases, a desired view is accomplished by routing the movement along certain ridge tops or valley bottoms (see as case studies Exon et al., 2000 for Stonehenge, England; Madry and Rakos, 1996 for Burgundy, France).

A separate set of cost-maps were created adding viewshed as a cost factor, as it was believed that constant visibility of the sky-scraping pyramid of Holmul's Group 1 may have had a symbolic significance in the choice of access routes leading to the center. A search algorithm in the GRASS-GIS program selected the least-resistance paths through the region originating from all directions at the margin of the map. All the known minor centers ringing Holmul appeared to be in close proximity to the predicted least-resistance access route (Figure 1). In the northeast and northwest quadrants, which were at the time unexplored, we predicted the existence of two new sites at the predicted 1-hour walk distance along the GIS-derived route. Remarkably, in the northwest, the site of Hahakab (Mayan for "looking at the earth") was found to be at 3.5 km from Holmul at the convergence of two predicted paths (viewshed and non-viewshed related) within meters of the target location. To the northwest, the site of Hamontun (Mayan for "stone house of the water macaw") was discovered along the predicted viewshed-related path, at 5 km from Holmul. This was the only path derived for the northeastern quadrant and stood in contrast to the eastern site of K'o, which is located on a non-viewshed related path. We concluded that the five largest centers with elite palatial and ritual architecture outside of Holmul were built in regularly spaced locations along major routes leading to Holmul from each cardinal direction. In addition, these elite centers combined a visual link with Holmul's

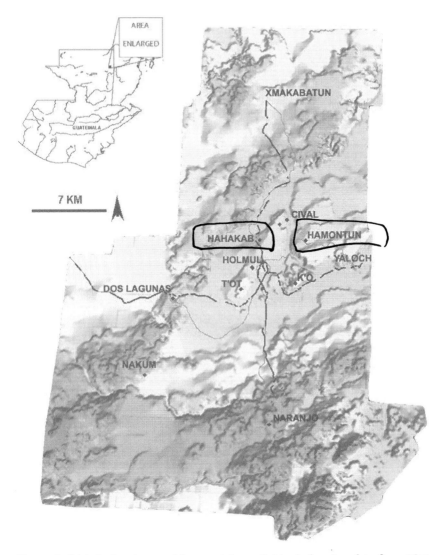

Figure 1. Digital elevation model created from digitized elevation data from IGM topographic maps at 1:50,000 scale showing paths of least resistance to Holmul and associated sites.

pyramids with a wide field of view outwards across the wetlands and ridges that ring the Holmul plateau. This arrangement may have served a variety of functions: 1) A visual link between the center of the ancient realm and its constituent parts may have served ritual functions at each locale and as a metaphor for cosmological order and for the secular power relations at work. 2) This disposition of elite palaces would have enabled effective control on

access to and from Holmul for security reasons. 3) From these locations at
the intersection of upland and lowland ecozones elite palaces could have had
optimal control of the widest variety of resources possible (Figure 1).

4. REMOTE SENSING ANALYSIS OF THE HOLMUL
LANDSCAPE

In addition to the interesting findings described above, the DEM was
highly instrumental for acquiring a first understanding of the geological
properties of the area surrounding Holmul, especially when combined with
LANDSAT imagery as overlays in three-dimensional renderings. These are
discussed below.

A set of images from optical (LANDSAT 5 and 7 TM/ETM+, IKONOS)
and microwave sensors (IFSAR) were acquired beginning in 2000. We
obtained from the U.S. Geological Survey (USGS, 2005; http://edc.usgs.gov)
a LANDSAT TM 5 image taken on December 27, 1989, which documents
the vegetation cover during the last part of the wet season (May-February),
and a LANDSAT 7 ETM+ taken on March 7, 2001, twelve years after the
first image and during the first part of the dry season (March-April). The
LANDSAT images provide coverage of a very wide area (185 km × 185 km)
well beyond the area of study, encompassing northeastern Guatemala and
most of Belize at a pixel resolution of 28.5 m and a spectral resolution spanning
three visible bands and three near- and mid-infrared bands, complemented by
a 15-m-resolution panchromatic band in the ETM+ dataset. We then acquired
an IKONOS image (Space Imaging, 2005; http://www.spaceimaging.com) of a
10 km × 10 km tile centered on Holmul. This image came with three visible
and one infra-red bands at 4-m resolution in addition to a panchromatic band
at 1-m resolution and dates to March 19, 2002. Lastly, in 2003 we acquired
through NASA's Scientific Data Purchase Program a dataset from the STAR3i
microwave sensor. This instrument was flown over a large portion of northeastern Petén in 1999 by NASA using the Interferometric Synthetic Aperture
Radar at 10,000 m altitude (Sever and Irwin, 2003). The dataset includes a
set of single-band orthorectified images (ORI) derived from the radar swath
at 2.5-m resolution and a digital elevation model (DEM) created from the
interferometry data at a 10-m resolution.

5. ENVIRONMENTAL SETTING OF HOLMUL
AND SURROUNDINGS

Our DEM dataset and the color-composite display of a subset of the LANDSAT
ETM+ image proved most helpful in characterizing the major geological and
hydrological features of the landscape in which Holmul is located. Figure 1

shows the location of Holmul at the center of an upland karst ridge trending southwest-northeast, surrounded by swampy flats locally known as *bajos*, which are in turn the center of a vast concave shelf bounded by two sharp karstic escarpments trending southwest to northeast, as well. The highest elevations in the area are found in the southernmost escarpment edge of the plateau, near the location of Holmul's largest neighbor, the Maya city of Naranjo (Figure 1). From these elevations the ground slopes gently toward north and northwest, rising again abruptly (200 m increase) on the northern escarpment. This trending slope determines the water flow direction of the main drainage system, the Holmul River, which enters the region from the west and sharply turns to the north, bisecting the karstic ridge where the ancient Holmul city is located into two halves. Numerous faults and joints criss-cross the limestone ridges draining most of the surface water, thus keeping the soils in elevated areas well drained. Straight sections of the Holmul River and its tributaries follow these structural weakness zones in the limestone and eventually debouch into large karstic depressions locally known as *bajos*, which remain flooded during the rainy season. To the west, a minor seasonal watercourse runs along the edge of the northern escarpment effectively draining the western *bajo* and eventually converging into the main branch of the river north of the site of Cival. To the south, an arroyo diverts from the Holmul River to the east towards the Yaloch lake, eventually reconnecting with the main branch immediately east of the Holmul site.

The LANSDAT 7 image in Figure 2 highlights the variety of forest cover and ecotones in the area with differences in tones and textures. The vegetation ranges from dense rainforest in the elevated areas to shrubs and grasses in the lowest areas (Fedick and Ford, 1990). In particular, in the well-drained uplands (Ford, 1986) are communities of tree species that include the ramon *Brosimum alicastrum,* cedar *Cedrela spp.*, ficus *Ficus glaucensces,* Chico Zapote *Manikara zapota* and mahogany *Swietenia macrophylla* (green-blue areas in Figure 2). In the transitional zones on the slopes of the uplands corozo *Orbignya cohune,* escoba *Crysophilia argents,* and guano palm *Sabal spp.* are more common (light green areas in Figure 2), and in seasonally inundated *bajos* several types of scrub trees are dominant, including in some areas the tinto redwood species *Haematoxylum campechanum* (magenta areas in Figure 2). In addition, along the watercourses are riparian forest areas containing a mix of palm, tall trees, and thorny bamboo groves (vibrant green areas in Figure 2). Finally, an interesting round feature can be seen in the upper portion of the image (in white in Figure 2) next to the Holmul River. This is a large sinkhole (doline) covered with grasses and reeds (white and pink areas) and shallow water pooling in the center (black dot).

Typical relationships among vegetation types, landscape forms, and settlement areas have been observed on the ground by a number of authors (Brewer et al., 2003; Puleston, 1983; Ford, 1986; Kunen et al., 2000). We intended to explore such relationships over large areas with the aid of

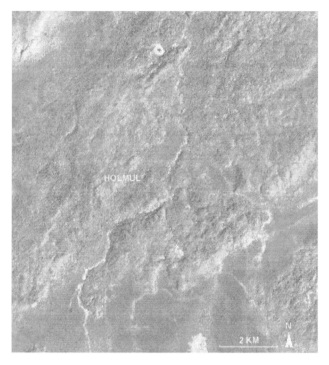

Figure 2. Landsat ETM+ color composite (bands 7, 4, and 2) highlighting major landscape features and showing the location of the Maya city of Holmul.

satellite imagery. This study was done through an unsupervised classification procedure using the *isodata* classifier performed on a subscene of the Landsat ETM+ image. The aim was to identify natural spectral classes that could be used to group vegetation species associated with specific land forms. The unsupervised classification method is a good approach to determine the number of spectrally contrasting land-cover classes that a sensor is capable of differentiating in a specific environment. After running the classification several times with different parameter settings (class numbers, iterations, maximum merged pairs), we isolated spurious pixels with a 3 × 3 kernel majority filter and included them in a larger class enclosing them. The result in Figure 3 shows that the diversity of vegetation species and the sensor's spectral (6 reflective bands) and spatial resolution (30 m) produced a mosaic of classes that are, at first, difficult to interpret in terms of associating vegetation assemblages with specific landforms. Nevertheless, some spatial patterns can be recognized, especially in the *bajo* areas (green and magenta classes) as well as in the uplands (blue, orange, and red classes) and along the slopes (yellow class). These classes may represent palm forest and grassy areas in the *bajos* and sinkholes, as well as tall to medium size forest in the elevated and sloping

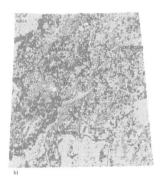

a)   b)

Figure 3. a) Isodata classification of ETM+ image (after applying a majority/minority filter; kernel: 3 × 3). b) The two most common classes identified as blue and orange in association with known settlement (in white) are overlaid on a DEM. Grid lines represent UTM grid at 2 km interval.

upland. According to Brewer et al. (2003), unique species tend to be found in the extremes of elevation, i.e., in ridges and valleys, while the highest diversity of trees is found in topographically transitional zones, i.e., along slopes and escarpments. In the classified LANDSAT image (Figure 3a) patches of blue-colored pixels may represent the tallest trees on the upland and the most vibrant vegetation of the riparian forest. Orange- and red-colored pixels are found in selected intermediate areas of the ridges and may represent a unique combination of trees and the low *escoba* palm species, whereas yellow pixels may correspond to mostly transitional plant communities located along the slopes where corozo and escoba palm are more common (Figures 3a, b).

In Figure 3b, the two most common classes in the ETM image classification are displayed against the terrain model and settlement areas showing not only that they are most common in upland and riparian ecozones, but also that they are mostly concentrated where the densest settlement occurs, especially in the central zones of Holmul and Cival. The most common of these two classes (blue in Figure 3) is associated with 38% of structures mapped at known sites in the region, while the second most common (orange) correlates with 26.7% of all structures. Together these two classes are found in connection with 64.7% of all structures. Further tests will be necessary to determine whether the association is caused by a certain type of soil most commonly found on the Holmul ridge on which both the most vibrant trees and the ancient settlers prospered, or if the vegetation clusters are in fact a result of their selecting of ruin areas because of the abundance of surface features such as the structures' rubble. It is worth noting, for instance, that ramon trees most commonly grow on ancient structures because their superficial roots can best wrap around the piles of rubble and walls of ruined buildings.

However, the Figure 3a image shows also a vast area to the northwest of Hahakab, which is a *bajo* or transitional zone, where class 1 (shown in blue) vegetation dominates. This anomaly will have to be investigated further. It is possible that the anomaly is the result of an incorrect pixel-class assignment by the algorithm or of an unusual abundance of certain upland trees in this low-lying area.

It is not unusual for some class confusion to occur in the classification process, because of the spatial and spectral resolution of LANSAT ETM+, which is obviously not suited to differentiate small-scale features and the wide range of vegetation composition and diversity found in tropical rainforests. To remedy at least some of the limitations of the LANDSAT sensor, we repeated the classification on the high-spatial-resolution image of the IKONOS sensor. Here the approach that was adopted was slightly different than with LANDSAT ETM+. A set of processing steps were necessary to extract the maximum information content from the IKONOS image and to test the sensor's discrimination power in rainforest environments. These steps included principal component analysis, vegetation index NDVI mapping, textural analysis by applying an occurrence filter, and both supervised (maximum likelihood) and unsupervised classifications (isodata). The supervised classification proved unsuccessful, probably because of the limited sample and diversity of the training sites selected (various archaeological sites of different epochs). The unsupervised classification, on the other hand, produced unbiased results. Only the most significant results are summarized below.

The first principal component (PC1) contains a high proportion of the non-correlated information found in the four bands and emphasizes mostly topographic features. The second component (PC2), however, highlights subtle variations contained in the multispectral bands, and in this case the information is related to contrasting upland and lowland vegetation types and soils (Figure 4). This interpretation is supported by the similarity found between PC2 and the vegetation index NDVI image. Both image products discriminate well areas in the rainforest with a dense canopy (light gray) from those with an open canopy (mid gray). They also distinguish well between the vegetated areas in the *bajos* (dark gray) and those with exposed soil and water (black). Unfortunately, the IKONOS image shows several clouds and shadows, which are shown in black in both image products and may lead to some confusion with respect to non-vegetated areas (i.e., soil and water). The remaining components do not provide much additional information on landscape features indicating that the first two components describe most of the image variance. Thenkabail et al. (2004) come to a similar conclusion in their study of African rainforests using IKONOS data.

The greatest advantage of the IKONOS image is its spatial resolution, which enables the detection of individual tree crowns in the panchromatic band. A disadvantage, however, is that the sensor lacks spectral bands in the mid-infrared region of the spectrum, thereby limiting its ability to discriminate

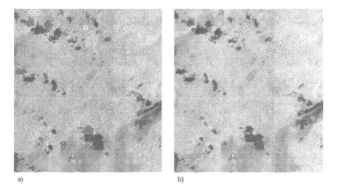

a)   b)

Figure 4. Principal component 2 (a) and NDVI product (b) of IKONOS image. PC2 and NDVI outputs look very similar and both stress differences in vegetated versus non-vegetated areas. The IKONOS image covers the same area as the Landsat subscene in Figure 2.

forest classes. Because the sensor's strength lies in the spatial domain, an attempt was made to classify rainforest vegetation based on canopy density and height by enhancing the textural component of the IKONOS image. An occurrence filter with a kernel size of 11×11 was applied to the panchromatic band in order to map canopy texture in forested areas and terrain surface roughness in forest-clear areas. This type of filter measures texture based on the number of occurrences of each pixel value (gray level) within the processing window (kernel). The IKONOS texture product was compared to a radar image of the same area taken with the airborne IFSAR sensor on March 1, 1999 (Figure 5). The IFSAR sensor imaged the study area with an X-band antenna and a spatial resolution of 2.5 m which was afterwards

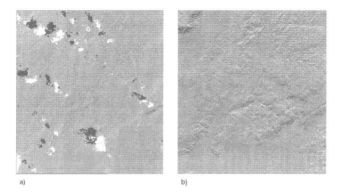

a)   b)

Figure 5. Comparison of texture components in IKONOS image (a) and IFSAR image (b). Images cover same area as in Figure 2.

downgraded to 10 m. Speckles in the image were largely removed by applying a Lee filter with a kernel size of 3 × 3. The IKONOS and IFSAR comparison in Figure 5 illustrates the advantages and disadvantages of optical and radar images. The most obvious one is that radar waves penetrate clouds, and thus are capable of providing complete coverage of the study area. Topographic relief is more pronounced in the radar image than in the IKONOS image even after enhancing its textural information. On the other hand, the orientation of landscape features with respect to the look direction of the IFSAR sensor will determine how well these features appear on the image. This problem can be illustrated with the Holmul River where segments of the river appear and disappear on the radar image but are mostly visible on the IKONOS image.

Finally, we performed an unsupervised classification of IKONOS multi-spectral bands using the same isodata classifier we used in the LANDSAT ETM+, but changing the parameter settings and applying a majority/minority filter until a satisfactory result was achieved. This time, a greater number of spectral classes needed to be used in order to separate clouds and shadows from other land-surface features (Figure 6a). The classifier produced good results in terms of mapping lowland features and differentiating them from upland ones. In Figure 6b, the most common class (shown in red) in the IKONOS image is singled out and displayed against the terrain and known settlement, thereby showing that it is most common in upland zones especially within the western part of the Holmul plateau. It should be noted that the artificial boundary that crosses the right-hand side of the image is caused by mosaicking two IKONOS scenes that were acquired under slightly different illumination conditions. However, the correspondence of the most common vegetation/soil class in the IKONOS image with upland terrain closely mirrors the pattern of the two most common classes in the ETM+

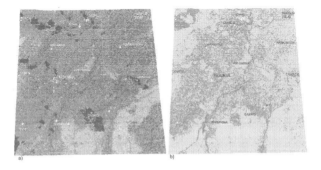

Figure 6. a) Isodata classification of IKONOS multispectral bands after applying a majority/minority filter with a kernel size of 5 × 5. Clouds and shadows (in black) are clearly separated from land-surface classes. b) IKONOS data overlaid on IFSAR DEM showing correlation of the most common class of vegetation, upland zones, and known settlement areas. Grid lines represent UTM grid at 2 km interval.

image and cannot be attributed to artifacts introduced by the mosaicking procedure. In addition, this vegetation/soil class is notably in association with 32% of all settlement known in the region in contrast to the second most common class, which is associated with 18% of settlement areas. This pattern also confirms the vegetation-to-structures association observed in the ETM+ classification.

6. CHANGE DETECTION

Looting and deforestation are two phenomena that threaten the integrity of the Maya Biopshere Reserve and often are found in close association. In order to monitor the good health of the forest, and identify stresses and illicit activities in and around our study area we performed a change-detection analysis on the two LANDSAT images at our disposal, the first of which was taken in 1989, the year prior to the institution of the Maya Biosphere Reserve, and the second in 2001.

Change-detection analysis is accomplished by using the following processing methods: 1) histogram matching for visual comparison of two images of the same area but of different dates, 2) change-detection technique using single-band-image algebra, and 3) classification of resulting change images to categorize and assess changes between the dates of image acquisition.

The first technique manipulates the image contrast by matching the histogram of a source image to that of a reference image (Figures 7a and b). This is mainly used for mosaicking adjacent images so that variations in brightness across the joint are minimal and the transition from one image into the other almost undetectable.

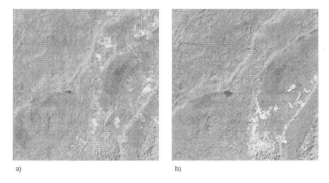

a)   b)

Figure 7. a) False-color composite using bands 7, 4, and 2 of the ETM+ image 2001. b) False-color composite using bands 7, 4, and 2 of the TM 1989 after matching its histogram to that of the ETM+ reference image. Image area is 17×17 km and lies east of Holmul.

The second technique performs the actual change detection analysis based on the image differencing method of each band in the ETM+ and TM image pair. In the resulting image, changed features are emphasized.

Individual single-band-difference images can be grouped or combined to increase the information content that will allow the separation of change versus no change features. One way of combining difference bands is by displaying them as false-color composites. Another way of distinguishing and categorizing change features on difference images is by performing an unsupervised classification (third technique) using several change bands together. Labeling these change classes is done afterwards by checking them in the field and/or surveying local land parcels to find the type of land-use change that may have occurred.

The change-detection procedure outlined above was performed on a subset of the LANDSAT TM/ETM+ image pair of December 27, 1989, and March 7, 2001 (Figures 7a and b). The image area lies east of the Maya city of Holmul and covers a variety of representative features, including seasonal variation in the *bajo* wetlands and human-induced changes in the forested area not far from the present-day town of Melchor. Most notable at the center of the image is the Yaloch Lake whose water appears in black pixels. Seasonally flooded and grass-covered areas around the lake appear as bright magenta. The eastern side of the image illustrates agricultural practices in the Belize region bordering Guatemala. The isodata classifier was used as the unsupervised classification method to group change versus no-change features into spectral classes (Figure 8). The change classes enclose areas of subtle to major changes and are color coded in Figure 8, whereas the unchanged areas are shown in gray. The light/dark blue classes are mainly associated with changes in the area surrounding the lake, and may represent a difference in soil moisture caused by more/less abundant rainy seasons and soil wetness (*bajos*). The yellow class represents new fields (forest clearing), whereas the green and magenta classes correspond to abandoned fields at different stages of re-growth. It is interesting to note that several small green/magenta speckles seen on the left margin of the 1989 image occur within the MBR protected area. These are possibly areas of clandestine clearings now reclaimed by the forest. Some of these were probably patches of forest cleared to grow marijuana. It is well known that in this area marijuana growing and looting of archaeological sites went hand in hand in the 1980s. According to local opinion, marijuana growers also dedicated themselves to looting nearby sites while waiting for their crop, thereby collecting two illegal harvests at the end of each season. It is likely that medium and small Maya mounds will be found near the clearings identified by this change-detection method when surveyed. Currently, our field surveys and the ETM+ data show that since the inception of the MBR protection laws and since our own project installed private guards at Holmul, these types of illicit activities have disappeared.

Figure 8. Isodata classification of ETM+'01 – TM'89 change image using three difference bands (respective band pairs of ETM+TM 7, 4, and 2). Classes were grouped to represent no-change classes in gray and change classes in color. Change areas are mainly associated with seasonal changes in the *bajos*, deforestation/reforestation.

7. CONCLUSIONS

While preliminary, the results presented in this report strongly encourage the implementation of a regional GIS and remote sensing methodology in the study of ancient settlement even in a highly forested region such as the Maya Biophere Reserve of Guatemala. The creation of a DEM from traditional paper maps has enabled us to obtain a coarse-grained database with which to identify the most significant geomorphological features of this region. In addition, we have been able to model optimal pathways across the hilly terrain and found correspondence between human settlement and least-resistance routes to Holmul. This result in turn has shed new light on the social, economic, and ideological factors influencing the ancient Maya choice of location for their settlement. Further analysis of higher resolution DEMs, such as the IFSAR data, and of spectral imagery will enable a fine-grain mapping of micro-ecotones available to each settlement unit.

The next generation of data will include a new AIRSAR dataset acquired in 2004 and being processed at NASA at the time of this writing (NASA JPL, 2004 http://airsar.jpl.nasa.gov/). This new foliage-penetrating radar sensor has the potential to identify medium and small ancient structures hidden below the forest canopy, thereby for the first time enabling mapping of Maya settlement over a wide continuous area of the Lowlands (Sever and Irwin, 2003). Our analysis of two different spectral sensors, the medium-resolution LANDSAT and the high-resolution IKONOS, gave interesting preliminary results that will be tested with further field surveys. We have identified a combination of soil and vegetation types that is most commonly associated with Maya settlement. This spectral signature appears to be concentrated on the western side of the Holmul upland ridge, including the upland and riparian forest ecotones. Although this feature is singled-out in both image types, it is in the IKONOS image that it appears to be more clearly associated with upland and riparian zones, as well as settlement. On the other hand, the LANDSAT classification includes also a broad area of what is believed to be a *bajo* area in the upper left quarter of the image. Since this area has not been documented by our field crews at the time of this writing, it is therefore difficult to pin-point the source of the spectral similarity between this vegetation zone and the uplands zone. This area, however, is known to be a *bajo* swamp dominated by the tintal tree, and we surmise that this scrub-type hardwood may have a spectral signature similar to the upland canopy forest. Further testing on the ground will ascertain whether ancient settlement is also present in this transitional-to-wetland zone.

As we have seen, the high-spatial resolution of the IKONOS image gave the best results in differentiating vegetation types, even with the limitations imposed by its low-spectral resolution. We look forward to performing further analysis using the next generation of spectral sensors which combine high-spatial resolution with hyper-spectral resolution, including hundreds of spectral bands. These data should enable us to differentiate vegetation types down to the individual tree species even in the most diverse tropical forest environment.

The change-detection analysis was presented here because of its straight-forward ability to monitor time-sensitive changes in the environment, including seasonal variations in moisture content, which may highlight buried features, and long-term land-use variations. The first LANDSAT satellite was launched in 1972 and imagery has been continuously stored for each returning pass on the same area, every 18 or 16 days (depending on the Landsat series), offering a wealth of time-sensitive data over the span of more than 30 years. Long-term variations in the forest cover may reveal deforestation trends associated with modern settlement and economic activities, licit or illicit; the general health status of the forest resources; and the efficacy of the policies implemented in the last decade.

We observed a general improvement in the integrity of the forest cover in the Holmul area of northeastern Petén since 1989, and this is undoubtedly a result of the effectiveness of the institution of Maya Biosphere Reserve and the implementation of its rules for a sustainable use of the forest resources.

We hope that the present archaeological investigation will contribute not only to a better understanding of the ancient settlement but also to the development of a wide range of cultural resources to complement the sustainable forestry with sustainable tourism. We believe the latter to be the only viable way to ensure the survival of the Maya tropical forest and its ancient cities into the future.

ACKNOWLEDGEMENTS

Permission to carry out this research was graciously granted by IDEAH of Guatemala. The Holmul Archaeological Project was initiated in 2000 with grants from National Geographic Society, the Foundation for the Advancement of Mesoamerican Studies Inc. to Estrada-Belli while at Boston University, and a grant from The Ahau Foundation to Professor Norman Hammond, Boston University. Since 2001 the remote sensing and field aspect of the project have operated with academic and financial support from Vanderbilt University, and additional grants from National Geographic Society, the Foundation for the Advancement of Mesoamerican Studies Inc., The Ahau Foundation, and private donations by ARB, Toyota Motors, Interco Tire co. Trailmaster, Warn, PIAA, Yamaha Motors, Rhino Linings, Borla, Garmin, Craftman Sears, Skyjacker, Eureka, Optima Batteries, ITP, Procomp, Painless, Molly Designs, Bearcom of Guatemala, Bill and Debbie McCanne, and an anonymous donor. Marc Wolf, Kristen Gardella, Justin Ebersole, and Jason Gonzalez have contributed the majority of the Holmul field mapping data over the years. Special thanks to the NASA Scientific Data Purchase Program for providing us access to the IFSAR 1999 Rio Bravo dataset from JPL. We also wish to thank Peter D. Harrison, George Stuart, Norman Hammond, Clemency Coggins, William L. Fash, Kenneth L. Kvamme, James Wiseman, Farouk El-Baz, and Marco and Inma Gross for their support and encouragement throughout this work.

REFERENCES

Adams, R.E.W., 1995, Introduction. In *The Programme for Belize Regional Archaeological Project: 1994 Interim Report*, edited by R.E.W. Adams and F. Valdez, Jr., pp. 1–15. University of Texas, San Antonio.

Adams, R.E.W., 1999, *Rio Azul: An Ancient Maya City*. University of Oklahoma Press, Norman.

Adams, R.E.W., Brown, Jr., W.E., and Culbert, T.P., 1981, Radar Mapping. Archaeology and Ancient Maya Land Use. *Science* 213:1457–1563.

Brewer, S.W., Rejmánek, M., Webb, M.A.H., and Fine, P.V.A., 2003, Relationships of phytogeography and diversity of tropical tree species with limestone topography in southern Belize. *Journal of Biogeography* 30:1669–1688.

Bullard, W., 1960, Maya Settlement Patterns in Northeastern Petén, Guatemala. *American Antiquity* 25:355–72.

Estrada-Belli, F., 2000, Archaeological Investigations at Holmul, Guatemala. Report of the first field season, May–June 2000. Report submitted to National Geographic Society and FAMSI. Online version, URL: http://www.famsi.org/reports/98010/index.html

Estrada-Belli, F., 2001, Maya Kingship at Holmul, Guatemala. *Antiquity* 75:685–686.

Estrada-Belli, F., 2002, Anatomía de una ciudad Maya: Holmul. Resultados de Investigaciones arqueológicas en 2000 y 2001. *Mexicon* XXIV(5):107–112.

Estrada-Belli, F., n.d. (in press), A Terminal Classic Ritualized Landscape At La Milpa, Belize. In *The 2000 Chacmool Conference Proceedings*. The University of Calgary, Calgary.

Estrada-Belli, F., Valle, J., Hewitson, C., Wolf, M., Bauer, J., Morgan, M., Perez, J. C., Doyle, J., Barrios, E., Chavez, A., and Neivens, N., 2004, Teledetección, patrón de asentamiento e historia en Holmul, Petén. In *XVII Simposio de Investigaciones Arqueológicas de Guatemala*, edited by J. P. Laporte, H. Escobedo, and Barbara Arroyo, pp. 73–84. Ministerio de Cultura Deportes, Instituto de Antropología e Historia, Asociacion Tikal, Guatemala.

Exon, S., Gaffney, V.,Woodward, A., Yorston, R., 2000, *Stonehenge Landscapes. Journeys through real and imagined worlds*. Archaeopress, Oxford.

Fedick, S. L., and Ford, A., 1990, The prehistoric agricultural landscape of the central Maya lowlands: an examination of local variability in a regional context. *World Archaeology* 22(1):18–33.

Ford, A., 1986, Population Growth and Social Complexity: An Examination of Settlement and Environment in the Central Maya Lowlands. *Anthropological Research Papers*, no 35. Arizona State University, Tempe.

Gardella, K., 2004, The settlement survey of Holmul: mapping the community and its landscape. Paper presented at the 69th Annual Meeting of the Society for American Archaeology, Montreal, April 4, 2004.

Garrett, W., 1989, La Ruta Maya. *National Geographic Magazine* 176(4):474–475

Kunen J. L., Culbert, T. P., Fialko, V., McKee, N., and Grazioso, Liwy, 2000, Bajo Communities: A case study from the central Petén. *Culture & Agriculture* 22:15–31.

Laporte, J. P., Mejia, H. E., Adanez, J., Chocon, J. E., Corzo, L. A., Ciudad Ruiz, A., and Iglesias, M. J., 2004, In *XII Simposio de Investigaciones Arqueologicas en Guatemala, 2003*, edited by J. P. Laporte, B. Arroyo, H. L. Escobedo, H. E. Mejía. Ministerio de Cultura y Deportes, Instituto de Antropología e Historia, Asociacion Tikal, Guatemala.

Lundell, C., 1937, *The vegetation of the Petén*. Carnegie Institution of Washington Publication 478. Washington, D.C.

Madry, S. H., and Rakos, L., 1996, Line-of Sight and Cost-Surface Techniques for Regional Research in the Arroux River Valley. In *New Methods, Old Problems: Geographic Information Systems in Modern Archaeological Research,* edited by H. D. G. Maschner, pp. 104–123. Center for Archaeological Investigations, Occasional Papers No. 23. Southern Illinois University, Carbondale.

Merwin, R. E and Vaillant, G., 1932, *The Ruins of Holmul. Memoirs of the Peabody Museum of American Archaeology and Ethnology*, vol. III no. 2. Harvard University, Cambridge.

Puleston, D. E., 1983,*The Settlement Survey of Tikal*. Tikal Report 13. University Museum, University of Pennsylvania, Philadelphia.

Sever, T. L., 1998, Validating Prehistoric and Cultural Social Phenomena upon the Landscape of the Petén, Guatemala. In *People and Pixels: Linking Remote Sensing and Social Science,* edited by Diana Liverman, pp. 145–163. National Academy Press, Washington, D.C.

Sever T. L., and Irwin, D. E., 2003, Landscape Archaeology. Remote Sensing Investigation of the ancient Maya in the Petén rainforest of northern Guatemala. *Ancient Mesoamerica* 14: 113–122.

Thenkabail, S. P., Enclona, E. A., Ashton, M. S., Legg, C., and DeDieu, M. K., 2004, Hyperion, IKONOS, ALI, and ETM+ sensors in the study of African rainforests. *Remote Sensing of Environment* 90:23–43.

USAID, 1996, *Plan maestro Reserva de la Biosfera Maya: consejo nacional de areas protegidas.* Centro Agronómico Tropical de Investigación y Enseñanza, Turralba, Costa Rica.

USGS (U.S. Geological Survey) EROS, Sioux Falls, SD, 2005, http://edc.usgs.gov/

Willey, G. R., Bullard, W. R., Glass, J., and Gifford, J., 1965, *Prehistoric Maya Settlements in the Belize River Valley. Papers of the Peabody Museum of American Archaeology and Ethnology*, vol. 54. Harvard University, Cambridge.

Chapter 12

Remote Sensing and GIS Use in the Archaeological Analysis of the Central Mesopotamian Plain

BENJAMIN F. RICHASON III AND CARRIE HRITZ

Abstract: The landscape of the southern portion of the Central Mesopotamian Plain is a unique maze of ancient settlement sites and canals. A number of these have been inventoried and mapped during the last century using ground surveys. Because of the size of the region and problems with accessibility, remote sensing and GIS mapping techniques have proven invaluable in our recent studies of this area. Specifically, we studied the area around the ancient site of Nippur and Lake Dalmaj, making use of a variety of sensor systems including Landsat, SPOT, Radarsat, ASTER, and Corona. A variety of enhancement and classification techniques were applied to these different image types to better understand patterns and distributions on the land, as well as the capabilities of different sensor systems. To aid in this investigation extensive GIS databases were created from a variety of field studies such as Robert Adams's work, the *Heartland of Cities*. These databases were used in conjunction with the imagery to isolate different age groupings of sites and canals and map their location and extent; areas potentially containing new sites were also identified. This study illustrates how distinctive this area is in terms of mounds and layers of canals as compared to areas farther north.

RODUCTION

Remote sensing imagery and associated analytical techniques have been employed in the discipline of archaeology for a number of years. The uses of remote sensing techniques in archaeology are varied. Some studies can be site-specific, meaning that just one individual site is being analyzed and inventoried. Quite often such a study will be done using high-precision orthophotography and supplemented, where possible, with detailed ground verification(Avery and Berlin, 1992:230). Remote sensing imagery can also be used to provide a panoramic overview of a study area in order to portray the regional characterization of the landscape. Of particular use for this application is de-classified Corona satellite photography, which has found widespread use in archaeological studies in the Near East (Kennedy, 1998). This archival photography can be used in conjunction with medium-resolution satellite imagery for verification and change detection. A further use of remote sensing imagery is in land-cover classification mapping. Such mapping may not be considered a specifically archaeological investigation, but the presence and pattern of certain types of land cover may indicate the location of ancient sites, thus providing information concerning site location. It is these uses of remote sensing that have been employed in recent archaeological investigations in central Iraq (Figure 1).

Figure 1. General reference map of central Iraq.

Current remote sensing technology does have the potential to present some problems in archaeological studies. For example, sensor resolution can be a difficulty when there is a need to study relatively small features on the earth's surface (Avery and Berlin, 1992; Holcomb, 1998). Most affordable civilian remote sensing imagery is considered to be mid resolution, that is, between 20 and 30 meters (Lillesand et al., 2004:398). When enlarged such imagery begins to "pixelate" with a resulting loss of detail. With the launch of new sensor systems that can provide resolutions of 1–5 m, and some at sub-meter accuracies, small ground features can be more effectively delineated. While remote sensing technology provides information about the pattern and surface configuration of features, it does not generally provide information about sub-surface features. The exception here would be ground penetrating radar and, in some cases, L-band and P-band SAR. Thus, studies in archaeology are similar to those in soils or geology where much of the information content is hidden from view. For this reason remote sensing imagery is not the primary tool of investigation in archaeology, but rather a secondary or supportive one behind on-site mapping and excavation. The main role of remote sensing is still basically one of reconnaissance mapping.

In conjunction with GIS coverages of several surveys in the region, remote sensing imagery was used in several archaeological studies in central Iraq's Mesopotamian Plain (Figure 2). Three separate areas were studied: the area around Hillah, the Abu Salabikh area and the Nippur-Lake Dalmaj area (Figure 3). These studies had two different emphases. First, were studies that

Figure 2. General outline of the extent of the Mesopotamian Plain. Modified from Buringh, 1960: Figure 13.

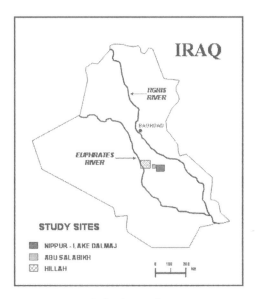

Figure 3. Study-area locations.

used imagery as a tool to aid in the identification and delineation of ancient occupation sites and the pattern of ancient canals, levees, and drainage ways. Second, studies were done to evaluate the potential of different sensor types and image processing techniques in terms of their utility in archaeological investigations in this region.

A number of different sensor systems were employed in these studies including the Landsat Thematic Mapper(TM) and Enhanced Thematic Mapper (ETM), SPOT panchromatic, ASTER, Radarsat, and Corona. SPOT imagery was used in all three investigations, while Corona and ASTER imagery were used for Hillah and Abu Salabikh, and Landsat TM and Radarsat were used in the Nippur-Lake Dalmaj area.

2. GENERAL INTRODUCTION TO THE MESOPOTAMIAN ALLUVIUM

The alluvial plain of southern Mesopotamia saw the rise of the earliest complex urban societies in the Near East. The plain can be characterized as a constantly aggrading geosyncline formed at the end of the Pleistocene period. The environment of the plain today consists of a number of units such as marshes, semi-arid dunal areas, and cultivable land (Verhoeven, 1998:161). The key forces that continue to reshape the landscape of the alluvial plain into these units are the Tigris and Euphrates rivers and human action.

These two rivers find their sources in the central Anatolian highlands. The Euphrates follows a meandering course on its way from these highlands, across Syria and into the Mesopotamian alluvium at which point it begins to split and re-join in a series of anastomosing branches. As the Euphrates enters the Syrian plain, it loses much of its volume to tributaries and evaporation so that when it reaches the alluvium at Falluja in Iraq it is a much less powerful river (Buringh, 1960). The Tigris enters the plain much earlier and is fed by many tributaries from the Zagros Mountains. It remains largely a single meandering course down the plain until it reaches Kut, Iraq, where it becomes more deeply incised. Both rivers meander over the length of the plain and join towards the head of the Persian Gulf to form the Shatt-al Arab waterway and expel their sediment load into the Gulf. Where natural vegetation is found on the plain, it consists primarily of small shrublets in the central portion. The climate is mostly hot and dry with winter lasting from December through February followed by a long, dominant summer. The plain lies outside the rainfed agricultural zone to the north and receives less than 100–200 mm of water per annum (Guest, 1966:17–18).

2.1. The Plain Dynamics

The present location of the Tigris and Euphrates rivers today is a result of the movement of both rivers (Figure 4). There is evidence, both textual and archaeological, for repositions of both of the rivers. Traces of the channel

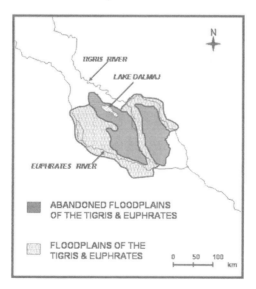

Figure 4. Locations of the floodplains of the Tigris and Euphrates rivers in central Iraq. Modified from Wilkinson, 2003: Figure 3.

movements of the Tigris are difficult to trace through time. Despite continued debate among scholars, it seems likely that extensive portions of the plain were drained by the Tigris or by some combined channel of the Tigris/Euphrates rivers (Algaze, 2001). Figure 1 shows how this joined course might have looked. In the case of the Euphrates, however, more evidence exists for continual movement of the Euphrates to the west over the past 3,000 years (Adams, 1981; Buringh, 1960; Gasche and Cole, 1998). The recent work of Gasche, Cole, and Verhoeven have shown that the textual, geomorphological, and archaeological evidence for river shifts can be correlated to better understand the movement of the Euphrates and the effects of this movement on settlement in specific periods (Gasche et al., 2002). Physical evidence for these shifts comes in the form of meander scars left in the modern landscape. These features are a result of the processes of erosion and deposition as the river moves across this relatively flat plain. Persistent erosion and stress from breaks in canal banks can cause river avulsions. This is the action by which a river or channel moves from its long-used path to form an entirely new path or rejoin the original path further downstream (Wilkinson, 2003:81). These elements of the natural landscape of the Mesopotamian plain enabled agriculture to spread from a narrow belt along a main channel of the Tigris or Euphrates river levee to wide stretches of the plain, thereby resulting in a shifting and complexly layered settlement- and canal-system through time.

2.2. Overview of Settlement Patterns

The settlement-pattern history in Mesopotamia is one of continuous change and seemingly delicate balance (Gasche et al., 2002). Given the natural environment, settlements are highly dependent on access to water. Settlements, in most periods, focused on taking canals from the Euphrates River rather than the Tigris (Adams, 1981:3–7). Because of the regimes of these two rivers, irrigation from the Tigris requires lifting devices which explains its much later role in large-scale irrigation systems on the plain. The Euphrates, with its numerous natural channels and their elevated levees, can be diverted with techniques such as sluices and breaks in levees. As the river moves, which is the tendency with the Euphrates once it enters the flat plain; settlements tend to shift as well.

For example, a significant shift in settlement pattern from beginning in the second millennium in southern Mesopotamia has been documented by both archaeological and textual research. This change is attributed to a major shift of the Euphrates to its western branches, similar to its present course. This shift is mirrored in changed settlement distribution. In general, the primary watercourses in the third-millennium B.C., mapped by archaeological survey, fell in lines that ran parallel with the north-northwest slope of the alluvium and down the center of the plain (Adams, 1981:158). Settlements were lined along the levees of these courses in a dendritic pattern. By the second millennium, this pattern shifted to the south and west. The canals that fed these southwest sites were larger and more artificial in their construction than the third millennium canals. Rather than following the slope, these new canals cut across the gradient (Adams, 1981:158). This shift to the west over time was complete by the later part of the second millennium. Settlement shifted to the west as well, which shift leads to the suggestion that significant settlement in this period may have been outside of the surveyed zone to the west (Brinkman, 1984). This series of changes provides an example of the dynamic nature of the plain.

The trend towards artificial canalization continues through time. It suggests an increasing agricultural intensity and exploitation of the productive potential of the plain as well as the development of a coping mechanism, i.e., artificial canalization. This canalization could be used to counter the Euphrates' movement without the loss of settlement in areas that were too far from the river course otherwise to be productive. The climax of the artificial channeling is the large transverse canals of the Sassanian period. These large, straight major canals fanned out across the plain in an attempt to draw most of the physiographic units of the plain into a single system under a centrally organized administration.

In summation, the plain can be characterized as a highly dynamic riverine system with a settlement pattern that fluctuates through time in response to watercourse fluctuations.

3. DEFINING THE STUDY AREA

The area for the studies discussed in this chapter is the central portion of the alluvial plain. This area, as outlined above, may be considered one of the most dynamic areas of the alluvial plain. During the period of intense archaeological survey of the 1960–1980s this area was subject to erosion as well as dune movements, obscuring archaeological sites. Today, this portion of the plain contains the overflow reservoir, Lake Dalmaj, and is an area that is under growing encroachment by modern cultivation, as may be seen in Figure 5. Amid the dunes and cultivation are hundreds of archaeological sites from all historical periods, relict canal traces that overlie each other to create a patchwork of landscapes of different time periods, and numerous sets of meander scars from the separation and movement of the Euphrates and Tigris rivers. Tells are visible from their elevation above the surrounding, generally flat, central portion of the alluvial plain. Relict canals can be traced by their patterns in modern fields and the remains of ancient levees which form wide elongated rises in the landscape. Meander scars are also traceable in the modern field patterns by their disruption of the cultivation pattern, sometimes as growth of deeply rooted shrubs that tap the water gathered in

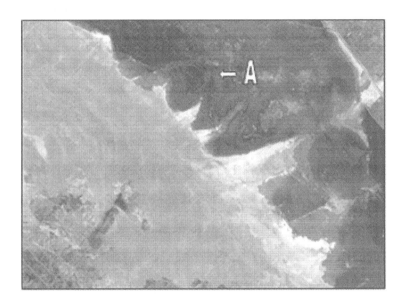

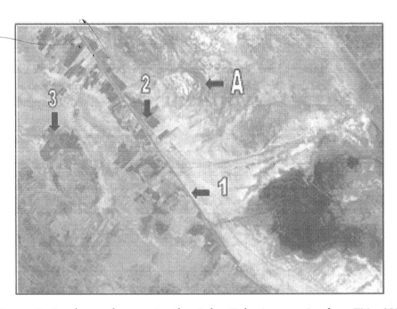

Figure 5. Land-use changes in the Lake Dalmaj area. Landsat TM, 1987 (1:52,600) (above). Landsat ETM, 2001 (1:52,600) (below). The arrows at "A" in both images point to a relict meander loop for reference. Note how the level of the lake has dropped between 1987 and 2001. Also note the creation of a new canal at "1" and the expansion of agriculture at "2" and "3."

ancient buried river beds. We undertook the studies described in this paper because we believed that deciphering this complex patchwork of landscapes will generate a better understanding of the dynamics of the plain through time, the history of settlement in southern Mesopotamia, and the role of irrigated agriculture in the development of urbanization.

3.1. History of Field Work in the Study Area

The settlement and irrigation patterns of southern Mesopotamia are illuminated through the combination of numerous disciplines and techniques, among which are historical geography, archaeology, geology, and geomorphology. Descriptions of the landscape in the central portion of the Mesopotamian alluvium, particularly in relation to the past, appear in earnest in the written record in the late 19th and early 20th centuries. There are three types of written sources: traveler's accounts and maps, technical geomorphological studies, and archaeological surveys. While the early travel accounts provide a crucial view of the landscape, the archaeological remains and the alluvial plain before modern industrial changes, travelers rarely ventured far into the central portion of the plain. Their accounts were often restricted to site descriptions, but in some cases included valuable accompanying maps locating ruins, villages, and modern and relict canals.

Technical descriptions from the 1960s provide important information on the ground conditions in this portion of the plain and were conducted at the same time as the majority of archaeological surveys in the region. In areas that have not been studied since by soil scientists, Buringh's study (1960) is the main source of information concerning the nature of the plain. Other soil surveys were done by prospecting groups, such as the Hunting group in the late 1950s and 1960s, which are almost totally unavailable as they either reside in an archive in Baghdad or were disposed of by the company's archivists. For example, in her unpublished dissertation, Pournelle describes the disposal of the Hunting Surveys KLM series of aerial photographs and soil data (Pournelle, 2003:32).

While all of these distinct disciplines have played a crucial role in our understanding of these patterns, archaeological survey is emphasized for its role in tracing the remains of these patterns through time. The systematic investigation of agriculture and irrigation systems through archaeological survey, as a means of understanding the cycles of development and decline of urban society, began with Jacobsen's and Adams' work in the Diyala Basin Archaeological Project in the 1950s (Adams, 1965; Jacobsen, 1957). The basic theme of Adams' work is the "...extensive, detailed inquiry into the forces responsible for the precocious early growth and those that later contributed to catastrophic decline and outright abandonment... (Adams, 1981:xvi). Adams attempted to answer the regional questions in Mesopotamian settlement history and understand the infrastructures that shaped Mesopotamian civilization. He viewed those structures as being mirrored "... in the pattern of agricultural land use and the hierarchical array of communities in which people lived" (Adams, 1965:3).

In the subsequent 30 years, Adams surveyed 1/3 of the central alluvial plain of southern Mesopotamia. His largest survey area was the central portion of the plain that he investigated from 1968 to 1975. He covered an area of hundreds of kilometers both on foot and in vehicles, and made use of aerial photographs and maps. Cultivation and ease of access dictated his boundaries. In areas where ground survey was difficult or not possible, Adams supplemented his study by using aerial photographs (Adams, 1981:28–37). Small subsidiary surveys helped to fill in more specific settlement questions aimed at understanding the development of settlement systems through time in this dynamic environment (Adams, 1972; Gibson, 1972; Wilkinson,1990).

More recent research includes the work of a team at the University of Ghent, Belgium, under the direction of De Meyer, Gasche, and Verhoeven (Gasche and Cole, 1998). This team used satellite images in conjunction with ground archaeological survey to re-evaluate the lines of watercourses in the second and first millennia in the area of the northern portion of the alluvium of near Sippar. By tracing ancient levees and incorporating textual data, as well as conducting geomorphological survey, the team has been able to show the importance of river avulsions in the area and to elucidate the interaction between cultural and physical transformations of the plain. Unfortunately, since the early 1990s no organized ground survey research has been done in the area. For this reason the analysis of remote sensing imagery has had to be relied on to a greater extent.

4. REMOTE SENSING APPLICATIONS

Employing remote sensing imagery in the analysis of the surface characteristics of earth objects can require using a number of different techniques. These techniques range from the preprocessing routines that prepare imagery for analysis, through the enhancement of imagery to improve its interpretability, to the actual classification of individual digital number (DN) values. Quite often these techniques are used in conjunction with one another.

Preprocessing routines are necessary in order to remove or reduce certain artifacts found in the raw datasets (Campbell, 2002:291). Some of these artifacts are the result of the way the sensor system operates. Circumstances such as line dropouts or striping may occur because of sensor malfunction and will have to be restored. Also, conditions in the atmosphere at the time the image was created, such as smoke or humidity, can result in the scattering or absorption of energy and thus affect the appearance of the image. Some effects tend to be specific to certain types of imagery, such as radar speckle that creates a sort of "salt and pepper" effect. In most instances imagery will have to undergo certain modifications to remove or reduce these artifacts.

Image enhancement techniques are a group of routines that are designed to make an image more visually interpretable (Jensen, 2005:255; Lillesand et al., 2004:509). One of the main goals in remote sensing is information extraction by either analog or digital means. During the last several decades the digital analysis of imagery has increased in significance, especially the interpretation of remote sensing imagery in the form of digital displays. While the interpretation of analog and digital images have a number of factors in common, digital images have their own characteristics that require special types of manipulations (Campbell, 2002:151).

Image classification includes a group of processes that segment or group image pixels into a defined set of categories such as land-cover classes. Such procedures, in effect, turn data into information to create a form of digital thematic map through the labeling of pixels (Mather, 1987:277). This labeling is accomplished through the numerical analysis of the pixel's spectral pattern. The term pattern, as it is used here, refers to a pixel's radiance measurements in each of the sensor's wavelength bands (Lillesand et al., 2004:551). For archaeological studies, having a classified land-cover map may facilitate the characterization of the present-day landscape in relation to ancient sites (Ebert and Lyons, 1983:1295).

4.1. Enhancement Techniques

Enhancement plays a very important part in digital image analysis. This is particularly true in archaeological investigations where remote sensing imagery is being employed. Many such studies are feature-specific for particular areas because only a single feature of interest is being isolated from the surrounding background cover. The extraction of certain objects such as canals, river levees, mound sites, ancient field boundaries, and transportation routes is thus facilitated through enhancement techniques. Indeed, enhancement operations may be all that is needed in certain types of studies where only specific sorts of features such as vegetation types or minerals need to be identified. Instead of attempting an overall land-cover classification, certain types of spectral or spatial enhancements may suffice (ERDAS, 1997:125).

There are a number of techniques that are available for enhancement. One of the most useful is contrast enhancement. In the case of the studies done in central Iraq such features as mound sites or tells, canals, and natural levees were the principal features being studied and highlighted (Figure 6). One of the simplest types of enhancements is to artificially color a gray-scale image such as a panchromatic band or a radar image (Figure 7). Because the human eye perceives more hues of color than shades of gray, details not readily apparent in a gray-toned image can be discerned in color (Avery and

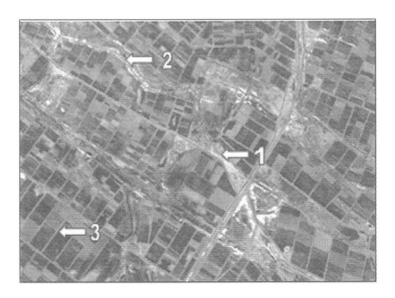

Figure 6. Examples of the types of features studied in the investigations. In the Landsat TM Band 3 image (1:7,000) above, various types of linear and curvilinear features denote relict drainage ways. The arrow at "1" indicates a relict river meander loop. The arrow at "2" shows an additional relict channel while the arrow at "3" points to a recent irrigation canal. (below) SPOT Panchromatic image (1:18,000) of the site of Abu Salabikh. The arrows point to two of the site's mounds. (SPOT image courtesy of NGA).

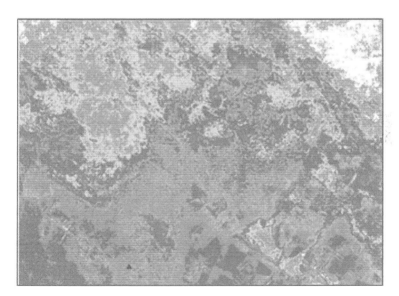

Figure 7. Corona Photograph, 1967 (1:30,500)(above). Color rendition of the same Corona photograph (1:30,500)(below). Note the increased detail of the dunes and sediment plains. (Courtesy of USGS).

Berlin, 1992:32; Sabins, 1997:52). In addition, altering the order in which different multispectral bands are displayed in different color planes on display devices can also enhance the detail of ground objects (Lillesand et al., 2004:435; Moik, 1980:153). The presentation of imagery in normal color is particularly advantageous in distinguishing subtle differences in rock and soil type, and can thus be beneficial in a number of uses in archaeology (Campbell, 2002:499).

With regard to the radiometric resolution or pixel depth of digital imagery, it is quite often the case that the pixel values do not fill the entire dynamic range of brightness levels (Sabins, 1997:266). For example, an 8-bit resolution image ranges from values of 0 (black) to 255 (white); a particular scene, however, may only have reflectance values between 12 and 117. This example shows that the image has a range of 105 brightness values, but, out of possible 256, 151 brightness levels are left unrepresented. If the number of levels of brightness the contrast can be stretched, thereby increasing the interpretability of an image. Once again, subtle variations in tone can be enhanced to bring out the detail of features that might otherwise be barely discernible.

There are three major types of contrast stretch: linear, piecewise and nonlinear (ERDAS, 1997:133). All three will spread or expand the original range of brightness values over the full dynamic range of possible values. It may be difficult, however, to predict exactly which type of stretch may provide the best results for specific features; a certain amount of adjustment by trial and error, therefore, may be needed. Before any enhancement technique is chosen, it is important to study the image-band histogram to see how the brightness values are distributed.

In the manipulation of the Thematic Mapper imagery used in the examination of archaeological sites in central Iraq, we found two enhancement techniques to be particularly beneficial. These techniques were histogram equalization and piecewise linear stretching. Simple linear and nonlinear stretching applies certain functions to the original data over the entire image. A piecewise stretch, on the other hand, utilizes a polyline function to enhance certain portions of the image to varying degrees (ERDAS, 1997:133). Such a stretch can be useful in isolating certain types of features while having the rest of the image displayed in more subdued tones (Figure 8). Once again, the gray-toned image can be further emphasized with a color rendition.

Another contrast stretch technique that is widely used is known as histogram equalization. This is a nonlinear stretch that redistributes individual pixel values into approximately an equal number of pixels within a specified range (ERDAS, 1997:138). The outcome of this enhancement is an essentially flat histogram where contrast is increased in that area with the greatest number of brightness values in the image while reducing the contrast in the very light and dark areas in the image (Jensen, 2005:272). The results of such equalization can be seen in Figure 9. The scene is a portion of a SPOT panchromatic image of the northern part of Lake Dalmaj that contains the site of Tell al-Arsan, which includes a circular citadel some 220 meters in

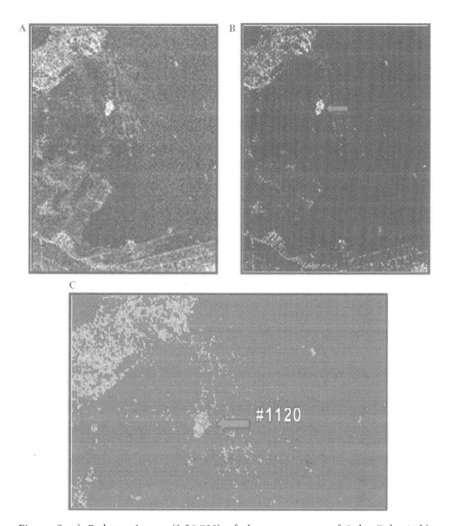

Figure 8. a) Radarsat Image (1:26,500) of the eastern part of Lake Dalmaj. b) Piecewise Linear Stretch image (1:26,500) of the scene in a). Note how the noise in the lake has been reduced, thereby bringing out the sites in the lake. c) Color Rendition (1:17,000) of the Piecewise Stretched image b). Site #1120 (Adams's survey) is labeled to show the additional detail that can be displayed. (Courtesy of EODS).

diameter (Adams, 1981). In Figure 9a the original SPOT image can be seen displayed in rather subdued tones. In Figure 9b the image has been enhanced through histogram equalization. In Figure 9b the outline of this site, as well as some of the smaller surrounding sites, can be seen in greater detail. The

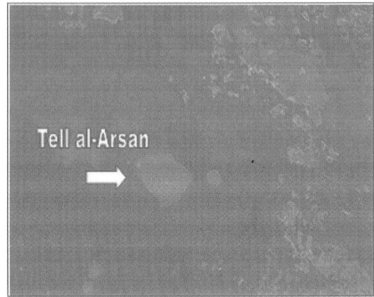

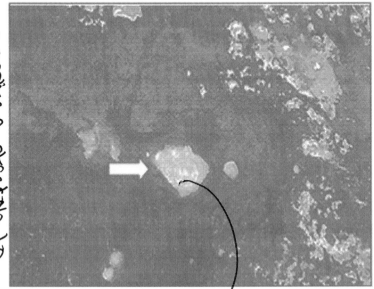

Figure 9. The effects of histogram equalization. (above) The original SPOT image
(1:6,400). (below) The same scene after histogram equalization. Note the increased
detail, especially the circular citadel of Tell al-Arsan. (Courtesy of NGA).

same is also true of some of the drainage features and dunes in portions of the dried up areas of the lake.

Another possible means of enhancement is through the use of various types of band ratios called indices. Many of these routines have been developed to enhance the appearance and detection of vegetation. Some of the more popular of these are the Normalized Difference Vegetation Index (NDVI) and the Tasseled Cap transformation (Deering et al., 1975; Crist and Cicone, 1986; Kauth and Thomas, 1976; Trishchenko, 2002). Other indices have also been developed specifically for mineral analysis and include those that attempt to isolate clay, ferrous, or hydrothermally altered minerals.

In terms of archaeological studies vegetation can be an aid or a hindrance. Dense vegetative cover can obscure the detail of underlying features such as ancient foundations (Adams, 1984: Avery and Berlin, 1992). On the other hand, vegetation type and orientation can also heighten details such as ancient canals or walls of old foundations. Also, the mixture of vegetated and nonvegetated areas could reveal occupation sites that have been taken over by modern agricultural development. By enhancing those areas where there is a break between the reflectance of healthy vegetation (current cultivation or natural vegetation) and bare rock/soil (which might indicate the works of humans) may become more apparent.

It should be noted, however, that such a result may not always be the case. These indices are ratios that are designed to enhance the reflectance of vegetation while other features are much more subdued. These bright tones may tend to mute the detail of surrounding nonvegetated areas. The exception would be where vegetation surrounds areas of bare soil such as those seen in Figure 10. This shows an area of irrigated agriculture just to the east of Lake Dalmaj. Within several of the agricultural fields small areas of soil reflectance can be seen. These are areas of small sites or mounds that have not been cultivated.

A more sophisticated vegetation index involves the application of coefficients to separate the regions of an image that represent the planes of brightness (bare soil/rock), greenness, and wetness. This technique is known as the Tasseled Cap transformation. By displaying band 1 (the brightness band), we can emphasize highly reflective areas of sites and ancient drainage patterns (Figure 11a). The area displayed in Figure 11 is just north of the extensively excavated site of Nippur. To bring out more detail further enhancement may be needed. For example, in Figure 11b histogram equalization was applied to the brightness band and then a gray-scale inversion was applied. The black line segments seen on this image indicate relict canals, some of which have not previously been recorded.

Another set of enhancements that can prove useful are those that emphasize the linear nature and orientation of features or to detect the edges of significant breaks in the reflectance pattern of the image (Sabins, 1997). Geologists and geographers employ such enhancements for detecting everything

Figure 10. A false-color Landsat TM composite of an area of irrigated fields just north of Lake Dalmaj. The black arrows point to the locations of some uncultivated tells and sites. Scale is 1:17,000.

from faults and jointing patterns to highways and boundary lines. In archaeology, and in the research noted here, edge enhancement was applied to detect edges and orientations of foundations and canals.

There are two main categories of edge enhancements, nondirectional and directional. Both types can be produced using convolution filters. As the name indicates, directional filters highlight a specific orientation, such as NE-SW, and are implemented using the correct filter coefficients. Only linear features with similar orientations will be emphasized. Nondirectional filters (also called Laplacian) do not have any particular directional bias, with practically all directions enhanced equally. These filters were used in the present study with varying degrees of success on Thematic Mapper and radar imagery.

One of the most useful types of directional filters results in the edges having the look of a shaded-relief map (Jensen, 2005). This process is generally referred to as embossing because of the raised appearance of the filtered images. This type of filtering can be found in most remote sensing image processing software, but similar, and just as effective, embossing filters can be found in programs such as Corel Photo-Paint and Adobe PhotoShop. Examples of such embossing can be seen in Figure 12. In Figure 12a a portion of a SPOT panchromatic image is displayed. In the upper right corner is an area of irrigated agriculture. The dark area is the water surface of Lake Dalmaj. In the center of the image various small white features can be seen. These are tells and other ancient occupation sites, some dating back to Uruk times. They stand out in stark contrast to the darker reflecting water. The arrows point to

Figure 11. A scene showing a Tasseled Cap enhancement of an area just north of Nippur. (above) This Tasseled Cap Band 1 image (1:82,500) displays only the brightness band or plane, i.e., one that emphasizes bare soil and rock. (below) This Inverted (Negative) Rendition (1:82,500) is the same scene as above, except that it is displayed as an inverted (negative) gray-tone image. Note how the meander loop at "1" and the relict canals at "2" are somewhat easier to distinguish.

a site designated as #1120 in Adams's survey and gives an indication of the reflectance of these sites (Adams, 1981). In Figure 12b an embossed rendition of the same scene can be seen. In this image note how the embossing filter causes these sites to stand out in "relief" as opposed to the flat surfaces of the

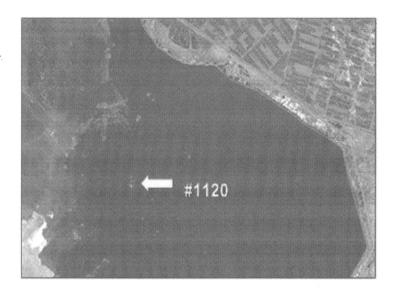

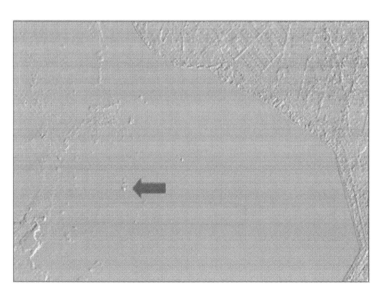

Figure 12. An example of embossing edge detection. (above) Landsat TM Band 4 image (1:37,000) of the eastern part of Lake Dalmaj. The location of site #1120 is indicated for reference. (below) The same scene as (above) after embossing. Note how sites in the lake now tend to stand out in "relief." Also note how irrigation canals and subsurface features are emphasized.

uniform reflectance of the water. In addition, to further highlight the "relief" of brightly reflecting sites, they can be displayed in a three-dimensional form or profile, as seen in Figure 13.

Another type of sensor used in our investigations was radar. Radar is a sensor that provides a unique perspective of the earth's surface (Holcomb, 1998). Through a burst of microwave energy it illuminates the earth's surface and records the strength of the return signal or backscatter. It has applications in the detection of soil moisture, linear features, and landform relief. In our study area of central Iraq where relief is extremely low, radar landform mapping is not as important as it would be farther east in the hills and mountains of Iran or even in northern Iraq. Radar did prove to be useful in central Iraq in the detection of linear features such as canals, irrigated agriculture, dunes and levees, and ancient drainage ways. It should be noted that when studying any radar image that the look direction of the sensor and the orientation of the feature on the ground has an important influence on the strength of the backscatter (Lewis et al., 1998:147).

Radar imagery does require some special preprocessing before image analysis. One of the problems with raw radar imagery is speckle, which is a type image noise or unwanted signal components in the image (Raney, 1998:67). Speckle appears on an image as random light and dark pixels, sometimes referred to as a "salt and pepper" effect. While speckle provides certain kinds of information about an image, it tends to be a distraction in image analysis. For this reason it is necessary to remove as much of the speckle as possible so as not to incorporate it into an enhanced or classified image. It should be noted, however, that while radar speckle can be significantly reduced, it cannot be completely eliminated (ERDAS, 1997).

To reduce speckle a series of special filters have been developed. These can be applied individually or in a sequence of combinations (Holcomb and Allan, 1992; Lee et al., 1994). It is not within the scope of this discussion to go into the specifics of all of these routines, although an example from our study is shown in Figure 14. Once the speckle noise has been reduced enhancement routines can be applied to the image just as they are done with other types of imagery. In the case of the radar imagery used in our investigation around Nippur, a Gamma MAP speckle-suppressed image was run through a Wallis Adaptive filter which enhances the image by adjusting the contrast stretch using a specified local region (ERDAS, 1997). Even with speckle suppression the enhanced radar imagery needed to be used in conjunction with SPOT panchromatic and Thematic Mapper imagery for the identification of new sites and the delineation of relict canals.

4.2. Image Classification

The mapping and analysis of the current regional land cover can be extremely beneficial in archaeological investigations (Sabins, 1997:387). It can point to

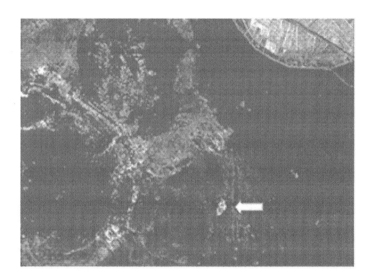

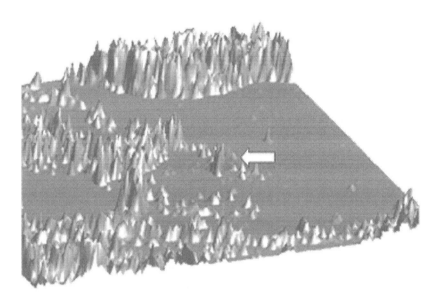

Figure 13. (above) Radarsat image of a portion of the scene in Figure 12. (below) Three-dimensional perspective rendition of the same scene. The perspective view can be used to enhance pixels with high brightness values. Once again, the arrows point to site #1120. Note also the high brightness values of aquatic vegetation and irrigated agriculture. (Courtesy of EODS).

Figure 14. An example of radar despeckle filter enhancement. (above) Original Radarsat Image (1:62,500). (below) Same scene after enhancement by GAMMA _ MAP Despeckle Filter. (Courtesy of EODS).

previous occupation sites and land uses, such as canals. In addition, mapping present-day patterns of land cover or change in land use and direction of growth can also be an aid in determining where there may be a potential for site destruction.

There are two categories of image classification, supervised and unsupervised (Lillesand et al., 2004:551–552). As the names indicate the main difference between the two is in the amount of prior knowledge of the area and the amount of analyst input into the classification routines. Because of the number of ground surveys that have been done in this area since the early 1960s there is some knowledge about the landscape that is available, but nothing current. For this reason, in the initial phases of our classification we decided to perform an unsupervised classification. One of the basic reasons for our using unsupervised classification in the investigation is the fact that most of the study area is not accessible because of the current political and military unrest.

An ISODATA unsupervised classification was the type of routine that was utilized. This is an iterative process that makes several passes through the dataset and clusters pixel values into categories based on spectral distance (Jensen, 2005: 385). This type of classification has the advantage of categorizing pixels without having to rely on ground-based training samples. Furthermore, all pixels in a dataset can be classified based on the statistical properties of their spectral distance from cluster means. The disadvantage is that once the pixels have been categorized, they still must be identified and named as to their land-cover types. Such a classification was performed on a Landsat Thematic Mapper subset of the Lake Dalmaj region (Figure 15a). The clustering was done based on 25 categories. The number of categories used was not based on any set figure, but instead on a certain amount of trial and error. It was found that from this particular scene a larger number of categories created a map that was difficult to interpret, while a smaller number did not bring out a sufficient amount of detail. Out of this classification three main groupings of land cover emerged; irrigated agricultural lands, water, and silt sediment flats or dunes. All of these class groupings trend in a generally northwest to southeast manner following the ancient patterns of the Tigris and Euphrates rivers.

Once the initial categorization was accomplished, the resulting map was studied and names were applied to the 25 categories (Figure 15b). This categorization was based on information from such sources as high-resolution SPOT panchromatic imagery, radar imagery, ground photography from earlier surveys, personal observations of field researchers, and the original Landsat image. Because of the discrimination provided by the 25 categories, general classes of land cover such as agriculture or silt were subdivided into separate class groupings such as Agricuture1, Agriculture2 and so forth based on their small spectral differences.

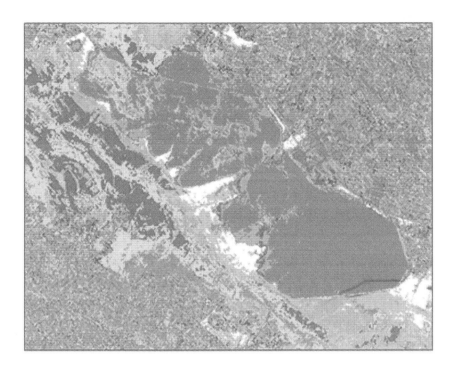

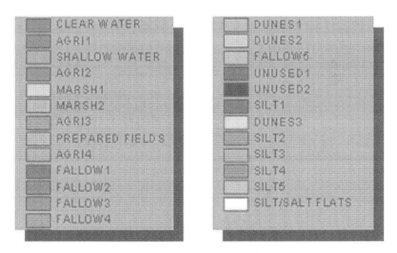

Figure 15. (above) Unsupervised land-cover classification map of the Lake Dalmaj Area. (below) An initial classified legend for the map above. Scale is 1:85:500.

With a general land-cover classification map the various human uses and natural landscapes can be studied and compared with current drainage patterns. Furthermore, the distribution of land-cover classes can be compared to the locations of known sites and canals. Buringh has stated, "ancient sites always occur on the best land" (Buringh, 1960:178). The land-cover types can be correlated with the satellite imagery to predict areas where additional sites not yet identified might be located. Just as image enhancement does, classification can bring out the subtleties in pattern that might not be apparent on the original satellite image. The unsupervised cluster categories thus cannot only be used as an aid in identifying undiscovered sites and canals, but also to confirm the locations of features already surveyed.

4.3. Remote Sensing and GIS Integration

Increasingly one of the key elements in the use of remote sensing imagery is its integration with Geographic Information Systems (GIS) (Ehlers et al., 1989; Westcott, 2000:1). Remote sensing imagery can be employed with GIS datasets in several different ways. First, the imagery can be used as a simple backdrop on which to plot various vector coverages. Second, rectified imagery can be used as a base on which vector digitizing can be done to delineate known archaeological features as well as updating existing surveys. Finally, remote sensing imagery can be used to create new coverages, such as the land-classification coverage previously discussed. Our studies in the central Mesopotamian Plain utilized all three of these applications.

One of the main uses of the survey GIS datasets is to display them on top of various types of registered remote sensing imagery (Lillesand et al., 2004:52). The location of known sites and canals can be correlated with their appearance on the imagery. In this way their types of spectral response can be extended into other areas not previously or intensively studied. By such methods new features can be plotted, or at the very least identified for potential ground observation and examination. With these new techniques for archaeological use and disparate datasets, we can begin to better understand landscape transformations of the plain in terms of watercourse changes and continue to add to the settlement record of archaeological sites.

5. CASE STUDIES

To illustrate how the techniques discussed above can be employed in actual investigations, three case studies are reviewed here. These case studies cover an area in the central portion of the Mesopotamian Plain (Figure 3). One goal of these studies was to test the use of various types of sensors (including Corona, Landsat, SPOT, ASTER, and Radarsat imagery) in archaeological research. In addition, these studies were undertaken to demonstrate the application of enhancement techniques and image classification in the detection and analysis of known sites and to extend the survey record of ancient sites and canals.

5.1 Abu Salabikh Study

The site of Abu Salabikh was surveyed by Adams (1981) and excavated by Postgate and Moorey (1976). A survey and geomorphological investigation was conducted by Wilkinson in 1988-1989 as part of the Abu Salabikh project in order to investigate the surrounding cultivated area for sites and canals, and to locate a branch of the Euphrates that ran through the site in antiquity (Wilkinson, 1990:75). The project identified sites in the surrounding area and augured along transects to determine the presence of ancient canals. The survey determined an "arc-shaped alignment of sites that ran NW to SE (Wilkinson, 1990: 82). The sites may have been situated on a canal or canal system that ran in this direction, but clear evidence for this system was difficult to discern during the survey because of heavy sedimentation (Wilkinson, 1990: 82) as well as increased cultivation in the area.

The 1968 Corona images of the area show a number of features that support the survey results and provide additional information. For example, they show the sites located during the survey as well as several more sites and the traces of a levee system that linked the archaeological sites. Both the ASTER and Corona images show tells that cast a shadow lining up in an arc as described by the surveyors (Figure 16). The additional information provided by the images is smaller sites that were later (at the time of survey) obscured by cultivation, thereby hiding the archaeological landscape around the main mounds of Abu Salabikh. The main canal system that would have fed these sites was not visible on the ground to the surveyors, but traces of it appeared in the corings (Wilkinson, 1990: 79–83). Low wide traces of levees between sites appear clearly on both the ASTER and Corona images, linking these fourth to third millennium sites. In this case, remote sensing has allowed for the addition of smaller satellite sites around the main site to be detected. The images have provided vital information on the canal system in the area that would have fed these sites and aided in answering questions of relationships of sites to canals that the ground survey could not properly address because of the constraints of archaeological survey and landscape transformations over time.

Adams mapped watercourses in the central portion of the plain in sections, because of cultivation and dune coverage. By georeferencing images of varying periods and overlaying them, we learned that these courses form a coherent system. For example, the Corona images that were contemporary with the surveys (late 1960s), show the courses in pieces with parts obscured by the dunes. When we compared the Corona, SPOT, and ASTER images for this area taken in different seasons over 30 years, the system became

Figure 16. An ASTER image of Abu Salabikh and the surrounding area. Scale is 1:10,500. (Courtesy of NASA/GSFC).

clear. From the 1960s to 1990s the dunes and cultivation boundaries shifted, exposing relict traces of levees once hidden by the dunes. Furthermore, large secondary branches, taking off from the main central branch which runs down the center of the plain, feed archaeological sites in the area. Analyzing the area as a whole would not be possible on the ground, even if political conditions permitted. In this case, using imagery in a GIS as well as techniques of remote sensing allow for a more inclusive analysis of watercourse patterns in particular periods.

In the case of archaeological sites, the integration of remote sensing and GIS has proved useful in numerous ways. The ASTER imagery proved to be very effective in the Abu Salabikh study. The Visible and Near Infrared

(VNIR) sensor provided multispectral imagery similar to that of Landsat TM, but at a resolution of 15 m. At this resolution it was much easier to pick out sites and to distinguish shadows. GIS also proved to be a valuable technique for the storage and analysis of spatial data. In the central portion of the plain, it has enabled the incorporation of many different datasets for comparison and spatial analysis. There are two major archaeological surveys that cover this area: Adams (1965, 1972 and 1981) and Gibson (1972). These surveys resulted in maps and site catalogues of information about individual sites. In order to integrate these survey maps into a study using remote sensing and GIS techniques, the maps were georeferenced to existing satellite imagery. In addition, a vector GIS database was created and the information from the site catalogues was placed in it. Each polygon or site had the following information attached: site number according to survey designation, site name, periods of occupation, shape, and additional information. Once referenced to the imagery, locations could be confirmed, errors in location or size noted, and the area calculated. Incorporation into a GIS for comparison allowed for more detailed analysis of spatial distributions of settlement in different periods and analysis of land-use patterns extrapolated from aggregate occupation areas.

These analyses resulted in our recognizing the tonal signature of an archaeological site (in this case tells) on a satellite image and comparing that signature with the signature of features in unsurveyed areas. Tells show a distinctive signature on satellite imagery. They appear as mounded colored features that cast a shadow to the north indicting a distinctive rise in elevation from the surrounding plain (Figure 16). Combining the characteristic signature of a tell to tells on imagery in surveyed areas, we may begin to fill in the gaps in the site record of Mesopotamia.

5.2. Hilla Study

Another case study of the use of remote sensing and GIS is presented in the region of the Hilla and Hindiyah River branches of the Euphrates (Figure 17). This area is typically indicated as blank and devoid of archaeological sites on maps of archaeological sites in Mesopotamia. Using remote sensing techniques and GIS, we analyzed imagery of different periods and seasons to detect the characteristic signatures of archaeological sites. It became clear to us that a number of new sites, perhaps as many as 33, do exist in this area. A comparison of Corona, SPOT, and ASTER imagery and British maps has suggested the existence of numerous archaeological sites in this area, as well as relict canal levees associated with these sites. Although dating of these sites is impossible without ground survey, integration of GIS and remote sensing has shown the need for continued work in this area to develop a more comprehensive settlement record.

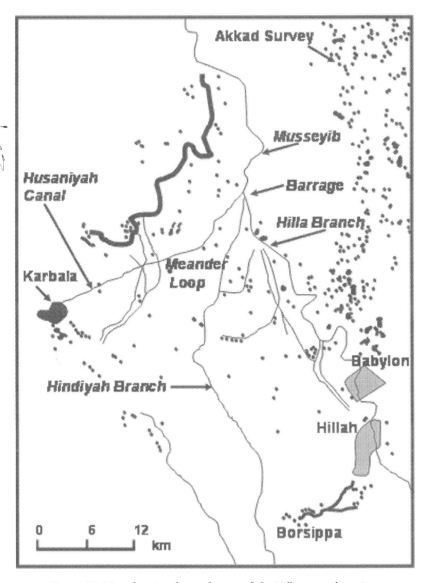

Figure 17. Map showing the study area of the Hilla research project.

5.3. Nippur – Lake Dalmaj Study

A study around the ancient site of Nippur up to Lake Dalmaj was undertaken to assess the capabilities of radar imagery in the delineation of relict drainage ways and in the identification of new sites. This area was chosen for several reasons. First, it contains a number of well-known sites that could be used for reference. Some of these sites are very prominent in shape and height while others are not (Figure 18). Also this is an area with a number of relict canals, as well as newly constructed ones (Figure 19). The radar imagery was used in conjunction with SPOT panchromatic and Landsat TM imagery. In addition, a GIS dataset of the survey coverage of Adams's catalog from the *Heartland of Cities* was used to verify currently mapped site locations. The SPOT imagery (WGS84 UTM Zone 38N) was used as the base against which the other imagery and databases were rectified.

At the time of our study relatively little commercial radar coverage of the study area was available, so we decided to contract to have a C-band Radarsat image of the Nippur-Lake Dalmaj area collected. At first we thought that L-band radar, with its greater surface penetrability in dry sands (see Holcomb and Shingiray in this volume) might be a better choice, but in the end concluded that the depth of alluvium in the study area and its silt/clay content would probably preclude successful use. Radarsat was also chosen because of its finer resolution, in this case approximately 9 m.

While the subsurface drainage patterns could not be detected to any great degree, the radar image proved quite effective in a number of ways. First, it brought out the detail of old meanders along the former course of the Euphrates river prior to its shift to the southwest. These can be seen in Figure 20b. In addition, old canal traces, as well as new canal construction can be differentiated on the Radarsat image in ways not presented on other types of imagery. These, too, can be seen in Figure 20b. Furthermore, sites and tells show up in a couple of different ways on the radar imagery as opposed to SPOT and Landsat. The look direction and angle can cause the radar backscatter to be recorded as either completely dark, irregular patches or as combinations of dark features with prominent white areas in a geometric form. The latter can also be seen in the form of Tell Al-Arsan in Figure 20b and in the site of Nippur with its walls and ziggurat in Figure 21.

On the other hand, many uniformly dark areas on the radar image 0can be noted, such as those in Figure 22. These dark areas stand out in comparison to the larger, rectilinear areas of light-toned reflectance. The lighter tones here represent the higher dielectric constant of high moisture levels in the irrigated field crops. The dark areas, in all probability, denote the flat surface of ancient occupation sites, and not merely areas where the seed did not germinate in the field. The lack of any other type of reflectance within these areas may indicate that they are of more recent origin in the archaeological record and do not have the kind of significant structures associated with them as major sites do, such as Nippur, Zibliyat, or Uruk. If such structures were present within these areas, there would be small bright spots or elongated bright streaks indicating the angular sides of structures or walls.

A despeckled, enhanced version of the Radarsat image (as discussed earlier) was subsequently used in conjunction with SPOT and Landsat imagery in an attempt to identify currently mapped sites and tells and to identify new

Figure 18. Ground photographs showing landscapes of (above) the Nippur tell and (below) the tell excavation of Abu Salabikh. Photos courtesy of John Sanders.

Figure 19. Ground photographs of two landscapes. (above) Modern irrigation canal near Nippur. Photo courtesy of John Sanders. (below) Portion of a relict canal near Abu Salabikh. The canal can be distinguished by differences in soil moisture content. Photo courtesy of Tony Wilkinson.

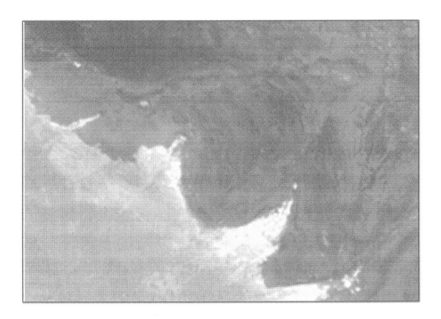

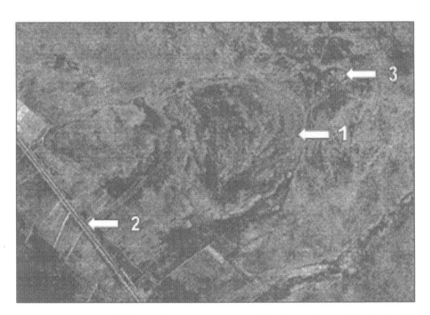

Figure 20. Images of the northwestern portion of Lake Dalmaj. (above) Landsat TM Band 1 image (1:15,800). (below) Radarsat image (1:16,150) of the same area. Note the relict meander loop at "1," the site of Tell al-Arsan at "2," and the new construction of canals at "3". (RADARSAT image courtesy of EODS).

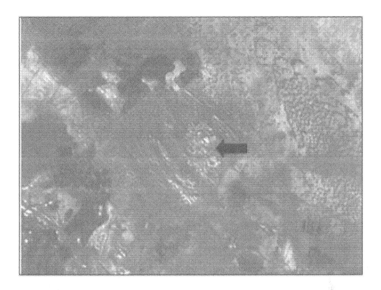

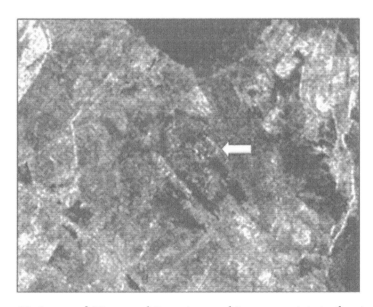

Figure 21. Images of Nippur and its region: a white arrow points to the ziggurat of Nippur in both scenes. (above) Corona photograph (1:5,000) of Nippur. (below) Radarsat image (1:6,270) of the same area. Note the high backscatter, represented as bright tones, of walls and foundations on the site. Also note the linear black area denoting small backscatter from more level drainage ways. (Corona image courtesy of USGS; RADARSAT image courtesy of EODS).

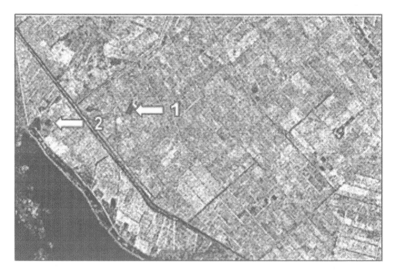

Figure 22. Area of irrigated agriculture east of Lake Dalmaj. (above) Landsat TM Band 3 image (1:20,000). A relict site is located at "1." Note the area at "2" with its bright reflectance, where other sites are located, b) Radarsat image (1:18,500) covering the same area. The black areas indicate small amounts of backscatter from level areas. These areas probably represent sites with little relief or prominent structures. Note how detail in the bright area at "2" in the image below can be discriminated better on the radar image. (RADARSAT image courtesy of EODS).

ones. All image sets were registered and a digital version of the *Heartland of Cities* site map was overlaid on top of the imagery for verification (Figure 23). Also overlaid was the land-cover map generated by unsupervised classification. The result of this processing was that 234 new potential sites were plotted (Figure 24). A new site was not created unless there was agreement between each image type as it was overlaid with the Adams' survey coverage to make sure there was not any duplication. The land-cover shapefile was also used to make sure that the area being classified as a new site agreed with current land-cover reflectances. Combining GIS coverages and remote sensing imagery proved to be a very effective means of verifying whether or not a site was unmapped or had already been identified and surveyed. In addition, we determined that a GIS can be an aid in studying specific surveyed sites that had already been mapped, as may be seen in the three-dimensional model created in Arcview 3D Analyst (Figure 25).

6. RESULTS AND CONCLUSIONS

It has been shown that the use of different types of remote sensing systems in the central Mesopotamian Plain of Iraq has definite advantages. In particular, higher-resolution imagery such as SPOT panchromatic was extremely useful because it enabled us to identify the shape, outline, and orientation of archaeological features. Another form of high-resolution imagery used to good effect was scanned Corona photography. This imagery has proven to be quite effective because it was contemporary with the major ground surveys of the time. Taken in the late 1960s, prior to much of the significant agricultural development in the region, Corona photography was able to be compared by us with more recent imagery. We were thus able to determine where former dunes had blown away, thereby exposing the surface, and also to detect the locations of ancient canals before newer canal construction had taken place. Multispectral imagery was also applied in several of our studies. Landsat TM and ASTER imagery provided the means by which natural vegetation and irrigated agricultural areas, especially in the infrared, could be differentiated from earth materials that might represent tells and drainage ways. Also, the ability to display the same imagery in a normal color rendition helped to distinguish subtleties in geomorphic and soil types in the study areas. Finally, radar imagery proved to be very beneficial in detailing level areas of terrain where additional sites might be located, as well as former channels of the Euphrates River. Radar imagery also aided in the differentiation between modern and relict canal systems.

Different dates of Landsat TM and ETM imagery helped to denote areas that had undergone the expansion of agricultural development into regions that have not yet been studied by detailed ground survey. The detection of areas that are being converted from natural cover to irrigated agriculture is of

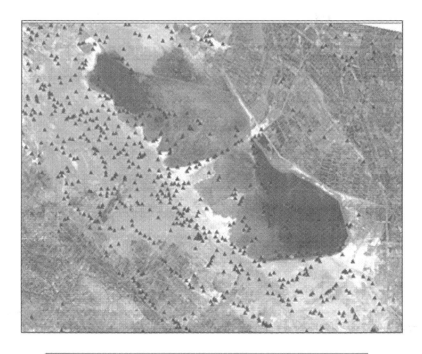

Number	Name	Site Shape	Length	Dimension	Width	Height	Dimension
825		cluster	0.0		0.0	2.0	110.00
826	Tell al-Arsan		0.0		0.0	0.0	0.00
827		circular	0.0		0.0	2.3	70.00
828		cluster	0.0		0.0	4.0	220.00
829		rectilinear	120.0	WNW	25.0	0.1	0.00
830		rectilinear	450.0	NNE	250.0	5.0	0.00
831		circular	0.0		0.0	2.8	300.00
832		rectilinear	350.0	NE	220.0	2.6	0.00
833		cluster	0.0		0.0	2.5	100.00
834		cluster	0.0		0.0	6.0	250.00
835		circular	0.0		0.0	1.2	170.00
836	Tell Dauran	rectilinear	180.0	N	100.0	6.5	0.00
837		circular	0.0		0.0	0.0	120.00
838		cluster	200.0	NE	90.0	2.8	0.00
839		rectilinear	150.0	NE	80.0	2.5	0.00
840		circular	0.0		0.0	0.8	100.00
841		circular	0.0		0.0	4.5	160.00
842		cluster	750.0	NE	200.0	2.8	0.00
843		circular	0.0		0.0	2.0	80.00
844		circular	0.0		0.0	0.0	60.00
845		rectilinear	320.0	NW	250.0	3.0	0.00
846		cluster	0.0		0.0	0.8	160.00
847		circular	0.0		0.0	4.0	450.00
848	Tell Abu Dhuwari	cluster	950.0	NW	200.0	3.5	0.00

Figure 23. (above) An example of the GIS database created from the Adams' survey. The small black triangles are the general locations of sites from this survey that have been overlaid on top of a Landsat TM image (1:138,600). (below) A portion of a few of the field types recorded in this database.

Figure 24. On this portion of a Landsat TM image (1:53,600) the blue triangle symbols represent the location of sites taken from the Adams' survey. The red circles represent additional potential sites that have been identified trough digital image processing and interpretation. It should be noted that none of these new sites has been verified on the ground.

tremendous importance. As agriculture expands in this region the potential for the destruction of known archaeological sites, as well as those that have not been inventoried and investigated, is great. The loss of such sites is incalculable. Remote sensing analysis can provide a means of identifying areas at risk to such devastating development and also predict the direction and rates of such growth.

The studies illustrated here also demonstrated the utility of various types of digital image enhancements. A number of these techniques are simple and straight-forward, yet can provide a significant amount of information. Being able to isolate and enhance specific types of landscapes or land cover features aids in the identification of various kinds of archaeological objects. Other types of techniques are more sophisticated and complex. For example, procedures designed to enhance the reflectance of vegetation or minerals can also be used for archaeological enhancements. It should be remembered that while enhancement can be done as a single operation, these procedures can also be used in conjunction with one another as a sequence of processes to maximize results.

In terms of enhancement techniques the application of color in the rendition of black- and white-imagery is one of the most basic. By being able to distinguish more hues of color than shades of gray, the analyst may perceive individual objects or patterns of reflectance that were not evident before. In addition, the use of contrast stretch techniques, especially histogram equalization, was invaluable in site delineation. The use of convolution filters

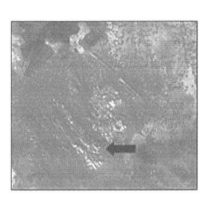

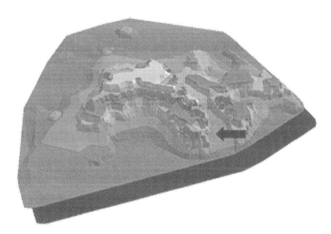

Figure 25. (upper left) Corona photograph (1:5,000) of Nippur. The black arrow locates a relict drainage way, the Shatt al-Nil. (upper right) Topography of Nippur: contour lines based on ground survey at Nippur. (below) Three-dimensional model of Nippur created in Arcview 3D Analyst based on contour lines generated from field surveys. (Courtesy Oriental Institute, University of Chicago).

for edge enhancements were also helpful in defining the linear orientation of canals and drainage ways. Particularly useful was a special type of edge enhancement known as embossing, which provided a kind of "raised relief" effect advantageous in identifying isolated sites and tells.

Employing various types of ratios or indices also proved effective in image analysis, notably the Tasseled Cap index. By being able to separate the brightness, greenness, and wetness components of an able to differentiate sites and canals from surrounding a.. Studying just the brightness band proved very effective in analyzing and potential sites. Displaying that band in an inverted gray-scale rendition further highlighted relict features.

Finally, the application of digital ground survey location data and land cover maps were shown to be worthwhile in the research. Landscape characterization through ISODATA clustering helped to isolate groups of statistically homogeneous pixels and confirmed site locations. It also aided in identifying areas where additional sites might be found.

The several studies reported here have shown the utility of remote sensing imagery in archaeological investigations in the central Mesopotamian Plain. Through various techniques of display, enhancement, classification and GIS integration the detection, delineation, and inventorying of ancient sites, tells, canals, and drainage ways is possible. Perhaps the real need for remote sensing lies in its ability to provide an image of inaccessible or dangerous locations, as is certainly the case in present-day Iraq. Because of political unrest and military operations, the access by archaeological ground-survey teams has been severely restricted or stopped. With no major on-site surveys done since the early 1990s, investigators have had to rely on a variety of remote sensing systems to provide the imagery necessary to maintain research projects. In the near future one of the main advantages of archaeological remote sensing will be to maintain the interest in and study of archaeology in this area of the world. Remote sensing will be the "bridge" technology for reconnaissance and research until archaeologists can once again return to the ground.

REFERENCES

Adams, R. E., 1980, Swamps, Canals, and the Location of Ancient Mayan Cities. *Antiquity* 54:206–214.

Adams, R. McC., 1965, *Land Behind Baghdad*. University of Chicago Press, Chicago.

Adams, R. McC., 1972, Settlement and Irrigation patterns in ancient Akkad. In *The City and Area of Kish*, edited by McG. Gibson, pp. 182–208. Field Research Projects, Miami.

Adams, R. McC., 1981, *Heartland of Cities*. University of Chicago Press, Chicago.

Algaze, G., 2001, Initial Social Complexity in Southwestern Asia: The Mesopotamian Advantage. *Current Anthropology* 42:199–233.

Avery, E., and Berlin, G., 1997, *Fundamentals of Remote Sensing and Airphoto Interpretation*. Fifth Edition, Macmillan Publishing Co, New York.

Brinkman, J.A., 1984, Settlement Surveys and Documentary Evidence: Regional Variation and Secular Trend in Mesopotamian Demography. *Journal of Near Eastern Studies* 43(3):169–80.

Buringh, P., 1960,*Soils and Soil Conditions in Iraq*. Republic of Iraq, Ministry of Agriculture, Baghdad.

Campbell, J., 2002, *Introduction to Remote Sensing*. Third Edition, Guilford Press, New York.

Crist, E., and Cicone, R., 1986, The Tasseled Cap De-mystified. *Photogrammetric Engineering and Remote Sensing* 55(5):601–610.

Deering, D., Rouse. J., Haas, R., and Schell, J., 1975, Measuring Forage Production of Grazing Units from Landsat MSS Data. In *Proceedings of the 10th International Symposium on Remote Sensing of Environment*, Vol. 2, pp. 1168–1178.

Ebert, J. and Lyons, T., Eds., 1983, Archaeology, Anthropology, and Cultural Resources Management. In *Manual of Remote Sensing*, edited by J. Estes, Vol. II, pp. 1233–1304. American Society of Photogrammetry. The Sheridan Press, Falls Church, VA.

Ehlers, M., Edwards, G., and Bedard, Y., 1989, Integration of Remote Sensing with Geographical Information Systems: A Necessary Evolution. *Photogrammetric Engineering and Remote Sensing* 57:669–675.

ERDAS, 1997, *ERDAS Field Guide*. Fourth Edition, Atlanta, GA.

Gasche, H. and Cole, S., 1998, Second and First Millennium B.C. Rivers in Northern Babylonia. In*Changing Watercourses in Babylonia: Towards a Reconstruction of the Ancient Environment in Lower Mesopotamia*, edited by H. Gasche and M. Tanret, pp. 1–64. University of Ghent and the Oriental Institute of the University of Chicago, Mesopotamian History and Environment Series II:V, University of Chicago Press, Chicago.

Gasche, H., Tanret, M., Cole, S., and Verhoeven, K., 2002, Fleuves du Temps et de La Vie: Permanence et Instabilitie du Reseau Fluviatile Babylonien Entre 2500 et 1500 Avant Notre Ere. *Annales Histoire, Sciences Sociales* 57(3):531–544.

Gibson, McG., 1972, *The City and Area of Kish*. Field Research Projects, Miami.

Guest, E., 1966, Flora*of Iraq*. Vol. 1 Introduction. Ministry of Agriculture, Baghdad.

Holcomb, D., 1998, Applications of Imaging Radar to Archaeological Investigations. In *Manual of Remote Sensing, Vol. 2, Principles and Applications of Imaging Radar*, edited by F. Henderson and A. Lewis, pp. 769–777. American Society of Photogrammetry and Remote Sensing, John Wiley & Sons, New York.

Holcomb, D., and Allan, J., 1992, Radar Data – An Important Source for the 1990's. In *18th Conference of the Remote Sensing Society*, pp. 408–413. University of Dundee.

Jacobsen, T., 1957, Salinity and Irrigation Agriculture in Antiquity. Diyala Basin Archaeological Project. Report on Essential Results, June 1, 1957–June 1, 1958. Unpublished.

Jensen, J., 2005, *Introductory Digital Image Processing*. Third Edition, Prentice Hall, Upper Saddle River, NJ.

Kauth, R., and Thomas, G., 1976, The Tasseled Cap – A Graphic Description of Spectral-Temporal Development of Agricultural Crops as Seen by Landsat. *Proceedings, Symposium on Machine Processing of Remotely Sensed Data*, pp. 41–51. LARS, West Lafayette, IN.

Kennedy, D., 1998, Declassified Satellite Photographs and Archaeology in the Middle East: Case Studies from Turkey. *Antiquity* 72:553–61.

Lee, J., Jurkevich, I., Dewaele, P., Wambacq, P., and Oosterlinck, A., 1994, Speckle Filtering of Synthetic Aperture Radar Images: A Review. *Remote Sensing Reviews* 8:313–340.

Lewis, A., Henderson, F., and Holcomb, D., 1998, Fundamentals: The Geoscience Perspective, In *Manual of Remote Sensing, Vol. 2, Principles and Applications of Imaging Radar*, edited by F. Henderson and A. Lewis, pp. 769–777. American Society of Photogrammetry and Remote Sensing, John Wiley & Sons, New York.

Lillesand, T, Kiefer, R., and Chipman, J., 2004, *Remote Sensing and Image Interpretation*. Fifth Edition, Wiley and Sons, New York.

Moik, J., 1980, *Digital Processing of Remotely Sensed Images*. Scientific and Technical Branch, National Aeronautics and Space Administration, Washington, D.C.

Postgate, J., and Moorey, P., 1976, Excavations at Abu Salabikh. *Iraq* 38:133–170.

Pournelle, J., 2003, *Marshland of Cities: Deltaic Landscapes and the Evolution of Early Mesopotamian Civilization*. Ph.D. dissertation, University of San Diego, Anthropology.

Raney, K., 1998, Fundamentals: Technical Perspective. In *Manual of Remote Sensing, Vol. 2, Principles and Applications of Imaging Radar*, edited by F. Henderson and A. Lewis, pp. 769–777. American Society of Photogrammetry and Remote Sensing, John Wiley & Sons, New York.

Sabins, F., 1997, *Remote Sensing: Principles and Interpretation*. Third Edition, W.H. Freeman Co, New York.

Trishchenko, A., Cihlar, J., and Li, Z., 2002, Effects of Spectral Response Function on Surface Reflectance and NDVI Measured with Moderate Resolution Satellite Sensors. *Remote Sensing of the Environment* 81:1–18.

Verhoeven, K., 1998, Geomorphological Research in the Mesopotamian Flood Plain. In *Changing Watercourses in Babylonia: Towards a Reconstruction of the Ancient Environment in Lower Mesopotamia*, edited by H. Gasche and M. Tanret, pp. 159–245. University of Ghent and the Oriental Institute of the University of Chicago Press, Mesopotamian History and Environment Series II:V.

Westcott, K. 2000, Introduction. In *Practical Applications of GIS for Archaeologists*, edited by K. Wescott, and J. Brandon, pp. 1–50. Taylor and Francis, London.

Wilkinson, T., 1990, Early channels and Landscape Development Around Abu Salabikh: A Preliminary Report. *Iraq* 52:75–83.

Wilkinson, T., 2003, *Archaeological Landscapes of the Near East*. University of Arizona Press, Tucson.

Wright, H., 1981, The Southern Margins of Sumer: Archaeological Survey of the Area of Eridu and Ur. In *Heartland of Cities*, edited by R. McC Adams, pp. 295–346. University of Chicago Press, Chicago.

Section **IV**

Geophysical Prospecting and Analytical Presentations

Chapter 13

Ground-penetrating Radar for Archaeological Mapping

LAWRENCE B. CONYERS

Abstract: Ground-penetrating Radar (GPR) is considered one of the more complicated of near-surface geophysical techniques, but also one of the more precise, because of its ability to map buried archaeological features in three-dimensions. Data from many two-dimensional reflections profiles within a tightly spaced grid, can be processed to remove noise, migrate reflections to their correct subsurface location, and then enhance important reflections from subsurface interfaces of interest. Three-dimensional images can then be constructed that produce realistic isosurfaces and amplitude slice-maps of buried features. When GPR reflections are incorporated with information derived from standard archaeological methods, and corrected to depth in the ground using velocity analysis, GPR maps can be used to display a large amount of information from limited excavations to produce a great deal of knowledge from a very large area. At the Albany, New York, town sites, historical maps of the city were compared to GPR images to determine neighborhood changes over time and the changing cultural landscape of one city block from early settlement through the early 20th century. At two sites in California and Colorado no reflections recognizable as cultural or geological were identified in reflection profiles, but amplitude slice-maps delineated spatial patterns that were found to be highly significant. Complex stratigraphy associated with buried cultural features can also be mapped, as illustrated in reflection profiles from aeolian dunes in coastal Oregon.

1. INTRODUCTION

Ground-penetrating radar is a near-surface geophysical technique that allows archaeologists to discover and map buried archaeological features in ways not possible using traditional field methods. By measuring the elapsed time between pulses of radar energy transmitted from a surface antenna, reflected from buried discontinuities, and then received back at the surface, two-dimensional profiles of buried stratigraphy can be produced. When the distribution and orientation of those subsurface changes can be related to certain aspects of archaeological sites such as the presence of architecture, use areas, or other associated cultural features, high-definition maps and images of buried remains can be produced. Ground-penetrating radar is a geophysical technique that is most effective with buried sites where artifacts and features of interest are located within 2–3 m of the surface, but has occasionally been used for more deeply buried deposits.

A growing community of archaeologists has been incorporating ground-penetrating radar (GPR) as a routine field procedure for many years (Conyers, 2004; Conyers and Goodman, 1997; Gaffney and Gater, 2003). Their maps and images act as primary data that can be used to guide the placement of excavations, or to define sensitive areas containing cultural remains to avoid. Archaeological geophysicists have also used the GPR method as a way to place archaeological sites within a broader environmental context and study human interaction with, and adaptation to, ancient landscapes (Kvamme, 2003).

Ground-penetrating radar data collection involves the transmission of high frequency radar pulses from a surface antenna into the ground. That energy travels at the speed of light in air, but quickly slows when it moves through the ground. At each interface where its speed changes, some of that energy is reflected back to the surface. The greater the velocity change, the higher the amplitude of the reflected radar-waves. The elapsed time between when radar waves are transmitted, reflected from buried materials or sediment, and soil changes in the ground, and then received back at the surface is then measured. When many thousands of radar-wave reflections are measured and recorded, as antennas are moved along transects within a grid, two-dimensional profiles of soil, sediment, and buried cultural feature changes can be created (Figure 1). When many tens or even hundreds of two-dimensional profiles are collected in a grid, three-dimensional maps can be constructed, making the GPR method one of the most precise tools for archaeological mapping.

The GPR method has recently become so accurate that the possibility now exists to test any number of working hypotheses concerning a broad range of anthropological, geological, and environmental questions important to archaeological interpretation. Some of those could be related to social organization and social change, when these cultural attributes can be directly related to the placement, orientation, size, geometry, or distribution of certain architectural and ancillary features on the landscape. Geological and environmental aspects of ancient landscapes such as soil changes and the nature of buried topographic features is also possible (Conyers, 1995; Conyers and Spetzler, 2002; Conyers et al., 2002a). Most importantly, the GPR method can gather a great deal of information about the near-surface in a totally non-destructive way, allowing large areas with buried remains to be studied efficiently and accurately, while at the same time preserving and protecting them.

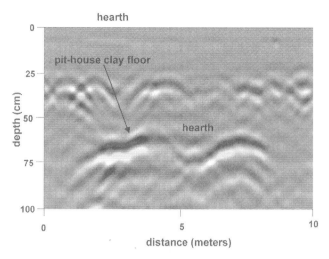

Figure 1. A two-dimensional GPR reflection profile showing a clay floor of a pit house buried in sand dunes.

The GPR method is especially effective in certain sediments and soils between about twenty centimeters and five meters below the ground surface, where the targets to be imaged are fairly large, hollow, linear, or have significant physical and chemical properties that contrast with the surrounding medium. Features as diverse as Maya house platforms and plazas (Conyers, 1995), burial tombs (Goodman and Nishimura, 1993), historical-period cellars, privies, and graves (Bevan and Kenyon, 1975), camp sites (Vaughan, 1986) and pit dwellings and kivas (Conyers and Cameron, 1998) have been discovered and mapped using the GPR method. Large-scale archi-tectural features such as stone walls and floors, surrounded by homogeneous soil or sediment, are especially visible as GPR reflections, and very suitable for three-dimensional mapping and image production (Conyers, 2004; Conyers et al., 2002a, Goodman et al., 2004, Neubauer et al., 2002).

Modern GPR systems are quite compact and easy to use. The typical system consists of surface antennas, a radar system to produce pulses, a computer to process and save the data, a video monitor, and a power source. This system can be easily transported to the field by plane, car, and backpack. Processing of data can often occur on a laptop computer after reflection data are downloaded, often within a few hours after data collection.

2. THE GPR METHOD

Ground-penetrating radar (GPR) is a geophysical method that can accurately map the spatial extent of near-surface objects and archaeological features or changes in soil media and ultimately produce images of those materials. Radar waves are propagated in distinct pulses from a surface antenna, reflected off buried objects, features, bedding contacts, or soil units, and detected back at the source by a receiving antenna. As radar pulses are transmitted through various materials on their way to the buried target feature, their velocity changes depending on the physical and chemical properties of the material through which they travel (Conyers, 2004a; Conyers and Goodman, 1997). The greater the contrast in electrical and to some extent magnetic properties between two materials at a subsurface interface (resulting in a stronger reflected signal), the greater is the amplitude of the reflected waves. When the travel times of energy pulses are measured, and their velocity through the ground is known, then distance (or depth in the ground) can be accurately measured to produce a three-dimensional dataset (Conyers and Lucius, 1996). Each time a radar pulse traverses a material with a different composition or water saturation, the velocity changes and a portion of the radar energy is reflected back to the surface, to be recorded at the receiving antenna. The remaining energy continues to pass into the ground to be further reflected, until it finally dissipates with depth.

The success of GPR surveys is to a great extent dependent on soil and sediment mineralogy, clay content, ground moisture, depth of burial, surface topography, and vegetation. It is not a geophysical method that can be immediately applied to any subsurface problem, although with thoughtful modifications in acquisition and data-processing methodology, GPR can be adapted to many differing site conditions. Although radar-wave penetration and the ability to reflect energy back to the surface is often enhanced in a dry environment, moist soils can still transmit and reflect radar energy, and GPR surveys can sometimes yield meaningful data even in totally saturated clay-rich soils. Recent research indicates that the amount and distribution of water in the ground is probably the most important variable that affects radar transmission and reflection (Conyers, 2004b). Often completely water-saturated ground will drastically slow radar wave velocity, but not attenuate the signal, if the material is not electrically conductive. Radar energy has been transmitted and reflected from depths approaching 3 m in totally saturated clay, when that clay is of a chemical composition that does not impede radar movement. In contrast totally dry sandy soil can potentially attenuate energy when it contains hydrous salts, as they are electrically conductive and will readily dissipate radar energy.

3. GPR DATA COLLECTION

To produce reflection profiles the two-way travel time and the amplitude and wavelength of the reflected radar waves derived from pulses generated at the antenna are amplified, processed, and recorded for immediate viewing or later post-acquisition processing and display. Most often this primary display is in the form of two-dimensional profiles. During acquisition of field data, the radar-transmission process is repeated many times per second as the antennas are pulled along the ground surface or moved in steps. Distance along each line is recorded for accurate placement of all reflections in space within a surveyed grid. Reflection profiles are usually spaced between 50 and 100 cm apart in a grid, but recent research has shown that greater resolution is almost always a function of a more densely spaced grid (Goodman et al., 2004; Neubauer et al., 2002).

The depth to which radar energy can penetrate and the amount of definition that can be expected in the subsurface is partially controlled by the frequency of the radar energy transmitted. Frequency controls both the wavelength of the propagating wave and the amount of signal spreading and attenuation of the energy as it travels in the ground. Commercial GPR antennas range from about 10 to 1500 megahertz (MHz) center frequency. Proper antenna frequency selection can in most cases make the difference between success and failure in a GPR survey and must be planned for in advance. In general the greater the depth of investigation, the lower the antenna frequency necessary. Lower-frequency antennas are much larger, heavier and more difficult to transport to and within the field than high-frequency antennas but can potentially transmit energy more deeply into the ground. In contrast, high-frequency antennas are quite small and can easily fit into a suitcase but are capable of shallower transmission.

Subsurface-feature resolution varies with radar-energy frequency. Low-frequency antennas (10–120 MHz) that generate long-wavelength radar energy can penetrate up to 50 m in certain conditions, but are capable of resolving only very large subsurface features. For example, dry sand and gravel, or unweathered volcanic ash and pumice are media that allow radar transmission to depths approaching 8–10 m, when lower-frequency antennas are used. In contrast the maximum depth of penetration of a 900-MHz antenna is about 1 m or less in typical soils, but its generated reflections can resolve features down to a few centimeters in diameter. A trade-off therefore exists between depth of penetration and subsurface resolution. These factors are highly variable, depending on many site-specific factors such as overburden composition and porosity, and the amount of moisture retained in the soil.

4. GPR DATA INTERPRETATION

Standard two-dimensional images can be used for most basic data interpretation, but analysis can be tedious if many profiles are included in the database. In addition, the origins of each reflection in each profile must sometimes be defined before accurate and meaningful subsurface maps can be produced. Often detailed image definition comes only with a good deal of interpretive experience, because the primary goal of most GPR surveys is to identify the size, shape, depth, and location of buried remains and related stratigraphy. The most straightforward way to identify the size, shape, depth, and location of buried materials is by first identifying their produced reflections and then correlating them within and between adjoining two-dimensional reflection profiles. A more sophisticated method of GPR processing is amplitude slice-mapping, which creates maps of amplitude differences of reflected waves in discreet horizontal slices within a grid. The result can be a series of maps that illustrate the three-dimensional location of reflection anomalies derived from a computer analysis of the two-dimensional profiles (Figure 2). This method of data processing can only be accomplished with a computer using GPR data that are collected and stored digitally.

Three-dimensional amplitude maps are possible with GPR data because all reflection data collected by GPR is nothing more than a collection of many individual traces along two-dimensional transects within a grid. Each of those reflection traces contains a series of waves that vary in amplitude depending on the amount and intensity of energy reflection that occurred at buried interfaces. An analysis of the spatial distribution of the amplitudes of reflected waves is important because it is an indicator of potentially meaningful subsurface changes in lithology or other physical properties. If amplitude changes can be related to important buried features and stratigraphy, then location of those changes can be used to reconstruct the subsurface in three-dimensions. Areas of low-amplitude waves usually indicate uniform matrix material or soils, while those of high amplitude denote areas of high subsurface contrast such as buried archaeological features, voids, or important stratigraphic changes. In order to be interpreted, amplitude differences must

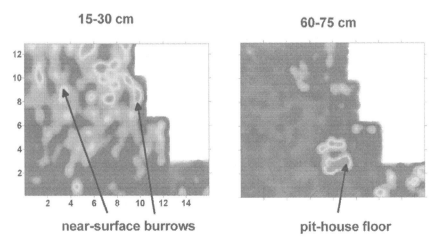

Figure 2. Amplitude slice-maps at two levels, showing in plan view rodent burrows near the surface and in the deeper slice the pit-house floor shown in one profile in Figure 1.

be analyzed in slices that examine only changes within specific depths in the ground. Each horizontal amplitude slice consists of the spatial distribution of all reflected-wave amplitudes, which are indicative of changes in sediments, soils and buried materials.

Amplitude slices need not be constructed horizontally or even in equal time intervals. They can vary in thickness and orientation, depending on the questions being asked. Surface topography and the subsurface orientation of features and stratigraphy of a site may sometimes necessitate the construction of slices that are neither uniform in thickness nor horizontal. To compute horizontal time-slices the computer compares amplitude variations within traces that were recorded within a defined time window. When this computation is done both positive and negative amplitudes of reflections are compared to the norm of all amplitudes within that window. No differentiation is usually made between positive or negative amplitudes in these analyses; only the magnitude of amplitude deviation from the norm. Low amplitude variations within any one slice denote little subsurface reflection and therefore indicate the presence of fairly homogeneous material. High amplitudes indicate significant subsurface discontinuities, in many cases detecting the presence of buried features. An abrupt change between an area of low and high amplitude can be very significant and may indicate the presence of a major buried interface between two media. Degrees of amplitude variation in each time-slice can be assigned arbitrary colors or shades of gray along a nominal scale. Usually there are no specific amplitude units assigned to these color or tonal changes.

The unique ability of GPR systems to collect reflection data in a three-dimensional package lends itself to the production of a number of other three-dimensional images not possible using other methods (Conyers, 2004a; Conyers et al., 2002a; Goodman et al., 1998; Goodman et al., 2004; Leckebusch, 2000, 2003). If reflection data are collected in a grid of closely spaced transects, and there are many reflection traces gathered along each transect, reflection amplitudes can be accurately placed in three-dimensions and then rendered using a number of visual display programs. In this way GPR data from archaeological sites become analogous to many other imaging techniques used in other disciplines, which rely on energy sources such as sonic waves and magnetic resonance. In medical imaging complex three-dimensional techniques can produce images of certain amplitudes derived from these waves to display internal body parts, or even electrical impulses in the brain as a function of different stimuli. In archaeology, radar reflections can be used in the same way, but instead produce images of buried archaeological features.

Using GPR data buried features can be rendered into isosurfaces, meaning that the interfaces producing the reflections are placed in three-dimensions and a pattern or color is assigned to specific amplitudes in order for them to be visible (Conyers et al., 2002a; Goodman et al., 2004; Leckebusch, 2003). In programs that produce these types of images certain amplitudes (usually the highest ones) can be patterned or colored while others are made transparent. Computer-generated light sources, to simulate rays of the sun, can then be used to shade and shadow the rendered features in order to enhance them, and the features can be rotated and shaded until a desired product results.

5. EXAMPLES OF GPR MAPS AND IMAGES

Often in complexly layered sites, where more deeply buried horizons contain materials of greater age, horizontal GPR amplitude maps can illustrate dramatic cultural changes over time. In this way the deeper slices will show older building activities, while the shallow slices show more recent ones. This difference was demonstrated at an historical-period site in Albany, New York, USA, where fire-insurance maps of the city were available showing the location of buildings present on town lots at specific time periods, going back to the year 1857. These maps showed dramatic building, demolition, and redevelopment episodes over a period of only 150 years. All buried materials from each construction episode are now located under a paved parking lot. The amplitude slice-maps were constructed in 25-cm-depth slices (after radar travel times were corrected for velocity), and images of the buried architectural features visible in the GPR amplitude maps were compared to the historical lot maps (Figure 3). The slice from 50–75 cm depth shows building foundations whose location compared almost exactly to domestic structures and a large kiln that were present in 1890. In progressively deeper slices, those 1890 buildings were still visible, but deeper foundations from older structures were also visible in the slice in the 150–175 cm depth (Figure 3). When the locations of those features were compared to the oldest maps from 1857, no correlation was found to any mapped structures. The deeper slices were therefore producing images of buildings that were present prior to the construction of any extant maps of the city. These very old building remains are awaiting excavation and their age and function remain unknown. The shallower slices and their structural remains correlated almost perfectly with the most recent fire-insurance maps.

In this example from historical New York the horizontal amplitude slice-maps can be a way not only to map building locations over time, but also, when integrated with enough other information such as historical maps and artifacts from excavations, determine their function as well. The changing make-up of neighborhoods can potentially be determined using these GPR amplitude maps, each of which is from a specific time period, if the sequential amplitude slices are roughly comparable to time periods. In this way GPR images can be much more than just a tool for finding and mapping buried features; they can also be a database from which to study social change and a wealth of other historic and anthropological questions.

Often reflection profiles can be difficult to interpret, even after filtering, post-acquisition processing, and the production of many profile views with

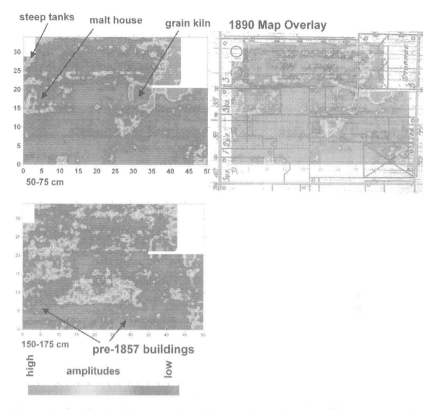

Figure 3. Amplitude slice-maps of an historical-period site in Albany, New York, preserved under a parking lot. The two slices illustrate very different preserved architecture at different depths. When the shallow slice from 50–75 cm is compared to the 1890 fire insurance maps, there is good correlation with what was known. The deeper slice shows smaller buildings from pre-1857.

differing vertical and horizontal exaggeration. There is often a temptation when one looks at particularly "noisy" reflection profiles such as the one in Figure 4 to give up and call the survey a failure as there are no amplitude changes readily visible in it. The profile in Figure 4 was collected with 900-MHz antennas in a boggy area in the California Sierra Nevada Mountains, USA. The goal was to map recently deposited sedimentary units in the hope of defining fluvial, marsh, and floodplain sediments that might have been present in the mid-19th century. Historical records indicated that the ill-fated Donner party, a wagon train that was immigrating to California and attempting to cross the mountains in November of 1846, camped near a creek in the study area and were stranded there all winter. Many eyewitness accounts reported that the survivors found themselves in a bog in the spring of 1847 when the snow melted. Finding the remains of that camp today is complicated

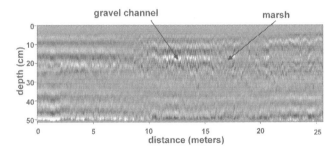

Figure 4. Reflection profile across a marsh which contains a small gravel channel. Reflection character is the only way to differentiate these lithological differences.

because the environment has changed a great deal, and is now on the edge of a reservoir that was flooded in the 1960s. It was hoped that an analysis of the environment as it existed during the time of the Donner Party encampment might yield clues to where the winter campsite was located, as it was known that there was a small creek nearby in the early winter, and the area became a bog in the spring. The GPR method was used because it had the ability to map in three-dimensions and, even though almost all the prospective area is today wet and boggy, because similar environments with an abundance of peat beds had proven excellent areas for GPR mapping in Scotland (Clarke et al., 1999).

The reflections in the 900-MHz profiles that crossed the present-day bog proved to be noisy and discontinuous, and few good reflections were visible that could be readily interpreted (Figure 4). There are, however, changes in the reflection character along profiles, with some areas containing few good reflections and others that appeared to contain many very small hyperbolic reflections, especially within the upper 20 cm of the reflection profiles. As little interpretation could be done using the individual profiles, it was decided to study the amplitude changes spatially within all the profiles in a grid to determine if there were any patterns to the distribution of these reflections either spatially or with depth.

When the amplitudes in all the profiles in the grid were studied in slice-maps, one sinuous area of higher-amplitude reflections was visible from the 10–20 cm slice (Figure 5). After further study of the reflection profiles in two-dimensions, it was hypothesized that the anomalously high amplitude area probably corresponds to the presence of many small gravel clasts deposited in a creek, each of which generated the small reflection hyperbolas at that depth. Auger holes were then dug on either side of the high amplitude feature, and within it (Figure 5). A study of the sediments recovered from them showed that holes 1 and 3 consisted of silt and peat, with abundant charcoal, while the sediment in hole 2 was mostly sand and gravel and contained very little peat. This subsurface information confirms that the sinuous anomaly in the 10–20 cm slice was likely a small sand- and gravel-filled creek. The areas

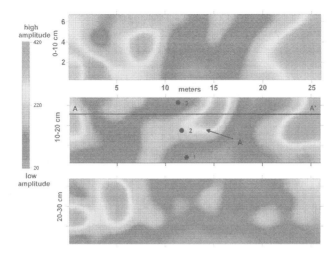

Figure 5. Amplitude slice-maps across the marsh and small channel seen in Figure 4. The sinuous channel (A) is readily visible in the middle slice and was confirmed in auger hole 2, while adjoining marsh sediment was recovered in holes 1 and 3.

adjacent to it, which are much lower in reflection amplitude, are areas where marsh and floodplain sediments were deposited.

Although remains of the Donner Party camp were not found in this immediate area, the study was successful in defining the shallow creek with adjacent marshy floodplain deposits, which can be used as a guide for further subsurface testing in a search for artifacts. Most importantly, this study illustrates how even reflection data that are difficult to interpret in two-dimensional profiles can produce useful data when studied in amplitude slices. When data of this type are then incorporated with standard archaeological and historical information, large areas of ground can be studied quickly and excavation efforts can be concentrated in the most prospective locations.

Often GPR amplitude slice-maps are capable of producing images that are not only almost invisible in reflection profiles, as shown in the example above, but the buried features that produced the reflection anomalies are also almost invisible to the human eye even when uncovered in excavations. A GPR survey was conducted in an orchard, where surface plowing had destroyed any indication of buried features likely to exist below. The area surveyed was the site of an early homestead in the mid-1800s in Denver, Colorado, USA, which was converted to a stage-wagon stop and finally reverted to a family farm in the 20th century. There were historical documents indicating that a number of buildings had been located somewhere in the orchard area, but their exact locations were unknown. The area had also been subjected to a number of floods, which buried any possible remaining features below more than a meter of sediment.

A grid of 400-MHz GPR reflection data was collected in the orchard, and slice-maps were constructed every 20 cm in the ground, after radar travel times were converted to depth (Figure 6). At the 75–100 cm depth a distinct linear feature was discovered, and in the deeper slices, another linear feature crossing it at an angle. Modern utility maps show plastic water lines cutting through the orchard, which generated the linear reflection anomalies. More interesting, however, was a 4-meter-square amplitude feature in the 75–100 cm depth, which was not correlative to any of the historical buildings that had been mapped in the area. This feature was hypothesized to be a buried building floor, because of its perfectly square geometry. Auger holes were dug both inside and outside the square feature, and no discernible difference could be seen in the two sediment samples from the depth indicated in the amplitude slice-map. Thinking that perhaps velocities, which had been estimated from hyperbola fitting of point source hyperbolas generated from the pipes, were

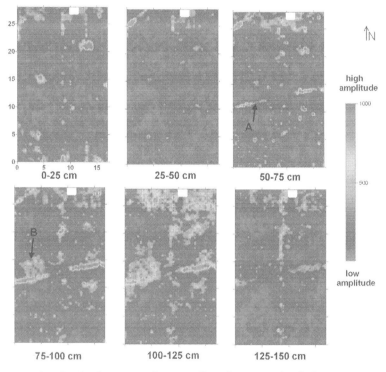

Figure 6. Amplitude slice-maps of GPR collected in an orchard. An irrigation pipe is visible as (A). The square feature (B) was excavated and found to be a very thin sand layer that was at one time associated with the floor of an historical-period building.

incorrect, a test excavation was placed directly on top of one of the pipes to uncover it in order to confirm the depth to a known object. The pipe was found at exactly the depth shown in the GPR maps, raising the confidence level of the velocity analyses and the slice-maps depths. A more careful analysis of the soil and sediment stratigraphy in the excavations adjacent to and just below the pipe was then made. A layer 2 cm thick that was just a little sandier than layers above and below was found within the square GPR feature. When excavations were extended outward, this sandy layer was found to be overlain by broken pieces of flat sandstone, which were probably used as pavers in the floor of a small house or shed.

It appears that there was a small building in the orchard at one time, whose floor was paved with sand and then covered with flat flagstone. When the building was abandoned, the usable sandstone pieces were probably salvaged for use elsewhere, and the remaining sandy sub-floor was covered by sediment during floods, and by the build-up of soil. All that remains of the house today is the very subtle sand layer. The feature is so subtle that normal excavation methods would have likely overlooked it, as it would have been interpreted as just another sandy layer in the orchard's sediment and soil package. Even once the feature had been discovered in excavations, it could still not be seen in GPR reflection profiles without using a good deal of imagination. Only subtle changes in radar reflection amplitudes, processed as amplitude slice-maps, were capable of finding and mapping this feature, which was only distinguishable by its distinctive square shape.

At the site of Petra, Jordan, Roman and Byzantine structures are buried in wind-blown sand (Conyers et al., 2002). Walls, floors, and standing columns are often visible in reflection profiles as distinct hyperbolic reflections. Amplitude slice-maps can be used to map the reflections from these architectural features in plan view, which is the usual way of imaging three-dimensional GPR reflections. Isosurfaces can also be created from these reflection data that reveal an image of the buried architecture in a very different way (Figure 7). In this rendering, the shallowest slices were not included, as they contained many small point source reflections from shallow rock rubble. The very deepest slices were also not included, as the data from that depth tended to be "noisier," and resulted in high amplitude streaking in the rendered images (Conyers et al., 2002). These reflection data were ideal for rendering because most of the reflections collected in the grid were generated from buried architecture, and the surrounding wind-blown sand matrix generated almost no reflections of any amplitude. Once the rendering was constructed on the computer it could also be rotated or tilted in a number of different orientations in a video display (Conyers et al., 2002).

Often in stratigraphically complex areas, reflections from geological layers can obscure and potentially clutter amplitude maps with reflections from non-cultural horizons. When this clutter occurs detailed analysis of reflections and the layers that produced them is necessary. Along the Pacific Coast near

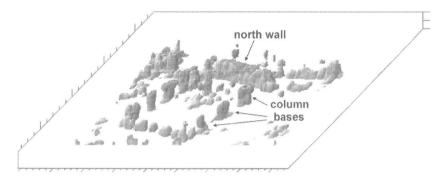

Figure 7. Rendering of high amplitude reflections from a buried Byzantine building at Petra, Jordan. Buried column bases and a distinct north wall are visible.

Port Orford, Oregon, USA, a buried pit-house village is preserved in stabilized coastal dunes. The dunes contain many large cross-bedded units of sand and silt, each of which produces distinct GPR reflections (Figure 8). A 400-MHz antenna was used in this area, which was capable of defining each of the major dune horizons as sloping reflections, with the pit-house floors as horizontal reflections (Figure 8). A reflection profile parallel to an erosional bank that exposed many of these geological units as well as one exposed pit-house floor, was used to determine the origin of reflections. When the GPR reflections profile was compared to the photograph of the natural and cultural layers, the origins of reflections were easily determined. A series of horizontal amplitude maps was then constructed that filtered out the steeply dipping beds (those produced from the sand-dune stratigraphy), and mapped only the horizontal reflections from the pit-house floors (Figure 2). Those floors could then be

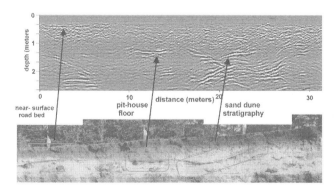

Figure 8. GPR reflection profile collected parallel to an erosional cutbank where a pit-house floor and complex sand dune stratigraphy are visible. These features are also visible in the reflection profile.

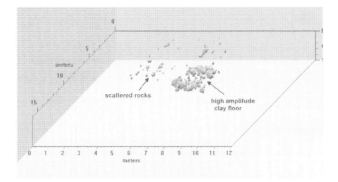

Figure 9. An amplitude isosurface of the pit-house floor visible in the amplitude slice-map in Figure 2.

discriminated from the surrounding sand matrix by filtering out the lower amplitudes, leaving only the high amplitudes from the compacted clay floors. When this was done, the remaining high-amplitude reflections were rendered in three-dimensions to produce an image of the buried floor, revealing two "benches" of compacted clay, separated by a less-compacted area, which might be a separate work area or partially disturbed floor surface (Figure 9).

6. CONCLUSION

Ground-penetrating radar has the unique ability of near-surface geophysical methods to produce three-dimensional maps and images of buried archi-tecture and other associated cultural and natural features. By using high-definition, two-dimensional reflection profiles, researchers can generate three-dimensional maps of amplitude changes that can define physical and chemical changes in the ground that are related to buried materials of impor-tance. Often it is how the physical and chemical changes in various buried materials affect the retention and distribution of water in the ground that is of highest importance. Subtle changes in the amplitude of radar reflection can often be related to the presence or absence of cultural materials, thereby producing maps of changes that would be difficult to see in normal excavation trenches or through other subsurface evaluation techniques.

In the processing of GPR reflection data, maps and images must be generated so that buried cultural materials of interest can be visualized and interpreted. This result can be accomplished by using horizontal amplitude maps, sliced in layers or three-dimensional isosurfaces, which produce images of only certain amplitudes within a three-dimensional volume of reflections. In all cases, the results of these amplitude images must be differentiated from the surrounding geological layers.

REFERENCES

Bevan, Bruce and Kenyon, Jeffrey, 1975, Ground-penetrating radar for historical archaeology. *MASCA Newsletter* 11(2):2–7.

Clarke, M. Ciara, Utsi, Erica and Utsi, Vincent, 1999, Ground-penetrating radar investigations at North Ballachulish Moss, Scotland. *Archaeological Prospection* 6:107–121.

Conyers, Lawrence B., 1995, The use of ground-penetrating radar to map the buried structures and landscape of the Ceren site, El Salvador. *Geoarchaeology* 10(4):275–299.

Conyers, Lawrence B., 2004a, *Ground-penetrating Radar for Archaeology*. AltaMira Press, Walnut Creek, California.

Conyers, Lawrence B., 2004b, Moisture and soil differences as related to the spatial accuracy of amplitude maps at two archaeological test sites. In *Proceedings of the Tenth International Conference on Ground Penetrating Radar: June 21–24, Delft, The Netherlands*, edited by Evert Slols, Alex Yaravoy and Jan Rheberger, pp. 435–438. Delft University of Technology, The Netherlands Institute of Electrical and Electronic Engineers, Inc., Piscataway, NJ.

Conyers, Lawrence B., and Cameron, Catherine M., 1998, Finding buried archaeological features in the American Southwest: New ground-penetrating radar techniques and three-dimensional computer mapping. *Journal of Field Archaeology* 25(4):417–430.

Conyers, Lawrence B., and Goodman, Dean, 1997, *Ground-penetrating Radar: An Introduction for Archaeologists*. AltaMira Press, Walnut Creek, California.

Conyers, Lawrence B., and Lucius, Jeffrey E., 1996, Velocity analysis in archaeological ground-penetrating radar studies. *Archaeological Prospection* 3:312–333.

Conyers, Lawrence B., and Spetzler, Hartmut, 2002, Geophysical Exploration at Ceren. In *Before the Volcano Erupted*, edited by Payson Sheets, pp. 24–32. University of Texas Press, Austin.

Conyers, Lawrence B., Ernenwein, Eileen G., and Bedal, Leigh-Ann, 2002, Ground-penetrating radar (GPR) mapping as a method for planning excavation strategies, Petra, Jordan. E-tiquity Number 1, *http://e-tiquity.saa.org/%7Eetiquity/title1.html*

Gaffney, Chris and Gater, John, 2003, *Revealing the Buried Past: Geophysics for Archaeologists*. Tempus, Stroud, Gloucestershire.

Goodman, D. and Nishimura, Y., 1993, A ground-radar view of Japanese burial mounds. *Antiquity* 67:349–354.

Goodman, Dean, Nishimura, Yashushi, Hongo, Hiromichi and Maasaki, Okita, 1998, GPR Amplitude rendering in archaeology. In *Proceedings of the Seventh International Conference on Ground- penetrating Radar, May 27–30, 1998*. University of Kansas, Lawrence, Kansas, USA. pp. 91–92. Radar Systems and Remote Sensing Laboratory, University of Kansas.

Goodman, Dean, and Piro, Salvatore, Nishimura, Yasushi, Patterson, Helen, and Gaffney, Vince, 2004, Discovery of a 1st century AD Roman amphitheatre and other structures at the Forum Novum by GPR. *Journal of Environmental and Engineering Geophysics* 9:35–42.

Kvamme Kenneth L., 2003, Geophysical surveys as landscape archaeology. *American Antiquity* 63(3): 435–457.

Leckebusch, J., 2000, Two and three-dimensional ground-penetrating radar surveys across a medieval choir: a case study in archaeology. *Archaeological Prospection* 7:189–200.

Leckebusch, J., 2003, Ground-penetrating radar: A modern three-dimensional prospection method. *Archaeological Prospection* 10:213–240.

Neubauer, W., Eder-Hinterleitner, A., Seren, S. and Melichar, P., 2002, Georadar in the Roman civil town Carnuntum, Austria: An approach for archaeological interpretation of GPR data. *Archaeological Prospection* 9:135–156.

Vaughan, C.J., 1986, Ground-penetrating radar surveys used in archaeological investigations. *Geophysics* 51(3):595–604.

Chapter 14

Integrating Multiple Geophysical Datasets

Kenneth L. Kvamme

Abstract: In the past two decades improvements in geophysical instrumentation, survey techniques, and computer methods for handling spatial data have yielded significant advances in the management, portrayal, and interpretation of subsurface data. Geophysical investigations on archaeological sites have long utilized multiple survey methods. The use of difference methods allows responses to a variety of physical properties and the possibility of confirmatory, complementary, or entirely new information from each device. Such datasets have conventionally been examined side-by-side allowing informative comparisons. With GIS and other computer methods data may now be co-registered and more fully integrated in composite graphics of multidimensional content. Several approaches to "data fusion" are investigated including mathematical-statistical techniques, GIS, and advanced computer graphics. High-resolution, large-area datasets from the historic commercial center of Army City (A.D. 1917–1921), in central Kansas, illustrate benefits of these approaches.

1. INTRODUCTION

Two recent developments in the practice of archaeological geophysics are investigated and combined. The first is the integration or "fusion" of geophysical results from multiple instruments into composite datasets, under the premise that combined information from disparate sources leads to improved insights. The second is the importance of large-area survey coverage, necessary to facilitate recognition of broadly distributed cultural patterns in anomaly maps. Taken together these approaches offer great potential for

345

identifying and interpreting significant cultural features of the subsurface, and for understanding relationships between multiple sensor outputs.

1.1. Anomalies

Useful geophysical results are obtained when (1) archaeological features possess physical or chemical properties different from the surrounding matrix, and (2) instrumentation capable of measuring those properties is utilized with sufficient precision and sampling density to detect contrasts against the natural background. Such contrasts, when mapped, are referred to as "anomalies." They arise from anthropogenic causes, the targets of archaeo-geophysics, but geological, pedological, and biological phenomena also create them (e.g., paleo-channels, tree throws, animal dens). Sources of anomalies may be identified through excavation or deductive reasoning.

Anomalous measurements, by definition, are extreme in value and differ from "normal" data from undisturbed deposits. Those larger than typical background values are referred to as "positive," while unusually small measurements are termed "negative." These concepts may be understood in terms of a statistical distribution where positive and negative anomalies exist on the right and left tails, respectively, while common background values define the central tendency (Figure 1). Robust, well-indicated anomalies tend

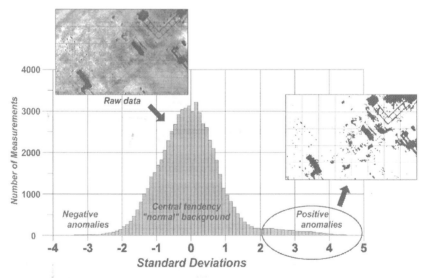

Figure 1. Two-dimensional mapping of an electrical resistivity dataset (top) and its one-dimensional histogram after a normalizing power transform and standardization of the data. Normal background measurements occupy the central tendency while values in the tails are anomalous, indicated by a mapping of positive anomalies beyond two standard deviations of the mean (right).

to be represented by measurements of large (absolute) magnitude, located in extreme portions of the distribution's tails relative to more subtly expressed anomalies lying closer to the center. In geophysics, unlike many applications, focus is therefore placed on one or both *tails* of data distributions, a fact that profoundly influences analysis and methods of integration.

1.2. Data Integration

Most instruments employed in geophysical surveys respond primarily to a single physical property—magnetometry to soil magnetism, electrical resistivity and electromagnetic (EM) induction to conductivity, and ground-penetrating radar (GPR) to dielectric contrasts. One may easily conclude that surveys of the same piece of ground utilizing multiple geophysical methods *must* offer improved insights because each yields information about a relatively independent aspect of the subsurface (Weymouth, 1986:371). Archaeo-geophysical prospection has therefore had a multidimensional focus since its inception (Clark, 2000:11–26), and multiple instrument surveys are commonly practiced (e.g., see Clay, 2001; Gaffney and Gater, 2003).

Multiple-method surveys offer a number of specific benefits. One is simple insurance. If buried archaeological features are not detected by one geophysical method, they may by another. A second benefit is complementary data. Anomaly distributions revealed by one technique can be very different from the output of another, potentially allowing a more holistic view of the subsurface. For instance, EM conductivity might reveal the foundations of an historical-period structure while a magnetic gradiometry survey might indicate its hearth, that the structure burned, and the loci of iron fittings. Multiple technique surveys therefore help us to sort out the puzzle of the subsurface by providing several lines of information (Clay, 2001). Van Leusen (2001:575) observes that when several archaeological datasets are integrated the "whole...is larger than its constituent parts." In addition, better geophysical detections reduce the amount of excavation needed to effectively evaluate a site. Field teams can be more informed about a site's content, precisely target features of interest for excavation, and realize significant cost savings (Hargrave, 2006). From a management perspective the increasingly refined application of geophysical and related methods ensures that a maximum of information about a site is gained non-destructively (David, 2001:525).

Data integration can mean many different things. The use of GIS technology—software appropriate for manipulating any information with a spatial component (Burrough and McDonnell, 1998)—is an essential step forward, but it should not be viewed as an end-point to integration. Even when several geophysical datasets from a site are incorporated within GIS, portrayals of results are typically represented side-by-side as distinct maps.

This arrangement was the common practice of display in the early liter-ature of archaeo-geophysics (Clark, 2000; Weymouth, 1986), and remains the standard form of presentation today (e.g., Gaffney and Gater, 2003). But is there another way?

A host of methods exists for creating composite imagery of high dimen-sionality and for generating mathematical and statistical combinations of data. Incipient expressions have already appeared in the archaeo-geophysical literature. Brizzolari et al. (1992:54) and Buteux et al., (2000:77) illustrate interpreted vectors representing cultural anomalies revealed by multiple geophysical surveys. Neubauer and Eder-Hinterleitner (1997) and Piro et al. (2000) advance a suite of simple mathematical operations that effectively combine geophysical data. A computer graphics solution is presented by Kvamme (2001) where a color composite of three geophysical datasets is offered. Johnson and Haley (2004) pioneer supervised classification methods to better define subtle anomalies expressed in multidimensional geophysical data. In each approach recognition of cultural patterns in anomaly distributions and relationships between detection methods is enhanced.

1.3. Large-area Datasets

Geophysical surveys of large areas increase the likelihood that culturally formed patterns in anomaly distributions may be recognized. A fundamental tenet in the field of aerial photo interpretation tells us, for example, that regular geometric shapes in the landscape—circles, ellipses, squares, rectangles, straight lines—are generally anthropogenic, arising much less frequently as products of nature (Avery and Berlin, 1992:52). Many cultural features exhibit repetitive organization, as illustrated by regular spacing between houses or graves (natural entities tend to be distributed much more irregularly), and relative size differences give important clues to the meaning of anomalies—public buildings tend to be much larger than dwellings, which in turn are larger than such outbuildings as privies or storage sheds. Moreover, associ-ations commonly exist between cultural features: houses combine to form villages, plazas and fortifications go with settlements, and trails or roads emanate from them. These characteristics all yield important clues that facil-itate the identification of cultural anomalies in remotely sensed imagery (Kvamme, 2007), and surveys of large areas are generally essential to make use of them (Buteux et al., 2000). When only a small area is surveyed, a segment of a linear anomaly could be interpreted as part of a room or house wall, the edge of a ditch, pavement, trail, or road, for example. Without imaging entire features and surrounding areas one cannot realize their shapes, relative sizes, and relationships, making it difficult to interpret their meaning or significance.

Many examples of wide-area surveys exist in archaeo-geophysics. Magnetic surveys at Wroxeter, England (Gaffney et al., 2000), and Kerkenes, Turkey

(Summers et al., 1996), are legendary for their sizes (nearly a square kilometer), but surveys of this type are often large owing to the speed of instrumentation (see Gaffney and Gater, 2003). Despite painfully slow data gathering, large electrical resistivity surveys using probe-contact instruments have also been reported (e.g., Dabas et al., 2000), but big EM surveys for the same phenomenon are relatively rare, even though field instrumentation allows rapid data capture (see Clay, 2001; Kvamme, 2003). GPR surveys of large areas are almost non-existent, largely because of the extraordinary volumes of data generated (upwards of 25,000 measurements/m^2) and the complexity of data processing. This situation is rapidly changing with improved software, however, and examples of large GPR investigations are beginning to creep into the literature (Neubauer et al., 2002; Leckebusch,2003).

The essential problem in pursuing integration of multi-sensor geophysical data over a large area lies in data acquisition—a region must be surveyed by several methods at compatible spatial resolutions, and such a procedure takes time and considerable effort. Spatial resolution determines the size of archaeological features that may be resolved, with a rule of thumb requiring measurement intervals to be no larger than *half* the size of the smallest feature of interest (so that multiple measurements may delineate it; Weymouth, 1986:347). Fortunately, a large study has recently been completed that was designed to integrate geophysical and other remote sensing data from several historical and prehistoric archaeological sites distributed across the United States (Kvamme et al., 2006). One of these sites is utilized here where six high-resolution geophysical datasets were acquired over a large area.

1.4. Goals

From the foregoing, it is apparent that archaeo-geophysical data integration is a subject worthy of study. Yet, because of limited prior work (cited earlier), there has been little effort to examine systematically and organize its legitimate domains and lines of inquiry. One goal seems to lie in simply developing superior ways to visualize the totality of subsurface evidence by portraying multiple dimensions simultaneously. Another concentrates on improved theoretical understandings of relationships between various types of surveys. Still another pursues more accurate means for anomaly identification and classification. The following pages examine these domains with a focus that is necessarily methodological. In each, a variety of integrating approaches is examined utilizing computer graphic, GIS, mathematical, and statistical solutions. Use of the same data throughout provides a control, facilitating comparisons between results. Detailed interpretations, evaluations of raw data contributions, theoretical digressions, and ground validation studies through excavation are beyond the scope of this effort, but are reported elsewhere (Kvamme et al., 2006).

2. CASE STUDY: ARMY CITY, KANSAS

Army City was a privately owned commercial complex established by local
entrepreneurs to provide entertainment and other services to troops at Camp
Funston (now part of Fort Riley) in the era of World War I (Figure 2). The
town, established in 1917, suffered from economic decline with the close of the
war in 1918, a major flood in 1919, and a fire that devastated its commercial
core in 1920. It was abandoned soon afterward when its remaining structures were dismantled, sold as scrap, or moved to a nearby town (Hargrave
et al., 2002). It rests at present under a hay field with few surface indications
of its presence, although architectural and other remains lie buried as little as
30 cm below the surface.

Six geophysical methods, five ground-based and one aerial, were investigated in a study block measuring 100 m × 160 m (1.6 ha) centered over the
town's former commercial core (Figure 2). Prior work (Hargrave et al., 2002)
showed good geophysical contrasts in electrical resistivity and magnetic
gradiometry surveys, with the former responding to highly resistant concrete
walls and floors and the latter to the many ferrous metal artifacts and areas
of burning. Test excavations and the period of occupation suggested the
likelihood of a network of underground pipes for water or sewage (the
earlier magnetic survey, which would have confirmed this, did not cover
this area). Although EM quadrature phase data are theoretically the inverse
of soil resistivity—measuring *electrical conductivity*—it was thought that EM
methods could potentially reveal new information because metal artifacts yield
pronounced anomalies that do not appear in probe-contact resistivity surveys
(Bevan, 1998:36–39). Moreover, the great fire of 1920 quite likely raised
magnetic susceptibility in the near surface, so the in-phase component of an
EM survey, limited to very shallow depths, was likely to prove insightful.
Despite the assertion by Hargrave et al. (2002:94) that "the high clay content
of the soil at Army City eliminated GPR as an effective technique" (clay is
generally conductive and disperses radar energy), prior experience, a long

Figure 2. Undated view of Army City, looking southeast down Washington Avenue.
The Hippodrome is the largest building in the center; Camp Funston lies to the right
(cropped from original panoramic view; photo credit: Kansas State Historical Society).

period of drought (dry soils lower conductivity), new software for signal processing, and other studies suggested the possibility of obtaining positive results. A full GPR survey of the study area was therefore pursued because this technique can measure yet a different physical property of the sub-surface (dielectric contrasts) and it offered the possibility of three-dimensional imaging at a variety of depths. Finally, concrete foundations immediately beneath the surface, filled cellars, pipe trenches, and street gutters recommended aerial thermography because significant thermal variations were likely to be generated by these types of features (see Scollar et al., 1990:601–605).

The five ground-based surveys were undertaken in a two-week period during the summer of 2002. All except GPR were conducted within survey blocks of 20 m × 20 m that served as control units allowing piecemeal coverage of the region. The GPR survey was conducted in larger blocks of 50 m × 50 m or 30 m × 50 m to avoid numerous field set-ups and data processing difficulties associated with concatenation of many small units. In all cases, instrument placement and spatial control were facilitated by meter marks on tapes or ropes placed within the survey units. The aerial survey, undertaken in an evening flight in the spring of 2004, utilized a slow, low-flying powered parachute as a platform at an altitude of 250–400 m above the surface. Thermal infrared imagery in the 10-micron band was acquired on digital video in an 8-bit format using a hand-held camera. Instrumentation, original spatial resolutions, and approximate prospecting depths for each survey are summarized in Table 1.

3. DATA PRE-PROCESSING

The goal of pre-processing geophysical data in most contexts is to remove survey defects (e.g., instrument drift, imbalances resulting from independent block surveys), reduce noise, and enhance the visibility of cultural anomalies through appropriate transformations (e.g., low- and high-pass filters). Since all of the ground-based data were collected within the same survey units they were naturally co-registered within a locally established coordinate system. Data from each of these units, however, had to be concatenated and placed in their correct spatial positions. Edge-matching, to adjust data means and variances between individual survey units, had to be performed on the EM and electrical resistivity data because of instrument drift in the former or variations in remote-probe placements in the latter. Extensive de-sloping was also applied to the EM and magnetic gradiometry data, which exhibited considerable drift owing to high temperatures (40° C) during the surveys or use of fluxgate technology, respectively. Pre-processing of the GPR data was much more laborious. Because of zigzag survey methods, even-numbered profiles had to be reversed and the ground-surface reflection identified in each. This procedure was followed by background removal of horizontal banding

Table 1. Geophysical methods, instruments, sampling densities, and principal anomaly types at Army City.

Geophysical Method	Instrument	Original resolution (m)	Prospection depth (m)	Unit of measurement	Principal indicated anomalies
Conductivity	Geonics EM-38	.5 × 1	1.5	mS/m	Metal pipes
Ground-penetrating radar	GSSI SIR 2000 & 400 MHz antenna	.025 × .5	2.3 (60 ns time window)	amplitude (16-bit scale)	Foundations, street gutters
Magnetic gradiometry	Geoscan Research FM-36	.25 × 1	1.5	nT	Iron pipes, large iron artifacts
Magnetic suscepti-bility	Geonics EM-38	.5 × 1	<.5	ppt received to transmitted signal)	Burned areas, foundations, street gutters, iron pipes
Electrical resistivity	Geoscan Research RM-15 twin-probe array	.5 × 1	.5	ohm/m	Concrete floors, foundations, building footers, gutters
Thermal infrared	Raytheon Palm IR-250 (10 micron band)	.1 × .1 (approx)	<1	brightness (8-bit scale)	Cellars, floors, foundations, streets, pipe trenches

common to GPR, and post-gaining to amplify time windows of anomaly prevalence. Time-slices (Conyers, 2004), consisting of maximum *positive* amplitudes within a larger voxel (the three-dimensional equivalent of a pixel, typically 0.5 m × 0.125 m × 15 nanoseconds), were defined that best characterized subsurface features of interest in the horizontal plane. These time-slices were then de-striped (striping resulted from slight amplitude changes caused by zigzag survey methods, coupling, and timing variations) and, along with the other data, were subjected to mild de-spiking of isolated extreme measurements (caused by small metal artifacts, rodent holes, and the like) followed by low-pass filtering to reduce noise. Finally, the variable spatial resolutions (Table 1) necessitated that each dataset be re-sampled to a uniform 0.5 m × 0.5 m to meet requirements of subsequent pixel-based data integrations.

Two of the geophysical datasets required additional processing for meaningful data integrations. The magnetic gradiometry and susceptibility surveys exhibited large dipolar values (dipolar anomalies that also occurred in the EM conductivity data were less extreme and therefore ignored). Each joint of the many buried pipes revealed by magnetic gradiometry exhibited adjacent positive and negative poles of extreme value that represented the same feature, which would negatively influence statistical or visual relationships in data integrations. Although an advanced technique known as "reduction to the pole" may theoretically be employed for reducing dipolar anomalies to source points (Telford et al., 1990:109), software reconstructions of linear features like pipes are difficult to achieve, so absolute values of the data were taken instead.

For the aerial thermal data, still frames were extracted from the digital video that showed ground control markers placed over the corners of the 20-m survey blocks of our ground-based data-collection units. These frames were subjected to standard affine transformations (Burrough and McDonnell, 1998:128) to rectify and register them to the coordinate base of the 1.6 ha study rectangle. In so doing the original spatial resolution of about 10 cm (Table 1) was re-sampled to a coarser 50 cm for compatibility with the other data and analysis needs. Considerable image noise from variations in vegetation cover was further reduced by low-pass filtering.

The six pre-processed datasets are illustrated in Figure 3.[1] They collectively indicate various aspects of Army City's commercial core that was centered on Washington Avenue (the large southeast-northwest trending "street" revealed by geophysics, particularly in Figure 3d, and photographed in the center of Figure 2). Building foundations, floors, footings, water pipes, sewage lines, and streets, are indicated to varying degrees. A collection of shops known as the Hippodrome (Figure 2, center; robustly revealed by the resistivity survey, Figure 3e) lies in the northwest corner of the study area, with the Orpheum Theater directly across the street to the east (Hargrave et al., 2002:100). Broad zones of high magnetic value may point to areas of intense burning during the 1920 fire (Figure 3c, d). Principal cultural anomalies revealed by each dataset are summarized in Table 1. The geophysical data of Figure 3 represent the start-point to data integrations that proceed from the simple to the complex, beginning with computer graphic techniques and moving through mathematical transformations and intricate statistical manipulations.

4. INTERPRETIVE APPROACHES TO DATA INTEGRATION

Geophysical findings are traditionally obtained through subjective interpretations of the data combined with deductive reasoning, whether the data are unidimensional or multidimensional. Successful interpretations of likely cultural features rely on expertise in the local archaeology and knowledge

---

[1] In all Army City imagery the grid represents 20 m squares, black is high in gray scale depictions (a geophysical convention because most significant anomalies are positive), and north is to the right. The GPR data represent a time-slice equivalent to approximately 20–40 cm below the surface.

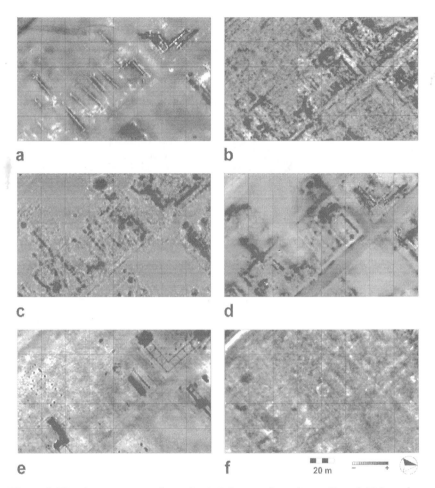

Figure 3. The six pre-processed geophysical datasets from Army City: a) EM conductivity, b) GPR (time-slice of maximum positive amplitudes from 20–40 cm below the surface), c) absolute magnetic gradiometry, d) absolute magnetic susceptibility, e) electrical resistivity, f) thermal infrared.

of corresponding archaeological signatures in geophysical data. The methodology rests largely on visual interpretation of geophysical maps followed by the laborious process of *vectorization*—the manual tracing of likely cultural anomalies on paper or a computer screen using GIS or equivalent digitizing software (Burrough and McDonnell, 1998:88). This subjective procedure is extremely time-consuming, and can be viewed as the conversion of continuous geophysical raster data to discrete point, line, or area vectors (in GIS parlance) that represent significant cultural features. In other words, the end-product is a series of interpreted maps depicting the locations of likely cultural features. If multiple geophysical surveys have been conducted then a separate map is produced showing interpreted cultural anomalies in each one. If combined and overlaid, a form of data integration is represented. Many examples exist in the literature (Buteux et al.,, 2000:77; Gaffney et al., 2000; Neubauer 2004:163) and the approach has justifiably been presented as a goal of archaeo-geophysical surveys (e.g., see Schmidt, 2001:23).

At Army City, each geophysical dataset (Figure 3) was carefully examined to define potential cultural anomalies based on their form, distribution, limited historical maps, photos (Figure 2), and excavation results. These interpreted anomalies were then vectorized and overlaid to indicate all anomalies thought or known to be culturally significant (Figure 4a). This integration represents a point of departure from methods that follow. In contrast to this manually intensive and subjectively driven technique, subsequent approaches present, for the most part, *algorithmic* and more automated means for data integration.

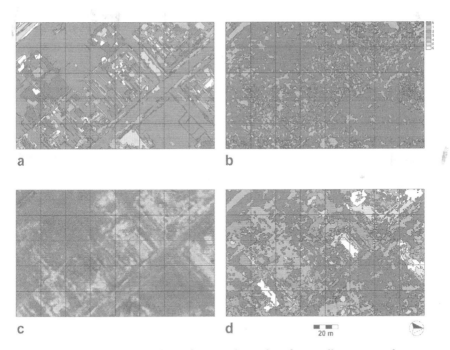

Figure 4. Simple and advanced data fusions. a) Overlay of manually interpreted vectors (gray=anomaly absence; bright blue=subtle anomalies; other colors correspond with robust anomalies indicated by specific geophysical methods). b) Sum of binary representations of robust anomalies in the six geophysical datasets (Figure 6a–f). c) Blue-green color composite of the two varimax rotated factors (blue=Factor 1; green=Factor 2). d) Results of the K-means cluster analysis with $k=6$.

5. COMPUTER GRAPHIC APPROACHES TO DATA INTEGRATION

Several computer methods of graphic portrayal have been applied for decades that yield effective data fusions. In fact, some are so common they may not be understood as fusions. While many display variations exist, only four principal ones are examined here.

5.1. Two-dimensional Overlays

Four methods are most commonly employed for displaying continuous map data: isoline contours representing measurements of equal value, gray- or color-scale variations signifying measurement magnitudes, pseudo-three-dimensional views where the vertical axis represents measurement size, and shadowing to indicate the steepness of gradients in three-dimensional surfaces. Each of these methods may be used to represent results of a single geophysical survey, but several datasets may be simultaneously portrayed through overlaying. An effective combination utilizes one data source as a color- or gray-scaled background image, with a second one overlaid with isoline contours. For instance, in Figure 5a the background electrical resistivity data tend to highlight Army City's buried concrete foundations, building footings, and other resistant features, while the overlaid EM conductivity data locates water or sewage pipelines associated with these structures (compare Figures 3a,e). A second useful display is illustrated in Figure 5b where magnetic susceptibility is portrayed as a three-dimensional surface that mostly reveals streets and some structural evidence (with variations emphasized through shading in gray), and GPR is overlaid in color indicating robust walls and floors (compare Figures 3b,d). These data fusions instantly convey profound similarities and differences in anomaly distributions and emphasize that various methods measure different properties of the ground.

Graphic overlays can obviously be carried to portrayals of more extreme dimensionality. For example, a surface might represent one dimension, shadowing a second, a color overlay a third, and contouring a fourth. Such maps rapidly become too complex for ready understanding, however, with two dimensions probably offering a practical limit to what the human brain can effectively process.

5.2. RGB Color Composites

Red-green-blue (RGB) color composites have been a standard means for displaying satellite imagery for decades (e.g., see Schowengerdt, 1997). In this visualization method, each of three datasets is assigned a primary color and the images are then combined to create color mixes that span the

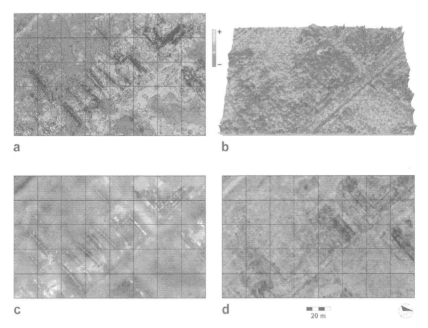

Figure 5. Computer graphic overlays and composites. a) EM conductivity contours overlaid on electrical resistivity data. b) GPR data (in color) overlaid on magnetic susceptibility surface with variations emphasized by shadowing. c) RGB color composite of electrical resistivity (red), EM conductivity (green), and magnetic susceptibility (blue). d) Overlay of six geophysical datasets, each represented by a different translucent color (conductivity=yellow, GPR=green, magnetic gradiometry=blue, magnetic susceptibility=cyan, resistivity=red, thermal infrared=purple).

full visual spectrum. When more than three datasets are available many primary color assignments become possible, and each can yield different insights. For instance, with six Army City surveys (Figure 3) there are $C_{6,3} = 20$ possible groups of three, with $3! = 6$ permutations of red, green, and blue for each, yielding 120 possible displays! One such RGB composite is illustrated in Figure 5c that well shows the site's commercial core and relationships between electrical resistivity, conductivity, and magnetic susceptibility.[2]

Knowledge of color theory and the RGB color model enhances interpretation. Bright red, green, or blue in a color composite points to high values of the variable assigned to the respective color, but low values of the other variables. White indicates large data values in all three datasets, while black

---

[2] Because EM conductivity is the theoretical inverse of resistivity its scale is *inverted* in order that parallel components of the data may complement, rather than cancel, each other when combined.

points to the reverse. Yellow means high values in the variables associated with green and red, magenta to elevated measurements in red and blue, and cyan to large values in green and blue. Consequently, at least eight simple interpretations are possible, plus numerous shades of gray and a myriad of colors between. In Figure 5c the Hippodrome, other foundations, and building footings are exclusively revealed by electrical resistivity (red), streets are indicated primarily by magnetic susceptibility (blue), and pipelines by EM conductivity (green); at the same time, certain walls and floors are robustly positive in the three primaries (making white) while a recent drainage ditch in the southwest corner shows a uniform negative result (creating black).[2]

5.3. Translucent Overlays

Recent advances in computer graphics allow color overlays of many more dimensions than the three of the RGB model through use of translucencies, but the approach lacks similar theoretical grounding. A very different color is assigned to each of k datasets; $k-1$ are then made translucent and overlaid on a kth opaque image beneath. Using this methodology, all six datasets of Figure 3 are simultaneously overlaid in Figure 5d, resulting in a more comprehensive depiction of anomalies than the three-dimensional RGB composite (Figure 5c). A general limitation is that high k tends to produce a "muddy" looking result, and layers lying beneath several others can be almost completely obscured.

5.4. Summary

Computer graphic solutions for data integration are easy to implement and effectively combine information from several disparate sources in easy-to-interpret displays. They offer a ready means to generate complex visualizations of the subsurface. Their weakness lies in their relatively low dimensionality— typically only 2–3 dimensions may be represented (or perhaps somewhat more using translucent overlays). Moreover, these methods are purely descriptive, capable of combining only what is mappable in contributing datasets.

6. MATHEMATICAL TRANSFORMATIONS FOR DATA INTEGRATION

Several straightforward operations may be employed to mathematically integrate multiple geophysical datasets. They may be divided into two classes: those that utilize binary data (information coded as zeroes or ones) and those that employ continuous measurements (the raw data of geophysics, generally represented by positive or negative real numbers with decimal points). Both require further pre-processing of the data in order for the methods to perform optimally.

6.1. Operations on Binary Data

Binary data lie behind a principal *modus operandi* in the realm of GIS modeling, forming the inputs to what is known as *Boolean* operations (Burrough and McDonnell, 1998:164–166). These approaches require the raw continuous geophysical data output by field instrumentation to be reclassified to a binary state. In the present context, a "1" may indicate the presence and "0" the absence of an anomaly. This binary representation is achieved simply by selecting appropriate thresholds near the tails of a statistical distribution and mapping only those locations with more extreme values as unity, zero otherwise (as in Figure 1).

Thresholding of the raw data might appear to be a straightforward operation, but problems arise in application. Broad regional trends arising from soil changes or underlying geology frequently can mask cultural anomalies. Removing these trends prior to binary classification is therefore warranted, and may be achieved through application of high-pass filters or subtraction of low-order polynomial trend-surfaces. A second problem occurs in relatively "quiet" datasets where the signal-to-noise ratio is low. In these contexts low-pass filters can help to consolidate and strengthen weak anomalies, making them more cohesive and apparent. These operations were applied to the data in Figure 3, appropriate thresholds were selected that visually appeared to best define cultural features in each, and binary datasets were generated (Figure 6a–f; a binary representation of the raw electrical resistivity data in Figure 1 may be compared against a binary map after data de-trending and low-pass filtering in Figure 6e). These results were then subjected to the following operations.

6.1.1. Boolean Union

A Boolean union occurs when *at least* one of the inputs is coded as "1." Its mapping shows all places where anomalies are indicated by one or more methods. In other words, the union simultaneously shows the loci of all defined anomalies in all datasets (Figure 6a–f). As such, the outcome of this data integration produces a primary map of interest (Figure 6g). As a combination of all anomalies from all geophysical sources, the union can map a relatively large area: 39.7 percent of the region (0.63 of 1.6 ha) in Figure 6g is anomalous.

6.1.2. Boolean Intersection

The Boolean intersection is also of interest because it reveals places where defined anomalies *simultaneously* occur in all geophysical datasets. In other words, for a result to be "true" a "1" must occur in *every* input data

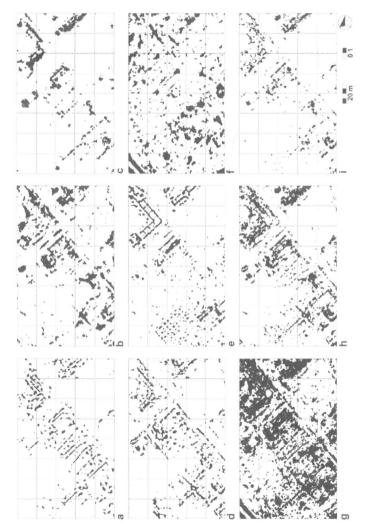

Figure 6. Binary representations of principal geophysical anomalies and Boolean results: a) EM conductivity, b) GPR, c) magnetic gradiometry, d) magnetic susceptibility, e) electrical resistivity, f) thermal infrared, g) Boolean union of a-f, h) anomalies indicated by at least 2 methods in a-f, i) anomalies indicated by at least 3 methods in a-f.

source. At Army City only 11.5 m² (.07 *percent* of the region's 16,000 m²) meets this condition, representing targets with very different properties in all dimensions (not illustrated). This restricted result is not surprising because subsurface entities that cause anomalies to be recorded will rarely yield strong contrasts concurrently in so many geophysical dimensions (magnetic, electrical, thermal). Studies using lower dimensionality (e.g., 2–3 methods) will typically yield less limited results.

6.1.3. Binary Sum

A *sum* of binary maps offers several distinct advantages over Boolean operations. With k inputs, the result can theoretically range between $0 - k$, yielding $k + 1$ classes. Zero occurs at locations where all binary geophysical inputs are coded zero while k (the maximum) will result where all inputs are coded as unity. Consequently, imbedded within this method are the Boolean intersection and union because locations achieving k represent the former and all non-zero loci the latter. The sum also represents a *ranking* of anomalies revealed by zero of k methods, by one of k methods, and so on up to k of k methods. As such, it can be interpreted as an anomaly "confidence" map, where larger values signify that more geophysical methods reveal the same anomaly (Figure 4b). This map is therefore much richer in content and detail than the Boolean union (Figure 6g) or intersection. It is noteworthy for its simplicity, and well indicates the layout of Army City's principal cultural features. It shows that many anomalies, particularly certain foundations, floors, pipelines, and gutters, are revealed by several geophysical methods, while other anomalies—representing floors, street gutters, walls—are indicated by only a single method.

6.1.4. Thresholded Binary Sum

The binary summation (Figure 4b) can itself be reclassified to indicate, for example, all places where two, three, or more (to $k - 1$) sensors reveal a major anomaly. This approach also yields binary outcomes, but ones that lie between the extremes of the Boolean union (one or more indications of anomalies) and intersection (k indications of anomalies). In other words, a threshold is selected in the binary summation map between 1 and k that may yield a more informative result than either Boolean outcome. To illustrate, mapping anomalies indicated by *two* or more geophysical methods (Figure 6h) well defines Army City's major features (foundations, floors, pipelines, street outlines) by eliminating less secure anomalies revealed by only a single method. The mapping of anomalies indicated by three or more methods shows particularly robust elements of the site (Figure 6i).

6.2. Operations on Continuous Data

Continuous data are naturally richer than categorized information, potentially enabling fusions with superior content. Neubauer and Eder-Hinterleitner (1997) examined sums, products, ratios, and differences between normalized resistivity and magnetometry results. Piro et al. (2000) investigated normalized sums and products of three geophysical datasets. In undertaking these operations one must first consider that various measurement scales, widely different data ranges, and even distributional forms exist under each geophysical method (datasets such as GPR generally illustrate extreme skewness, for example). In order to circumvent difficulties that such a mixed bag might introduce to mathematical manipulations (e.g., one variable might dominate because of larger measurements), a series of power or logarithmic functions was applied to approximately normalize each distribution (e.g., see Figure 1). The data were then standardized with: $z = (x_i - \mu)/\sigma$ (where μ is the mean and σ the standard deviation), yielding $\mu = 0$, $\sigma = 1$, and about 99 percent of the data vary between -3 and $+3$ standard deviations.

6.2.1. Data Sum

The sum of the standardized data illustrates anomalies from all sources simultaneously, including those of large and small magnitude (Figure 7a).[2] Much as in the previous binary results, strong indications of foundations, floors, cellars, pipelines, and street gutters are seen throughout the town's core, but a plethora of subtle anomalies may also be discerned that help to "fill out" entire buildings, add partitions to structures, and complete the outlines of streets, producing a richer outcome. Given the simplicity of this operation the result is particularly pleasing because it compares closely with outcomes derived by more complex means below. Variations of this method are possible by assigning more or less weight to individual sensors.

6.2.2. Data Product

The data product was also investigated under the theory that cross-multiplication should make anomalies of extreme value more pronounced. A constant of 10 was first added to each standardized dataset (resulting in $\mu = 10$ and $\sigma = 1$) to insure positive distributions and avoid problems caused by multiplying negative and positive values. The product of the six datasets is illustrated in Figure 7b.[2] Although the result appears broadly similar to the data sum, close inspection reveals that major anomalies tend to be better contrasted and slightly exaggerated while subtler anomalies are more subdued or absent.

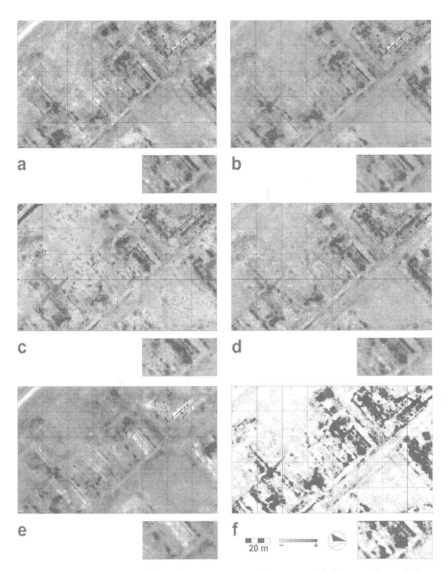

Figure 7. Continuous geophysical integrations: a) data sum, b) data product, c) data maximum, d) Factor 1, e) Factor 2, f) logistic regression probability surface for anomaly presence. Inset shows close-up detail for comparison.

6.2.3. Data Maximum

A maximum of data inputs should theoretically map the most important *positive* anomalies in a multidimensional collection of data. At any locus, assuming commensurate measurement scales, the largest value in all inputs is taken. In other words, $x_{out} = MAX(x_1, x_2, \ldots, x_k)$, where x_{out} is the transformed result, $x_1 - x_k$ are the $k$input datasets , and "MAX" is the maximum function. A similar operation can be taken for negative anomalies (using a "minimum" function) or the maximum of absolute values might be considered. At Army City negative anomalies were eliminated by pre-processing of both magnetic datasets and GPR (see above), strong negative anomalies are nearly absent in the resistivity data (most likely because of drought conditions at the time of acquisition; Figure 3e), and the same may be said of positive anomalies in EM conductivity owing to their theoretical relationship with resistivity (Figure 3a). Only the thermal infrared data expressed significant positive and negative anomalies (Figure 3f), so absolute values of this dataset were computed and the conductivity data were again inverted.[2] The *maximum* function was then applied (Figure 7c) enabling definition of the most robust anomalies—principal foundations, pipes, and street boundaries. A number of unanticipated artifacts are introduced by this procedure, however, that occur in relatively "quiet" open areas with few prominent anomalies. In these regions random noise (probably stemming from minor natural soil variations) is apparently maximized yielding a poor effect.

6.3. Summary

Relatively simple mathematical transformations offer a rich suite of methods for integrating any number of geophysical datasets. Discrete, binary methods offer the advantage of clear-cut maps of anomaly presence or absence and allow use of readily available and understandable Boolean operations. Continuous data integrations, on the other hand, allow robust and subtle anomalies to be simultaneously expressed, producing composite imagery with high information content. Although they allow complex visualizations of the multidimensional character of the subsurface, these approaches are descriptive in nature and ultimately represent only a combination of their individual parts.

7. STATISTICAL METHODS FOR DATA INTEGRATION

Multivariate statistics offer several mechanisms for data integration. They may broadly be divided into two classes: those that reduce dimensionality and classification methods. The former includes principal components and factor analyses that produce linear combinations of the original data. The significant components that result, fewer in number than the original variables, represent data fusions based on statistical correlations. Classification methods, on the other hand, attempt to define "groups" or "classes" in bodies of data through two different methods. One employs a form of *cluster analysis* where observations are grouped into "natural" classes based on their proximities in a measurement space. The other includes statistically optimized *discriminant functions* derived from data patterns in samples from known classes. These functions are applied to data of unknown membership to assign them to likeliest classes. In satellite remote sensing the former is referred to as "unsupervised classification," because the algorithm is data driven to find naturally occurring classes. The latter is known as "supervised classification," because functions are developed from statistical characteristics of samples provided by a supervisor (Schowengerdt, 1997). It is emphasized that classification models of any kind may be regarded as having a predictive capacity when mapped classes indicate not only robust and well-recognized anomalies, but vaguer or nearly invisible ones as members of the same class. In these cases, such methods can indeed predict from the known (anomalies of certain or nearly certain identity) to the unknown (anomalies of uncertain membership), thereby augmenting prospecting possibilities.

It is prudent to examine first a fundamental characteristic of the Army City geophysical data. It has been indicated that many anomalies do not recur in different datasets, thereby suggesting independence or a lack of correlation (compare elements of Figure 3). Yet, it is the assumption of correlation that lies at the very heart of such methods as principal components analysis. The Pearsonian correlation matrix (Table 2) is therefore both insightful and alarming. Noteworthy are the generally low levels of correlation, a circumstance that is surprising because, theoretically, one might suspect resistivity and conductivity to be highly correlated, as well as (absolute) magnetic gradiometry and susceptibility. Yet, the highest absolute correlations in the entire dataset only approach $|r|=.3$ (about $100r^2=9$ percent of variance in common), with the thermal data being almost completely independent.

7.1. Principal Components and Factor Analysis

Principal components analysis (PCA) linearly combines multivariate data based on the correlation structure between k variables (Table 2). The resulting k components are uncorrelated and ordered in such a way that the first

Table 2. Pearsonian correlation matrix between the six normalized geophysical variables.

	Cond	GPR	Mag	MS	Res	Therm
Cond	1.000	−.160	.073	−.286	−.137	−.081
GPR	−.160	1.000	.231	.304	.218	−.053
Mag	.073	.231	1.000	.299	.277	−.019
MS	−.286	.304	.299	1.000	.073	.022
Res	−.137	.218	.277	.073	1.000	−.035
Therm	−.081	−.053	−.019	.022	−.035	1.000

KEY:**Cond**=EM *conductivity;* **GPR**=*ground-penetrating radar (maximum positive amplitude);* **Mag**=*absolute magnetic gradiometry;* **MS**=*absolute magnetic susceptibility;* **Res**=*electrical resistivity;* **Therm**=*thermal infrared.*

represents more of the total variance in the data than the second, which in turn carries more than the third, and so on, with the first few components typically characterizing the bulk of the information content in most datasets (Davis, 2002:509–525). When standardized, each variable contributes a variance of unity. A PCA of the Army City data reveals that only the first two components account for more variance than any single variable alone (eigenvalues larger than unity), with the first accounting for nearly 30 percent of the total variance, and the second about 19 percent (Table 3). These low values result from the lack of correlation between the variables (Table 2). Yet, that the first two components account for nearly half the total variance offers encouragement that useful fusions of data commonalities are achieved. The *loadings* (correlations of variables with a component) on the first component indicate a linear combination of all variables except *thermal*, with moderate contributions from each. The second component primarily reflects *conductivity*, with minor contribution from the thermal data. The remaining components more or less represent individual variables except the last, which contains remaining noise (Table 3).

While the foregoing components are informative, the variance they represent can be better distributed through a *varimax* rotation, sometimes referred to as factor analysis (Davis, 2002:533). This method rotates the two component axes of interest within the six-dimensional measurement space to

Table 3. Results of principal component and factor analyses of the geophysical data. Principal loadings in columns are indicated in bold typeface.

Component	PCA Eigenvalues			Factors after Rotation		
	Total	% Variance	Cumulaive %	Total	% Variance	Cumulative %
1	1.791	29.86	29.86	1.712	28.54	28.54
2	1.153	19.23	49.07	1.232	20.53	49.07
3	.964	16.06	65.13			
4	.899	14.99	80.12			
5	.714	11.91	92.02			
6	.479	7.98	100.00			

	PCA Loadings (correlations with components 1–6)						Factor Loadings after Rotation		Factor Score Coefficients	
	1	2	3	4	5	6	1	2	1	2
Cond	−.423	.701	.313	.284	.219	.321	−.150	−.805	−.007	−.652
GPR	.681	.042	−.169	.073	.704	−.072	.652	.210	.369	.100
Mag	.611	.467	.388	.217	−.259	−.379	.737	−.222	.462	−.259
MS	.694	−.258	−.060	.477	−.249	.397	.559	.486	.284	.346
Res	.540	.280	.130	−.732	−.095	.262	.604	−.072	.367	−.121
Therm	−.021	−.545	.816	−.045	.182	.030	−.212	.503	−.177	.439

KEY: **Cond**=*EM conductivity*; **GPR**=*ground-penetrating radar (maximum positive amplitude)*; **Mag**=*absolute magnetic gradiometry*; **MS**=*absolute magnetic susceptibility*; **Res**=*electrical resistivity*; **Therm**=*thermal infrared*.

more equally distribute the variance they represent; in so doing, the loadings
with the original variables are altered, a circumstance that can improve interpretability. This was undertaken and, while the variance accounted for by the
second component (or factor) after rotation is only slightly increased to 20.5
percent, the loading structure is significantly altered yielding clearer results.
The contribution of *conductivity* is nearly removed in the first rotated factor,
while *magnetic gradiometry* is greatly reduced in the second, although the
contribution from *magnetic susceptibility* increases. The first factor represents
a linear composite of *GPR*, *magnetic gradiometry*, a portion of *magnetic susceptibility*, and *electrical resistivity*; the second combines *conductivity*, an equal
portion of *magnetic susceptibility*, and *thermal* (Table 3). These uncorrelated
factors are illustrated in Figure 7d, e and represent independent underlying
dimensions of the data, with the first pointing to foundations, floors, and
gutters, and the second emphasizing such negative anomalies as pipelines,
cellars, and a modern drainage ditch (near the southwest corner of the study
area). The dual results, each representing a fusion of only some of the data,
present something of a quandary because only some of the anomalies are
indicated in each. A variety of tactics, presented earlier, might be employed
in conjunction with this method to resolve this issue; a color composite is
offered here (Figure 4c).

7.2. K-means Cluster Analysis (Unsupervised Classification)

Cluster analysis includes a series of algorithms designed to define natural
groupings in a body of multivariate data such that each one is more or less
homogeneous and distinct from others (Davis, 2002:487). With a goal of
defining natural classes the result is discrete. K-means cluster analysis was
selected because it operates rapidly on large datasets (here, $n = 64,000$ in
each of six dimensions). Beginning with a user-specified number of clusters,
k, the algorithm places k arbitrarily located means in the six-dimensional
measurement space, computes the Euclidean distance between each point
and the nearest mean (hence the importance of commensurate measurement
scales), and computes a sum of squared error (SSE) statistic, the sum of
squared distances from the respective means. At each step the algorithm
iteratively moves the k-means about the measurement space until the SSE
is minimized, which indicates that an optimal partitioning into k classes is
achieved with each case assigned to the closest cluster mean. Of course,
one often has no idea about how many classes might truly exist, a circumstance that poses a real problem in the present context. One might make an
argument for only two classes ("anomaly" versus "background"), three classes
("positive anomaly," "negative anomaly," "background"), or many classes,
perhaps hoping the algorithm can differentiate among such different anomaly
types as "wall," "floor," "building footing," "pipeline," "gutter," "street," and
"background," for example.

The standardized data were subjected to k-means cluster analyses, with $k = 2 - 10$. It was found that with $k \geq 7$ small and insignificant clusters containing few pixels were defined (small micro-groups). The $k = 6$ solution might be most defining of a myriad of anomaly types (Figure 4d), but it is difficult to assign specific interpretations to many of the classes (space does not permit detailed comparisons against the primary data of Figure 3). Several other solutions with smaller k, however, are informative and easier to interpret. The $k = 2$ solution tends to show anomalies versus the natural background, as predicted (Figure 8a), while $k = 3$ splits the anomaly class of the $k = 2$ result largely along the lines seen in the two factor analysis dimensions (Figure 8b). The four-cluster solution is also insightful (Figure 8c), because at this level of partitioning the "background" of the previous iterations is divided according to thermally indicated anomalies (see Figure 3f), shown previously to represent an independent dimension (Table 2).

7.3. Binary Logistic Regression (Supervised Classification)

A large number of algorithms exist in satellite remote sensing for developing supervised classifications (Schowengerdt, 1997). Many are based on a multivariate normal model (e.g., maximum likelihood, Fisher's discriminant), while others apply simple geometric operations to the multidimensional measurement space (e.g., minimum distance and parallelepiped classifiers). A particularly robust discriminant funtion is logistic regression, a nonparametric classifier (Hosmer and Lemeshow, 2000). Although a multi-class solution defining several *types* of anomalies might ultimately be attempted using multinomial logistic regression, a simple binary model is pursued to produce a continuous probability surface for anomalies as a single class (where low probabilities suggest anomaly absence).

The selection of "training sites"—locations of known class membership provided to the algorithm—is vital to a good result because the classification function that results is optimized to patterns in these samples. Training-site selection was investigated using a number of approaches. One that works well for modeling robust indications of anomalies, and is related to other fusions investigated here, was a five percent random sample taken from the binary result showing anomalies indicated by at least two techniques (Figure 6h). Approximately 3,200 sample points were drawn of which 16 percent fell in the robust anomaly class. No interaction terms were selected and all variables were forced into the model, resulting in *pseudo-R^2*=.67. The following function was derived that maximizes differences between the classes: $L=-2.984-.224*Cond+1.184*GPR+1.194*Mag+.778*MS+.667*Res-.547*Therm$ ("*" signifies multiplication; see Table 2 for explanation of abbreviations). All parameters in this model are significant contributors (at $\alpha = .01$) and they offer interpretive potential. Since the data are standardized, the absolute sizes of the coefficients indicate that *GPR*

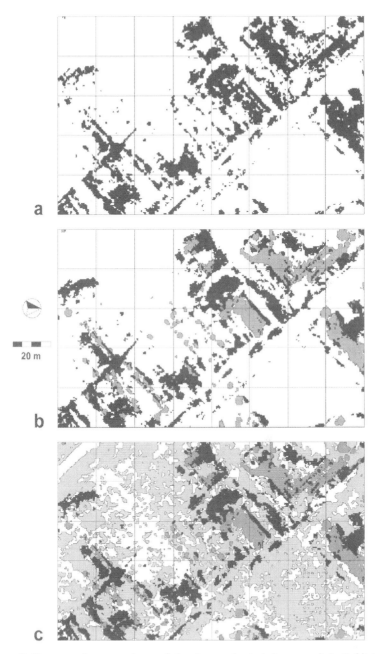

Figure 8. K-means cluster analyses of the six geophysical datasets: a) $k=2$, b) $k=3$, c) $k=4$.

and *magnetic gradiometry* contribute most to discrimination, *conductivity* and *thermal* least, and the signs inform us that negative values of *conductivity* and *thermal* and positive values of the other variables are related to anomaly presence. The logistic transformation, $p=(1+exp(-L))^{-1}$, conveniently rescales this axis to a 0-1 range, which may be interpreted as a probability surface for robust anomaly presence (Figure 7f). This map well represents not only the training sites (Figure 6h), but many other anomalies illustrated previously, forming an effective fusion.

7.4. Summary

Statistically based approaches to data integration are extremely powerful and yield insights beyond the capabilities of other methods. While they, too, offer descriptive potential in more complete visualizations of the subsurface, interpretive data are also generated in the form of principal component scores, factor loadings, or regression weights that add considerably to understanding of interrelationships and underlying dimensionality. Predictive aspects are also noteworthy—whether in the form of probability surfaces or discrete classified maps—because data patterns may point to less visible anomalous conditions that might otherwise be overlooked and ignored.

8. CONCLUSIONS

Although the approaches to geophysical data integration examined here span a wide range of commonly available techniques, they are by no means exhaustive. A host of other supervised and unsupervised classification algorithms exists (Schowengerdt, 1997) and new context-based and image-segmentation methods for classification (van der Sande et al., 2003) have not yet been explored. If the foregoing results can be generalized, it is that moderate and robust anomalies exist in the data and tend to dominate any fusion, regardless of the method employed. The consequence is amazingly parallel results between widely different forms of integration (compare elements of Figure 7). A correlation matrix demonstrates this fact for the quantitative results (Table 4). Given that many integrating approaches are linear combinations of the geophysical variables, with similar weights, high correlations may not be surprising.[3] Discrete data fusions may also

---

[3] Factor 1 is especially correlated with the binary logistic regression ($r=.985$) because the factor score coefficients (Table 3) show an uncanny proportional resemblance to the regression coefficients (given in text). With Factor 1 characterizing the principal axis of variation in the data, from anomalous through common conditions, this relationship is reasonable. The relatively low relationship of Factor 2 with logistic regression and the other fusions arises because it characterizes variation independent of the main axis represented by Factor 1. This result suggests that a three or more class solution that includes specific features represented by Factor 2 may be necessary for a *multinomial* logistic regression to capture this other dimension of variation.

Table 4. Relationships between the continuous (top) and discrete (bottom) data integrations.

PEARSONIAN CORRELATIONS BETWEEN CONTINUOUS INTEGRATIONS

	Sum	Product	Maximum	Factor 1	Factor 2	Regression
Sum (Figure 7a)	1.0	.902	.628	.795	.543	.802
Product (Figure 7b)	.902	1.0	.755	.887	.369	.913
Maximum (Figure 7c)	.628	.755	1.0	.751	.059	.750
Factor 1 (Figure 7d)	.795	.887	.751	1.0	0.0[a]	.985
Factor 2 (Figure 7e)	.543	.369	.059	0.0[a]	1.0	.082
Regression (Figure 7f)[b]	.802	.913	.750	.985	.082	1.0

PERCENT CORRECT BETWEEN DISCRETE INTEGRATIONS

Method 0 = anomaly absent; 1 = present		Anomalies by 2+ Methods (Fig. 4h)		Overall Accuracy/ Kappa
		0	1	
Robust Interpreted	0	82.5%	23.0%	81.6%.462
Vectors (Figure 4a)	1	17.5	77.0	
Boolean Union	0	71.6	0.0	76.1/.442
(Figure 6g)	1	28.4	100.0[c]	
K-means ($k=2$)	0	89.6	19.7	88.2/.611
(Figure 8a)	1	10.4	80.3	
K-means ($k=3$)	0	85.3	15.1	85.2/.558
(Figure 8b)	1, 2	14.7	84.9	
K-means ($k=4$)	0, 1	88.3	21.8	86.7/.570
(Figure 8c)	2, 3	11.7	78.2	
Number of cases/pixels		$n_0=53,924$ $n_1=10,076$		$N=64,000$

[a]Factors are independent. [b]The raw regression function is used here because the logistic transformation is non-linear. [c]Anomalies shown by 2+ methods are a subset of the Boolean union.

be compared. The binary mapping of anomalies indicated by at least two methods (Figure 6h) cross-tabulated against the manually interpreted vectors (Figure 4a) and the cluster analyses (Figure 8) shows that here, too, results tend to be similar, often approaching 80 percent agreement or more between indicated anomalies (Table 4).

Which methods are best? The answer may depend on purpose. Some yield visually pleasing results that appear to well integrate available information while others may seem less revealing but offer interpretive or predictive potential. If a goal is to define discrete anomalies that may be labeled and subsequently interpreted, then categorical methods may be best. If a goal is merely a continuous-tone image that represents most of what is known about the subsurface, then a composite graphic or mathematical-statistical integration may be most suitable. Classification models, weights, scores, or loadings produced by statistical methods offer added insights and the results of any method can be correlated with the primary geophysical data to understand better relationships between the inputs and anomaly types (a capability that space does not permit here). In this age of high technology, itself responsible for the geophysical data, it is apparent that approaches to data presentation and integration have generally been neglected. The computer graphic, GIS, mathematical, and statistical solutions offered here provide a glimpse of what is possible in today's software environments.

ACKNOWLEDGEMENTS

This work was made possible by a grant to the Center for Advanced Spatial Technologies (CAST) and Department of Anthropology, University of Arkansas, from the Strategic Environmental Research and Development Program, U.S. Department of Defense. W. Fredrick Limp, Director of CAST, was project co-Principal Investigator with the author. Michael L. Hargrave of the U.S. Army Corps of Engineers Research Laboratory served as coordinator with Fort Riley and contributed to fieldwork. Scott Hall and the support staff of Fort Riley are thanked for their cooperation. Lawrence B. Conyers and an anonymous reviewer provided helpful comments. The data were collected by the author, Jo Ann C. Kvamme, and University of Arkansas students Jennifer R. Bales, Eileen G. Ernenwein, Charles K. Kvamme, and Dorothy Neeley. Tommy Ike Hailey, of Northwestern State University, Louisiana, piloted the powered parachute for the aerial thermography. Much of the data pre-processing, especially of GPR, was performed by doctoral student Eileen G. Ernenwein. Data processing was accomplished using Geoplot 3 by Geoscan Research, Radan 5 by Geophysical Survey Systems Inc., Idrisi Kilmanjaro 14 by Clark Labs, SPSS 12 by SPSS Inc., Surfer 8 by Golden Software Inc., and Photoshop 7 by Adobe Systems Inc. This research is dedicated to the memory of my grandfather, Harold E. Theobald, 41st U.S. Infantry, who received basic training at Camp Funston in 1918.

REFERENCES

Avery, T.E., and Berlin G.L., 1992, *Fundamentals of Remote Sensing and Airphoto Interpretation*, 5th edition. Macmillan, New York.
Bevan, B.W., 1998, *Geophysical Exploration for Archaeology: An Introduction to Geophysical Exploration*. Midwest Archeological Center Special Report 1. U.S. National Park Service, Lincoln, Nebraska.
Brizzolari, E., Ermolli, F., Orlando, L., Piro, S., Versino, L., 1992, Integrated Geophysical Methods in Archaeological Surveys. *Journal of Applied Geophysics* 29:47–55.
Burrough, P.A., and McDonnell, R.A., 1998, *Principles of Geographical Information Systems*. Oxford University Press, Oxford.
Buteux, S., Gaffney, V., White, R., and van Leusen, M., 2000, Wroxeter Hinterland Project and Geophysical Survey at Wroxeter. *Archaeological Prospection* 7:69–80.
Clark, A., 2000, *Seeing Beneath The Soil: Prospection Methods in Archaeology*. Routledge, London.
Clay, R.B., 2001, Complementary Geophysical Survey Techniques: Why Two Ways are Always Better than One. *Southeastern Archaeology* 20:31–43.

Conyers, L.B., 2004, *Ground-penetrating Radar for Archaeology*. AltaMira Press, Walnut Creek, California.

Dabas, M., Hesse, A., and Tabbagh, J., 2000, Experimental Resistivity Survey at Wroxeter Archaeological Site with a Fast and Light Recording Device. *Archaeological Prospection* 7:107–118.

David, A., 2001, Overview—the Role and Practice of Archaeological Prospection. In *Handbook Of Archaeological Sciences*, edited by D.R. Brothwell and A.M. Pollard, pp. 521–527. John Wiley, New York.

Davis, J.C., 2002, *Statistics and Data Analysis in Geology*, 3rd edn. John Wiley, New York.

Gaffney, C., and Gater, J., 2003, *Revealing the Buried Past: Geophysics for Archaeologists*. Tempus Publishing, Stroud, England.

Gaffney, C., Gater, J.A., Linford, P., Gaffney, V., and White, R., 2000, Large-Scale Systematic Fluxgate Gradiometry at The Roman City of Wroxeter. *Archaeological Prospection* 7:81–99.

Hargrave, M.L., 2006, Ground Truthing the Results of Geophysical Surveys. In *Geophysical and Airborne Remote Sensing Applications in Archaeology: A Guide for Cultural Resource Managers*, edited by J. Johnson. In press, University of Alabama Press, Tuscaloosa.

Hargrave, M.L., Somers, L.E., Larson, T.K., Shields, R., and Dendy, J., 2002, The Role of Resistivity Survey in Historic Site Assessment and Management: An Example from Fort Riley, Kansas. *Historical Archaeology* 36: 89–110.

Hosmer, D.W., and Lemeshow, S., 2000, *Applied Logistic Regression*, 2nd edn. John Wiley, New York.

Johnson, J.K., and Haley, B.S., 2004, Multiple Sensor Applications in Archaeological Geophysics. In *Proceedings of SPIE Vol. 5234, Sensors, Systems and Next Generation Satellites VII*, edited by R. Meynart, S.P. Neeck, H. Simoda, J.B. Lurie and M.L. Aten, pp. 688–697. SPIE, Bellingham, Washington.

Kvamme, K.L., 2001, Archaeological Prospection in Fortified Great Plains Villages: New Insights through Data Fusion, Visualization, and Testing. In *Archaeological Prospection: 4th International Conference on Archaeological Prospection*, edited by P.M. Doneus, A. Eder-Hinterleitner, W. Neubauer, pp. 141–143. Austrian Academy of Sciences Press, Vienna.

Kvamme, K.L., 2003, Geophysical Surveys as Landscape Archaeology. *American Antiquity* 68:435–457.

Kvamme, K.L., 2007, Remote Sensing: Archaeological Reasoning through Physical Principles and Pattern Recognition. In *Archaeological Concepts for the Study of the Cultural Past*, edited by A.P. Sullivan III. University of Utah Press, Salt Lake City, in press.

Kvamme, K.L., Ernenwein, E., Hargraave, M., Sever, T., Harmon, D., and Limp, F. 2006, *New Approaches to the Use and Integration of Multi-Sensor Remote Sensing for Historic Resources Identification and Evaluation*, SERDP Project CS-1263. Final project report submitted to the Strategic Environmental Research and Development Program, U.S. Department of Defense, Washington, D.C.

Leckebusch, J., 2003, Ground-penetrating Radar: A Modern Three-dimensional Prospection Method. *Archaeological Prospection* 10: 213–240.

Neubauer, W., 2004, GIS in Archaeology—the Interface between Prospection and Excavation. *Archaeological Prospection* 11:159–166.

Neubauer, W., Eder-Hinterleitner, A., 1997, Resistivity and Magnetics of the Roman Town Carnuntum, Austria: An Example of Combined Interpretation of Prospection Data. *Archaeological Prospection* 4:179–189.

Neubauer, W., Eder-Hinterleitner, A., Seren, S., Melichar, P., 2002, Georadar in the Roman Civil Town Carnuntum, Austria: An Approach for Archaeological Interpretation of GPR Data. *Archaeological Prospection* 9:135–156.

Piro, S., Mauriello, P., and Cammarano, F., 2000, Quantitative Integration of Geophysical Methods for Archaeological Prospection. *Archaeological Prospection* 7:203–213.

Schmidt, A., 2001, *Geophysical Data in Archaeology: A Guide to Good Practice*. Oxbow Books, Oxford.

Schowengerdt, R.A., 1997, *Remote Sensing: Models and Methods for Image Processing*. Academic Press, San Diego.

Scollar, I., Tabbagh, A., Hesse, A., and Herzog, I., 1990, *Archaeological Prospection and Remote Sensing*. Cambridge University Press, Cambridge.

Summers, G.D., Summers, M.E.F., Baturayoglu, N., Harmansah, Ö., and McIntosh, E., 1996, The Kerkenes Dag Survey: An Interim Report, *Anatolian Studies* 46:201–234.

Telford, W.M., Geldart, L.P., Sheriff, R.E., 1990, *Applied Geophysics*, 2nd edn. Cambridge University Press, Cambridge.

van der Sande, C., de Jong, S.M., and de Roo, A.P.J., 2003, A Segmentation and Classification Approach to IKONOS-2 Imagery for Land Cover Mapping to Assist Flood Risk and Flood Damage Assessment. *International Journal of Applied Earth Observation and Geoinformation* 4:217–229.

van Leusen, M., 2001, Archaeological Data Integration. In *Handbook of Archaeological Sciences*, edited by D.R. Brothwell and A.M. Pollard, pp. 575–583. John Wiley, New York.

Weymouth, J.W., 1986, Geophysical Methods of Archaeological Site Surveying. In *Advances in Archaeological Method and Theory, Vol. 9*, edited by M.B. Schiffer, pp. 311–395. Academic Press, New York.

Chapter *15*

Ground Penetrating Radar Advances in Subsurface Imaging for Archaeology

Dean Goodman, Kent Schneider, Salvatore Piro,
Yasushi Nishimura, and Agamemnon G. Pantel

Abstract: Advances in imaging software for Ground Penetrating Radar (GPR) have greatly enhanced the utility of this geophysical remote sensing tool for archaeological discovery. Time-slice analysis, isosurface rendering, and "overlay analysis" are among several image analyses used to identify subsurface archaeological remains. Static corrections, in which the tilt of the transmitting antenna is accounted for over areas with significant topography, are presented here for the first time. GPR-GPS surveying to facilitate and automate remote sensing, are presented as a method for accurate identification of unmarked grave sites. An example of the use of GPR imaging in surveying ancient Roman sites and the application of GPR to help guide the stabilization and restoration of the second oldest church in the New World, are presented. GPR imaging is also summarized in several case histories involving detection of burials and other structures at Japanese, Byzantine, and Native American sites.

1. INTRODUCTION

The earliest application of Ground Penetrating Radar (GPR) technology dates back to the late1920s and was employed in Europe for measurement of glacier thickness. The first data outputs were limited to examining radar pulses on an oscilloscope and measuring two-way travel times of reflections to the bottom of ice flows. Since these early days, advances in GPR data visualization and processing have significantly increased the utility of this geophysical remote sensing tool, particularly for the field of archaeological prospection. GPR has enabled archaeologists to view a subsurface image prior to any excavation, allowing them to better understand the internal structure of a site and potentially to avoid destruction of the most important materials and evidence of a site. In some cases, the completeness of GPR images has allowed archaeologists to understand fully the essential secrets of a buried archaeological site, without having to perform any "destructive" excavation. In the same sense that modern medical doctors would never do exploratory surgery without using imaging technology for the human body, archaeologists are beginning to use GPR imagery to enter a site surgically to unearth only the key areas which may have the most likelihood of yielding valuable information. In the case of Roman sites, GPR images can reveal, with great detail, the entire design of buildings that are buried beneath the ground surface. The subsurface ruins of the villa of Emperor Trajan have been nearly completely discovered by GPR, and the building designs documented before any major excavation of the site has begun (Piro et al., 2003). The surveys at the Forum Novum site in Vescovio, Italy, show that a complete town, including previously unknown amphitheaters, Roman funerary complexes, and other buildings were accurately imaged prior to extensive excavation (Goodman et al., 2004a). Some of the results from this study are presented below.

The key ingredient that has unleashed the utility of GPR technology has been the manipulation of the raw radargram profiles, which in themselves are often difficult to discern and interpret. Time-slice imaging is a process that compares and maps changes in reflected amplitudes across a site. Small changes in the recorded reflections, which could never be deciphered or seen within the pulse waveforms of the vertical slice radargrams, become markedly obvious reflections when horizontal time-slice imaging is applied. Although the seismic industry has been doing these kinds of three-dimensional analyses for over 60 years, it was late to be implemented for GPR. One of the first-ever reviewed journal papers referring to time-slice analysis for GPR was cited as recently as late 1993 (Goodman and Nishimura, 1993). In the mid 1990s the utility of advanced imaging of GPR datasets began to be recognized. GPR images of archaeological surfaces that are buried in the ground now include 3D rendering where multiple isosurfaces can be displayed within a single volume (Grasmuck, 1996).

Examples of GPR imaging and visualization by our research team at a variety of important archaeological sites are presented in this article. We identify them briefly before presenting a more detailed report on each.

Roman. The results of GPR imaging at ancient Forum Novum, Italy, are presented in which an unknown amphitheater and funerary precinct were discovered. Forum Novum is a site being studied by the British School at Rome and is one of the most important urban Roman sites in the Tiber Valley.

Native American. An Abenaki site in Highgate, Vermont, was studied with GPR in hopes of detecting burials and/or dwellings. The site is planned for development, but any burial remains found would have a great impact on future land use in the area. Typical analysis of the site would usually involve only "blind" test pits. Using GPR in conjunction with test pits is shown to be a much better means to locate potential archaeological hotspots, and adds value to test-pit extrapolations. At another site in the Kistachi National Forest, Louisiana, a GPR-GPS experiment was initiated to locate unmarked burials from a Jena-Choctaw Tribal cemetery with great success. Using 3D volumes of data that are generated from "random" GPR tracks monitored from real-time differential GPS, isosurface renders of marked and unmarked burials are clearly imaged.

Japanese. Several Japanese burial mounds from the Kofun (tumulus) period (300–700 A.D.) are presented. At sites in the Saitobaru National Burial Mounds in Miyazaki, new static corrections, which account for the tilt the GPR antenna incurs as it is pulled over large mounded tombs, are introduced for the first time.

Spanish Site in the New World. Early 16th-century remains on the site of the second-oldest standing church in the New World, the Iglesia San Jose in San Juan, Puerto Rico, are also studied. This church is listed on the World Monuments Watch List of the 100 Most Endangered Sites for 2004 and is undergoing extensive stabilization and restoration. GPR was used to assist in these conservation measures by mapping subsurface water damage, unknown crypts and voids, as well as in the examination of the constructive evolution of the building from the 16th to 19th centuries. The application of GPR survey methods also assisted in the detection of early Christian burials within a contemporary urban context, specifically within a historic Puerto Rico town center.

Byzantine. A final example of GPR imaging details the search for an ancient necropolis at Chersonesos in Ukraine.

2. ROMAN: FORUM NOVUM

2.1. GPR 3D Time Slices

The search for Roman ruins is often aided by the search for crop marks (Sever, 2000). Buried walls often affect the surface vegetation by giving a different color to the leaves, by changing vegetation density, or possibly altering the color of the ground soil. Leaching of minerals within the cemented Roman walls is the primary cause of the crop marks that can be seen from aerial photographs. The marks have often been used by Roman archaeologists to map the subsurface buildings with great precision. Near-surface wall-foundations, however, do not always produce a surface effect that can be detected. One such area in which the shallow foundations below ground could not be revealed by aerial photography is the Forum Novum marketplace.

The Forum Novum is located in the Sabine hills at the present town of Vescovio at the head of the valley of the Aia river, which flows into the Tiber river northeast of Rome. The site was established as a new marketplace center sometime during the 2nd century B.C. By the 1st century A.D. the town received the status of a Roman municipium (Pliny, NH 3.107). The town was very active throughout the Imperial period and the marketplace was still functioning through the 4th century A.D., and to a lesser degree through medieval times. The site is being investigated as part of the Tiber Valley Project of the British School at Rome (Gaffney et al., 2001; Patterson, 2004).

The site was extensively surveyed with a magnetometer with spectacular results: many subsurface structures were discovered. In several areas, however, there were no magnetic signals detected because of highly magnetic soils. Examination of survey results from magnetometry in adjacent areas suggested to the archaeologists that buildings probably continued, even though nothing could be seen in the aerial photographs. Archaeologists Helen Patterson (British School at Rome) and Vince Gaffney (University of Birmingham, England) invited us to do a GPR survey in spite of little hope of finding any significant buried structures (Gaffney et al., 2001).

Shown in Figures 1 and 2 are time-slices made at the surface and at 18.2 ns. On the surface time-slice, which actually contains a slicing window from the surface down to 21 cm several geometric features are imaged, including cylindrical and circular structures, as well as a rectangular structure with rounded sides. This last feature proved to be buried foundations of a destroyed mausoleum. On the 18.2 ns time-slice (corresponding to a depth of approximately 60 cm), an oval structure was discovered (Goodman et al., 2004a). This structure is a Roman amphitheater that had gone undetected for centuries prior to the 500-MHz GPR survey. The amphitheater was constructed with 8 entrances and has an inner wall cemented with a limestone mortar. Pottery sherds from subsequent excavations indicated that the amphitheater was constructed in the 1st century A.D. After discovery of the amphitheater was made, some of the geometric features imaged on the surface time-slice (which could also be seen by the naked eye during the field survey as crop marks), were identified as destroyed structures associated with the training of gladiators.

It is quite remarkable that the shallow-buried amphitheater walls did not have an effect on the surface vegetation at the site, which would have allowed for easy detection of the site by aerial photography. It was also very fortuitous that the archaeologists recommended surveying a site which had a less than "favorable" rating in terms of yielding significant Roman structures. The GPR discovery of the amphitheater dramatically changed archaeologists' understanding of the Forum Novum as an important Roman urban center (Gaffney et al., 2001).

GPR also helped in the discovery of Roman burials at Forum Novum, or Vescovio, as it was called from at least the 10th century. The GPR image

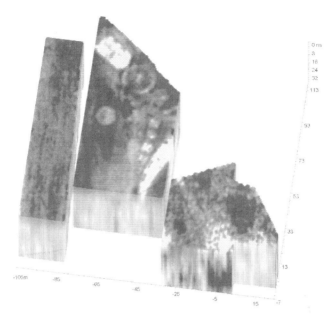

Figure 1. A surface time-slice map of Forum Novum shows several geometric features which also correspond to crop marks that can be visually seen on the surface of the site. The geometric shapes imaged are believed to be reflections recorded from destroyed structures that were once used for the training of gladiators.

Figure 2. A deeper time-slice map at 20 ns shows an oval structure that subsequent excavation revealed to be the remains of a Roman amphitheater of the 1st century A.D. with eight entrances (photo linked to time-slice after Goodman et al., 2004).

of one area to the north of the Romanesque church of Santa Maria shows
two long walls meeting to form a 45-degree angle (Figure 3). GPR reflections
at the apex of the area where the walls joined indicated the existence of a
semi-circular structure at depth. This structure was subsequently identified as
buried foundation walls of a destroyed mausoleum, which still has portions
that remain above ground. Located on the outside flanks of both walls are
small rectangular structures, which are very clearly imaged outside the western
wall. The eastern area was excavated to reveal a Roman funerary precinct.
Various burial artifacts were later recovered by the excavations (Figure 3).

GPR surveys were thus able to map buried buildings at Forum Novum
with remarkable accuracy, as well as at other Roman ruins in other parts

Figure 3. GPR imaging (bottom image) at Forum Novum helped to discover the
remains of Roman burial crypts flanking angular garden walls (upper right). The partial
rectangular anomalies imaged led archaeologists to excavate these areas, which yielded
a multitude of burial artifacts (photo, upper left).

of Europe (Nishimura and Goodman, 2000). The benefits of remote sensing prior to excavation, at least in the case of Forum Novum, are that areas which have the most potential for artifact recovery can be identified. This capability has helped archaeologists to manage more efficiently the limited resources for excavation.

3. NATIVE AMERICAN

3.1. GPR and Test Pits

3D imaging with GPR was recently applied by our team in the study of Abenaki sites in Highgate, Vermont. These sites, unlike the "well groomed – high contrast" Roman sites of Forum Novum, are areas where subtle ground features need to be extracted from the datasets. Soil-to-soil contacts rather than stone and soil, are the primary components that archaeologists need to distinguish. One objective of the survey was to identify any features which may be related to Indian habitation on the site. The area is planned for development, and any Indian remains or ancient dwellings might severely limit the future land-use. One of the advantages of surveying the site was that any anomalies detected could be quickly followed up by test-pit excavations. Had GPR not been available, coarse test-pit excavations would have been the only means by which this particular site could have been surveyed because of limited resources. Applying GPR with follow-up excavations would help archaeologists better understand the nature of detected anomalies and possibly allow for "calibrated extrapolations" into areas which would not be excavated.

A GPR survey was made with a 400-MHz antenna that provided about 2 m of ground penetration for this particular site. 3D time-slice images were developed from radargram profiles that were spaced 0.5 m apart. Shown in Figure 4 is a time/depth slice made at about 50 cm. One anomaly was chosen on this site because of its discreet nature and was followed up by test-pit excavation. This particular anomaly was then attributed to reflections from the floor of a fire hearth which contained some charcoal. According to the site archaeologists, the amount of charcoal was the most found in any single test pit for several years. Nonetheless, the amounts seemed to us quite small to account for the strength of the anomaly detected. The main energy that was recorded was most likely not from the charcoal, but from the bottom of the fire hearth which was dug and reworked by the ancient Abenaki inhabitants. The bottom of the fire hearth has a clear concave shape to it. The concavity of the hearth floor is a good reflector of radar waves. The structural shape of this particular hearth can be likened to a parabolic satellite dish which focuses the energy back, thereby allowing for detection and measurement. For similar discussions of ray focusing, see Conyers and Goodman (1997) and Goodman (1994).

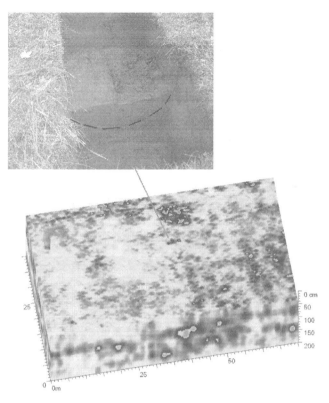

Figure 4. GPR images made at an Abenaki site in Vermont helped to guide archaeologists in making informative test-pit excavations. Shown in the figure is a small discreet anomaly on the 3D time-slice (lower image) that was later dug to reveal a buried fire hearth (photo at top).

3.2. GPR and GPS

Proper navigation of ground-penetrating radar equipment is a tedious chore in the field; it is, however, the most essential factor in dictating the accuracy of the data. With advances in global positioning system technology, the potential for using this instrumentation for navigation of GPR survey remains to be evaluated in a variety of field conditions. In November 2003 the United States Forest Service sponsored studies to implement GPS navigation for GPR surveying (Goodman et al., 2004b). The purpose of the GPS integration was to automate the collection of GPR datasets, particularly at sites where establishing grids or using survey wheels may also be problematic.

      One study to implement GPS recording was done at the Whiterock Cemetery, in the Kisatchi National Forest in Louisiana. The head archaeologist for the forest, Velicia Bergstrom, had received many requests for assistance from local tribes in locating lost burial remains. The implementation of GPR would provide a good non-invasive method and, combined with GPS, would allow for a high degree of accuracy in determining the location of graves. The cemetery is a burial ground for members of the Jena Choctaw tribe. The purpose of the GPR study was to detect modern unmarked burials in the cemetery, which had been lost from improper maintenance of the site over the last 40 years.

To expedite the study, navigation of the GPR antenna's course across the Jena Choctaw Tribal Cemetery was recorded using a Trimble Pro XRS GPS system with real-time differential correction capabilities. Differential GPS data were available from a CORS station in the field and recorded at 1-second intervals. A total of 11 minutes of continuous data was recorded in survey. An isosurface render of the top 70% strongest reflectors in a 3D volume of data collected at the Jena Choctaw cemetery indicates the location of known graves. Using the isosurface shapes detected from known graves at the site, in which longitudinal anomalies can be seen, we determined that several similar reflections were indicative of unmarked graves, which were later excavated (Figure 5). Although there may be some multi-pathing of the GPS signals from trees in the woods nearby, the accuracy of the navigation was better than 70 cms on average. The use of GPS with GPR, therefore, has been shown to add great flexibility and accuracy in locating sites that need to be restored and protected.

4. JAPAN: KOFUN (TUMULUS) BURIALS

4.1. GPR Overlay Analysis

More often than not, archaeological features that are recorded at different levels in the ground are usually displayed on separate time-slice maps. The interpreter may thus have difficulty in piecing the essential information together between maps. In this instance, overlay analysis can be applied to create comprehensive images of subsurface anthropogenic features. Overlay analysis is different from compositing or addition of grid information. Kvamme (2003) has recently looked at compositing of information from various geophysical methods including resistivity, magnetics, and GPR to create comprehensive figures that show many physical properties of the subsurface that may be related to archaeological features. These are "additive" in that RGB colors are assigned and then color combinations from overlaying these results can be viewed.

In overlay analysis the time-slice images at user-chosen levels have gains individually adjusted. Next, an overlay map is generated by extracting the relative-strongest reflectors from each individual map, and then posting them on a single map. The method of overlay analysis is not an addition of grids;

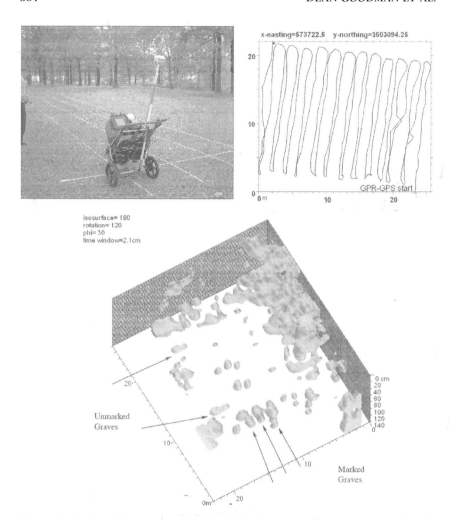

Figure 5. An isosurface render of the top 70% strongest reflections measured at the Whiterock (Jena Choctaw) Cemetery shows the location of marked and unmarked graves at the site. The 3D volume was generated from GPR tracks (upper right) that were navigated using a Trimble PRO-XR GPS system. A 400-MHz GPR antenna was used in the profile collection (photo, upper left).

it is a binary operation where only the relative-strongest reflectors within set time/depth levels are collected for a single grid location.

The method was first applied in 1991 and an extremely successful application was made in 1998 to study Kofun burial #100 at the Saitobaru National Burial Mounds in Miyazaki Prefecture, Japan (*www.GPR-SURVEY.com*). Overlay analysis has been applied to another 6th-century burial mound in the Saitobaru National Burial Mound Park. Since 1992 the park has been extensively studied by GPR at the insistence of the museum director, Hongo Hiromichi, and the curator, Noriaki Higashi, who were early to realize the importance of this geophysical remote sensing tool. In one area (Figure 6) where a partially destroyed burial existed, extensive excavation was implemented. The purpose of the excavation was to discover the location of the moat surrounding this partially intact burial mound, so that the mound could be accurately rebuilt to its former specifications of the 6th century A.D. Most of the mound had been removed by farming activities on the site. Six excavation trenches were placed near the remaining mound. The profile of the moat could not been seen on the walls of the excavation trenches as there were no detectable contrasts between soils within and outside of the moat. Because of the unsuccessful attempt by excavation to detect the outside moat boundary, which determination was necessary for reconstructing the mound to its ancient size, a GPR survey was conducted to study the area.

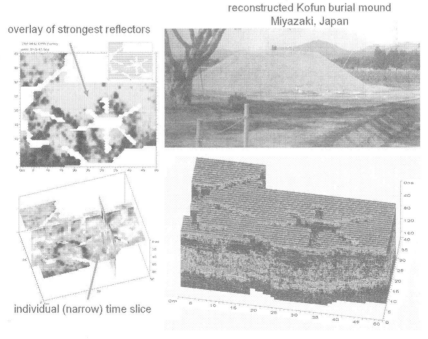

Figure 6. A 3D radagram of Kofun burial mound is at lower right. A star-shaped survey grid in which numerous excavation trenches were placed before the GPR survey yielded no information about the location of an ancient burial moat. Overlay analysis was used to create comprehensive images of subtle reflections at variable depths, which, when "put together," revealed a circular moat. The photograph of the burial (upper right) is a reconstruction made using the size specifications found from GPR overlay analysis (Saitobaru National Burial Mounds, Miyazaki).

an horizontal slice

Using time-slice analysis, a circular shape for one of two burial-mound moats present could be easily seen in the dataset on the individual–narrow time-slice (Figure 6, bottom time-slice image). This reflection is from a mound moat (located to the left of the tree in the photo (Figure 6, upper right), but not shown in the photograph). However, the mound where the excavations were conducted showed no continuous reflection data to indicate that an even, circular structure was buried at 1.4 m depth. Using time-slices that were individually adjusted for gain within a 30-ns time window, a composite overlay (Figure 6, upper left) indicating a mound approximately 12 m in diameter could be "synthesized." The mound shown (Figure 6, upper right) is a reconstructed one that was rebuilt in 2002 using the specifications obtained from the overlay time-slices.

4.2. GPR Static Corrections for Antenna Tilt

In addition to frequent studies of moats surrounding Japanese burials, GPR surveys are often conducted on the tops of protected burial mounds to detect the presence or absence of primary graves. In such field situations, special corrections need to be applied to the data since the topography of the mounds is quite significant compared to the depth of GPR penetration. Shown in Figure 7 is the generalized effect that topography has on the tilt of the GPR antenna and the transmitted wave. The top diagram shows the effects of mild topography. In this field situation, the downward vertical rays emanating from the antenna do not cross each other. If the topography changes abruptly, the rays transmitted into the ground near adjacent locations can traverse the same swath of ground; such a return traverse becomes more probable the deeper

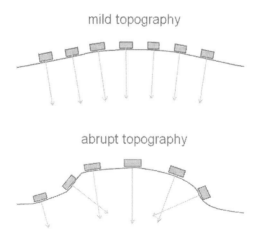

Figure 7. A diagram showing the effect that abruptly and gradually changing topography of a site can have on the geometry of the transmitted GPR microwave.

the penetration is. The amount of shifting of the location of a recorded radar reflection, however, can be significant if the antenna is tilted. For instance, reflections at 100 ns, where an antenna is tilted 30 degrees and assuming a 6 cm/ns microwave velocity, can be shifted 1.5 m horizontally at depth. If the velocity were 12 cm/ns the reflections at 100 ns could be shifted by 3 m. Archaeologists using GPR data to detect targets in narrow test pits could miss these anomalies entirely.

In typical topographic corrections, the radargram scans are simply shifted up and down in the binary files to match the topography. The amount to shift radar scans is a function of the speed of the microwave velocity across the survey area. In general, a single velocity is assumed when applying the topography to a radargram. Shown in Figure 8 (top diagram) is an example of a GPR profile corrected for topography using the normal correction, which is a simple shift vertically—up or down—to the radar scan. The correction for topography and tilt of the antenna is shown in the lower diagram. The tilt of the antenna is assumed to be perpendicular to the ground surface and was computed from the first derivative of the topography profile. The topography for this particular site does not have the condition of abruptly changing topography and no crossing of the antenna beam at different locations is registered for the depths traverse. The image is taken across a subterranean stone chamber of the 6th century A.D. beneath a Kofun in the Saitobaru National Burial Mounds in Miyazaki Prefecture. The stone chamber may have housed the remains of an important military person, since armor and swords were discovered with this ancient burial.

5. SPANISH NEW WORLD

5.1. GPR for Evaluation and Discovery at Historic Buildings

GPR surveying is not limited to subsurface detection but can also be applied in the study of important historic buildings and standing infrastructure above ground (e.g., Lualdi and Zanzi, 2002). We present here one example of such a survey, which was conducted for the purpose of helping with building restorations in Puerto Rico. The Iglesia San Jose in Old San Juan is the second oldest standing structure built by Europeans in the Western Hemisphere. The church was constructed by the Spanish in 1532 and is identified by a unique series of New World Gothic vaulted ceilings, and some of the earliest known European murals in the Americas. Ponce de León's coat of arms hangs above the main altar in the church. The church is the oldest surviving building in Puerto Rico, and was originally part of the 16th century Dominican Convent built on the island. Over the years, the structure has been debilitated by structural deficiencies resulting from poor interventions during the 20th century. Water infiltration and monitoring and maintenance issues have aggravated

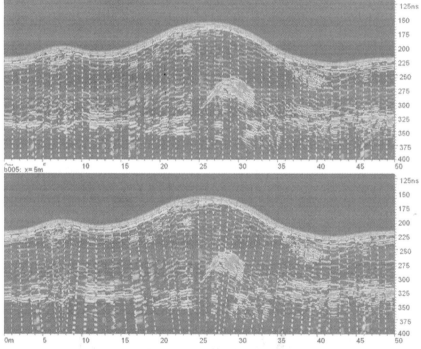

Figure 8. Topographic corrections (bottom diagram), which account for the tilt of the GPR antenna, are compared with traditional static corrections for Kofun burial #111 (Saitobaru National Burial Mounds, Miyazaki). In this particular example, the size of the reflection from the top of a burial chamber is found to be much narrower beneath the burial mound, when tilt-statics are applied. A photograph of the excavated chamber that was imaged is also shown. (At the time of the photograph, a synthetic resin was being applied to the burial chamber for the purposes of reconstructing it for a museum display.)

the state of conservation of this monumental building. As a result, the World Monument Fund included Iglesia San Jose on its World Monuments Watch list of the world's 100 Most Endangered Sites for 2004. A GPR survey was initiated to assist in ascertaining the degree of foundation degradation, and to help in locating of underground crypts. There was also a possibility of detecting original foundations prior to remodeling of the church which were initiated after several earthquake episodes centuries earlier.

A GPR survey was made by our team within the church at 25-cm profile spacing using a 400-MHz antenna (Figure 9). The time-slice image from about 50 cm indicates locations of two known crypts below the church floor. A previously unknown crypt was also "rediscovered." This crypt is located between two pillars on the east side of the survey area. Within the background reflection is a large area on the east side of the church in which slightly stronger amplitudes are recorded. Two possible scenarios that can account for the reflections have been considered. In one interpretation the GPR is simply responding to dampness of the tiles and subsurface soils as a result of water infiltration. During the survey, although there was no standing water in the church, areas of extended dampness were visible. (Recently, steps have been taken to repair roof drains to divert rainwater as one of the first conservation measures). In another scenario, the dampness of the floor on the east may be responsible for "illuminating" structural differences beneath the floor. Several structural features that are linear (some curvilinear) can be seen extending

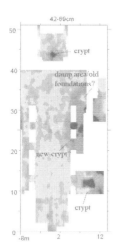

Figure 9. GPR survey results from the Iglesia San Jose in Puerto Rico. The time-slice image (north is at the top) of the main floor of the church indicates the location of several known crypts and a rediscovered crypt. Areas on the east side of the church show moderate reflected energies corresponding either to water dampness or to old foundations that had been buried in the course of renovations. The altar visible in the background of the photo is above the crypt shown in the time-slice near y=43 m.

into the central church and are not at right angles to the present structure. The church has undergone remodeling in centuries past, and we believe that earlier walls or passageways into the church may have been removed and tiled over. These features may only have been detected because of the introduction of water into the base of the floor. The dampness may have helped to create a contrast in electrical properties of the building foundations, which might have little or no contrast with surrounding materials when they are dry.

5.2. GPR and Burials in Urban Environments

Over twenty historic town centers in Puerto Rico are being remodeled. Utilities are being buried underground and new concrete streets and sidewalks are being laid. Many of these downtown areas have undergone several reconstructions over the last 200 years. Early 20th-century urban expansion, earthquakes, and other natural disasters have also modified street configurations, destroyed buildings in some areas, and buried rubble piles beneath the present-day streets. Subsurface structures are a mixture of many different materials, new and old, making a difficult environment for subsurface remote sensing.

One of our recent GPR surveys was of the historic downtown center including the main Spanish plaza and Cathedral in the historic town of Mayagüez on the west coast of Puerto Rico. This work was made possible by the support of the Directorate of Urbanism of the Puerto Rico Department of Transportation and Public Works. Nearly 40 linear kilometers of GPR data was collected at 0.5 meter intervals along key sections of nine streets.

Modern roads have replaced church grounds which once were often relegated to cemeteries. The possibility of discovering burials beneath roads that cut through the church yards was to be tested in our survey.

Figure 10 relates a time-slice image to a photograph of burials discovered and excavated along one street adjacent to a church. More than thirty burials were eventually excavated here, some of which had been accounted for and seen within the time-slice dataset resulting from the survey that took place before excavation. The burials appear to have been put into the ground without coffins and the deceased were placed directly in the soil. The degree of preservation of the skeletal material of the burials was extremely poor. The deteriorated nature of the bone, indeed, practically mimics the surrounding soil matrix, a condition that made the detection of these features as "anomalies" a challenging task. The anomalies at these levels had lower overall reflections from soils within the burial pits.

Although the complex subsurface images contain some noise from utilities traversing the street, the subtle reflections from the burials can be discerned. Burial identification in urban environments, however, may prove to be a difficult task in a typical blind survey. The subtle anomalies can be easily overlooked when the nearby ground has many features that overshadow reflections from the target areas. The advantage of having the post-excavation

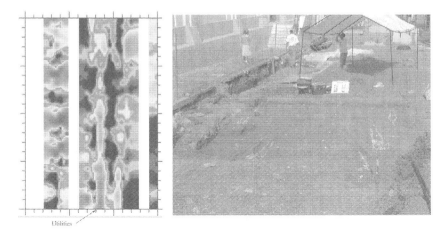

Utilities

Figure 10. Photograph and time-slice map of burials in downtown Mayagüez, Puerto Rico. Several burials were identified on a shallow time-slice image at about 50 cm below the ground surface.

results recorded with the GPR images, as in Figure 10, lies in providing a source of base-line data for making comparisons of future sites in which similar subsurface features are present.

6. BYZANTINE PERIOD IN UKRAINE

6.0.1. GPR and Necropolis Tomb Burials

Extensive ruins from Byzantine period buildings of ancient Chersonesos can be seen just on the outskirts of Sevastopol, Ukraine, on the Black Sea. The location of the necropolis to entomb the elite of this ancient city, however, still remains an elusive question. Under grants obtained by Joseph Carter, a professor at the University of Texas in Austin, a GPR survey was requisitioned to aid in the search for the necropolis. The suspected necropolis site was unfortunately the location of fierce battles during the Crimean War and World War I. Many bomb craters and shrapnel are still clearly visible on the surface even after a century and a half since the first battles were fought there. Because of the metallic interference, magnetic surveys were impractical. In addition, the hard limestone outcroppings and soil would have been difficult if not impossible for typical electrical resistivity other than for shallow probing contact resistivity to work. For these reasons, GPR was the most likely candidate for remote sensing of this area.

Shown in Figure 11 are some results obtained at one area studied that had clear subsurface anomalies. The strong reflections measured on a 300-MHz radargram profile show a parabolic like anomaly appearing near the 64 m

DEAN GOODMAN ET AL.

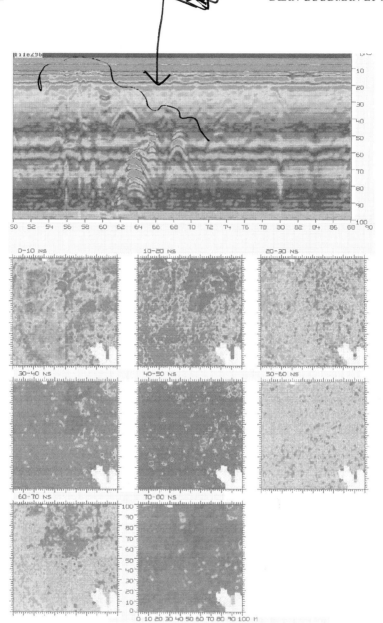

Figure 11. A 300-MHz GPR image was collected at a suspected necropolis site at Chersonesos, Ukraine. The radargram (image at top) shows a strongly contrasted anomaly. Time-slice results (lower eight images) show the location of several strong and narrow reflections at depth.

range and at a depth beginning at about 50–60 ns. This interesting anomaly can also be seen on the time-slice maps, and is particularly distinct on the 70–80 ns map at the location x = 98 m, y = 64 m. The time-slices generated at 10 ns intervals (∼ 30 cm depth slice intervals) show many strong and distinct anomalies. Several test drillings were made at some of the identified anomalies at the site. Unfortunately, the test drilling to a depth of 2 meters did not reveal any subsurface disturbances that appear to be manmade. The strong anomalies that were identified on the time-slice maps are possibly a result of subsurface voids in the limestone rocks at the site, which occur naturally in these formations as they weather. A remote chance that the drilling did not penetrate deep enough to detect the necropolis also remains a low-probability scenario. Nonetheless, GPR was able to clearly differentiate the most likely areas in which strong subsurface anomalies existed. To this date, however, the location of the ancient necropolis has yet to be discovered.

7. CONCLUSIONS

As illustrated in the examples provided by this overview of some of our recent projects, GPR is an invaluable tool for modern archaeology. The ability not only to detect targets in the ground, but also to yield size, shape, and depth information as well, is shown in direct comparison with excavations. Essential images are not always readily derivable from simple time-slice analysis, but require additional processes. Overlay analysis is a very useful tool which is not practiced much in the prospection community yet, but needs to be employed if more useful and comprehensive images are to be extracted from GPR datasets. With the introduction of GPS into GPR, the acquisition of survey data will become more automated and make the process of surveying prior to study or any archaeological excavation more desirable. The ability to synthesize large sets of GPR transect data into visual graphic images of subsurface features, resolves the human-interpretation error inevitably inherent in the analysis of a large number of GPR radargrams. The application of time-slice analysis, isosurface rendering, and various processes to adjust data for unique field conditions will make the value of GPR imaging more useful for the archaeologist. GPR imaging software presents a significant advance in resolving subsurface structures with radar equipment, and has a tremendous potential for expanded uses in archaeology around the globe.

8. REFERENCES

Conyers, L. B., and Goodman, D., 1997, *GPR: An Introduction for Archaeologists*. AltaMira Press, Walnut Creek, California.

Gaffney, V., Patterson, H., and Robert, P., 2001, Forum Novum–Vescovio: Studying urbanism in the Tiber Valley. *Journal of Roman Archaeology* 14:59–79.

Gaffney V., Patterson, H., and Roberts, P., 2004, Forum Novum (Vescovio): a new study of the town and bishopric. In *Bridging The Tiber: Approaches To Regional Archaeology In The Middle Tiber Valley*, edited by H. Patterson, pp. 237–251. Archaeological Monograph 13. British School at Rome, London.

Goodman, D., Piro, S., Nishimura, Y., Patterson, H., and Gaffney, V., 2004a, Discovery of a 1st century A.D. Roman amphitheater and other structures at the Forum Novum by GPR. *Journal of Environmental and Engineering Geophysics* 9:35–41.

Goodman, D., Schneider, K., Barner, M., Bergstrom, V., Piro, S., and Nishimura, Y., 2004b, Implementation of GPS navigation and 3D volume imaging of ground penetrating radar for identification of subsurface archaeology. *Proceedings of the Symposium on the Application of Geophysics to Engineering and Environmental Problems.* pp. 806–813. Environmental and Engineering Geophysical Society, Colorado Springs, Colorado.

Goodman, D., 1994, Ground penetrating radar simulation in engineering and archaeology. *Geophysics* 59:224–232.

Goodman, D., and Nishimura, Y, 1993, Ground radar view of Japanese burial mounds. *Antiquity* 67: 349–354.

Grasmuck, M., 1996, Ground-penetrating radar applied to fracture imaging in gneiss. *Geophysics* 61(4):1050–1064.

Lualdi, M., and Zanzi, L., 2002, GPR investigations to reconstruct the geometry of the wooden structures in historical buildings. In 9th *International Conference on Ground Penetrating Radar*, edited by S.K. Koppenjan and H. Lee, *SPIE* 4758:63–67.

Kvamme, K., 2003, Multi-dimensional prospecting in North American Great Plains village sites. *Archaeological Prospection* 10:131–142.

Nishimura, Y. and Goodman D., 2000, Ground penetrating radar survey at Wroxeter. *Archaeological Prospection* 7:101–105.

Patterson, H., and Rovelli, A., 2004, Ceramics and coins in the middle Tiber valley from the fifth to the tenth centuries A.D. In *Bridging The Tiber: Approaches To Regional Archaeology In The Middle Tiber Valley*, edited by H. Patterson, pp. 269–284. Archaeological Monograph 13. British School at Rome, London.

Piro, S., Goodman, D., and Nishimura Y., 2003, The study and characterization of Emperor Traiano's villa using high-resolution integrated geophysical surveys. *Archaeological Prospection* 10(1):1–25.

Sever, Thomas L., 2000, Remote Sensing Methods. In *Science and technology in historic preservation*, edited by R.A. Williamson and P.R. Nickens, pp. 21–51. Kluwer Academic/Plenum Publishers, New York.

Chapter 16

Landscape Archaeology and Remote Sensing of a Spanish-Conquest Town: Ciudad Vieja, El Salvador

WILLIAM R. FOWLER, JR., FRANCISCO ESTRADA-BELLI,
JENNIFER R. BALES, MATTHEW D. REYNOLDS,
AND KENNETH L. KVAMME

Abstract: The *villa de San Salvador*, founded in 1525 and refounded in 1528 as a Spanish-conquest town, had a resident indigenous (Pipil and Tlaxcallan) population that was perhaps twenty times greater in number than its Spanish population. The town was abandoned in 1545, and its 17-year permanent occupation spans the crucial years of the Conquest period in Central America. The well preserved ruins of this town, known today as the site of Ciudad Vieja, afford a rare opportunity for archaeological study of the dynamics of early Spanish-Indian culture contact. Archaeological research at the site emphasizes spatial study of the town, viewing it as a cultural landscape, and focusing on the mutual interaction of the different cultural groups that shared the terrain. Approximately two dozen Spanish cities were founded in Central America during the Conquest period. Very few of them have been investigated archaeologically, and Ciudad Vieja is unique among them for its exposure, preservation, and ease of access. The cultural landscapes of these settlements formed the spatial matrix within which social and physical relations were enacted. These relations are amenable to archaeological investigation because they were the product of the behavior enacted within a social space which is in turn reflected in the material remains of the site. Geophysical surveys, conducted at the site in November–December 2002 and in March 2003,

395

have formed a critical part of the investigation of the cultural landscape of Ciudad Vieja. Instruments used in these surveys were the Geometrics G858 cesium gradiometer and the Geonics EM38/EM38B electromagnetic induction meter. Initial results obtained in the fall of 2002 indicated that the latter, operating in in-phase or magnetic-susceptibility mode, provided the most interpretable results. The 2003 survey concentrated on widespread magnetic susceptibility coverage of the site. The results of the remote sensing survey depict the locations of probable buried stone foundations because of the high iron-oxide content of volcanic stones used in the built environment.

1. INTRODUCTION

This paper addresses Spanish-American urbanism in the Conquest period and the application of subsurface remote sensing to investigate the spatial distribution and variation of a Spanish-conquest town in Central America. The ruins of the first *villa* of San Salvador form the archaeological site of Ciudad Vieja, El Salvador. The town was first founded in 1525, hastily abandoned, refounded in 1528, and permanently abandoned in 1545, after only 17 years of permanent occupation. A comprehensive, multidisciplinary archaeological investigation and full-scale analysis of Ciudad Vieja under Fowler's direction has been ongoing at the site since 1996, in collaboration with and by permission from the Consejo Nacional para la Cultura y el Arte (CONCULTURA) of El Salvador (Fowler and Gallardo, 2002).

The conquest of Guatemala and El Salvador was an extension of the conquest of Mexico. After the fall of the Mexica Aztec capital of Tenochtitlan in 1521 (Hassig, 1994), Hernán Cortés dispatched Pedro de Alvarado to lead the conquest of unknown lands to the south. A small contingent of Spanish conquistadors led by Alvarado and accompanied by several thousand Tlaxcallan and other indigenous Mexican auxiliary forces invaded Guatemala and western and central El Salvador in June 1524 (Escalante Arce, 2001:20–31; Fowler, 1989:135). They met fierce resistance by the native Nahua-speaking Pipils of the region who, after two major battles, forced the Spaniards to return to their new capital at Iximche, Guatemala. The first attempt at settlement in Pipil territory came the following year. Dispatched from Guatemala under the command of Gonzalo de Alvarado, a small group of Spaniards founded the first *villa* of San Salvador in 1525, probably on the same site as the later 1528 settlement (Barón Castro, 1996:41–42; although Lardé y Larín [2000:80, 102] insisted that the 1525 settlement was "in or near" the Pipil center of Cuscatlan; cf. Escalante Arce, 2001:34). The town was built in a small valley to the north of Cuscatlan Pipil territory that had little or no indigenous settlement at the time of the Conquest (Fowler and Earnest, 1985), but it was still subject to attack. The Pipils rebelled and drove out the Spaniards in 1526 (Barón Castro, 1996:39–44).

Pipil resistance waned by early 1528, however, allowing the Spaniards to return and found a permanent settlement, the second founding of the villa of San Salvador on 1 April 1528 (Barón Castro, 1996:87–91, 197–202). The second foundation involved 73 conquistadors, all of whom became residents of the town (Lardé y Larín, 2000:108–110). All original documents concerning the foundation and growth of the town have vanished, but the Dominican chronicler Antonio de Remesal (1964–66: vol. 2, bk. 9, ch. 3, p. 201), who apparently consulted some of these documents, reported that the residents spent 15 days laying out the streets, plaza, and the church, and in building a few residences. After completing the layout of the town, each Spanish resident was assigned a *solar* or house lot within the town, following Spanish custom.

The ruins of this first permanently settled villa of San Salvador now form the archaeological site known as Ciudad Vieja, located in a rural area 10 km south of Suchitoto, El Salvador (Figures 1,2). Ciudad Vieja is the best preserved Spanish conquest town on the American mainland. Its location has been known for centuries; indeed, one could argue that it was never forgotten (Barón Castro, 1996). But until 1996 no systematic archaeological research had been conducted at the site. The site is totally accessible and completely exposed with very light vegetation cover and no modern occupation to obscure surface features. It has suffered very little damage from agricultural disturbances, and it has not been prone to illicit digging by looters. The site was built on a grid plan with a core area covering 45 ha (Figure 3), virtually all of

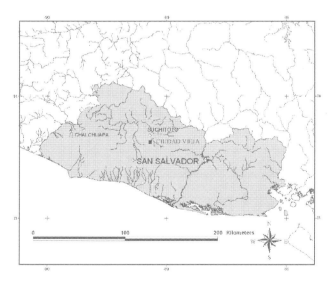

Figure 1. Map showing location of Ciudad Vieja within El Salvador. Map by Francisco Estrada-Belli, based on data distributed by the National Aeronautics and Space Administration (http://servir.nsstc.nasa.gov).

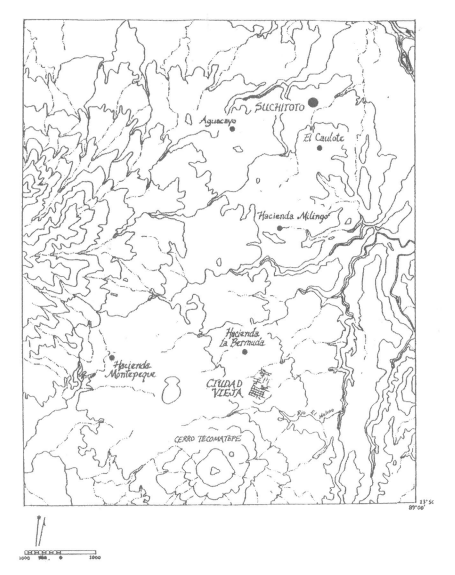

Figure 2. Location of Ciudad Vieja in the valley of La Bermuda, central El Salvador. Map by Rebecca Cutler.

which was artificially levelled and filled with various types of densely packed constructions, thereby making it truly an urban landscape.

Ciudad Vieja is located in central El Salvador at 13° 51' 33.07692" N latitude, 89° 01' 58.30929" W longitude, at an elevation of 534 m above sea level. The town was built on a small *meseta* formed by an extrusive andesite outcrop rising above a small natural basin south of the

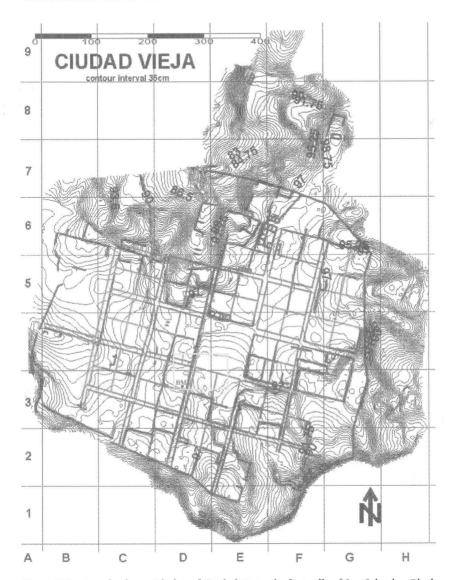

Figure 3. Projected urban grid plan of Ciudad Vieja, the first *villa* of San Salvador. Black lines represent features visible on the surface, blue lines are based on partial visibility, while red are projected. The yellow rectangle indicates the area in Figures 6-9, on the south and southwest sides of the main plaza with the survey benchmark (BM) indicated. Map by Conard Hamilton.

middle reaches of the Lempa River known as the Paraíso Basin (Fowler and Earnest, 1985). Before construction of the town, the top of this *meseta* would have been irregular and craggy with many andesite outcrops and boulders, thus requiring extensive leveling and terracing. The dominant natural features of the surrounding landscape are Cerro Tecomatepe, a small remnant volcanic cone to the southwest, and the extinct Guazapa Volcano to the west. The natural vegetation is tropical deciduous forest of the seasonal formation series (Daugherty, 1969:49; Fowler, 1989:82). Some characteristic tree species of this formation are the *ceiba* (*Ceiba pentandra*), *amate* (*Ficus* spp.), and *conacaste* (*Enterolobium cyclocarpum*). The area was probably very thickly wooded at the time of the Conquest, thus requiring a great deal of clearing for the construction of the town. The labor for clearing and leveling and for construction of the town was provided by Pipil commoners from towns in the Cuscatlan polity. Agricultural tribute commodities from the same communities supplied the town with food.

2. THE URBAN CONQUEST OF AMERICA

Cities and towns played a crucial role in the Spanish enterprise of conquest in America. Since Spanish society of the sixteenth century was essentially urban, it followed that urban places provided the institutional framework for taking possession of the land, for launching further explorations and conquests, and consolidating imperial control over colonized peoples and territories. Cities offered focal points for the development of integrated social, economic, political, and ideological networks essential to the conquistadors and the crown. Thus, early colonial Spanish cities functioned overtly as vehicles of conquest and played a crucial part in the radical transformation of the cultural and physical landscape of Spanish America (Morse, 1962:325).

Conquistadors endeavored to build cities in the New World almost immediately after entering a region, in some cases even before the region's native societies were subdued (San Salvador is a good example), dutifully following the dictum of the chronicler Francisco López de Gómara: "[he] who fails to settle fails to conquer properly" (Morse, 1984:77). The founding of a New World city was both a duty and a privilege for colonists, and the formal act of foundation transformed conquistadors into *vecinos* (residents or householders). The *municipio* held the power to assign land, and thus it functioned as the main instrument of colonization in America as it had done during the reconquest of Spain (Domínguez Compañy, 1984:29; McAlister, 1984:133).

3. THE GRID-PLAN CITY

All early Spanish-American cities were built on a grid plan which varied in form, structure, and internal arrangements from place to place, but the general grid-plan or checkerboard layout was inviolable (Aguilera Rojas, 1994:66;

Brewer-Carías, 1997; Butzer, 1992:554-555; Crouch, Garr, and Mundigo, 1982; Deagan, 1995; García Zarza, 1996; Hardoy, 1975; Kagan, 2000:33; Kubler, 1978; Morris, 1994:302–306; Stanislawski, 1946, 1947; Tejeira-Davis, 1996). As George M. Foster (1960:34) observed, this urban grid plan represents one of the most striking examples of new Spanish-American cultural forms. It is innovative in the Spanish-American context because Spanish medieval cities grew by organic accretion rather than by deliberate, geometric planning. The only predecessors in Spain to the American urban grid-plan layout were all in Andalucía: Chipiona, the Puerto de Santa María, Sanlúcar de Barrameda, Puerto Real, and Santa Fe, the latter a military town established during the siege of Granada (García Zarza, 1996:62–63; Jiménez Martín, 1987:72–76; Morse, 1962:319; Marrinan, 1995:168–169; Muro Orejón, 1950).

As to the origins of this plan, scholars have speculated that it was derived from the ancient Roman city, the medieval Bastide towns of southern France and northern Spain, the revived classicism of the Italian Renaissance, influence from indigenous cities, pragmatic solutions implemented more or less spontaneously, and a form of cultural expression embedded in a matrix of deep traditions (Morse, 1984:68–69). These are not mutually exclusive possibilities; all are worthy of consideration, and they can be reconciled. As Morse (1984:69) put it, "what began as a debate over the genealogy of urban design has evolved into a discussion of larger historical process." Clearly, the layout of early Spanish-American urban centers followed widely held precepts of urban planning that had existed since classical antiquity and which were reformulated in a very complex tradition by Spanish-American settlers (Crouch, 1991; Solano, 1990:36–39). Central among these principles were (1) the idea of the city as emblematic of the imperial will to conquer and dominate, and (2) a reflection of the bureaucratic need for order and symmetry (Morse, 1984:68). The plan of the city, symbolizing civilization itself, embodied a propagandistic statement concerning the power of the empire that translated into very specific notions of spatial patterning (Crouch, Garr, and Mundigo, 1982:xx; Kagan, 2000:34; Ricard, 1950:325; Robinson, 1989). This last point, we believe, is extremely important because it allows us to understand the Spanish-American grid-plan town simultaneously from the emic and etic perspectives. This urban model called for a grid of parallel streets crossing at right angles to form square or rectangular blocks, central plazas, and the siting of the church and municipal government buildings on the plaza (Foster, 1960:34; Markman, 1978:475; Solano, 1990:39–40). In a very real geopolitical sense, the process of conquest emanated outward from the plaza and its surrounding grid (Domínguez Compañy, 1984:30; Palm, 1992).

Concerning proximate sources of the Conquest-period grid-plan city in Mexico and Central America, Rochelle Marrinan (1995:169–170) pointed out that Nicolás de Ovando, appointed governor of Hispaniola in 1501, had been present at the siege of Granada and was familiar with the grid-plan layout of

Santa Fe. After the original settlement of Santo Domingo was destroyed by a hurricane in 1502, Ovando rebuilt the town in the grid-plan layout, and he supervised the construction of a network of 14 other grid-plan cities during the eight years of his governorship. To be sure, the layout of Santo Domingo (see Deagan, 1995a:Figure 13.2; Marrinan, 1995:Figure 7.2), while it was built on a grid, is not rigid; the streets are straight but not parallel, and the polygonal blocks vary in size. Likewise, the main plaza is polygonal and off-center (Tejeira-Davis, 1996:33). Concerning the implementation of this particular plan, although Ovando was acting on instructions from the crown to found new *villas* for Spaniards throughout the island, his choice of the off-center grid-plan layout appears to have been based on his own discretion and his experience in Granada rather than explicit royal orders. Ovando's implementation of the plan did, however, inform the crown's very explicit instructions to Pedrarias Dávila for the colonization of Castilla del Oro (Panama) in 1513 (Morse, 1984:73). Nevertheless, Eduardo Tejeira-Davis (1996:33) reminded us that the crown's instructions to Pedrarias did not explicitly call for a rigid grid plan and that there is no requirement of a strictly orthogonal grid with a central plaza.

Of the half-dozen cities founded by Pedrarias, the only ones known and available for study today are Panamá la Vieja (1519) and Natá (1522). Tejeira-Davis (1996) compared the plans of the two cities and found that the layout of Panamá la Vieja, a port city, was quite irregular and polycentric with the plaza being off-center, the lots long and narrow, and the streets unparallel. He concluded that the layout of Panamá la Vieja was closer to the medieval conception than to the novel Conquest-period grid-plan city (Tejeira-Davis, 1996:43). On the other hand, Natá, an inland site located 15 km up the Río Chico from the Gulf of Parita, near the populous sixteenth-century chiefdom center of the same name, was laid out as a spacious, orthogonal grid with a central plaza, (probably) square lots, and four-lot blocks enclosed by parallel streets (Tejeira-Davis, 1996:45). Tejeira-Davis (1996:52) suggests that Natá was the first example of a large-scale, orthogonal, grid-plan city in Spanish America and that it served as the model for similar plans of later date, echoed by later cities built in Central America and Mexico, especially Santiago de Guatemala in Almolonga (1527), San Salvador (1528), and Oaxaca (1529). In this connection, it is intriguing that the *jumétrico* who laid out Mexico City, Veracruz, and Oaxaca, Alonso García Bravo, first went to the New World with Pedrarias in 1513, later leaving for Mexico in 1520 (Tejeira-Davis, 1996:54). It would not seem at all unlikely that one or more individuals who had worked with García Bravo in Mexico, or who at least were familiar with his plans, directed the layout of both Santiago de Guatemala and San Salvador. Lending support to the suggested "Natá connection" is the fact that the *acta de fundación* of Santiago de Guatemala in Almolonga is very similar to that of Natá (Tejeira-Davis, 1996:56). The *acta de fundación* of San Salvador unfortunately has not survived, but since it was founded by conquistadors from Mexico and Guatemala, sent out directly from Santiago, its layout almost certainly reflected that of Santiago (Domínguez Compañy, 1978:195-196; Sáenz de Santa María, 1991:xxi-xxiii).

4. A LANDSCAPE OF CONQUEST

The concepts of landscape and cultural landscape have gained great currency in archaeology in the past decade or so. A recent synthesis by Anschuetz et al. (2001) is very useful because it is comprehensive, incorporating the perspectives of many other specialists who have also written recent syntheses (for example, Crumley and Marquardt [1990], Fisher and Thurston [1999], Kelso and Most [1990], Knapp and Ashmore [1999], and Yamin and Metheny [1996]), offering original ideas on the history of the concept, and suggesting directions for the further development of the approach. Anschuetz et al. suggest that the power of the "landscape paradigm" lies in its potential to connect patterns of human behavior with particular places and times. They outline the following four interrelated premises that provide the foundations for a landscape paradigm in archaeology. (1) Landscapes are not synonymous with natural environments. (2) Landscapes are worlds of cultural product (not merely the world we see and not the same as built environments). (3) Landscapes are the arena for all of a community's activities. (4) Landscapes are dynamic constructions, with each community and each generation imposing its own cognitive map on an anthropogenic world of interconnected morphology, arrangement, and coherent meaning (Anschuetz et al., 2001:160–161). Note that by these premises, especially the first one, the concept of cultural landscape is conflated with landscape in general: "Landscapes are synthetic, with cultural systems structuring and organizing peoples' interactions with their natural environments. Landscapes mediate between nature and culture" (Anschuetz et al., 2001:160).

In defining precisely what a landscape is, Anschuetz et al. (2001:164) take as a point of departure the following definition by Carl Ortwin Sauer (1925:25).

The cultural landscape is fashioned from a natural landscape by a culture group. Culture is the agent, the natural area is the medium, the cultural landscape is the result. Under the influence of a given culture, itself changing through time, the landscape undergoes development, passing through phases, and probably reaching ultimately the end of its cycle of development. With the introduction of a different—that is, alien—culture, a rejuvenation of the cultural landscape sets in, or a new landscape is superimposed on the remnants of an older one.

While this definition certainly seems relevant to a broad conception of landscape, Sauer's latter point makes his definition especially *a propós* to the present case. The Spanish conquest resulted in the creation of highly structured cultural landscapes in which conquerors and conquered interacted, each bringing their own distinctive cultural attitudes toward the organization and use of space. At Ciudad Vieja, we are attempting to research the making of the cultural landscape of a Conquest-period town through the dynamic interaction of its European and Native American inhabitants. Of vital importance to the Ciudad Vieja research is the idea that cultural landscapes not only reflect the factors that led to their formation, but they are also spatial arenas in which social and physical relations are enacted (Anschuetz et al., 2001:161; Orser, 1996:138; Unwin, 1992:195–196).

5. THE VANDERBILT UNIVERSITY CIUDAD VIEJA PROJECT

As mentioned previously, the Vanderbilt University Ciudad Vieja Archaeological Project began research at the site under Fowler's direction in 1996 with subsequent field seasons of investigation and analysis in 1998, 1999, 2000, 2001, 2002, and 2003. The Vanderbilt project is conducted by permission from the Consejo Nacional para la Cultura y el Arte (CONCULTURA), an agency of El Salvador's Ministry of Education, which owns and manages the site. The Department of Archaeology of CONCULTURA also carried out investigations and consolidation of two structures in 2001, 2002, and 2003, primarily with a view toward developing the site for tourism. In the remainder of this chapter, we offer a brief summary of archaeological research on the Ciudad Vieja site fosusing on mapping and subsurface remote sensing, excavations and architecture, and interpretation of the urban landscape.

5.1. Mapping

The primary goal of the research at Ciudad Vieja has been to document and explore the spatial patterning and community organization of this remarkable urban landscape through precise mapping, subsurface remote sensing, surface collection, and excavation. The great potential for identifying and distinguishing between the indigenous and Spanish material culture components of the town demanded that a highly accurate map of the site be made. We also expect eventually to be able to identify different activity areas and spatial patterning correlated with socioeconomic status differences of the residents. To enable the kind of spatial analysis of the settlement that is required if it is to be studied as a cultural landscape with a view toward internal variation and patterning that might be correlated with ethnic or socioeconomic distinctions, in the 1998 and 1999 field seasons a map of the site was made by Conard Hamilton of Tulane University using a Nikon DTM-420 total station (Hamilton, Gallardo, and Fowler, 2002). Many features visible on the surface such as terrace walls, *solar* boundary walls, and street edges were clearly documented by thousands of data points shot in the survey (Figure 3). These points tend to be very closely spaced, which when plotted result in a representation of linear architectural features. Peripheral portions of the site were added to the map with data gathered in surveys using a Trimble Geoexplorer II GPS unit. The use of this sophisticated mapping technology has enabled us to produce a map of the site that is accurate to within a few centimeters. Such highly accurate and detailed mapping allows us to study the size of each house lot and the size and distribution of structural remains from surface data.

The first methodological step in producing the map was to impose a coordinate grid system over the entire site which is used for designating and identifying surface collection units and excavation units. The archaeological grid was designed independently of the urban grid plan of the town in order to provide maximum control in locating and referencing the natural physical and social divisions within the site. An arbitrary grid of 100 m squares oriented to magnetic north was imposed over the site. Each grid square received a binomial designation with numerals *1* to *8* along the *y* axis and letters *B* to *G* along the *x* axis. Grid squares receive a numeral and letter designation, and structures are numbered sequentially within each square, for example, Structure 6F1.

Concurrent with the mapping in 1998, a 100% systematic surface collection of the entire site was conducted within standardized collection units of 50 m × 50 m (dividing each 100 m × 100 m grid square into quarters). In addition, the surface collection units were treated as quadrats within which randomly placed shovel-test units were excavated in the south zone of the site where structures are scarcely visible on the surface. These test units consisted of 100 squares 50 cm × 50 cm in size, placed within the quadrats on a stratified, systematic, unaligned basis, plus an additional five units 2 m × 2 m. They were excavated to sterile subsoil which is encountered in most places throughout the site within a depth of about 40 to 50 cm from the surface. The systematic surface collection and the shovel testing provided important data on internal variability within the site.

5.2. Excavations and Architecture

Excavations of residential and nonresidential structures and activity loci have been conducted in each of the 1996, 1998, 1999, 2000, 2001, 2002, and 2003 field seasons. A total of 18 structures or activity areas have been excavated. These include a large Spanish residence, several indigenous residences, a kitchen structure, two blacksmith's workshops, a commercial structure, an observation post, and several special-function structures. With very few exceptions, the orientations of the buildings follow the general site grid of 12°. The multicourse stone wall foundations are usually 80–85 cm in width, or approximately one Spanish *vara* of 83 cm, although some foundations were thicker, in the range of 100–120 cm. The foundations run very deep, usually to at least

1 m below the top row of stones. The width and depth of foundations were correlated with the width and height of the walls supported. The stones were generally carefully cut with at least one dressed face, laid carefully to form noticeably straight foundations. Walls were generally constructed of *tapia* or rammed earth, but in some cases adobe bricks were used. Floors were either earthen or covered with *baldosas* (brick floor tiles); occasionally cobblestones arranged in decorative patterns were used. Roofs were thatched or covered with *tejas* (ceramic roof tiles) laid over a wooden framework (Fowler and Gallardo, 2002).

5.3. Subsurface Remote Sensing

Kvamme (2003:438) recently noted that "current geophysical surveys now can map entire villages and surrounding landscapes, allowing examination of interrelationships between such individual site components as houses or house clusters, the lanes between them, dumping grounds, public structures, storage and borrow pits, gardens, plazas, fortifications, and the like." The research reported here represents the initial steps in applying this approach to Ciudad Vieja. From the start of our research there, the site presented itself as an excellent environment for the application of a comprehensive geophysical survey.

Because of its exposure, visibility, and preservation, the site is an ideal environment for total-station mapping and subsurface remote sensing. The site is vegetated in pasture grass and weeds. The latter posed a slight hindrance to the geophysical survey and were burned off midway through the work. The geophysical targets at the site consisted mainly of stone building foundations buried slightly below the ground surface. The foundations were made of the natural andesite stone that is present on the site, a highly magnetic volcanic rock. The stones ranged in size from nearly a meter in diameter to cobble size. The foundations were located in a dry sandy soil matrix. With the benefit of the geophysical data, in some cases, we have even been able to examine the size, dimensions, and floor plans of structures prior to excavations.

Subsurface remote sensing involving magnetic gradiometry, electrical conductivity, and magnetic susceptibility was conducted in November–December 2002 and in March 2003 by Jennifer R. Bales of Ethnoscience, Inc., and Matthew D. Reynolds of the University of Arkansas (Bales, Reynolds, and Kvamme, 2003; Reynolds and Bales, 2003). This phase of the project has been guided by Francisco Estrada-Belli of Vanderbilt University and Kenneth L. Kvamme of the University of Arkansas. During the first test we used the Geometrics G858 cesium gradiometer and the Geonics EM38B electromagnetic induction meter (Figure 4). The former is a type of magnetometer and the latter simultaneously yields measurements of soil conductivity and magnetic susceptibility. Our initial results indicated that the electromagnetic induction meter operating in in-phase, or magnetic-susceptibility mode, provided the

Icanic rocks →

Figure 4. Jenny Bales operating the Geonics EM38B electrical conductivity meter at Ciudad Vieja, view to north. Photograph by Francisco Estrada-Belli.

most interpretable results (Bales, Reynolds, and Kvamme, 2003; Reynolds and Bales, 2003).

Although it was not possible to conduct specific field or laboratory tests on soil and rock magnetism at Ciudad Vieja, we do know that andesite, forming the bedrock and common building stone, possesses high magnetite content and exhibits high magnetic susceptibility (informal field observations with available instruments confirmed this assessment). Of volcanic origin, andesite must also contain a large remanent magnetism, undetectable by electromagnetic induction methods. In volcanic regions with surface rock and rising bedrock, this remanent magnetism can be an impediment to magnetometry surveys, saturating the signal and obscuring more subtle anomalous features. Indeed, the cesium gradiometry survey did not prove to be very effective in this environment. The high magnetite and iron content of the bedrock produced very skewed readings and a very noisy background against which it was difficult to distinguish the architectural building stones of the same locally available volcanic material. In these contexts magnetic susceptibility surveys can be advantageous because only the induced component of the magnetic field is relevant, and even the slightest variation in magnetic susceptibility could be recorded with a very minimal amount of noise by the EM38B. When the survey continued in March 2003, we employed two Geonics EM38/EM38B meters. The conductivity and magnetic susceptibility surveys obtained by these instruments have now covered an area of 5.5 ha, or approximately 12% of the site, with some very intriguing results, especially in the area on the south side of the plaza.

5.3.1. Magnetic Susceptibility

Magnetic susceptibility is a measure of the ability of a material to become magnetized in the presence of an external magnetizing field, such as the earth's geomagnetic field (Telford et al., 1995:73). Features are identified in magnetic-susceptibility data when they possess higher or lower magnetic susceptibility than the non-target, or background, environment. The Geonics EM38/EM38B electromagnetic induction meter is primarily intended to record soil conductivity, but it also has a mode that quantifies magnetic susceptibility. The EM38/EM38B has a coil separation of one meter and operates at a frequency of 14.4 kHz. The rear-mounted transmitter coil transmits a continuous sine wave of electromagnetic energy into the ground. As the energy enters the ground it sets up eddy currents in the subsurface conductors, which in turn transmit weak electromagnetic energy to the receiver coil located at the front of the instrument (Kearey et al., 2002:209). The signal received by the instrument contains three important components. The first is the primary signal transmitted directly to the receiver by the transmitter. This component is made null in the instrument setup procedure. The second component is made up of a wave 90° out of phase with the transmitted signal. This is the quadrature-phase signal, related to the electrical conductivity of the soil. The final component is the in-phase signal, related to the magnetic susceptibility of the soil (Clark 2000:105). It represents the ratio in strength of the induced to transmitted fields, quantified as "parts per thousand" (ppt). Unfortunately, magnetic-susceptibility surveys are capable of only limited prospecting depths because the instrument's active signal is attenuated going into the ground *and* on its return to the receiver, causing sensitivity to fall off at a rate of $1/d^6$ (where d is distance), in contrast to the magnetometer's $1/d^3$ (Clark 2000:102). Prospecting depth for magnetic susceptibility with the EM38B is therefore less than 50 cm, making it suitable only for very shallow investigations, but ideal in the context of Ciudad Vieja.

5.3.2. Field Methods

The magnetic susceptibility survey was conducted in 20 m × 20 m sections aligned with the extant archaeological site-grid system. A permanent concrete survey monument located at N4860 E4944 was used as a benchmark to tie all survey grids back to the site grid, where it is designated BM in grid 3D (Figures 3, 6). Each 20 m × 20 m grid square was surveyed using north-south trending transects. These transects were 1 m apart along the east-west axis of the grid. Forty readings were recorded along each survey transect, spaced 50 cm apart. This resulted in 800 data points per 20 m × 20 m square. By the end of the spring 2003 survey, 137 squares were completed. Two large areas in the northern half of the survey area were skipped. The southern of these is the location of a modern, but razed, house. Brick, metal, and other debris would have made the survey data in this location extremely noisy. The second, northern, area was skipped because it contained a metal storage shed adjacent to an open archaeological excavation. Geophysical survey was physically impossible at the location of the shed and data collected over the open excavation, if interpretable, would have been redundant.

5.3.3. Surface Conditions and Data Quality

A certain amount of unwanted noise is normal in geophysical data. Noise refers to variations in the strength of the geophysical signature that are unrelated to the actual properties of the deposit. Excessive amounts of noise can potentially mask features of interest and can introduce an additional level of difficulty to anomaly interpretation. There are several potential sources of noise in geophysical data. First, there is a certain amount of noise that is inherent to the instrument and cannot be controlled. A second source of unwanted noise is the operator. The person operating the geophysical instrument might produce noise by varying the height and/or orientation of the instrument while moving it over the survey area; it is often attributable to the operator's natural gait. Finally, surface conditions of the survey area, such as thick vegetation, uneven terrain, or obstacles, can also be a source of unwanted noise. These surface conditions produce noise by causing variation in the height and/or orientation of the instrument, particularly for geophysical instruments that must be pulled along the ground surface. An area of 2,400 m^2 in extent was surveyed in the fall of 2002 that was heavily vegetated with thick weeds. This area was resurveyed in the spring of 2003 after the vegetation had been cleared. There is a remarkable difference between the two datasets (Figure 5). The 2003 data provide much better feature definition. Several linear anomalies, which may represent building foundations, toward the south end of the 2003 data are completely invisible in the 2002 data because of noise.

5.3.4. Results

The magnetic-susceptibility survey of Ciudad Vieja resulted in the identification of numerous linear and rectilinear anomalies. Figure 6 shows an unprocessed magnetic-susceptibility map of anomalies of a zone on the south and southwest sides of the main plaza. The top image in Figure 7 depicts the filtered and processed results of the field work in the same zone. The bottom image provides interpretations of the data. Visible in these images are the surface expressions of the *solar* boundaries, which are depicted in magenta. These consisted of surface piles of stone which represented topographic obstacles to the survey. Their signature on the map is both positive and negative. These features contain strong positive signals because the high susceptibility stones were very close to, or even touching, the instrument during the survey. The negative values are the result of having to lift the instrument over stone walls and other topographic obstacles. In these cases the meter was reading only free space, resulting in a zero or sometimes negative reading. The linear anomalies indicated in red represent the potential building foundations or walls that are beneath the surface. The

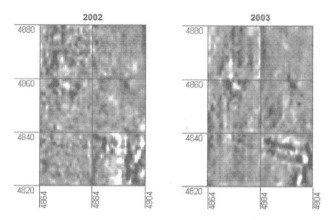

Figure 5. A 2,400m² area that was surveyed in 2002 (left) and again in 2003 (right) after the removal of heavy vegetation. Linear anomalies appear in the 2003 data that are invisible in the 2002 data. Images by Jennifer R. Bales, Matthew D. Reynolds, and Kenneth L. Kvamme.

ground verification

ground-truthing excavations conducted during the 2003 field season, discussed below, indicated that most of these anomalies represent buried stone foundations of structures. The proximity of the anomalies to the *solar* boundaries reinforces

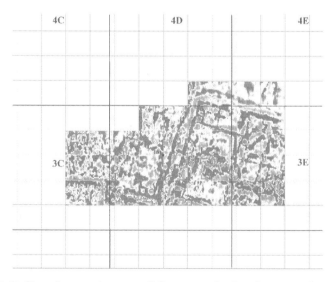

Figure 6. Unfiltered magnetic-susceptibility map of subsurface anomalies on south and southwest side of main plaza, identifying archaeological grid (100 m × 100 m), geophysical survey grid (20 m × 20 m) squares, and benchmark point (BM) in grid 3D. Image by Francisco Estrada-Belli.

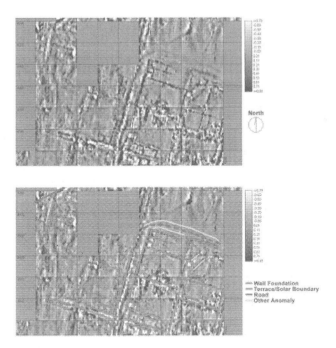

Figure 7. Geophysical results, filtered and processed, in the central portion of the survey area on south and southwest sides of main plaza. Top: Processed magnetic-susceptibility data. Bottom: Magnetic-susceptibility data with superimposed interpretive vectors. Images by Jennifer R. Bales, Matthew D. Reynolds, and Kenneth L. Kvamme.

the interpretation of a perimeter-courtyard arrangement of the structures. In many locations the anomalies are discontinuous. These may be locations where the stone has been removed in post-abandonment agricultural clearing. The anomalies depicted in blue are caused by roads. The solid blue line indicates a modern road. The dashed blue lines are the locations of roads that are associated with the Conquest-period town. Most of these roads are visible on the surface. Yellow is used to indicate anomalies of an unknown source or that are oriented in a different direction than the majority of the features at the site. Two of these anomalies, located near N4900, outline a possible curve in the road that is not visible on the ground surface. A discrete anomaly located at N5035 E5055 is similar in appearance to the blacksmith's furnace, discussed below, located at N4975 E4899, but it is of a lower magnitude. This anomaly has a maximum value of 2 ppt compared with the furnace, which has a maximum value of 6 ppt.

5.3.5. Ground Truthing

The subsurface remote sensing conducted in 2002 indicated numerous linear and rectilinear anomalies representing a street and a number of structures on the south side of the main plaza (Bales, Reynolds, and Kvamme, 2003; Reynolds and Bales, 2003) (Figures 6 and 7). Recognition of these features substantially changed our view and understanding of the plaza, and, more importantly, none of them was visible on the surface. Especially interesting was an apparent structure with three rooms in a north-south alignment forming the interior southwest corner of the plaza and a series of attached rooms to the east.

Excavations directed by Fowler began here in January 2003, leading to the discovery of a blacksmith's shop on the southwest corner of the plaza, an adjacent food-preparation area and charcoal manufacturing area to the east, and a street that enters the southwest corner of the plaza, turns at the corner of the blacksmith's shop, and runs toward the southwest corner of the church platform. This cluster of rooms was designated Structure 3D2 (Figures 8 and 9). The blacksmith's shop on the corner measures approximately 18.5 m north-south and 6.7 m east-west. The furnace was located in the north room. It was associated with a long, narrow tempering tank constructed in the floor, apparently for tempering swords. Adjoining this structure was a large, open room measuring 12.2 m east-west and 7.1 m north-south. Here we found a number of partial and complete cooking and serving vessels. Smaller rooms continue to the east. The magnetic-susceptibility data clearly indicated the presence of wall foundations in this area. Figure 8 depicts the geophysical data and the subsequent excavation results. Figure 9 is a plan view of the structure after excavation. Excavation revealed that the large, high magnetic-susceptibility anomaly in the center of the north room is a black-smith's furnace. Chunks of iron debris were discarded along the west exterior wall of the building, which helped to define the feature in the magnetic susceptibility data. We interpret this cluster of rooms on the south side of the plaza as a market-industrial area producing and selling iron tools and weapons, food, and probably many other items as well that were used in the Conquest.

5.3.6. Intensive Survey

In the spring of 2003 a 400 m² area was selected for a special intensive survey. Designated Structure 3D1, it had been excavated and backfilled in 2001 by Fowler and Jeb Card and was interpreted by Fowler as the location of a commercial building. Textual evidence suggests the presence of *tiendas* or commercial structures "on the plaza" in San Salvador (Altman, 1991:47–51), and we propose that they were located in this area. Located approximately 150 m south of the *cabildo* ("municipal council building") and just off the southwest corner of the plaza, this structure was initially identified on the surface by a very dense concentration of roof-tile fragments. Approximately 50% of the structure was excavated. While the exact dimensions of the full

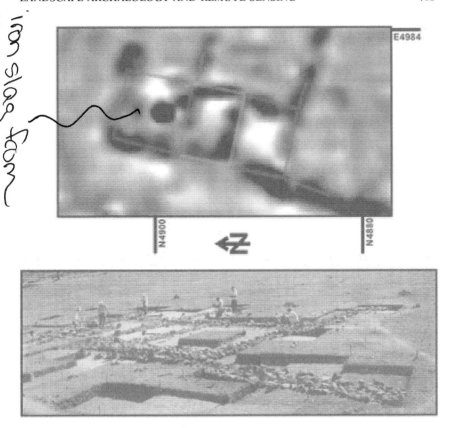

Figure 8. Ground-truthing, southwest corner of plaza, Structure 3D2: A portion of the magnetic-susceptibility data compared with excavation results. Three corresponding rooms are color coded in the geophysical data and the excavation photo for comparison. Compare with Figures 6, 7, and 9. Images by Jennifer R. Bales, Matthew D. Reynolds, and Kenneth L. Kvamme.

structure are unknown at this point, the excavated portion measures approximately 12 m east-west and 20 m north-south.

The 2002 magnetic-susceptibility survey detected the foundation of the former building, but the detail was somewhat lacking. There are two ways in which the level of detail in geophysical data can potentially be increased during data collection. The first method is to increase the sampling interval. Decreasing the distance between measurements will allow smaller features to be detected, increasing the overall detail of the data. The second method is to collect the data using multiple transect directions. Linear features that are oriented in the same direction as the survey transects have the potential to fall in between transects and not be detected at all. Therefore,

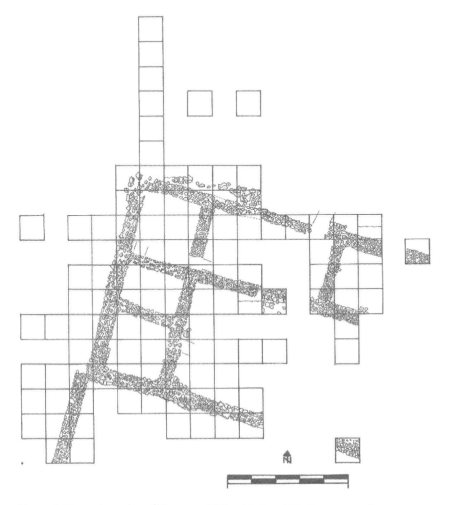

Figure 9. Excavation plan of Structure 3D2, a blacksmith's shop on southwest corner
of Main Plaza, each grid square measures 2 m on a side. Compare with Figures 6, 7,
and 8. Map by Francisco Galdámez.

linear features are best detected when they are perpendicular to the survey
direction. Surveys conducted multiple times using varying transect directions
will produce increased detail because they will detect subtle linear features
that are oriented in a variety of directions. The 2003 intensive survey utilized
a sampling interval of 0.5 × 0.5 m, while in 2002 the area was surveyed at
a sampling interval of 0.5 × 1.0 m. In 2003 the area was surveyed twice,
once with north-south oriented transects and once with east-west oriented
transects. Figure 10a shows the poorly defined data that were collected in 2002
using north/south transects and a sampling interval of 0.5 × 1.0 m. The data

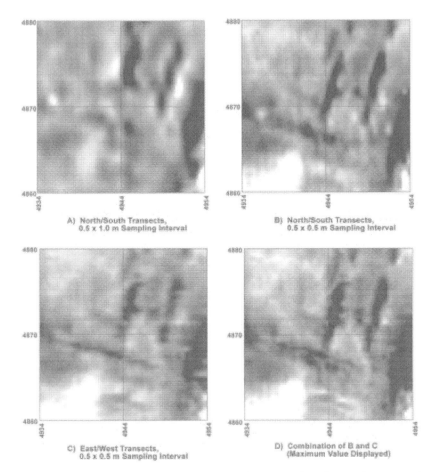

Figure 10. Effects of transect direction and sampling interval on feature clarity. Results of a 400 m² intensive survey over the location of Structure 3D1. a) Poorly defined data collected in 2002; b) Data collected in 2003 using north-south oriented transects and an increased sampling interval; c) Data collected using east-west oriented transects and the increased sampling interval; d) Combination of b and c depicting the maximum values from the two datasets. Images by Jennifer R. Bales, Matthew D. Reynolds, and Kenneth L. Kvamme.

in Figure 10b were collected in the same direction but at a sampling interval of 0.5 × 0.5 m. The increase in detail with this method alone is remarkable. The data in Figure 6c were collected using east-west transects and reveal slightly different information than Figure 6b. By combining the information from data collected with north/south transects and data collected with east-west transects, as in Figure 10d, a highly detailed image of the feature can be obtained.

5.3.7. Discussion and Conclusions

The results of the magnetic susceptibility survey provide a detailed map of the subsurface features in the central portion of the site (Figures 11 and 12). The utility of such a map was demonstrated through ground-truthing excavations

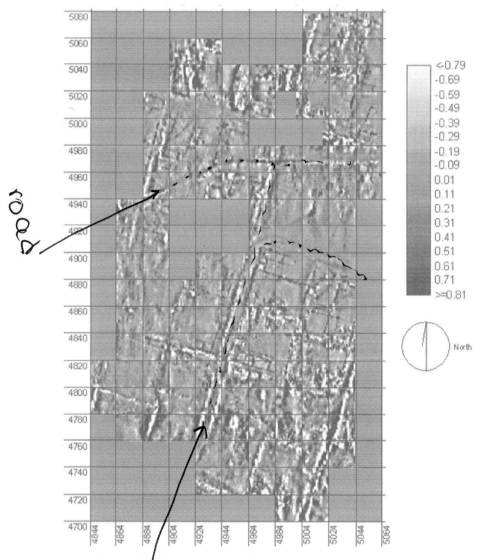

road

main avenue

Figure 11. Results of the 2002 and 2003 magnetic susceptibility surveys at Ciudad Vieja, El Salvador. Map by Jennifer R. Bales, Matthew D. Reynolds, and Kenneth L. Kvamme.

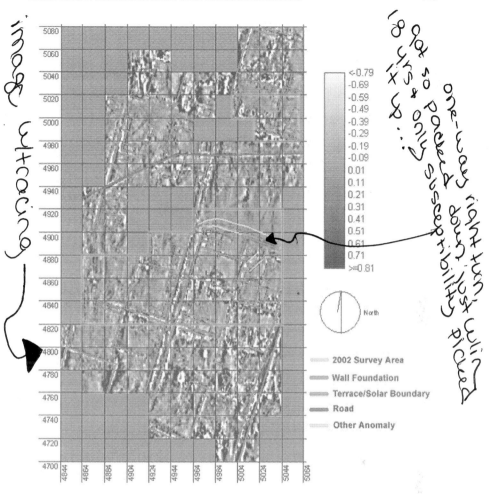

Figure 12. Interpreted results of the 2002 and 2003 magnetic–susceptibility surveys at Ciudad Vieja, El Salvador. Map by Jennifer R. Bales, Matthew D. Reynolds, and Kenneth L. Kvamme.

that were conducted in the spring of 2003. The excavations exposed building foundations that were exactly as they appeared in the geophysical map of subsurface anomalies. In addition, through excavation, a large, high magnetic-susceptibility anomaly in one of the rooms was identified as a blacksmith's furnace. The intensive survey conducted over Structure 3D1 indicates that a greater level of detail can be obtained by increasing the sampling interval during data collection. Although this level of detail may be ideal, it comes with increased time and cost invested in data collection. The goals of the survey must be considered when deciding on geophysical data-collection methods. Although the sampling interval used for the bulk of the survey does not provide the best feature definition, it allows for rapid coverage of the site at a level of detail that does not sacrifice the utility of the data. The best possible feature definition, no matter what the sampling interval, can be ensured by taking steps to avoid excessive amounts of noise in the data. Data from the portion of the site that was surveyed in 2002 and then resurveyed in 2003 after the removal of heavy vegetation attests to the difference this type of preparation can make.

Complete geophysical coverage of the site of Ciudad Vieja would provide valuable new information about the landscape of an entire Spanish American conquest town. This information simply could not be obtained without this type of technology because it would be prohibitively expensive and time consuming, and also very damaging to the site itself. The dataset discussed here, 5.5 ha in area (12% of the defined area of the site), acquired over the course of just eleven days in the field (Figure 13), is already large enough to begin studying the distribution of various building forms and the function of different areas within the site without invasive and costly excavation. In general, with few exceptions, the orientation, size, and shape of the subsurface anomalies conform to expectations of *solar* and structure layout and distribution derived from our knowledge of the Ciudad Vieja landscape based on mapping and excavations. Additional, future ground-truthing excavations will continue to aid in the interpretation of the geophysical data by revealing the source of various types of anomalies that are common to the landscape of Spanish-conquest towns.

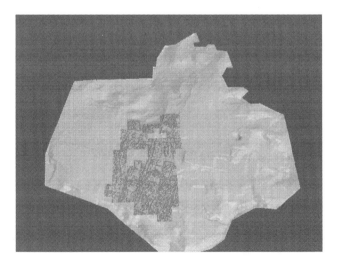

Figure 13. Three-dimensional model, bird's-eye view of Ciudad Vieja, from south to north, based on total-station survey (see Figure 3), with geophysical results superimposed. Image by Francisco Estrada-Belli.

REFERENCES

Aguilera Rojas, Javier, 1994, *Fundación de ciudades hispanoamericanas*. Editorial MAPFRE, Madrid.

Altman, Ida, 1991, A New World in the Old: Local Society and Spanish Emigration to the Indies. In *"To Make America": European Emigration in the Early Modern Period*, edited by Ida Altman and James Horn, pp. 30–58. University of California Press, Berkeley and Los Angeles.

Anschuetz, Kurt F., Wilshusen, Richard H., and Scheick, Cherie L., 2001, An Archaeology of Landscapes: Perspectives and Directions. *Journal of Archaeological Research* 9:157–211.

Ashmore, Wendy, and Knapp, A. Bernard, eds.,1999, *Archaeologies of Landscape: Contemporary Perspectives*. Blackwell Publishers, Oxford.

Bales, Jenny, Reynolds, Matthew D., and Kvamme, Kenneth L., 2002, Final Report of Geophysical Investigations Conducted at Ciudad Vieja, El Salvador. Unpublished manuscript on file, Department of Anthropology, Vanderbilt University.

Barón Castro, Rodolfo,1996, *Reseña histórica de la villa de San Salvador desde su fundación en 1525, hasta que recibe el título de ciudad en 1546*, 2nd ed. Consejo Nacional para la Cultura y el Arte, San Salvador.

Brewer-Carías, Alan R., 1997, *La ciudad ordenada*. Instituto Pascual Madoz, Universidad Carlos III de Madrid, Madrid.

Butzer, Karl, 1991, Spanish Colonization of the New World: Cultural Continuity and Change in the New World. *Erdkunde* 45:204–219.

Clark, Anthony, 2000, *Seeing Beneath the Soil: Prospecting Methods in Archaeology*. Reprinted. Routledge, London. Originally published 1990, B.T. Batsford Ltd., London.

Crouch, Dora P.,1991, Roman Models for Spanish Colonization. In *Columbian Consequences*, vol. 3: *The Spanish Borderlands in Pan-American Perspective*, edited by David Hurst Thomas, pp. 21–35. Smithsonian Institution Press, Washington, D.C.

Crouch, Dora P., Garr, Daniel J., and Mundigo, Axel I., 1982, *Spanish City Planning in North America*. MIT Press, Cambridge.

Crumley, Carole L., and Marquardt, William H., 1990, Landscape: A Unifying Concept in Regional Analysis. In *Interpreting Space: GIS and Archaeology*, edited by Kathleen M. S. Allen, Stanton W. Green, and Ezra B. W. Zubrow, pp. 73–79. Taylor & Francis, London.

Daugherty, Howard E., 1969, *Man-Induced Ecologic Change in El Salvador*. Ph.D. dissertation, University of California, Los Angeles. University Microfilms, Ann Arbor.

Deagan, Kathleen, 1995, After Columbus: The Sixteenth-Century Spanish-Caribbean Frontier. In *Puerto Real: The Archaeology of a Sixteenth-Century Spanish Town in Hispaniola*, edited by Kathleen Deagan, pp. 419–456. University Press of Florida, Gainesville.

Dominguez Compañy, Francisco, 1978, *La vida en las pequeñas ciudades hispanoamericanas de la conquista*. Ediciones de Cultura Hispanica del Centro Iberoamericano de Cooperación, Madrid.

Dominguez Compañy, Francisco, 1984, *Política de poblamiento de España en América: La fundación de ciudades*. Instituto de Estudios de Administración Local, Madrid.

Escalante Arce, Pedro Antonio, 2001, *Los tlaxcaltecas en Centro América*. Consejo Nacional para la Cultura y el Arte, San Salvador.

Fisher, Christopher T., and Thurston, Tina L., eds., 1999, Special Section: Dynamic Landscapes and Socio-Political Process: The Topography of Anthropogenic Environments in Global Perspective. *Antiquity* 73:630–631.

Foster, George M.,1960, *Culture and Conquest: America's Spanish Heritage*. Viking Fund Publications in Anthropology, No. 27. Wenner-Gren Foundation for Anthropological Reserach, New York.

Fowler, William R., Jr., 1989, *The Cultural Evolution of Ancient Nahua Civilizations: The Pipil-Nicarao of Central America*. University of Oklahoma Press, Norman.

Fowler, William R., Jr., and Earnest, Jr., Howard H., 1985, Settlement Patterns and Prehistory of the Paraíso Basin of El Salvador. *Journal of Field Archaeology* 12:19–32.

Fowler, William R., Jr., and Gallardo, Roberto, eds., 2002 *Investigaciones arqueológicas en Ciudad Vieja, El Salvador: La primigenia villa de San Salvador*. CONCULTURA, San Salvador.

García Zarza, Eugenio, 1996, *La ciudad en cuadrícula o hispanoamericana: Origen, evolución y situación actual*. Instituto de Estudios de Iberoamérica y Portugal, Universidad de Salamanca, Salamanca.

Hamilton, Conard C., Fowler, Jr., William R., and Gallardo, Roberto, 2002, El levantamiento topográfico de Ciudad Vieja. Introducción. In *Investigaciones arqueológicas en Ciudad Vieja, El Salvador: La primigenia villa de San Salvador*, edited by William R. Fowler, Jr. and Roberto Gallardo, pp. 33–37. Consejo Nacional para la Cultura y el Arte, San Salvador.

Hardoy, Jorge, 1975, La forma de las ciudades coloniales en la América española. In *Estudios sobre la ciudad iberoamericana*, edited by Francisco Solano, pp. 315–344. Consejo Superior de Investigaciones Científicas, Madrid.

Hassig, Ross, 1994, *Mexico and the Spanish Conquest*. Longman, London and New York.

Jiménez Martín, Alfonso, 1987, Antecedentes: España hasta 1492. In *Historia urbana de Iberoamérica*, vol. 1: *La ciudad iberoamericana hasta 1573*, edited by Francisco de Solano and María Luisa Cerrillos, pp. 23–79. Consejo Superior de los Colegios de Arquitectos de España, Madrid.

Kagan, Richard L., 2000, *Urban Images of the Hispanic World, 1493–1793*. Yale University Press, New Haven.

Kearey, Philip, Brooks, Michael, and Hill, Ian, 2002, *An Introduction to Geophysical Exploration*, 3rd ed. Blackwell, Oxford.

Kelso, William M., and Most, Rachel, eds., 1990, *Earth Patterns: Essays in Landscape Archaeology*. University Press of Virginia, Charlottesville.

Knapp, A. Bernard, and Ashmore, Wendy, 1999, Archaeological Landscapes: Constructed, Conceptualized, Ideational. In *Archaeologies of Landscape: Contemporary Perspectives*, edited by Wendy Ashmore and A. Bernard Knapp, pp. 1–30. Blackwell Publishers, Oxford.

Kubler, George, 1978, Open-Grid Town Plans in Europe and America. In *Urbanization in the Americas from Its Beginnings to the Present*, edited by Richard P. Schaedel, Jorge E. Hardoy, and Nora Scott Kinzer, pp. 327–341. Mouton, The Hague.

Kvamme, Kenneth L., 2003, Geophysical Surveys as Landscape Archaeology. *American Antiquity* 68:435–457.

Lardé y Larín, Jorge, 2000, *El Salvador: Descubrimiento, conquista y colonización*, 2nd ed. Consejo Nacional para la Cultura y el Arte, San Salvador.

Markman, Sidney David, 1978, The Gridiron Town Plan and the Caste System in Colonial Central America. In *Urbanization in the Americas from Its Beginnings to the Present*, edited by Richard P. Schaedel, Jorge E. Hardoy, and Nora Scott Kinzer, pp. 471–489. Mouton, The Hague.

Marrinan, Rochelle A., 1995, Archaeology in Puerto Real's Public Sector: Building B. In *Puerto Real: The Archaeology of a Sixteenth-Century Spanish Town in Hispaniola*, edited by Kathleen Deagan, pp. 167–194. University Press of Florida, Gainesville.

McAlister, Lyle N., 1984, *Spain and Portugal in the New World, 1492–1700*. University of Minnesota Press, Minneapolis.

Morris, A. E. J., 1994, *History of Urban Form Before the Industrial Revolutions*, 3rd ed. Longman Scientific & Technical, Essex.

Morse, Richard M., 1962, Some Characteristics of Latin American Urban History. *American Historical Review* 67:317–338.

Morse, Richard M., 1984, The Urban Development of Colonial Spanish America. In *The Cambridge History of Latin America*, vol. 2: *Colonial Latin America*, edited by Leslie Bethell, pp. 67–104. Cambridge University Press, Cambridge.

Muro Orejón, Antonio, 1950, La villa de Puerto Real, fundación de los Reyes Católicos. *Anuario de Historia del Derecho Español* 20:753.

Orser, Charles E., Jr., 1996, *A Historical Archaeology of the Modern World*. Plenum Press, New York.

Palm, Erwin Walter, 1992, La ville espagnole au Nouveau Monde dans la première moitié du XVI siècle. In *Heimkehr ins Exil: Schriften zu Literatur und Kunst*, edited by Helga von Kügelgen and Arnold Rothe, pp.129–141. Böhlau Verlag, Cologne and Vienna.

Remesal, Antonio de, 1964–1966, *Historia general de las Indias Occidentales y particular de la gobernación de Chiapa y Guatemala*, 2 vols.. Biblioteca de Autores Españoles, vols. 175 and 186. Ediciones Atlas, Madrid.

Reynolds, Matthew D., and Bales, Jennifer R., 2003, Magnetic Susceptibility Survey at Ciudad Vieja, El Salvador. Unpublished manuscript on file, Department of Anthropology, Vanderbilt University.

Ricard, Robert, 1950, La plaza mayor en España y en América española. *Estudios Geográficos* 11:321–327. Madrid.

Robinson, David, 1989, The Language and Significance of Place in Latin America. In *The Power of Place: Bringing Together Geographical and Sociological Imaginations*, edited by John A. Agnew and James S. Duncan, pp. 157–184. Unwin Hyman, Boston.

Sáenz de Santa María, Carmelo, ed., 1991, *Libro viejo de la fundación de Guatemala*, edición crítica de Carmelo Sáenz de Santa María; confrontación de la paleografía María del Carmen Deola de Girón. Academia de Geografía e Historia de Guatemala, Guatemala City.

Salcedo, Jaime, 1990, El modelo urbano aplicado a la América española: Su génesis y desarrollo teórico-práctico. In *Estudios sobre urbanismo iberoamericano*, edited by Ramón Gutiérrez, pp. 9–85. Seville.

Sauer, Carl Ortwin, 1925, The Morphology of Landscape. *Publications in Geography*, vol. 2, no. 2, pp. 19–53. University of California, Berkeley.

Solano, Francisco de,1990, *Ciudades hispanoamericanas y pueblos de indios*. Consejo Superior de Investigaciones Científicas, Madrid.

Stanislawski, Dan, 1946, The Origin and Spread of the Grid-Pattern Town. *Geographical Review* 36:105–120.

Stanislawski, Dan, 1947, Early Spanish Town Planning in the New World. *Geographical Review* 37:94–105.

Tejeira-Davis, Eduardo, 1996, Pedrarias Dávila and His Cities in Panama, 1513–1522: New Facts on Early Spanish Settlements in America. *Jahrbuch für Geschichte von Staat Wirtschaft und Gesellschaft Lateinamerikas* 33:27–61.

Telford, William M., Geldart, Lloyd P., and Sheriff, Robert E., 1995, *Applied Geophysics*, 2nd ed.. Cambridge University Press, Cambridge.

Unwin, Tim, 1992, *The Place of Geography*. Longman, London.

Yamin, Rebecca, and Metheny, Karen Bescherer, 1996, *Landscape Archaeology: Reading and Interpreting the American Historical Landscape*. University of Tennessee Press, Knoxville.

Chapter 17

Urban Structure at Tiwanaku: Geophysical Investigations in the Andean Altiplano

Patrick Ryan Williams, Nicole Couture, and Deborah Blom

Abstract: Fieldwork at Tiwanaku sites in recent decades has opened new avenues of inquiry into the characteristics of the Tiwanaku polity, but archaeology has been hampered by the inability to make large-scale characterizations of the organization of urban and monumental space because of the depositional processes that interred its cities and monuments. In this paper, we use several geophysical techniques, including magnetometry, electrical resistivity, and ground-penetrating radar to begin to visualize some of these spaces. In doing so, we outline a methodology for recreating the urban spatial structure of the six-square-kilometer Tiwanaku capital. Our preliminary work compares excavations at the Putuni, Mollo Kontu, and Akapana East sectors with geophysical data that has expanded our understanding of both residential and monumental spaces to provide a first glimpse at the potential of imaging Tiwanaku's buried urban structure.

1. INTRODUCTION

As one of the longest-lived and more extensive pre-Inka polities, Tiwanaku flourished in the south-central Andes from approximately A.D. 600 to 1000 (Figure 1). Situated near the southern shore of Lake Titicaca in the highlands of modern-day Bolivia, Tiwanaku had a sphere of power and influence that

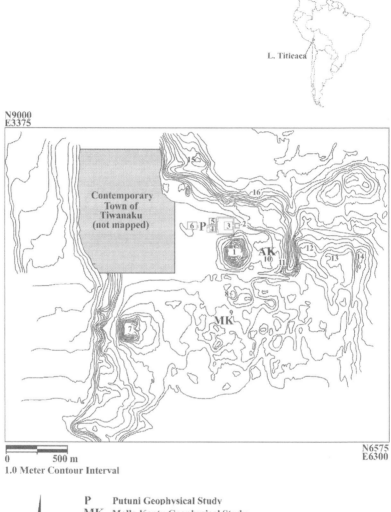

P   Putuni Geophysical Study
MK   Mollo Kontu Geophysical Study
AK   Akapana East Geophysical Study

1 Akapana	9 Mollo Kontu South
2 Semi-subterranean Temple	10 Akapana East 1 Mound
3 Kalasasaya	11 Akapana East 1
4 Putuni	12 Akapana East 2
5 Chunchukala	13 Marcapata
6 Kheri Kala	14 Ch'iji Jawira
7 Pumapunku	15 Kk'araña
8 Mollo Kontu Mound	16 La K'araña

Figure 1. Map locating principal sectors of Tiwanaku. Large-format letters indicate locations of geophysical survey. Map from Kolata, 2003a.

extended to the coastal regions of northern Chile and southern Peru, and to the tropical lowlands of eastern Bolivia. In recent years, extensive ethnohistorical and archaeological research has resulted in several competing interpretations to explain Tiwanaku sociopolitical complexity. Our research builds on these competing ideas by addressing patterns of social differentiation and urban identity through an analysis of architectural and social space, focusing on both elite and non-elite residential complexes and monumental areas. With data from several short field seasons at three different areas of Tiwanaku, we establish the potential for geophysical survey to elucidate broad- and fine-scale urban and residential patterns at the site.

These preliminary results of the geophysical survey, in turn, provide a critical database from which to plan the logistics of major research projects at these areas in the future, including large-scale excavations at Mollo Kontu. For example, the collection of geophysical data will enhance our understanding of subsurface remains, including house foundations, boundary walls, drainage canals, wells, and tombs. Since most of the ancient architectural remains at Tiwanaku are not visible on the surface, geophysical prospection is one of the only means of obtaining information about the large-scale patterns of spatial organization at the site. A geophysical database provides a powerful resource for targeting optimal excavation areas, identifying architectural features and activity areas, and making informed decisions regarding sampling strategies.

Most archaeologists now agree that the Tiwanaku polity was composed of multiple social and/or ethnic groups. Debate, however, is still underway as to the nature of this polity, be it a centralized state (Kolata, 1993, 2003a; Goldstein, 1989, 1993; Ponce Sanginés, 1981; Stanish, 2003) or a loose federation organized along segmentary lines (Albarracín Jordan, 1996). Others argue that while there did exist a broadly shared "state culture," it co-existed alongside expressions of local identity in Tiwanaku regional centers or at more distantly located Tiwanaku-related sites (Bermann, 1994; Janusek, 2002, 2003; Seddon, 1998). While these studies have contributed to a richer and more balanced understanding of the overall Tiwanaku polity, many of them fail to take into consideration the social diversity and internal complexity of Tiwanaku's urban capital (Blom, 1999; Couture, 2002; Janusek, 2003). That is, hampered by a paucity of available comparative datasets from the center, many analyses have relied upon a vision of a homogeneous, if not monolithic, population residing at Tiwanaku itself. We argue, therefore, that issues of ethnic diversity, resistance, and the articulation of state power—particularly when making core-periphery comparisons—can only be properly addressed if researchers take into account the social diversity associated with Tiwanaku's complex urban landscape.

2. URBAN ARCHAEOLOGY AT TIWANAKU

Until the late 1980s, our understanding of Tiwanaku's urban history was based primarily on investigations of the site's most prominent ceremonial structures (i.e., Kalasasaya, Semi-subterranean Temple, and Akapana), located within the site's central monumental district (Figure 1; see Ponce Sanginés, 1981,2003). This emphasis led a number of scholars to conclude that Tiwanaku was an impressive, though circumscribed religious or pilgrimage center with a small permanent population (Bennett, 1934; Menzel, 1964; Sehaedel, 1988). Given that Tiwanaku's residential architecture was built almost entirely of adobe whose eroded remains leave few readily detectable traces on the surface, the characterization of the site as an "empty ceremonial center" is not entirely without foundation if one cannot visualize the built space beneath the surface (Figure 2).

More recent research at Tiwanaku by the joint North American/Bolivian *Proyecto Wila Jawira*,[1] including a pedestrian survey, systematic surface collection, and extensive excavations, has provided conclusive evidence that the site extended well beyond the central monumental district and maintained a large, socially and economically diverse urban population. Current data indicate that at its peak, in the Late Tiwanaku IV (ca. A.D. 600–800) and Tiwanaku V (ca. A.D. 800–1150) Periods, the site measured an estimated 6 square kilometers and housed a projected urban population of 15,000–20,000 (Figure 1; Kolata, 2003a:15).[2]

Figure 2. View to north of two monumental structures in Tiwanaku's central civic-ceremonial district: Kalasasaya Platform (left) and the Semi-Subterranean Temple (right). Both were constructed of finely worked sandstone blocks. Remains of residential structures elsewhere at the site are not visible on the surface today.

---

[1] The *Proyecto Wila Jawira* was directed by Dr. Alan Kolata from the University of Chicago and Lic. Oswaldo Rivera Sundt, then director of the Bolivian Instituto Nacional de Cultura. Results of the *Wila Jawira* research have recently been published in an edited volume by Smithsonian Press (Kolata, 2003b).

[2] Kolata was not the first to argue for the presence of a large residential population at Tiwanaku. Based on a two-day surface survey of the site in 1966, Parsons (1968:243) estimated that Tiwanaku covered an area of 2.4 square kilometers and housed a population between 5,200 and 10,500, and possibly as high as 20,000. Somewhat later, Ponce Sanginés (1981:62) concluded that Tiwanaku measured approximately 4.2 square kilometers and maintained an urban population of 46,800.

Proyecto Wila Jawira excavations in eight different residential sectors at the site, located both inside and outside of the central district, have shown that Tiwanaku maintained a dense urban population residing in well-defined, spatially segregated neighborhoods, or *barrios*, bounded by massive adobe compound walls. These walls were typically set on cobblestone and field stone foundations measuring between 0.50 m and 1 m in width (Couture, 2003; Janusek, 1994, 2003). Excavations in different parts of the site have shown that the stone foundations are often well preserved, and in a few cases may be found associated with a matrix of eroded adobe (Figure 3). In addition, all of the compound walls are oriented to the cardinal directions, indicating that Tiwanaku's residential spaces were established according to a well conceived and centralized urban plan (Couture, 2002, 2003; Couture and Sampeck, 2003; Kolata, 1993; Janusek, 1994, 2003). Internally, residential neighborhoods were characterized by multiple clusters of domestic structures (e.g., kitchens, sleeping quarters, storage facilities), some of which were apparently organized around a small private patio. Though no neighborhood has been mapped or excavated in its entirety, it appears that the inhabitants of these house clusters may also have had access to larger, shared outdoor plaza areas utilized for communal ceremonial events (Janusek, 1994, 2003). Our future research will integrate the first detailed community-scale maps, focusing on the internal organization of these neighborhoods, in relation to the broader Tiwanaku site.

Drawing on ethnographic and ethnohistoric descriptions of residential and urban spaces in Colonial period and contemporary Andean communities, Janusek (2002, 2003) has argued that each compound at Tiwanaku served as the residential focus of multiple household units related by kinship and

Figure 3. Excavated architectural features of the Early Tiwanaku V Period in the Mollo Kontu South residential area; view to south. The cobblestone foundation on the right (ca. 0.75 m wide) once served as the base for a major adobe compound wall. To the left are the stone foundations and flagstone entrance of an adobe residence. In the lower left corner of the photo are remains of an earlier compound wall, also oriented north-south, that runs directly below the habitation structure.

ritual ties, that is, along the lines of lineages or *ayllus* (Abercrombie, 1986; Platt, 1982, 1987; Rasnake, 1988). Furthermore, based on partial excavations of residential neighborhoods at both Tiwanaku and Lukurmata (a major Tiwanaku administrative and ceremonial site in the neighboring Katari valley), he notes that some compounds were also associated with occupational specialization (Janusek, 1994, 2002). Finally, significant differences in the distribution of non-local ceramics and imported food crops (principally maize, which comes from the lowlands) across the site, has led Janusek (2002, 2003) and Rivera (2003) to suggest that residents of specific neighborhoods maintained strong ties with ethnically distinct groups from distant regions of the Tiwanaku realm.

As noted above, while several neighborhoods, or *barrios*, located in different areas of the site have been partially excavated (Couture, 2002; Couture, 2003; Couture and Sampeck, 2003; Escalante, 1992, 2003; Janusek, 1994, 2003; Rivera, 1994, 2003), to date no residential complex has been excavated in its entirety. As a consequence, we do not know the size of a "typical" Tiwanaku neighborhood, nor do we know how different household clusters within a single compound relate to each other socially and architecturally. In addition, we lack concrete data on the spatial articulations between compounds. The orientation of *barrio* walls to the cardinal directions indicates that residential space was built according to an overarching grid plan. However, the extent to which social space within individual neighborhoods also conformed to a generalized architectural template is yet to be determined. The urban homogeneity imposed by the grid plan may in fact be masking important differences in the organization of domestic space and patterns of economic and social diversity.

3. GEOPHYSICAL RESEARCH AT TIWANAKU

Geophysical prospection enhances our understanding of the subsurface remains at Tiwanaku sites, including house foundations, boundary walls, drainage canals, wells, and tombs. Since most of the ancient architectural remains are not visible on the surface, geophysical prospection is an important means of obtaining information about the large-scale patterns of spatial organization at the site. Based on the magnetic and electrical properties of the archaeological features under consideration and their contrasts with the surrounding soils, we obtain distributional information on the spatial organization of the last architectural phase at the ancient capital. Geophysical investigations designed to extract information on subsurface structures involve a multiple sensor approach (Weymouth and Huggins, 1985) that includes ground-penetrating radar, electrical resistivity, and gradient magnetometry, all of which have been tested at Tiwanaku with promising results in recent years (Williams et al., 2001, 2003).

3.1. The Putuni Sector at Tiwanaku

Some of our earliest geophysical work in the summer of 2000 in the Tiwanaku core took place in the Putuni sector, which had just been excavated extensively by Nicole Couture in 1999–2000 (Figure 1). The Putuni is a large elite residential and ceremonial complex located in the western part of the site's central monumental district. The easternmost, public side of the complex, known as the Putuni Platform (Figure 4), was excavated almost in its entirety in the early 1970s (Ponce Sanginés, 1981). More recent research at Putuni has focused on the more private, residential component of the complex known as the "Putuni Palace," which is composed of a series of elite residential structures organized around an open patio (Couture, 2002; Couture and Sampeck, 2003). We conducted a GPR survey and a magnetometer survey over Couture's recently backfilled excavations in a section of one of these residential structures (the "West Palace") to test the accuracy of geophysical equipment in defining buried architectural features (Figure 5). While a backfilled area will not respond exactly the same as an area buried for many years to geophysical devices, especially GPR, the differences may not be that substantial. Returns in a backfilled area can be caused by packing and the discontinuity in the backfill to intact earth interface. While ideally one would conduct a geophysical survey, then dig, the reverse methodology provided an adequate database to assess the versatility of the instruments and to follow features found in the excavations beyond the boundaries of the dig.

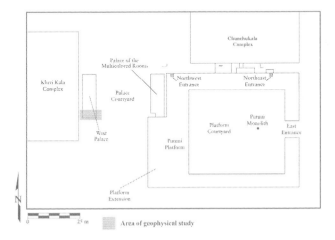

Figure 4. Simplified ground plan of the Putuni complex showing the location of excavations and geophysical studies in the West Palace structure (after Couture and Sampeck, 2003).

Figure 5. The Putuni West Palace during 1999 excavations; view to west. In the foreground are the F63 stone canal (left); sections of the F37 cobblestone wall foundation (center); and the F39 stone canal with capstones still in place (right). Part of the massive F173 lintel can be seen in the center of the photograph, where it lies on top of the remains of the F37 cobblestone foundations.

3.1.1. Ground-Penetrating Radar (GPR)

A GPR pilot study at Tiwanaku using a GSSI SIR-2000 ground-penetrating radar with a 400 MHz antenna indicated a field-estimated dielectric constant for the soil A Horizon of 17 (Williams et al., 2002). While this is rather high for depth penetration, work along open profiles was able to identify the presence of a buried wall to a depth of almost 1.5 m. Based on previous excavations of residential architecture in neighborhoods outside of Tiwanaku's central monumental district, including Mollo Kontu South (Couture, 1993, 2003) and Akapana East (Janusek, 1994), most subsurface remains from the Late Tiwanaku IV and Tiwanaku V Periods are found at a depth of 0.40 to 1.0 m below the modern surface. Therefore signal attenuation is only an issue for more deeply buried structures or those in areas where the water table is especially high. Resolution is also fairly good, with an estimated footprint radius of 30 cm at a depth of 1 m in the Tiwanaku soil medium. Radar work has indicated that the dielectric contrast between the natural soil matrix and stones used as construction material at Tiwanaku provides adequate radar reflections to be easily identifiable.

Radar data are ideally collected in transects oriented E-W and N-S in order to provide cross-polarized returns. Data are collected in a continuous manner at a velocity of 1 m/sec with a recording rate of 30 pulses/second (or one pulse every 3 cm). Given the approximate 30 cm radius footprint for a 400–500-MHz antenna, radar transects are ideally conducted every 50 cm (25 cm radius on either side) in order to provide optimal survey at the target depth of 1 m. Data can be processed in the field, taking advantage of amplitude slice maps (Conyers and Goodman, 1997) to assist in interpretation. These slice maps can also serve to integrate the radar data as a layer in the site-wide GIS. In areas where the water table is close to the surface, radar signals are further attenuated, making subsurface features difficult to detect. Therefore, we employ complementary geophysical techniques, including electrical resistivity and magnetometry, to compensate.

3.1.2. GPR Results at Putuni

The individual cobbles in the F37 cobblestone wall foundations appear quite clearly at the 10 and 14 nanosecond time-slices (ca. 30 and 50 cm below surface) as does the F39 canal line, as shown in Figure 6. The same 14 ns time-slice also corresponds to some features appearing in the southern part of the excavations corresponding to Tiwanaku V constructions, particularly the F63 canal. Other features are also corroborated in other time-slices and using different filtering techniques on the complex data. In all cases, however, the data are extremely complex. The features of interest are not homogeneous structures that lend themselves easily to interpretation. Locating a few cobbles aligned as part of a wall foundation is extremely tricky business without ground truthing. Thus, for this level of high-resolution survey in complex archaeological strata, the GPR works best interpolating between excavated (ie ground truthed) areas.

3.1.3. Magnetic Gradiometry

Magnetometers measure the strength of the magnetic field over a point on the Earth's surface. This field contains many components including underlying geology, surface materials, and natural diurnal variations and in the Titicaca Basin has one of the lowest mean values on the planet – approximately 26000 nT. To remove the effect of the above components and enhance highly local sources, two separate sensors are aligned vertically with a known separation to provide a measure of gradient measured in nT/m (nano-Tesla / meter).

   These instruments detect two forms of magnetism – that induced in a material by the Earth's own magnetic field and proportional to the magnetic susceptibility of the material, and remanent magnetism. In an archaeological context the latter is frequently thermoremanent magnetism (TRM), caused by the heating of certain materials in a suitable environment so that they become weak magnets in their own right (Breiner, 1973). At Tiwanaku, the andesite building materials (which are imported from outside the site) have a high TRM as a result of their volcanic origins and provide a strong contrast to the relatively weak induced magnetism of the native soil matrix.

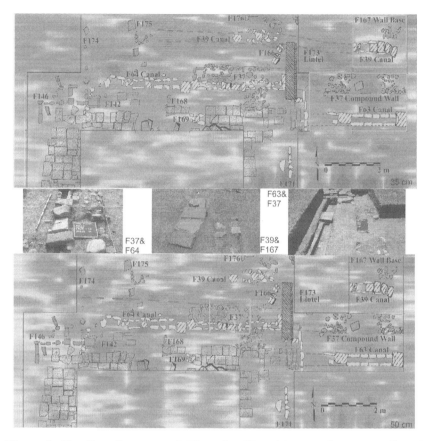

Figure 6. The Putuni's excavated Tiwanaku IV architecture (upper panel) and Tiwanaku V architecture (lower panel), each overlayed on a GPR slice at 14 nanoseconds. Green boxes highlight features detected by GPR. Photographs of the same features are displayed on the central panel; all views are to the west. From left to right the photos show 1) F37 compound wall and F64 canal (corresponding to the green box from the lower panel of GPR data), 2) F39 canal and F167 wall base (corresponding to the northernmost green box from the upper panel of GPR data), and 3) F63 canal and F37 compound wall (corresponding to the other green box in the upper panel of GPR data). Architectural map fromCouture and Sampeck, 2003.

Pilot gradiometer surveys carried out during three days in 2000 (Williams et al., 2001) with a Geometrics G-858G gradiometer belonging to Boston University revealed that measured variations in natural soil magnetism in areas bereft of archaeological remains at the site of Tiwanaku average less than 2 nT/m at 1 m above ground. Measured variations of anomalies of likely archaeological origin range from 3–5 nT/m up to 250 nT/m in the case of some of the largest andesite monoliths. These variations may have included hearths, smaller stone features, lintels, and perhaps cobblestone wall foundations. Gradiometer work complements the radar and resistivity surveys to provide additional information on subsurface structure and especially building material composition.

3.1.4. Magnetometry Results at Putuni

Magnetic gradiometer survey of the same area in the Putuni Palace was also very enlightening. Throughout this survey region, Total Field measurement varied on the order of 150 nT while gradient measurements maxed at ca. $+/-45$ nT/m. For reference, this is the same magnetic field a kilogram of iron would induce at 4–6 feet from the sensor (Breiner, 1973). Notable is the general elevated field over the main central trench of the excavation unit that indicates the position of the andesite paving stones used to cap the F39 canal and to build the F169 flagstone patio (Figure 7). The largest anomaly of 45 nT/m is generated by the F173 andesite door lintel that marks the east entrance to the palace structure. Several other notable magnetic anomalies on the order of 25 nT/m are noted outside the main excavation area and are likely generated by TRM from other building materials of volcanic origin.

3.2. The Akapana East Sector at Tiwanaku

Opposite the Akapana from the Putuni, on the easternmost edge of the site's monumental core, lies the Akapana East sector (Figure 1), excavated by John Janusek in 1990 (Janusek, 1994, 2003). Akapana East is composed of residential compounds, or *barrios*, containing multiple households and semi-public spaces for the communal use of a social group. Excavations in one of these large residential compounds, designated Akapana East 1 Mound, revealed residential room groups containing kitchens with hearths, patio space, living areas, storage pits, and refuse zones dating to Tiwanaku IV (Figure 8). New construction activities associated with a massive episode of urban renewal in the subsequent Tiwanaku V Period transformed these spaces. Although many of the domestic activities and architectural orientations remained the same, ceramic assemblages and spatial density of occupation, including distinct households, were no longer visible in the excavated areas.

3.2.1. Resistivity

While resistivity studies are more time consuming than either GPR or magnetic gradiometry, field testing at Tiwanaku carried out in July 2003 was very successful in identifying buried *barrio* wall foundations and other architectural features. Electrical resistance is dependant on changes in the electrical properties of buried materials. The predominant effect on soil

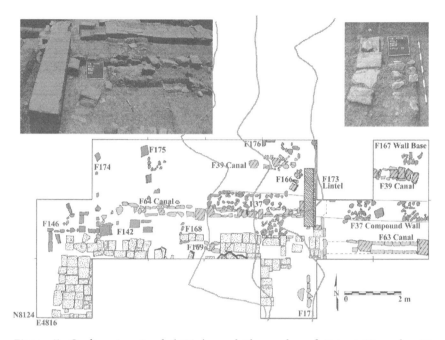

Figure 7. Gradiometer signal (nT/m) overlaid on plan of Putuni Tiwanaku IV and Tiwanaku V architecture; the graphic illustrates how excavated features were detected by magnetometry. Particularly noteworthy is the strong negative signal at the south end of the F173 lintel, the positive signal over the F39 canal in two transects, and the slightly negative signal over the F169 pavement. Photographs of the features highlighted by magnetometry are illustrated at the top of the figure. From left to right they are 1) F173 lintel and F169 pavement (view to south), and 2) F39 canal and F167 wall base (view to west). Architectural map from Couture and Sampeck, 2003.

resistance is soil moisture, such that archaeological remains can be detected if the amount of moisture they retain is significantly different from that retained by the surrounding soil matrix. Soil permeability, ion content, and temperature will also influence the electrical properties of soils. The method is effective for identifying buried walls, voids, solid floors, and some pits and ditches given the proper natural soil conditions, and is of particular benefit when the magnetic susceptibility contrast between archaeological features and the soil is low. Stone foundations exhibit marked resistivity contrasts with the surrounding soil in moist soil conditions. In some areas where the water table is very close to the surface, resistivity surveying is the only means of identifying wall foundations because of the effects of moisture on attenuation of the radar signal.

Figure 8. Structure 1, a residential structure of Tiwanaku IV Period located at Akapana East 1 Mound; view to north. The stone foundation of one wall can be seen along the upper edge of the image; a nearly intact adobe block is visible in the left half of the photograph. Note also the well preserved cobblestone foundation of a wide compound wall (upper right).

3.2.2. Resistivity Results at Akapana East

Several resistivity transects were located within the vicinity of the Akapana East 1 Mound excavations in 2003 in order to delimit the area of the compound and assess the nature of urban renewal in Tiwanaku V. These preliminary data support contentions that each of these residential compounds contained several households. More importantly, the resistivity survey delimited the size of this compound to ca. 30 m on a side by locating its southern and eastern compound walls (Figure 9). It also identified a possible drainage feature outside those compound walls, other potential internal wall features which could help elucidate the nature of household composition in Tiwanaku V, and internal voids that likely represent open plaza space. The latter lend support to Janusek's (2003) interpretation of these compounds as the loci of social-group residence in an embedded system of kinship, perhaps along the lines of the segmentary *ayllu* organization described in ethnohistoric and ethnographic accounts. As such, elucidating the overall urban plan of this sector yields new data on the nature of corporate integration in Tiwanaku society.

3.3. The Mollo Kontu Sector at Tiwanaku

Located immediately south of Tiwanaku's central civic-ceremonial core (Figure 1) and measuring approximately 8 ha in area, the Mollo Kontu sector will be a focus of future research, having already been preliminarily tested in 1991 and 2001. Earlier investigators have suggested that parts of Mollo Kontu served as dense cemeteries (Ponce Sanginés, 1961, 1981), and test excavations

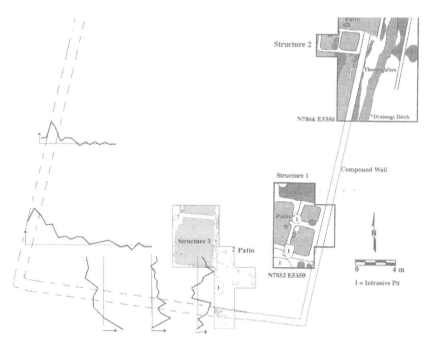

Figure 9. The remains of the Akapana East 1 Mound compound walls, Structures 1 and 2 (both dating to Tiwanaku IV), and Structure 3 (Tiwanaku V) exposed through excavation are shown with solid lines. Walls revealed by resistivity data are illustrated with dashed lines. Plots of the actual resistivity data from which compound walls were interpolated are displayed as line graphs superimposed over the architectural map with arrows indicating the direction of higher relative resistivity values. Architectural maps modified from Janusek 2003, Figures 10.4 and 10.16.

in 2001 have confirmed the sugestion (Blom, Couture, and Mendoza, 2004).[3] Preliminary studies at Mollo Kontu South (located toward the southern edge of the Mollo Kontu sector) have also established that this sector maintained at least three superimposed residential occupations dating to the Late Tiwanaku IV and Tiwanaku V Periods. Excavations here showed that household groups in this area were organized into discrete neighborhoods, or *barrios*, bounded by massive adobe walls set on cobblestone foundations (Couture 1993, 2003).[4]

---

[3] Funding for a systematic survey and surface collection in the Mollo Kontu sector, carried out in July and August 2001, was funded by an $8000.00 grant from the H. John Heinz III Fund for Archaeological Field Research in Latin America.
[4] Preliminary excavations in Mollo Kontu South were conducted over the course of six weeks in 1990 and 1991 as part of Couture's (1993) M.A. thesis research and were carried out under the umbrella of the *Proyecto Wila Jawira*. A description of this research was recently published by Couture (2003) as a chapter in an edited volume on Tiwanaku archaeology (Kolata, 2003b).

While the preliminary excavations at Mollo Kontu South were limited in scope, the presence of well preserved architectural remains, including the foundations of two north-south compound walls, makes this an ideal starting point for geophysical studies. In addition, the presence of a nearby cemetery—identified during excavation of test units approximately 10 m to the north in 2001—offers a unique opportunity to study the articulation between residential space and a highly specialized ritual space at Tiwanaku. The Mollo Kontu South mortuary area represents the first securely identified and the only systematically excavated cemetery at the site of Tiwanaku.

3.3.1. Magnetometry Results at Mollo Kontu

While all three geophysical techniques have been tested in Mollo Kontu, only magnetometry has proven successful. Resistivity survey was limited because it was too difficult to insert the probes and establish contact resistance in the hard packed soil often associated with the *altiplano's* dry season.[5] Survey in the plowed zone at Akapana East indicated that shallow plowing can increase the soil moisture and establish low contact resistance, so this technique will be applied in the future. GPR also had some problems in the cemetery area since tombs were often located beneath the 1.5-m depth window. A lower frequency antenna will be utilized in the future to address this issue.

Pilot magnetometry surveys were somewhat successful, identifying several anomalies, including one immense anomaly just south of the Mollo Kontu mound (Figure 10). This behemoth reading had a gradient of −100 nT/m, reflecting an equatorial field paramagnet of giant proportions. Depth estimate techniques (cf. Breiner, 1973) indicate the anomaly was produced about 2.5 m below the sensors, or about 1.5 m beneath the surface. Given this distance from the sensors and the magnitude of the magnetic field produced, we can only interpret this as a highly susceptible material, with a magnetic moment the equivalent of a fifty-kilogram mass of pure iron. More than twice the anomaly size of the Putuni lintel, we interpret this anomaly as a massive andesite monolith, perhaps another stone stelae as represented by the famous Bennett and Ponce monoliths. Only excavation will confirm our suspicions.

---

[5] The Mollo Kontu sector of Tiwanaku was purchased by the Bolivian government in the early 1990s as part of an ongoing effort to preserve the integrity of the site. While local farmers are still allowed to graze their cattle and sheep in Mollo Kontu, they no longer engage in heavy plowing. As a consequence, the topsoil in Mollo Kontu is significantly more hard-packed than in other parts of the site, such as Akapana East, where more intensive agricultural activity still takes place. The authors have been granted permission to plow shallow sections of Mollo Kontu (where the surface has already been disturbed by modern plowing) — to facilitate future geophysical survey.

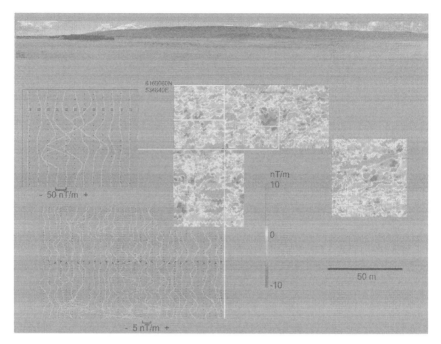

Figure 10. One hectare of gradiometer data from Mollo Kontu is shown in the blue-to-red color gradation plot; north is at the top. Magnifications of the actual survey lines highlight unexcavated anomaliesof -100 nT/m (north-center of color gradation plot) and -6 nT/meter (northeast corner of color gradation plot). Landscape photograph of the area surveyed with the Mollo Kontu mound and the Akapana mound in the background is displayed in the upper register of the figure. (COLOR).

4. CONCLUSIONS

This project presents some of the first intensive studies of overall urban space at the site of Tiwanaku using geophysical techniques. Given the nature of the site's depositional history, including the burial of domestic features under thick deposits of silt and eroded adobe, these techniques provide an invaluable means to study the organization of urban spaces across the city, ranging from the intimate spaces within individual domestic compounds to the organization of entire residential neighborhoods. As such, this approach presents a powerful lens through which to visualize Tiwanaku's built environment at both the large and small scale.

The long-term results of this research will contribute to the way archaeology is conducted at Tiwanaku, as well as at other sites in the Titicaca Basin and beyond. First, this project provides an important oppor-tunity for collaboration with Bolivian colleagues, including students from the Universidad Mayor de San Andrés and local collaborators from the community of Tiwanaku, in geophysical field methods. The Bolivian *Unidad Nacional de Arqueología* currently owns both a ground-penetrating radar and a magnetometer; international collaboration enhances understanding of the techniques and promotes advances in methodology for both sides. As this project advances, our collaborations will promote additional non-destructive geophysical research at Tiwanaku, as well as at other archaeological sites.

Second, the results of the geophysical research program will provide an important comparative dataset on the urban landscape, which can be used to identify patterns in social space not only between different sectors of Tiwanaku's diverse urban population, but also between Tiwanaku and other Tiwanaku-related sites located in the *altiplano* heartland and the more peripheral regions of the state. As such, our research goals and questions will nicely complement those of other ongoing Tiwanaku research in the Titicaca Basin heartland and in the more distant reaches of the Tiwanaku realm (e.g., in Moquegua, Peru, and eastern Bolivia). When placed within the context of these broader investigations, the geophysical research program contributes new data relating to issues of ethnic diversity, local resistance, and articulations of state power in the Pre-Columbian Andes.

ACKNOWLEDGEMENTS

The geophysical research was carried out in 2000–2001 and 2003 under the direction of the authors. Couture's excavations at the Putuni undertaken in 1999-2000 were funded by a PRA Fellowship from the Organization of American States and a grant from the Bolivian Archaeological Research Fund at The University of Chicago.

We wish to thank Licenciado Javier Escalante, the Director of the Unidad Nacional de Arqueología in Bolivia for granting permission to undertake this work. The John H. Heinz III Fund of the Heinz Family Foundation under-wrote the work at Mollo Kontu in 2001, and Boston University supplied the use of a SIR-2000 ground penetrating radar unit and a Geometrics G-858 gradiometer. Kenneth Sims assisted in data collection.

The 2003 resistivity test at Akapana East was undertaken in collaboration with John Janusek, who excavated that compound under the auspices of *Proyecto Wila Jawira* in 1990. The Field Museum of Natural History provided the resistivity meter manufactured by TR Systems. Ben Vining and Claire Sammels assisted in the field.

This paper benefited from the collaboration of John Janusek, Ben Vining, Chris Dayton, and two anonymous reviewers. We wish to extend our gratitude to Jim Wiseman and Farouk El-Baz for their encouragement and assistance in preparing this chapter. All errors or omissions remain our sole responsibility.

REFERENCES

Abercrombie, Thomas A., 1986, The Politics of Sacrifice: an Aymara Cosmology in Action. Unpublished Ph.D. dissertation, Department of Anthropology, University of Chicago.

Albarracín Jordan, Juan, 1996, Tiwanaku Settlement System: the Integration of Nested Hierarchies in the Lower Tiwanaku Valley. Latin American Antiquity 7(3):183–210.

Alconini, Sonia, 1995, Rito, Simbolo e Historia en al Pirámide de Akapana: un Análisis de Cerámica Ceremonial Prehispánica. Editorial Acción, La Paz.

Bennett, Wendell C., 1934, Excavations at Tiwanaku. Anthropological Papers of the American Museum of Natural History, No.34. American Museum of Natural History, New York.

Bermann, Marc, 1994, Lukurmata: Household Archaeology in Prehispanic Bolivia. Princeton University Press, Princeton, New Jersey.

Blom, Deborah E., 1999, Tiwanaku Regional Interaction and Social Identity: A Bioarchaeological Approach. Unpublished Ph.D. dissertation, Department of Anthropology, University of Chicago.

Blom, Deborah, Couture, Nicole, and Mendoza, Velia, 2004, Informe de Labores en Mollo Kontu, 2001–2002. Informe de Investigación Arqueológica del Proyecto Jach'a Marka, 2001. Unpublished report presented to the Unidad Nacional de Arqueología, La Paz, Bolivia.

Breiner, Sheldon, 1973, Applications Manual for Portable Magnetometers. San Jose: Geometrics.

Conyers, Lawrence B., and Goodman, Dean, 1997, Ground Penetrating Radar: An Introduction for Archaeologists. AltaMira Press, Walnut Creek, CA.

Couture, Nicole C., 1993, Excavations at Mollo Kontu, Tiwanaku. Unpublished M.A. thesis, Department of Anthropology, University of Chicago.

Couture, Nicole C., 2002, The Construction of Power: Monumental Space and Elite Residence at Tiwanaku, Bolivia. Unpublished Ph.D. dissertation, Department of Anthropology, University of Chicago.

Couture, Nicole C., 2003, Ritual, Monumentalism, and Residence at Mollo Kontu, Tiwanaku. In Tiwanaku, and Its Hinterland: Archaeology and Paleoecology of an Andean Civilization, Volume 2, Urban and Rural Archaeology, edited by Alan L. Kolata, pp. 202–25. Smithsonian Institution Press, Washington, D.C.

Couture, Nicole C., and Sampeck, Kathryn, 2003, Putuni: History of Palace Architecture at Tiwanaku. In Tiwanaku and Its Hinterland: Archaeology and Paleoecology of an Andean Civilization, Volume 2, Urban and Rural Archaeology, edited by Alan L. Kolata, pp. 226–63. Smithsonian Institution Press, Washington, D.C.

Escalante Moscoso, Javier,1992, Arquitectura Prehispanica en los Andes Bolivianos. Producciones CIMA, La Paz.

Escalante Moscoso, Javier, 2003, Residential Architecture in La K'araña, Tiwanaku. In Tiwanaku and its Hinterland: Archaeological and Paleoecological Investigations of an Andean Civilization, Volume 2, Urban and Rural Archaeology, edited by Alan L. Kolata, pp. 316–26. Smithsonian Institution Press, Washington, D.C.

Goldstein, Paul, 1989, Omo, A Tiwanaku Provincial Center in Moquegua, Peru. Unpublished Ph.D. dissertation, Department of Anthropology, University of Chicago.

Goldstein, Paul, 1993, Tiwanaku Temples and State Expansion: a Tiwanaku Sunken-Court Temple in Moquegua, Peru. Latin American Antiquity 4(1):22–47.

Janusek, John W., 1994, State and Local Power in a Prehispanic Polity: Changing Patterns of Urban Residence in Tiwanaku and Lukurmata, Bolivia. Unpublished Ph.D. dissertation, Department of Anthropology, University of Chicago.

Janusek, John W., 2002, Out of Many, One: Style and Social Boundaries in Tiwanaku. Latin American Antiquity 13(1):35–61.

Janusek, John W., 2003, The Changing Face of Tiwanaku Residential Life: State and Local Identity in an Andean City. In Tiwanaku and its Hinterland: Archaeological and Paleoecological Investigations of an Andean Civilization, Volume 2, Urban and Rural Archaeology, edited by Alan L. Kolata, pp. 264–95. Smithsonian Institution Press, Washington, D.C.

Kolata, Alan L., 1993, *The Tiwanaku: Portrait of an Andean Civilization*. Blackwell Publishers, Cambridge, M.A.

Kolata, Alan L., 2003a, The Proyecto Wila Jawira Research Program. In *Tiwanaku and its Hinterland: Archaeological and Paleoecological Investigations of an Andean Civilization*, Volume 2 *Urban and Rural Archaeology*, edited by Alan L. Kolata, pp. 3–17, Smithsonian Institution Press, Washington, D.C.

Kolata, Alan L., 2003b, *Tiwanaku and its Hinterland: Archaeological and Paleoecological Investigations of an Andean Civilization*, Volume 2, *Urban and Rural Archaeology*. Smithsonian Institution Press, Washington, D.C.

Kolata, Alan and Ponce Sanginés, Carlos, 1992, Tiwanaku: the City at the Center. In *The Ancient Americas: Art from Sacred Landscapes*, edited by Richard E. Townsend, pp. 317–334. The Art Institute of Chicago, Chicago.

Menzel, Dorothy, 1964, Style and Time in the Middle Horizon. *Nawpa Pacha* 2:1–105.

Parsons, Jeffrey, 1968, An Estimate of Size and Population for Middle Horizon Tiahuanaco, Bolivia. *American Antiquity* 33(4):243–45.

Platt, Tristan, 1982, *Estado Boliviano u Ayllu Andino: Tierra y Tributo en el Norte de Potosí*. Instituto de Estudios Peruanos, Lima.

Platt, Tristan, 1987, Entre ch'axwa y muxsa: para una Historia del pensamiento político aymara. In *Tres Reflexiones sobre el Pensamiento Andino*, pp. 61–132. Hisbol, La Paz.

Ponce Sanginés, Carlos, 1961, *Informe de Labores*. Centro de Investigaciones Arqueológicas en Tiwanaku, Publicación No.25, La Paz.

Ponce Sanginés, Carlos, 1981[1972], *Tiwanaku: Espacio, Tiempo y Cultura*. Los Amigos del Libro, La Paz.

Ponce Sanginés, Carlos, 1995, *Tiwanaku: 200 Años de Investigaciones Arqueológicas*. Producciones CIMA, La Paz.

Rasnake, Roger, 1988, *Domination and Cultural Resistance: Authority and Power among an Andean People*. Duke University Press, Durham.

Rivera Casanovas, Claudia S., 1994, *Ch'iji Jawira: Evidencias sobre la Producción de Cerámica en Tiwanaku*. Unpublished Licenciatura thesis, Universidad Mayor de San Andrés, La Paz.

Rivera Casanovas, Claudia S., 2003, Ch'iji Jawira: a Case of Ceramic Specialization in the Tiwanaku Urban Periphery. In *Tiwanaku and its Hinterland: Archaeological and Paleoecological Investigations of an Andean Civilization*, Volume 2, *Urban and Rural Archaeology*, edited by Alan L. Kolata, pp. 296–315. Smithsonian Institution Press, Washington, D.C.

Schaedel, Richard P., 1988, Andean World View: Hierarchy or Reciprocity, Regulation or Control? *Current Anthropology* 29(5):768–775.

Seddon, Mathew H., 1998, *Ritual, Power, and the Development of a Complex Society: the Island of the Sun and the Tiwanaku State*. Unpublished Ph.D. dissertation, Department of Anthropology, University of Chicago.

Stanish, Charles, 2003, *Ancient Titicaca: the Evolution of Complex Society in Southern Peru and Northern Bolivia*. University of California Press, Berkeley, CA.

Weymouth, J.W., and Huggins, R., 1985, Geophysical Surveying of Archaeological Sites. In *Archaeological Geology*, edited by G. Rapp and D.A. Gifford, pp. 191–235. Yale University Press, New Haven, CT.

Williams, Patrick, Blom, Deborah, Couture, Nicole, and Sims, K., 2001, Geophysical Survey and Tiwanaku Urbanism. Paper presented at the 29th Annual Meeting, Midwest Conference on Andean and Amazonian Archaeology and Ethnohistory. Ann Arbor, MI.

Williams, Patrick Ryan, Blom, Deborah, Couture, Nicole, Dayton, Christopher, Janusek, John, and Vining, Benjamin, 2003, Visualizing the Urban and Monumental Components of the Tiwanaku State: New Perspectives from Geophysics in the Andean Altiplano. Paper presented at the 22nd Annual Northeast Conference of Andean Archaeology and Ethnohistory. Harvard University, Cambridge.

Chapter 18

Geophysical Archaeology in the Lower Amazon: A Research Strategy

A. C. ROOSEVELT

Abstract: Geophysical survey has been integral to the archaeological research strategy
of the Lower Amazon Project in Brazil from 1983 until the present. In our
project, the systematic integration of multiple-instrument geophysical survey
with traditional archaeological approaches in three-stage, problem-oriented
research has produced unprecedented information about the organization of
prehistoric Amazonian societies, their history, and their ecological adaptations. In
our results, there was a consistent correlation of the geophysical anomaly patterns
from survey with the patterning of stratigraphy and distribution of objects in
the sites, as determined by excavation. This correlation between survey and
excavation results has allowed us to map theoretically important aspects of the
structure and composition of the archaeological sites in advance of excavation.
The resulting information about the nature of the human occupation of the
tropical rainforest has changed theoretical understanding of human history in
the New World. In this article I explain why and how geophysical methods have
been used in the project, illustrate the explanation with three case studies, and
assess the potential of the methods for archaeology in the future.

1. INTRODUCTION

Across the globe, geophysical survey has not been part of research strategy
for most archaeologists and had no role in my dissertation on prehistoric
subsistence and population in the Orinoco, Venezuela. Why have we not

used geophysical survey more? Three possible reasons come to mind. Reports on geophysical archaeology often use jargon—not very effective advertising. Also, few give excavation data to demonstrate that geophysical techniques can detect features of archaeological interest. Finally, the results often seem limited in scope and detail compared to what archaeologists want to know about ancient sites and cultures. It is not surprising, then, that archaeologists might be skeptical of the approach, given its apparent costs. My own experience in Amazonia, however, is that geophysical techniques can indeed produce results of great archaeological interest.

1.1. The Relevance of Geophysical Methods to Archaeologists' Interests

Archaeologists' research interests vary, but one enduring goal is manifest. We hope to understand the reasons for the changing patterns of human culture and biology in time and space. For some, this goal means looking at what the past shows about the evolution of human nature and culture. For others, the research has narrower, more practical goals, such as constructing regional cultural histories relevant to the interests of certain ethnic, economic, or national groups. Whatever the case, our interests lead us to investigate such facets of the past as humans' changing environments and economic systems, residence patterns, health status and genetic variation, social and political organization, systems of ritual and symbolism, and achievements and styles in architecture, art, technology, and public works.

How can geophysical survey help us pursue such investigations? The main way, I believe, is that geophysical survey can quickly, inexpensively, and harmlessly reveal some of the underground patterning that archaeologists want to detect and sample. Archaeologists' traditional methods for understanding sites as wholes are not very effective for this purpose. The methods fail either because they are do not give a useful picture of what is underground or because they do so too slowly, expensively, and destructively. For example, many long-term human-occupation sites are vertical, layer-cake configurations whose surfaces lack detectable remains of earlier occupations. These tell-like mound sites, common in Amazonia, frequently have many meters of occupation strata underground, built up over thousands of years. Later building stages at such sites completely obscure earlier occupations. Only the topmost levels are accessible to surface surveys at such sites, while the rest are fully buried.

As another example, so-called large-scale excavation really is not a good technique for uncovering site layouts. Many sites in Amazonia are much larger than the "wide areas" that archaeologists customarily dig, leaving most of the site area unexcavated. And, as Brazilian geophysicist Jose Seixas Lourenco (1985) has pointed out, excavation without prior geophysical survey is digging blind. One can only dig a minority of the site area, and there is no way of knowing what lies outside that area. It could be similar to what was dug or could be entirely different. Furthermore, if we use blind excavation as our routine discovery method we lose the opportunity to choose what to dig in an informed way. Digging up large areas of sites just to find where certain features of interest are ends up destroying areas that, had one known the layout ahead of time, one would not have needed to dig. Conversely, since one cannot dig the whole site, one frequently misses excavating the areas of greatest research interest.

What geophysical survey can do at archaeological sites to help with such problems is to give quite specific information in either 2- or 3-D of some of the very types of things archaeologists want to know about: archaeological layers, features, structures, facilities, activity areas, certain kinds of objects, and disturbances that have affected these. Having such information ahead of digging can help archaeologists get what they need from the archaeological deposits to pursue their practical or theoretical research goals.

Geophysical surveys work best in integration with traditional archaeological research methods, to my mind. Like the observations from archaeological surface survey, the findings from geophysical survey usually have to be evaluated with excavations. Surveys may obviate extensive excavations in some situations, but I cannot think of any situations that would not require some type of subsurface testing and analysis at least to assess the archaeological nature of the patterns produced by geophysical survey. Thus, although geophysical methods can greatly assist archaeological research, I believe that for most research problems they do not eliminate the need for some sort of traditional archaeological investigations.

Over and above how geophysical surveys can help archaeologists' research strategies, the methods have the potential to open up new lines of inquiry that archaeologists have not yet thought of because of the limitations of strictly archaeological methods. That potential is an exciting possibility to ponder for future research. Having developed more effective ways to do research within the box, we can now start thinking outside of it. But I will have more on that issue in the conclusions.

1.2. Archaeological Research Problems and Site Composition in Amazonia

How one might apply geophysical methods in a region depends on what research problems are there. In Amazonia, a group of questions about long-term human-environment interactions have been defined. How long have humans been in the humid tropics? What role have such habitats had in shaping human evolution? How have preindustrial humans managed tropical forests as habitats? How have they subsisted there and how did those adaptations influence lifestyles and cultures? Is there anything unique about the trajectory of prehistory in that habitat? Did people there take a different path than those in other habitats? How have Amazonians interacted with people in other areas? What was the region's role in the culture history of the hemisphere?

To approach these overarching, comparative issues in our project, we needed to recover standard archaeological information that would allow local prehistoric occupation sites of different periods to be compared with those elsewhere in the world. For the evolutionary questions, we needed to establish, through stratigraphy, artifact analysis, and dating, sequences of the development of human societies through time. To be able to compare community organization, we needed information about the size, layout, and composition of sites at different times and places. To assess the long-term human-environment interaction, we needed to collect and analyze biological and geological materials in stratigraphic and cultural context at those sites.

The research problems of our project grew out of what I perceived as a misfit between consensual notions about human culture history in the Americas as applied to Amazonia. Much of this culture history was summarized in Gordon Willey's excellent book on South America (Willey, 1971) and is a theoretical formulation important in scientific American archaeology. In a version still popular, big-game hunters from Asia entered the Americas across the Bering Straits before 12,000 years ago, passing quickly via the cool grassy uplands of the hemisphere to reach the tip of South America as the Ice age ended. With climates warming, oceans rising, game going extinct, and wetlands and forest habitats expanding, people began to settle down and diversify subsistence, adding small fauna and plants. In propitious areas, such as the lake and river basins of Central Mexico and the Central Andes, people domesticated local plants. The agricultural Formative villages that then developed became the context for the rise of the first "high" cultures: the urban civilizations. People only ventured outside those nuclear areas late in prehistory because of a lack of resources for hunting and agriculture. In the tropical forests people were limited to slash-and-burn horticulture and foraging and lived in small, shifting settlements. They never could develop agricultural civilizations, with the great art and monuments typical of such societies.

When I began a project of surface surveys and excavations in the tropical lowlands of the Middle Orinoco, Venezuela in the mid-seventies, I focused on the contradictions of this culture history's view of the relationships among resources, agriculture, and population (Roosevelt, 1980). By the time I started in Brazil in the early eighties, I realized that archaeologists' consensus about settlement and social organization in the lowlands also might be wrong (2000b). Several archaeologists had disputed the consensus about Amazonia (Lathrap, 1970), but had not been able to disprove it. Despite common assumptions that tropical sites would be slight and poorly preserved, early researchers had recorded stratigraphy with artifacts, plant remains, and bones as well as substantial built environments and cultural landscapes. They described large constructions of earth, refuse, adobe, and plant materials. It seemed clear that Amazonian sites held evidence relevant to the theoretical issues.

Accordingly, it became a priority to investigate a series of multicomponent sites that could reveal differences among prehistoric societies over time and space. It seemed especially important to learn more about the structure, composition, and history of sites, their habitats, and subsistence remains. Obviously, geophysical methods that might reveal stratigraphy, structures, and object distribution ahead of excavation had appeal for me.

1.3. The Development of Geophysical Archaeology for the Lower Amazon Project

It was by accident that I learned of the utility of incorporating geophysical methods in a research project.

One of the first people to follow up on earlier insights on the complexity and richness of Amazonian sites was J. S. Lourenco, mentioned above, who had received his Ph.D. in geophysics at Berkeley. As a museum director, he was familiar with archaeology collections and became interested in the mounds of the famous Polychrome Horizon Marajoara culture at the mouth of the Amazon, near Belem. He realized right away that geophysical survey had the capacity to help discover and map features, layers, and structures, so he began preliminary research at a mound, Teso dos Bichos (Alves and Lourenco et al., 1981). When we met during my first visit to Brazil, we exchanged publications, and he invited me to join him on Marajo Island. I did so for two seasons, after which he left to head the state university and joined the Federal government. I then continued the research on Marajo and expanded to other areas of the Amazon.

Although Lourenco and his colleagues were actively applying geophysical methods in the Amazon, the fact that you could use such technology in a humid tropical lowland was not widely recognized. Several reviewers of my National Science Foundation research proposals were convinced that such methods were unworkable there, because of excessive water, acid soils, and thick vegetation. Those reviewers could not, however, cite evidence, and fortunately experienced archaeologists of tropical lowlands—such as the late Irving Rouse of Yale and Wesley Hurt of Indiana University—supported the geophysical research strategy. As a result I got grants and was able to start field work in Brazil in 1983, on Marajo.

2. GEOPHYSICAL TECHNIQUES FOR ARCHAEOLOGY

2.1. The Challenges

In order to engage in geophysical archaeology, we archaeologists had to try to learn those rudiments of the geophysics relevant to our goals. In order to help us, the geophysicists had to learn what it is that archaeologists want to know. They also had to learn about archaeological site patterning and figure out what methods to best detect and map it. As Principal Investigator, I had to define research questions suitable for the region and operationalize them for specific site investigations. To put together a comprehensive surveying system, I had to review the different instruments with the geophysicists and choose recording hardware and software with our computer specialists. I also had to plan the excavations and analysis of material to take advantage of survey results and get the needed information from the sites.

What I found out about the geophysical techniques is that each one had some disabilities but that some technique could fill in for another's deficiencies. By applying them together one could get the information to plan archaeological excavation. Some techniques turned out to be unsuitable for some sites, but what made one technique not useful at a site tended to favor another one. Accordingly, I invited a versatile, experienced archaeological geophysicist, Bruce Bevan, Director of Geosight, to be project geophysicist, in order to expand the number of techniques that we could apply to sites. Our assessment of the value of the techniques for our sites has been presented in detail elsewhere (Bevan, 1986, 1989; Bevan and Roosevelt, 2003; Roosevelt , 1991:188–342), so I will only summarize below.

2.2. Site Investigation Goals Relevant to Geophysical Surveying

In the site research in Amazonia, I wanted to assess the larger-scale stratigraphy and layout in advance of excavation. This information could help us understand the history of occupation (single or multiple, continuous or discontinuous) and the types of activity involved (domestic, ceremonial, both, etc.). It also was desirable to try to identify the number, size, and functions of structures and larger features throughout the site. Traditional surface survey by itself could not do so because of the lack of surface material. There was no hope of excavating even one entire Amazonian site in a lifetime, and one did not want to destroy deposits unnecessarily. With results from an initial geophysical prospection, one could excavate at a sample of the geophysical patterns recorded and then extrapolate the results to areas of similar anomalies that were not excavated. To furnish the data for such reconstructions, we focused on electrical resistivity, magnetometry, electromagnetic conductivity, radar, and seismic surveying (Figure 1). Below, I briefly evaluate each technique, based on our experience with them.

2.3. Electrical Resistivity Surveying

This technique (Figure 1A) is one of the most versatile, and advances in instrumentation and computerization keep increasing its value. Electrical resistivity measures how quickly current travels through underground sediment between electrode pairs on the surface. Conductivity varies in strata, because of different grain size, density, chemistry, moisture, etc., allowing reconstruction of stratification. The depth of measurement varies according to the electrode

Figure 1 (Continued). Geophysical equipment used in the Lower Amazon Project.
A. Proton magnetometer, Teso dos Bichos mound. B. Ground-penetrating radar, Teso
dos Bichos mound. C. EM 31 Conductivity meter. D. Seismic refraction, Guajara
mound. E. Electrical resistivity, Guajara mound. Photos by A. C. Roosevelt.

spacing, so a whole profile can be reconstructed by changing electrode spacing
in successive measurements. Over and above the delineation of vertical stratification in such "pseudosections," the technique can make a contour map of
different buried layers. The major internal morphology of sites thus can, in
theory, be traced. Most sites are apt for survey by resistivity. Dampness and
high conductivity are no barriers, although changing environmental moisture
requires calibrations from year to year. During drought electrodes may not
have sufficient ground contact and may need to be wet for each measurement,
leading them to rust. Dry, well-drained deposits may lack sufficient moisture
to conduct electricity at all, but such deposits are rare in Amazonia. The main
limitation of the technique is that many archaeological objects and features,
such as artifacts or post-molds, are too small to map.

2.4. Magnetometry

This highly useful technique lacks electrical resistivity's ability to map deep
vertical stratification, but in the average site it can easily and rapidly map
a common, important type of archaeological feature: burned areas of iron-
bearing sediment. Because many archaeological sediments and objects contain
iron and were burned, there is much that can be mapped: ovens and hearths,
burnt floors and walls, and burnt rock or metal concentrations. One only
can measure the magnetic variation within a meter or two from the surface
of the ground, but even then magnetic features under a meter in size are
mappable. A site totally lacking magnetic material is not mappable, obviously,
and a site full of modern ferrous metal rubbish can have so much magnetic
"noise" that the slighter and vaguer archaeological patterns get submerged.
Changing local, global, and solar magnetic fields also can swamp archaeological patterning, but can be subtracted by use of data from a stationary

magnetometer measuring at regular intervals. Soil moisture and high conductivity are no barriers, though shell and sand may lack measurable magnetic material. The proton magnetometer is light and easily carried while walking rapidly (Figure 1B).

2.5. Conductivity Surveying

Electromagnetic conductivity surveying, the technical alter-ego of electrical resistivity surveying, allows finer, more rapid measurements of electrical current through sediment than does resistivity surveying. Thus, common archaeological features with contrasting conductivity, such as hearths, filled pits, earth or stone constructions, etc., can be mapped. Depending on the instrument used, one can measure the conductivity of the ground over a half-meter to a meter-and-a-half's depth (EM 38 instrument) or between three to six meters' depth (EM 31 instrument). Each measurement, however, is an average, although one can detect some vertical changes by comparing surveys with different instruments. With conductivity, as with magnetometry, one can better map discrete horizontal areas, features, and objects than multiple superimposed vertical layers at a site. Like magnetometers, conductivity meters are easily and rapidly deployed while the operator is walking (Figure 1C). But, as with resistivity, sediments can occasionally be too dry or well-drained to register significant differences. Also, electrical interference, even by distant lightening, can obscure data patterns of archaeological scale.

2.6. Ground-penetrating Radar

Radar has the potential to produce a subsurface image comparable to the stratigraphy exposed by excavation. Its antennae can pick up the same variations seen in a well-differentiated stratigraphic profile or plan map. Layers, interfaces, walls and foundations, tombs or containers, decayed organic objects, and other features thus show up clearly in radar profiles. Only microstrata and small objects are too small to be mapped. Like resistivity, radar gives vertical as well as horizontal maps. Unlike magnetometry, radar has the potential to image through an entire site profile, depending on the sediment. Even remote, shuttle-deployed radar has imaged geological stratigraphy faithfully as deep as 15 m (McCauley et al., 1982). It is most useful in the dry, grainy sediments that are difficult for resistivity or conductivity. Radar, however, cannot penetrate deeply in very moist or conductive sediments, in which its signal is attenuated. It is not helpful for imaging alluvial clays, limestone, or mafic volcanic ash. Radar antenna sleds can be drawn as rapidly as one can walk or drive (Figure 1D), but some clearing may be necessary in disturbed undergrowth. Additionally, radar may encounter interference from other large equipment.

2.7. Seismic Refraction Surveying

Seismic refraction surveying is an excellent method for mapping earth masses, thick layers, and their interfaces. Similar to resistivity, it requires cumbersome laying out of sensors, and one must repeatedly bang the ground to elicit an average of the wave patterns that reflect variations in density (Figure 1E). The technique is oversensitive: a grazing cow or breeze through a tree can add confounding "noise" to seismic data. But major structures and the overall patterning of large deep sites can be reconstructed from seismic data, as long as they contrast in density within the site matrix. Measurements can go as deep as one requires, but, as with resistivity, important smaller architectural features, such as house floors or walls, really cannot be mapped by this method at present.

2.8. Methods for Collecting and Analyzing Data

Our research goals and the nature of humid tropical sites influenced our approach to data gathering. In carrying out the work and analyzing results, we have learned how better to apply and combine the different techniques of geophysics and archaeology. Our tinkering with methodology has involved both research design and methods: choice of instruments, collecting and recording methods, and analyses. It has been a priority to try to detect and prevent errors. Early on we learned the rule of security by duplication. Any truly indispensible instrument is brought in multiples so we can continue if an instrument malfunctions and all records are recorded multiple times on different media and stored in different places.

We map topography in detail early on. We need ground control to correct geophysical results for topography and to overlay geophysical and archaeological results for interpretation. Topography also can give information about history of disturbance. We use automated infrared laser theodolites. The large sizes of the sites and our need for good spatial control on sites with significant topography meant that we cannot rely on Brunton compass maps, sketch maps, or maps based on "chaining" measuring tapes.

Fortunately, soon after we started, laptops became available. We use computers in many ways: collecting, organizing, checking, analyzing, contouring, and printing data in the field. For maps we have used Cadd programs, Golden's Surfer, and Mapviewer. For databases we have used DBase, Access, Filemaker, and spreadsheet programs. To detect and prevent mapping errors, we remeasure points many times during a season and again in later seasons. At the suggestion of John Douglas, the archaeologist in charge of our topographic mapping and statistics, we make maps of the density of surveyed points, to avoid permitting spurious contouring to become artifacts of inadequate coverage.

We use portable printers to print maps at night or on the weekend to make interpretive comparisons and scan for striations and anomalies from instrument malfunctions or software bugs. Thus we can correct or re-take data in the field rather than face problems back home after the research season. Our initial automation was guided by David Stephan at Pima Community College in Tucson. Linda Brown and John Douglas, now at the University of Montana, Bruce Bevan, and the staff with the Pima Community College Archaeology Department did much of our software writing for the instruments, but now most manufacturers have developed their own (Roosevelt, 1988a).

Being far from civilization at most sites, we rely on solar power to run equipment and lights. Some heavy equipment, such as the radar, requires a gasoline generator. In addition, we can recharge batteries at outlets at night when staying at houses or hotels that have electricity. The Archaeology Department of Pima Community College designed our solar collecting system, which we still use. We now supplement those large, heavy panels and the heavy batteries with smaller, lighter panels directly attached to the instruments.

In the archaeology operation our data needs led to somewhat different approaches from those usual in the tropical lowlands. To be able to compare the object contents of different layers and features, we fine-screen all deposits, not just a sample, and we retain screen contents for analysis in the laboratory. We find that most of the identifiable plant and animal remains are found in those residues. Having a special focus on human ecology and subsistence, we work with local informants and biologists in the field to collect biota to help identify ancient remains. To help define cultural sequences for the region, we run numerous dates on different materials with different methods at different labs. To further develop sequences, we use attribute analysis (sometimes called modal analysis) of pottery and lithics, rather than type-variety, to create fine-grained periods. Knowing that stable isotope ratios varied greatly in the humid tropics, we have the stable isotope ratios analyzed for both paleoecological and subsistence information. Much of that work was begun in collaboration with the late Hal Krueger of Geochron.

3. RESEARCH DESIGN

3.1. The Three Stages

We combine archaeology and geophysics in research that we carry out in three stages. We want to be critical of both geophysical methodology and traditional archaeological methods. Thus, the stages involve explicit testing of our and others' interpretations. We have had to continually refine methods and interpretations because we began in relatively new territory for both archaeology and geophysics. Late 20th century scientific archaeology had hardly been applied in Amazonia in 1980 when we started; the cultural sequence was not yet established; and assumptions on prehistoric human and natural ecology were theoretically, not empirically, derived. Most of our and others' assumptions were bound to be disproven.

To help create a regional sequence, we mostly have chosen for investigation the multicomponent sites with superpositions. To be able to study site layout, we have focused on sites that had preserved facilities and structures to map and excavate. To be able to infer the nature of activities, we have chosen sites with abundant ancient biological and cultural remains that could be dated relatively or absolutely, and analyzed for cultural and habitat information. And finally, we chose sites that exemplified some phenomenon that for theoretical reasons we wanted to know more about and that realistically could be investigated with the time and resources available to us.

Among the sites we chose were the following: a habitation of early sedentary foragers, namely, a large shellmound; two mid-prehistoric tell-like earth platform-mounds of moderate size; and a center of a complex society. At the shellmound, we found and dated Formative deposits appropriate for research in the future. We felt that the shellmound could show how early villagers gained their subsistence and help date the beginning of ceramic-making. We felt that the earth mounds could reveal the subsistence and community organization of an early complex society and that the large cultural center could reveal the ecological adaptations and organization of a late prehistoric complex society.

3.1.1. Stage 1

First we survey the site in detail with different techniques, checking results to assess accuracy.

We initially walk the surface. Because of active soil deposition, little appears on the surface, but at erosion features or disturbances one can see both strata and objects. At some sites, too, the vegetation reflects the sediments beneath. We record such patterns on the map or in the data base site form.

Next comes topographic survey. The shape and size of sites give evidence of their function, composition, and scale, relevant to theories about demography and cultural development. We also need fine-grained site maps for laying out and interpreting the geophysical surveys. The geophysical surveys are conducted upon grids and transects staked on the site and recorded in the map so that a geophysical pattern can be referenced to its archaeological manifestation. We map topographic anomalies, any visible disturbances and modern occupation features, which affect both geophysical and archaeological results.

Map in hand, we carry out the geophysical surveys to get information about site layout, stratigraphy, and contents in advance of excavation. The entities of interest include major use areas of the site, major layering, structures and facilities, use areas, features, and objects. We need geophysical observations in both two and three dimensions where possible, to try to get information about both the most recent occupation and earlier occupations. We deal with the uncertainties and limitations of each type of geophysical survey or instrument by using multiple techniques. The particular combination we use was suggested by Bruce Bevan. We get more definitive information because a question raised by results from one instrument could be solved with data from another. The complementary geophysical information we get, compared to excavation results, helps compare the efficacy of different techniques for different questions and site conditions. To carry out and correct geophysical measurements, we conduct survey transects in directions contrary to what we know of the geometry of site patterning to detect striations in data from malfunctioning instruments so we can redo the transects after repairing or changing the instrument.

After the survey results are checked, the geophysicists make their interpretation of the stratigraphic meaning of the geophysical patterns. The archaeologists then make provisional conclusions about the archaeology.

3.1.2. Stage 2

For the second stage of research, we carry out a test-excavation plan to assess the preliminary conclusions from the geophysical and topographic surveys and provide archaeological data for an initial reconstruction of the occupation.

Our interpretations of the geophysical results allow us to divide the site into sampling areas for excavations. The aims of the text excavations are to evaluate the conclusions from the surveys and to locate, map, and sample entities of particular archaeological interest for future broad-area excavations. We need to acquire datable samples and well-referenced cultural and environmental materials in spatial control. We devise a preliminary sampling plan for each different area or type of geophysical feature. The plans could involve either stratified random sampling or "purposive" sampling. Based on the research questions and the nature of the deposits, as judged from surface and geophysical examination, we choose digging, soil-processing, and analytical methods. The dense, hard, and nearly sterile earthwork layers require heavy digging tools and have few objects to map and collect. Their recordable layering, however, ranges from simple, massive, and regular to small-scale and varied. The smaller or softer layers of hearth features, floors, and burials require delicate hand tools and very detailed plan and profile maps. They produce hundreds of objects to uncover, number, map, and photograph in place.

In this second stage, we make preliminary analyses of the archaeological results. We interpret stratigraphic sections in the light of the geological, biological, and cultural remains. We review the pottery and lithics for both chronological and functional information. Comparing objects by successive levels gives relative chronology to define cultural phases. Comparing objects in areas of the same age helps us define and interpret peoples' different activities. We also sometimes carry out geophysical measurements on excavated strata and materials during and after excavation, when this work might add information. Collected materials are archived for future stages and to address new questions.

We then compare the archaeological results with the predictions from the geophysical surveys and revise the latter, if necessary. The end-results of these comparisons are conclusions about the history and nature of the prehistoric occupations. We make conclusions in the light of what is being learned at other sites and in the context of the larger "developmental sequence" of the research region: the Lower Amazon. These conclusions include uncertainties and questions about facets that could not be assessed in the first two stages but then help shape the operations of the third stage.

3.1.3. Stage 3

In the third stage of our research, we use the geoarchaeological results from the first two stages to decide what to open in wide excavations to gain deeper or broader knowledge of the site and its successive human communities and landscapes.

The interpretive goals of our stage-3 research include the following. Although the first two stages help visualize the layout of sites as wholes, that information can be too gross or incomplete to give us a clear understanding of the specific characteristics of the geoarchaeological features. Stage-3 real excavations are needed to get adequate exposures and sufficient examples of strata and constructions and sufficient samples of objects relevant to the understanding of the human-environmental sequence in Amazonia.

Having carried out the first two stages of research at four sites by now, we have defined goals for stage-three research at the three sites where we will continue research work, and we have begun stage-3 excavations at one of them, Santarem. Next, I will summarize the research and results at the four sites.

4. THE CASE STUDIES

4.1. Marajo

On Marajo Island our research was aimed at understanding the famous archaeological earthen mounds in the seasonally-flooded eastern part of the island. The Marajoara culture was hypothesized to have been a short-lived colony from a late prehistoric agricultural civilization in the Andes foothills, which decayed into a horticultural village society in the environment. The mounds were assumed to have been purely ceremonial sites, not population centers.

Our work at two mounds on Marajo was our first in Brazil, and much of our methodology was developed there (Bevan, 1986, 1989; Bevan and Roosevelt, 2003; Roosevelt, 1987, 1988a and b, 1989, 1991, 1999a-e). Teso dos Bichos is a c. 3 ha. 7 meter high mound near Lake Arari, and Guajara is a ca. 1 ha. 7 m high in the Anajas river drainage. We have carried out stage-1 and -2 research at both mounds. The purpose of the stage-1 surveys was to assess site layout. The stage-2 test-excavations were carried out to test our interpretations of the surveys and to allow preliminary reconstruction of the chronology, ecological adaptation, and local artifact culture.

We worked first at Teso dos Bichos, then at Guajara (details in Bevan, 1986, 1989; Bevan and Roosevelt, 2003; Roosevelt, 1988a, 1989, 1991). In stage 1, after making detailed topographic maps, examining the mound surfaces, and laying out staked grids, we geophysically surveyed the mounds with multiple techniques. The topography showed that the mounds were flat-topped and had eroded sides, colluvial aprons, and borrow pits nearby. The tops of the mounds were mostly bare of artifacts, but the surface surveys revealed strata, features, and objects exposed at erosion features. These exposures suggested the presence of earthworks, house floors, grouped adobe hearths, and burial urns.

Our magnetometer surveys were the most effective operations to reveal in detail the distribution of important occupational features in the earth mounds (Figure 2A). Conducted at meter intervals across the mounds and beyond, they revealed many large magnetic lows dispersed through the mounds except at their centers and outside the mounds. Most of the lows were encountered in the top two meters, because of the depth limitation of magnetometer measurements. But where earlier levels of occupation had been exposed by erosion at the sides of Guajara, the magnetometer survey encountered comparable magnetic lows.

Our selective electrical resistivity surveys showed that the areas with magnetic lows were comparatively resistant to electrical current and that the edges of the mounds were much less resistant. Similarly, the detailed EM-38 conductivity surveys we conducted at 1-m intervals at both mounds showed low-conductivity areas at the magnetic lows, intermediate-conductivity areas outside them, and high-conductivity areas around the perimeter of the mounds except where erosion had exposed less conductive sediment. At the second mound, Guajara, a deeper EM-31 conductivity survey showed a major stratigraphic interface several meters down from the surface. Thus, conductivity survey was able to define four geophysically distinct major site areas and also buried interfaces. The seismic survey carried out at Guajara in selected transects suggested multiple major interfaces at different depths.

Figure 2. Teso dos Bichos mound, Marajo Island. A. Magnetometer map of mound. By B. Bevan. B. Photograph of hearths and floors excavated at the magnetic anomaly at grid-point N175/E333. Photo by A. C. Roosevelt.

Our radar survey, conducted in linear transects 50 cm (at Guajara) to 2 m (at Teso dos Bichos) apart over the sites, was only able to map near-surface disturbances, due to the comparatively high conductivity of mound sediments in which radar could only penetrate about a half-meter deep. Radar did produce profiles and maps containing interpretable anomalies but, being relatively superficial, these turned out to be recent disturbances, when excavated.

Our interpretation of these stage-1 results in the light of geophysical technology and of known Majoara archaeology was that the magnetic lows could be hearth groups, filled pits, or urn burials. We felt that the higher-resistivity/low-conductivity results in and around the magnetic lows could reflect sand-tempered floors, porous burnt adobe, or burnt soil. We expected the conductive/low-resistivity areas at mound edges to be exposures of the earthworks supporting the sites. We interpreted the intermediate geophysical results as middens or trampled plazas. We guessed that the radar anomalies reflected near-surface burials, other large ceramic vessels, adobe hearths, or disturbances. We interpreted the seismic results as indicative of several major building stages and, at the mound's base, a possible paleosurface on which the mound had been erected.

Our stage-2 test excavations showed the different geophysical areas differed in archaeological stratigraphy, features, and objects. At each site, we excavated a random selection of magnetic lows. The excavations at the magnetic lows uncovered many groups of fired-adobe domestic hearths set in structure floors (Figure 2B). The only magnetic low lacking hearth group was a filled-in, burned floor of a structure at Teso dos Bichos. The excavated hearth groups correlated with the higher-resistivity and lower-conductivity areas. Contrary to expectation, no cemeteries or pits have turned up at magnetic lows, though the excavations encountered occasional urn burials intruded into the remains of an earlier house from above. The excavation and survey results and the stratigraphy at the hearths and floors suggest that houses built onto low earthen foundations were placed in a rough oval around an open area on top of a larger platform supporting the whole village. Both hearths and associated floors manifest numerous successive plasterings and repairs, and the radiocarbon dates that we and others have had run from sites range over a long period, from A.D. 400 through 1300.

We also excavated test pits in the intermediate-conductivity/resistivity areas away from magnetic lows at both sites and at selected radar anomalies and high-conductivity areas at Teso dos Bichos. Where resistivity and conduc-tivity were intermediate, the stage-2 excavations uncovered deep black-soil garbage fill. The middens sometimes contained urn burials and feasting features with ritually-broken special foods and objects. The radar anomalies, on excavation, turned out to be intrusions of loose or grainy material into the mound construction layers. At high-conductivity anomalies, we found mound-construction layers.

The excavations showed that the artifacts differed among the strata and features of the different geophysical areas. Pottery was encountered in most geophysical features but was rare or absent in the earthworks at high-conductivity anomalies. Purposely-broken cooking bowls lay on top of many hearth groups. The rare domestic pottery in the associated house floors had been trampled into small pieces. In the primary feasting features in the middens at geophysically intermediate areas, vessels were larger than those in the domestic hearths and houses and much more elaborately shaped, finished, and decorated. Both kinds were mixed up in the middens in intermediate-size fragments.

Biological remains also contrasted among the different geoarchaeological areas, with more large faunal species and special fruits in the feasting areas than in the houses and hearths, where tiny fishbones predominated. The peripheral earthworks and the hearth adobe were poor in biological remains. The food and environmental remains in the residential, ritual, and refuse areas included diverse fruits, seeds, fishbones, turtle shells, and rare mammal and reptile bones, from lush tropical forest and riverine microenvironments now diminished by cattle ranching.

Research at Teso dos Bichos was discontinued after stage 1 and 2 because erosion appeared to have removed most of the cemetery areas of the occupation under investigation, making it unsuitable for a holistic investigation of the ancient community. Guajara (Figure 3A) was chosen instead for further research because many urns (Figure 3b) had been found at the site. We encountered cemeteries in or next to intermediate geophysical areas flanked by magnetic lows. In addition to the magnetic lows and the strata visible in erosion features lower down, the excavations showed at least two different occupation and building stages in the upper meter and one-half of the mound, each with different placement of structures, midden fill, and urn burials. These multiple mound occupation stages fit the predictions from the seismic and resistivity results.

For the third stage at Guajara, we plan to excavate the entire footprint of two or more of the structures with hearths, to assess their function and size. We also wish to excavate at least two cemeteries, in order to assess the differentiation within the population in access to culture and resources. Our study will be based on analysis of the skeletons and burial accompaniments in stratigraphic and chronological context. We also will expand excavations in black soil middens to fill in information about subsistence, environment, and crafts, and at one or more of the feast areas, to learn more about the size of dishes, the nature of special food, and the ceremonial activities. We also plan to extend excavation to the base of the mound to trace building stages from the beginning of the occupation.

The main revelation from the stage-1 and -2 work on Marajo is that the mounds were sizeable, long-lived villages built on top of large platforms. They were not just empty ceremonial centers. The research suggests that each

A

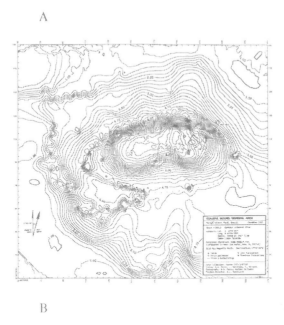

B

Figure 3. Guajara mound, Marajo Island. A. Topographic map of mound showing borrow pit at the southeast and recent erosion features on the south. Map by J. Douglas. B. Illustration of burial urn from the mound. S. Catalano.

mound's upper occupations had multiple large residential structures arranged around an open area. The houses had grouped hearths of adobe, clean earth floors, and possibly thatch roofs. Their artifacts were mostly simply made and often worn, indicating a residential function. Adjacent to the houses we found secondary garbage fill, cemeteries, and the in situ remains of feasts. The surveys, test excavations, and analyses suggest that both mounds held multiple stages of building, successive villages, and multilevel cemeteries, evidence for a stable, long-lived culture with extended periods of occupation, contrary to the interpretation of the culture as an invasion from the Andes foothills which rapidly collapsed in the tropical environment (e.g., Meggers and Evans, 1957). The radiocarbon dates collected in stage-2 research showed the cultures to be almost a thousand years too old to be offshoots the Andean culture. No obvious administrative buildings have been identified so far at the mounds, contrary to the idea that the mound cultures were state societies (e.g., Meggers and Evans, 1957). However, the significant similarity of mound structure and material culture across Marajo suggests strong regional cohesion, as in regional tribal organizations. The elaboration and complexity of the material culture at some mounds and the monumentality of mound constructions suggest the existence of a "high" culture, occupational specialization, and social ranking (Schaan, 1997, 2004). Investigating the nature and history of social differentiation will be a priority of the stage-3 excavations at structures and cemeteries.

The most useful general methodological result from the research is that site areas different from one another geophysically also contrasted in stratigraphy and contents. Thus, we believe we can now locate hearth clusters or burnt house-floors, earthworks, and garbage areas in sites without excavating by mapping with geophysical methods. But one disappointing finding was that our geophysical surveys apparently cannot detect large pots or burials, two theoretically important types of archaeological objects. Unfortunately, we still can only detect these indirectly with geophysical methods, by finding the house floors or garbage dumps in which they were placed.

4.2. Taperinha

Taperinha is a ca. 3,000-ha property about 60 km downstream from Santarem on the south bank of the Amazon. Nineteenth-century geologists had identified a large shellmound there (Figure 4A) as a possible campsite of early Holocene ceramic-age fisherpeople, but it had not been professionally excavated, dated, or reported (Hartt, 1886). If their observations were valid, then the site and others like it constituted further evidence against the short chronology for Amazonia. In preliminary museum research, we verified the presence of pottery sherds at the shellmidden and got a conventional radiocarbon date of about 5703 B.P. from shell that the early geologists had excavated with the pottery (Roosevelt, 1995; Roosevelt, 1991). Our research problem, then, was to verify the site's structure, stratigraphy, age, material culture, subsistence, and state of preservation. Part of it was destroyed by commercial shell mining in the 1970s, according to oral history.

We carried out three seasons of stage-1 and -2 research at the shellmound, revealing its topography, its minimum size, aspects of its cross-section, and the presence of hearths, post-holes, refuse, and burials (Bevan, 1989; Roosevelt, 1991). The mound turned out to be a habitation of broad-spectrum foragers of the Pottery Archaic period. Only foraged foods were identified in the refuse: primarily fish, shellfish, turtles, and other small animals, such as

A

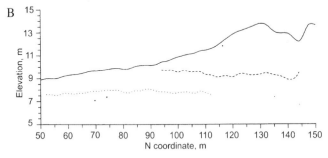

B

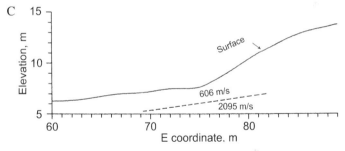

C

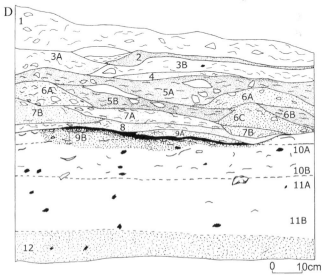

D

E

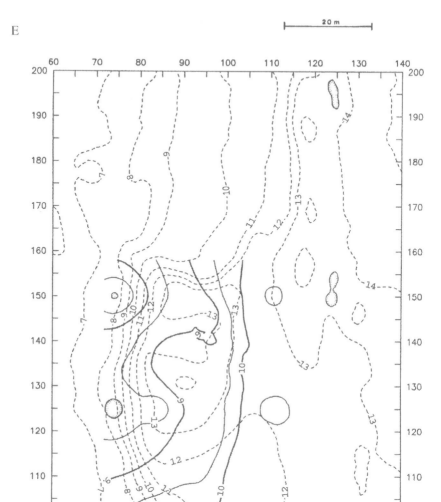

Figure 4 (Continued). Taperinha shellmound. A. Mound. Photo by A. C. Roosevelt. B. Interpretive cross-section from radar data. C. Interpretive cross-section from seismic data. Section by B. Bevan. D. Stratigraphic cross-section of dated hearth feature at shellmound base. Section by A. Roosevelt. E. Map showing interpretive basal cross section of shellmound. By B. Bevan.

amphibians. Our radiocarbon dates on charcoal, shell, and pottery and our TL dates on pottery bracketed the Archaic occupation between ca. 7,100 and 6,000 uncalibrated radiocarbon years ago. We named this culture Taperinha after the property to which it belongs.

Our surface walkovers of the site revealed a discrete c. 6-m-tall, 1-ha. shellmound on sloping terrain overlooking river channels and swamps. On one side was a large pit cut though the mound to the sand under it. Members of the property owners informed us that the Ludwig Jari plantation company had bulldozed this pit and carted the shell away for fertilizer. Our detailed topographic map revealed that the original shellmound was a topographic anomaly that extended about two ha. One-fourth to one-half of the original archaeological mound might have been removed by the quarrying.

Our geophysical surveys were aimed at estimating the area, thickness, and main layering of the shellmound and at locating middens, structures, burials, and disturbances. They were conducted in the cardinal directions north-south and east-west. Electrical resistivity, radar, and seismic refraction surveying methods worked well. In all three surveys, shell layers contrasted readily with the sand. Electrical resistivity transects produced useful predictive pseudo-sections of the mound. They indicated a superficial low-resistivity layer between about 40 cm to 1 m below the mound's surface; a very high resistivity mass within the body of the mound, just below the surface layer; and a moderate resistivity layer beneath the shellmound. Based on the geophysical results and our surface observations, the team interpreted the upper resistivity stratum as the late prehistoric black soil midden; the intermediate strata block as the main body of the shellmound; and the lowest resistivity layer as either the sand or water below the mound.

Radar was carried out in transects across the mound, the bulldozed quarry area, and the surrounding topography. It produced excellent, deep, detailed profiles in both the shell deposit and the underlying sand. The radar profiles indicated that the base of the shellmound lay about 5 m below the present surface of the mound (Figure 4B) and that there were 5- to 10-m long, oblong features within the mound, possible structure floors or large pits. In addition to the layering, the radar profiles recorded numerous small features that we interpreted as small, discrete occupation features, such as hearths, minor structural elements, or burials. Seismic soundings were carried out in selected transects through the same area as the radar. They recorded the same interface at the base of the midden, and an additional one, 2–3 m below the base, possibly indicating the water table or bedrock, in the view of the geophysicists (Figure 4C).

Thus, radar, seismic, and resistivity surveys had roughly similar results regarding the main layering of the mound. At least two major occupation stages were detected in the shellmound and distinct interfaces at and below its base.

Electromagnetic conductivity surveying results were not as useful as radar, seismic, and resistivity results. They were marred by sferic disturbances. Also, the loose sediment at the site made readings difficult to make, and conductivity variation within the site was too low at the shallow depths measurable to produce interpretable patterns. The conductivity results, however, did support the topographic evidence for the original size of the mound. The conductivity maps revealed a large area of greater conductivity on and around the mound and within the quarry pit. Magnetometer and self-potential were also attempted along some of the same transects, but they did not produce readily interpretable results. Unfortunately, the shell itself had no magnetism, though surprisingly all the rocks within the mound were magnetized. Similar iron-rich, lateritic rocks, when found outside the archaeological site were not magnetic, so it is probable that the site rocks got their magnetism by being burned during human activities. But our stage-2 excavations showed that the rocks in the site were mostly too small to be detectable from the surface by magnetometers, but the few magnetic anomalies detected in the mound may be concentrations of such rocks. As at our other sites, self-potential survey recorded little variation on and around the site.

Our stage-2 excavations accomplished so far were aimed at exposing stratigraphy, sampling some smaller radar features, verifying the artifact associations of strata and features, assessing disturbances, and establishing the chronology. Based on the surface remains and geophysical results, we expected to uncover remains of structures, hearths, datable materials, food remains, significant artifact groupings, burials, and garbage in the anomaly areas. We placed a long trench down the cut face of the mound and several test-pits on top, below, and beside it. These excavations indeed uncovered abundant artifacts, identifiable biological remains, samples for dating, hearths (Figure 4D), human remains, and many postholes.

Our excavations have confirmed the early Holocene age of the mound and revealed information about its ceramics and residential character and subsistence by foraging. Palynological research by others at Prainha, opposite Taperinha, documented multiple indicators of human disturbance of vegetation but no proof of plant cultivation (Piperno and Pearsall, 1998). Thus, our work at Taperinha has revealed the existence of a culture new to Amazonian scientific archaeology, one with unexpected characteristics in terms of earlier theories. These river foragers existed long before Amazonia was thought inhabited by humans, and they attained a degree of residential permanence and a level of cultural innovation—early decorated pottery—thought impossible there.

The original area of cultural occupation, according to the topography, geophysical surveys, and excavations, covered as much as 5 ha., more than expected. The shellmound and its vicinity held at least five cultural strata: the superficial later prehistoric Santarem phase deposit, traces of Polychrome Horizon and Formative occupations just above the Archaic shellmound occupation, and a non-shell occupation layer beneath it. Thus, the shell midden itself does not represent the only cultural occupation at the site nor the whole of the Taperinha phase archaeological settlement. It is merely a topographic eminence in the middle of a very complex ancient settlement.

The Santarem thick black-soil midden that we excavated on top of the mound was between 50 and 100 cm thick. This midden matched the top layer in the resistivity pseudosections. The excavations located the base of the shellmound between 4 and 5 m down from the surface, approximately the depths predicted (Figure 4E). In addition, within the sand under the mound, we found a 1- to 2-m earlier cultural stratum, ending in very moist sand. This cultural stratum may be the lower of the two basal radar interfaces. We have not yet dug deep enough to ascertain whether the basal seismic interface is water-table or bedrock. Outside of the shellmound and the survey area, under recent colluvium, our test pits uncovered a wide area rich in archaeological remains, including garbage, post holes, and burials.

Because of the size and complexity of the ancient site, we plan to extend stage-1 and -2 research beyond the areas covered so far. We plan to dig down to the lowest apparent interface. We also need to excavate more types of occupational features in and around the shellmound and further sample both the pre- and post-shellmound occupations. Of special interest is to identify archaeologically the broader discrete geophysical anomalies within the mound. If these are ceremonial structures and facilities like those found in South Brazililan coast shellmounds, we may have to interpret the Archaic societies as complex hunter-gatherers, rather than simple village societies (Gaspar, 1999).

With this knowledge, we can generate a model of the site at different times. Then in stage-3 research we will carry out wide-area excavation at the above significant activity areas and any that further stage-2 research can uncover. We hope to ascertain the original size of the mound, the size and nature of structures, the health status of the ancient populations, and their ecological and economic adaptations. This final research stage should be able to refine our interpretation of the nature of the settlement and the people's way of life.

4.3. Santarem

The Santarem culture is known for elaborate pottery vessels decorated with representational and geometric plastic and painted designs, large ceramic human figures, and varied polished-stone tools and sculptures. The culture was assumed to have been a chiefdom of the conquest period. Santarem city, for which the culture is named, may have been the central site of the culture. The city lies in sandy uplands about half way between Manaus and Belem on the south bank of the Lower Amazon. Like other Santarem sites, the city is covered with a thick layer of archaeological black earth. To judge from inhabitants' occasional finds, ancient Santarem appears to have been a large indigenous settlement of ca. 4 sq km. Though no scientific excavations had been carried out before, over the centuries the town inhabitants had found abundant decorated pottery and stone artifacts in *bolsoes*, described as large, deep pits filled with black soil, smashed pots, stone artifacts, and human statues. Additionally, locals noted the existence of low earth mounds in the site (Sr. Laurimar, personal communication, 1987).

A major issue has been whether Santarem was organized as a chiefdom or state and whether the European conquest influenced its formation. Many researchers have assumed that tropical rainforest could not support complex, sedentary, agricultural societies. Our goal at the site was to date the culture, evaluate its subsistence, environment, and crafts, and to uncover structures and activity areas with the layout of the site. Their patterning could give insights into the nature and history of socio-political and socio-economic organization. Our project's stage-1, -2, and preliminary -3 research has dated the culture and uncovered stratigraphy, structures, ritual facilities, and in situ cultural and biological material. Although magnetic, resistivity, and conductivity surveys were not helpful for mapping site patterning, the radar surveys produced superlative results, and during excavation we uncovered distinct stratigraphy and materials at the different types of radar anomalies.

Since much of the site now lies under modern paving and structures, we are focusing stage-3 research in the open area at the port. In stage-1 and -2 work in the city proper, work spaces were tight and recent stratigraphic disturbances frequent. In stage 1, we made detailed topographic maps of the entire port, an area of about 8 ha., and of a backyard in the city. The topographic map of the Port near some oil tanks showed two parallel rows of low mounds and depressions (Figure 5A). We completely radar-surveyed this area and another near a building owned by the Port stevedores, together amounting to about 550 sq m. Radar worked especially well in the dry, sandy sediment of Santarem. We attempted magnetometer, resistivity, and conductivity survey also, but these methods did not produce patterns of archaeological scale. In some areas of the port, the large amount of modern metal trash obscured archaeological magnetic and conductivity patterns, which, in any case, were minimal. The resistivity generally was high, increasing with depth, and indicated sediment horizons 1–3 m deep were slightly more conductive than the subsoil. We hypothesized that these intermediate horizons might be the prehistoric black or grey soils exposed at disturbances in the site.

In the high-resistivity sediment, radar worked well. The stage-1 radar research, run in north-south lines 50 cm apart, recorded three kinds of distinctive geophysical features. The first were numerous single-point anomalies about 1 m in diameter (Figure 5B), dotted about the surveyed areas mostly between 1 and 2 m in depth (Figure 5C). We hypothesized that these might be *bolsoes* (pockets of black earth), buried urns, or large human statues. The second were larger, linear or oblong areas of greater impedence of microwaves, interspersed among the point anomalies. We hypothesized that these could be stratigraphic interfaces with more conductive sediment, such as the edges of house foundations or platforms cemented with clay or silt or, alternatively, lines of the type of buried objects marked by the point anomalies. Some of these linear anomalies were narrow, slanting lines of

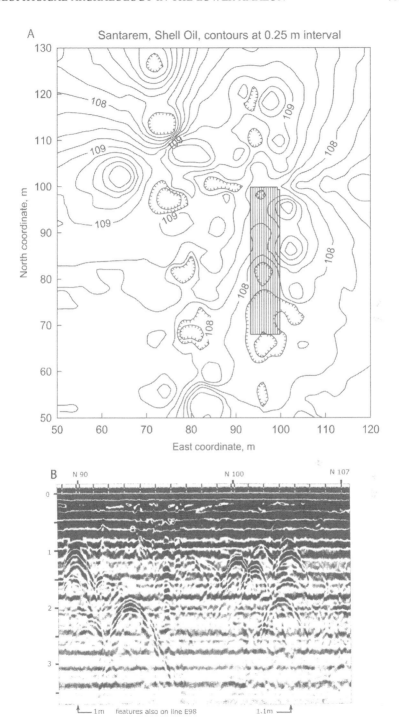

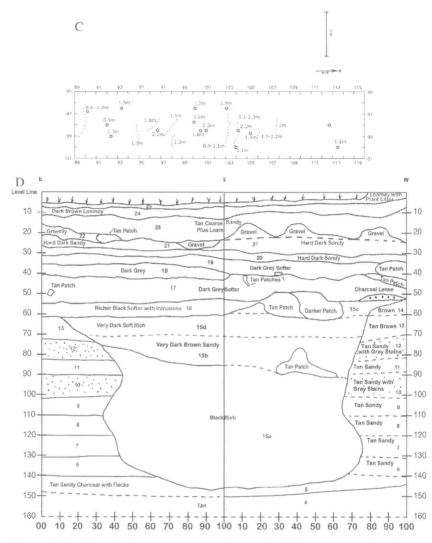

Figure 5 (Continued). Santarem Port site. A. Map detail showing mounds, adjacent borrow pit depressions, and project radar survey tract, Shell Oil locality. Map by M. Perry and B. Bevan. B. Radar profile anomalies at the Stevedore locality. Readout by B. Bevan. C. Geophysical interpretation map of radar anomalies, same locality. By B. Bevan. D. Stratigraphic profile of ritual pit (*bolso*) at the point anomaly at the far right in Figure B. Profile by A. C. Roosevelt, E. Quinn, L. Brown, and P. Lane.

greater impedence. The geophysicists hypothesized these to be modern roots or buried tree trunks. (Acidic, sandy tropical soils can indeed preserve ancient "organic horizons" of tropical forest litter, and some local trees have deeply penetrating roots.) The third type of anomaly was a distinct, broad interface at depths of 6m or more. This interface we guessed marked bedrock, a lateritic conglomerate.

Our stage-2 test-excavations uncovered both domestic and ritual deposits in a dark archaeological soil deposit 1–3 m below the surface. All the cultural layers and features we encountered within the black deposits pertained to the Santarem culture. These deposits arguably are the less resistant layers detected by survey. Below these features and layers was a diffuse layer of dark grey sandy sediment containing only scarce objects consisting of carbonized tree fruits, food bone, and very rare small stone flakes and sherds. This lower, less organic deposit could be the more resistant horizon detected below the less resistant horizon by the surveys.

In the excavations at the point radar anomalies, particularly, we uncovered large ritual pits filled with black soil, special food remains, broken decorated pottery and statues, and stone carvings and tools (Figure 5D). These indeed were *bolsoes*. The upper parts of the pits were surrounded or flanked by layers of comparable soils and materials, representing areas of ritual activity where people prepared and ate special foods and made and used fancy objects. These layers associated with the *bolsoes* seem to correlate with the larger, more linear anomalies. The artifacts in the ritual-activity areas were typical of classic Santarem culture (Quinn 2004): parts of highly decorated representational ritual vessels; parts of large and small human figures; large, broken, incised spindle whorls; and tools for making them. The food remains included bones of very large fish (> 1 m long), shells and bones of large river turtles (*Podocnemis*), numerous pits of Amazonian fruits now valued as ritual food, and a small amount of maize, which we had also found at other Santarem sites. The purposeful breakage of the fine artifacts in these features suggest termination ceremonies.

Adjacent to the ritual deposits, we encountered hard-packed black earthen floors containing scarce cultural and biological remains: small red-washed sherds, tiny stone artifact fragments, small charcoal pieces, and bones of small fish, representing intensive communal fishing. The floors were composed of many thin superimposed clayey-silty laminations, occasionally associated with a post base. We interpret the floors as the foundations of domestic houses maintained for appreciable periods of time. The interface between these blocks of floors and adjacent stratigraphy may have caused some of the larger, longer radar anomalies detected in the surveys.

Below the structure floors and the ritual features next to them, was an earlier deposit of dark grey, sandy sediment. This layer may represent a Formative-age activity area where people planted and utilized tree crops. We associate this earlier deposit with ancient "agroforestry" activity, since there were abundant useful plants but only rare cultural materials. In and below this layer, we uncovered modern tree roots and ancient carbonized logs. Our survey appears to have detected this grey layer as a zone of decreased resistivity.

The radiocarbon dates from our excavations show that classic Santarem culture began at least two hundred years before the European conquest and thus could not have been elicited by contact with Europeans, who arrived in the Amazon about 1540. Eleven tree-ring-calibrated radiocarbon dates on carbonized wood excavated with diagnostic potttery fell between A.D. 1300 and 1500. Two dates on charcoal from the interface of a Santarem ritual pit and the earlier dark grey, sandy layer were 2329 +/− 63 (379 B.C.) and 3003+/− 56 B.P. (1095 B.C.), making the "agroforestry" stratum that the pit cut into Formative in age. The stable isotope ratios of the Santarem culture dates averaged −25.6., indicating a thinned tree canopy during the occupation, compared to the earliest Formative date's stable isotope ratio of −29.2, which is significantly more negative, suggesting a closed-canopy forest then (Roosevelt, 2000a, appendices). (Today Santarem is surrounded by savanna woodland. Forest has been eliminated by cutting, burning, and bulldozing.)

To summarize the accomplishments of our stage-1, -2, and -3 research, the surveys and excavations have successfully pinpointed the important *bolsoes* features, platform structures, and adjacent craft and feasting activity areas. Our continuing stage-3 excavations at Santarem are investigating further the ancient site plan. We are exposing groups of house platforms and ritual facilities by excavating fully the two Port areas surveyed and tested in stages 1 and 2. We will use dating and stratigraphy to assess how structures and activities were related and how long they were used; we can compare the artifact technology and style between them to assess cultural diversity and change through time; and we can further investigate the earlier Formative culture and land use. With a larger sample of dated and identified biological remains we can better trace patterns of environment and human subsistence through time and space. We can, in addition, dig to bedrock to test the radar survey's predictions. We also hope to find deposits of the historical period, which could reveal what happened to Santarem during the conquest. The combined information will give quite a full picture of the neighborhood of ancient Santarem now in its Port.

5. CONCLUSIONS

As a result of the geophysical archaeology carried out by our project so far in Amazonia, some theories about human biological and cultural evolution need to be revised.

Our results from Taperinha show that the humid neotropics were not off-limits to early sedentary foragers (Roosevelt , 1991; Roosevelt et al., 1996), contrary to theories that tropical forest wild foods are too limited to sustain humans (Bailey et al. 1989). The Taperinha foragers got most of their food not by big-game hunting or agriculture, but by collecting fish, shellfish, plants, and small animals such as turtles, frogs, and rodents. The early Holocene pottery cultures at Taperinha and nearby sites were by far the earliest in the Americas, as early as 7,500 radiocarbon years ago (Roosevelt, 1995; Roosevelt, 1991), contrary to the idea the first cultures appeared in late prehistory. Our results from Marajo show eastern Amazonian Polychrome Horizon mounds were platforms for large, sedentary villages, not ephemeral, empty ceremonial centers, as had been assumed (Meggers and Evans, 1957). They did not live in temporary camps or ephemeral hamlets, as do modern Amazonian Indians, but in large, long-term, architecturally complex villages or towns. The site layouts that we investigated do not resemble those of recognized states or maximal chiefdoms but rather those of prosperous tribal societies. The large well-built villages appear to lack palaces and civic structures indicating centralized hierarchical rule. Contrary to the Andean invasion idea, the culture's start-date early in the Christian era is nearly a millennium earlier than the Andean Polychrome cultures. And finally, our surveys and excavations at late prehistoric Santarem uncovered artificial house mounds and adjacent special ritual facilities. How the people were organized within the site and the society we do not know, as we only know about part of a single neighborhood so far. Our results, however, show that the elaborate culture found at Santarem and other sites is a prehistoric culture, and so its origin cannot be attributed to influence by Europeans, who arrived later. Like the Polychrome culture, then, this complex culture, too, was indigenous to Amazonia.

Such findings show that, at times, Amazonia was a leader in South American cultural developments, not just a passive recipient of innovations from elsewhere. Amazonian ecology, then, was not such a barrier to indigenous development, and it stands to reason that other tropical rainforests would not have been barriers to prehistoric human development, either. Our Amazonian findings therefore encourage the rethinking of human evolution in other parts of the world.

The research in Amazonia brings up other provocative general questions amenable to geophysical archaeological approaches. For example, our and others' research show prehistoric human activities that had impacts on the environment: damage to forest vegetation near habitation sites occupied for hundreds of years, transformation of landscapes with large earthworks and middens. From these examples, we can understand the tropical forests and savannas of the globe neither as natural settings for human events nor as inexorable determinants of human behavior, but rather as humanized habitats that coevolved along with their inhabitants. And we now know that archaeologists hold unique information about the nature and history of tropical forests because the sites yield abundant datable biota not found in paleontological or palynological sites.

Given the apparent dynamism of human cultures and human ecology in Amazonia, there are many questions to resolve, and geophysical archaeology can help. To better document the diversity of the first peoples, we need to look at a wider range of late Pleistocene and early Holocene sites.

Remote sensing can record the noon-to-midnight thermal anomalies that show where caves exist, the soil chemistry and vegetation anomalies at early shellmounds, and the sonar anomalies at sites now under rivers and coastal waters. With geophysical techniques we also can investigate the internal organization of human sites. Scientists with the Center for Remote Sensing at Boston University have shown that simple balloon-lofted systems sensitive to heat, moisture, topography, and albedo are more cost-effective than on-land prospection to investigate site plans. We also can economically assess the magnitude of ancient human settlement on regional scales. Existing remote sensing images can be used to map the topography of ancient earthworks or the unique vegetation on the huge garbage dumps of Amazonian cultural centers.

But to take best advantage of both airborne and ground-based geophysical techniques for archaeology, whatever the habitat, we need a cooperative geoarchaeology institute that can maintain-state-of-the-art equipment and computers, foster needed expertise, teach methodological innovations, organize conferences and workshops, and disseminate results worldwide.

REFERENCES

Alves, J. de A., and Lourenco, J. S., 1981, Metodos geofisicos aplicados a arqueologia no estado do Para. *Boletim do Museum Paraense Emilio Goeldi [serie] Geologia [N.S.]* 26:1–52.

Bailey, R. C., Head, G., Jenicke, M., Owen, B., Rechtman, R., and Zechenter, E., 1989, Hunting and gathering in a tropical forest: Is it possible? *American Anthropologist* 91:59–81.

Bevan, B. W., 1986, *A geophysical survey at Teso dos Bichos*. Report on the 1985 fieldwork on Marajo Island.

Bevan, B. W., 1989, *Geophysical Surveys at Three Sites along the Lower Amazon River*. Pitman, NJ: Geosight. Unpublished report.

Bevan, B. W. and Roosevelt, A. C., 2003, Geoarchaeological Exploration of Guajara, A Prehistoric Earth Mound in Brazil. *Geoarchaeology* 18(3):287–331.

Gaspar, M. D., 1999, Os Pescadores-Colectores-Cacadores do Litoral. In *Prehistoria de Terra Brasilis*, edited by M. C. Tenorio, pp. 159–170. Universidade Federal de Rio de Janeiro.

Hartt, C. F., 1886, Contribuicoes para a ethnologia do valle do Amazonas. *Archivos do Museu Nacional* 6:1–174. Rio de Janeiro.

Lathrap, D. L., 1970, *The Upper Amazon*. Thames and Hudson, London.

Lourenco, J. S., 1985, *Anomalias magneticas de feicoes arqueologicas na Ilha de Marajo, Para, Brazil*. Manuscript.

McCauley, J. F., Schaber, G. G., Breed, C., Grolier, M. J., Haynes, C. V., Issawi, B., Elachi, C., and Blom, R., 1982, Subsurface valleys and geoarchaeology of the Eastern Sahara revealed by Shuttle radar. *Science* 218(4576):1004–1020.

Meggers, B. J., and Evans, C., 1957, Archeological Investigations at the Mouth of the Amazon. *Bulletin of the Bureau of American Ethnology* 167.

Piperno, D., and Pearsall, D., 1998, *Origins of Agriculture in the Lowland Tropics*. Academic Press, San Diego.

Quinn, Ellen R., 2004, *The Age and Archaeological Context of Ceramics at Santarem*. Ph.D. dissertation, Department of Anthropology, University of Illinois, Chicago.

Roosevelt, A. C., 1980, *Parmana: Prehistoric Maize and Manioc Subsistence along the Amazon and Orinoco*. Academic Press, New York.

Roosevelt, A. C., 1987, Chiefdoms in the Amazon and Orinoco. In *Chiefdoms in the Americas*, edited by R. Drennan and C. Uribe, pp. 153–185. University Press of America, Lanham, MD.

Roosevelt, A. C., 1988a, Microcomputers in the Lower Amazon Project. *Advances in Computer Archaeology* 4:41–53.

Roosevelt, A. C., 1988b, A Arqueologia Marajoara. *Revista do Museu Paulista* 34:7–40. Sao Paulo.

Roosevelt, A. C., 1989, Resource Management in Amazonia before the Conquest: Beyond Ethnographic Projection. *Advances in Economic Botany* 7:30–62.

Roosevelt, A. C., 1991, *Moundbuilders of the Amazon: Geophysical Archaeology on Marajo Island, Brazil*. Academic Press, San Diego.

Roosevelt, A. C., 1995, Early Pottery in the Amazon: Twenty Years of Scholarly Obscurity. In *The Emergence of Pottery: Technology and Innovation in Ancient Societies*, edited by W. Barnett and J. Hoopes, pp. 115–131. The Smithsonian Press, Washington, D. C.

Roosevelt, A. C., 1997, The Excavations at Corozal, Venezuela: Stratigraphy and Ceramic Seriation. *Yale University Publications in Anthropology*, No. 83. New Haven.

Roosevelt, A. C., 1999a, The Maritime-Highland-Forest Dynamic and the Origins of Complex Societies. In *History of the Native Peoples of South America*, edited by F. Salomon and S. Schwartz, pp. 264–349. Cambridge University Press, Cambridge.

Roosevelt, A. C., 1999b, The Development of Prehistoric Complex Societies: Amazonia, A Tropical Forest. In *Complex Polities in the Ancient Tropical World*, edited by A. A. Bacus and L. J. Lucero, pp. 13–33. *Archaeological Papers of the American Anthropological Association*, No. 9.

Roosevelt, A. C., 1999c, Twelve thousand years of human-environment interaction in the Amazon floodplain. In *Diversity, Development, and Conservation in Amazonia's Whitewater Floodplains*, edited by C. Paddoch, J. M. Ayres, M. Pinedo-Vasquez, and A. Henderson, pp. 371–392. *Advances in Economic Botany*, Vol. 13.

Roosevelt, A. C., 1999d, The role of floodplain lakes in human evolution in Amazonia and beyond. In *Ancient Lakes: Their Cultural and Biological Diversity*, edited by H. Kawanabe, G. W. Coulter, and A. C. Roosevelt, pp. 87–100. Kenobi Publications, Ghent, Belgium.

Roosevelt, A. C., 1999e, The development of prehistoric complex societies: Amazonia, A tropical forest. In *Complex Polities in the Ancient Tropical World*, edited by E. A. Bacus and L. Lucero, pp. 13–14. *Archaeological Papers of the American Anthropological Association*, Number 9.

Roosevelt, A. C., 2000a, The Lower Amazon: A Dynamic Human Habitat, The Case Study of Santarem. In *Imperfect Balance: Landscape Transformations in the Precolumbian Americas*, edited by D. Lentz, pp. 455–491. Columbia University Press, New York.

Roosevelt, A. C., 2000b, New Information from Old Collections: The Interface of Science and Systematic Collections. *Cultural Resource Management* 23(5):25–30.

Roosevelt, A. C., Douglas, J., and Brown, L., 2002, Migrations and Adaptations of the First Americans: Clovis and Pre-Clovis Viewed from South America. In *The First Americans*, edited by Nina Jablonski, pp. 159–236. *Memoirs of the California Academy of Sciences* No. 27. University of California Press.

Roosevelt, A. C., Lima da Costa, M., Lopes Machado, C., Michab, M., Mercier, N., Valladas, H., Feathers, J., Barnett, W., Imazio da Silveira, M., Henderson, A., Sliva, J., Chernoff, B., Reece, D. S., Holman, J. A., Toth, N., and Schick, K., 1996, Paleoindian Cave Dwellers in the Amazon: The Peopling of the Americas. *Science* 272:373–384.

Roosevelt, A. C., Housley, R. A., Imazio da Silveira, M., Maranca, S., and Johnson, R., 1991, Eighth Millennium Pottery from Prehistoric Shell Midden in the Brazilian Amazon. *Science* 254:1621–1624.

Schaan, D. P., 1997, A Linguajem iconografica da Ceramica Marajoara: Um Estudo da Arte Prehispanica da Ilha de Marajo (400–1300 A.D.). Edipucrs, Porto Alegre.

Schaan, D. P., 2004, The Camutins chiefdom: rise and development of social complexity on Marajo Island, Brazilian Amazon. Ph.D. dissertation, Anthropology, University of Pittsburgh.

Willey, G. R., 1971, *An Introduction to American Archaeology*. Vol. 1. Prentice-Hall, Englewood Cliffs, N. J.

Section V

Maritime Setting Applications

Chapter *19*

Archaeological Oceanography

ROBERT D. BALLARD

Abstract: Recent advances in deep-submergence technology as well as the discovery of a growing number of cultural sites in the world's oceans have led to the emergence of a new field of research, Archaeological Oceanography. The purpose of this paper is to review the short history of this newly emerging field of research including the evolution of remotely operated vehicle systems, various examples of their application in field programs, and the challenges future programs face in gaining access to the resources and tools needed to carry out such work in the great depths and remote locations of the world where submerged cultural resources may be found.

1. INTRODUCTION

Marine archaeology and deep-submergence science and technology are relatively young efforts, each beginning the early 1950s (Ballard, 2001). During the intervening years they have run along parallel paths, periodically intersecting but in large part these two marine programs have been operating in two separate worlds: one shallow, the other deep; one largely supported by private sources, the other by the Federal Government.

The field of marine archaeology has been dominated by SCUBA diving technology while deep-submergence programs have by their very nature required researchers to encapsulate themselves in small expensive submersibles to guard against the crushing deep of the abyss or to use sophisticated unmanned vehicle systems to avoid human risk altogether and to increase their bottom time.

There are past examples where these two worlds have worked together. The pioneering work of George Bass with the development of the ASHERAH submersible and the use of deep-towed side-scan sonars, or the more recent development and use of the submersible *Carolyn* are excellent cases in point. But by in large, these early efforts were exploratory in nature, were limited to relatively shallow depths, and did not become mainstream tools, particularly manned submersibles, for the marine archaeological community that continues to this day to rely heavily upon SCUBA diving technology. Conversely, oceanographers like K.O. Emery and Robert McMaster (Ballard and Hively, 2000) used manned submersibles to investigate Neolithic sites on the shallow continental shelf off New England, but they also quickly returned to their primary studies in marine geology and did not mount sustained archaeological programs under the oceanographic banner.

The initial lack of major cooperation between these two worlds was based upon not only fundamental cultural and academic difference but also the lack of a common technological base to bind them together. Geological oceanographers like Emery and McMaster were interested in mapping undersea terrains to varying degrees of detail. As a result, the oceanographic community has developed a broad array of mapping sonar systems including multi-narrow-beam systems like SEABEAM and phased-array sonars like the MR-1 system. But earth scientists have always had to play the range versus resolution game. SEABEAM and the MR-1 have an operating frequency of 12 kHz that makes it possible to map large expanses of the ocean floor, but such a frequency lacks the ability to detect smaller objects like ancient shipwrecks.

When the oceanographic community began to develop higher frequency sonar systems, in particular side-scanning sonars with 100–200 and now 500 kHz, they lost range but picked up important resolving power that made them of interest to the archaeological community since they could now detect ancient shipwrecks. In a similar fashion, navigating accuracy and mapping precision were initially crude when oceanographers relied upon Loran for ship navigation and long baseline transponder networks for vehicle and submersible tracking, far too inaccurate for the fine-scale mapping of a submerged cultural site. Submersibles also lack precision control, thereby making it next to impossible to work near a cultural site for fear of causing damage.

In the mid-1980s, our Deep Submergence Laboratory (DSL) at the Woods Hole Oceanographic Institution (WHOI) began to develop a high-frequency navigation system called SHARPS (Ballard et at., 1991; Ballard, 1993). Our desire for precision navigation grew not out of a need to make highly accurate maps but a desire to precisely control our new remotely operated vehicles or ROVs. A by-product of such precision control, however, was precise navigational accuracy. All of sudden, we could determine the three-dimensional orientation of our vehicle 10 times a second to a few centimeters of accuracy in three-dimensional space. With such positioning accuracy for our sensors, we were able to create micro-topographic maps of the bottom with a contour interval of centimeters that could clearly detect and delineate shipwrecks and their associated features. Such accuracy in navigation also permitted precise manipulator control, thereby making it possible to pick up small and delicate objects without landing on a site as well as making it possible to quickly construct photo-mosaics of bottom features; all to the precision of traditional archaeological mapping efforts. It was the emergence of this technology base that not only served the needs of the oceanographic community, but also the archaeological community that caused our DSL Group to begin working with marine historians, archaeologists, and anthropologist in the 1980s.

Ironically, it was our quest to find the sunken British luxury liner RMS TITANIC in the mid-1980s (Ballard, 1987) that led to the emergence of what is now being called Archaeological Oceanography. At that time, all three major "Blue-water" oceanographic institutions—the Scripps Institution of Oceanography, Columbia University's Lamont Geological Laboratory, and the Woods Hole Oceanographic Institution—sought to find TITANIC's final resting place. These three efforts were lead by Fred Spiess, William Ryan, and the author respectively. All three are marine earth scientists who shared an interest in deep-submergence engineering as well as human history beneath the sea. It was at these three oceanographic institutions were deep-submergence science and technology flourished in the 1980s with Federal funding from the Office of Naval Research and later the National Science Foundation. At that time, Scripps and Lamont focused upon unmanned vehicles while the focus of Woods Hole was on manned submersibles, although it was common for scientists from all three of these institutions to use both manned and unmanned deep-submergence assets.

Project FAMOUS, the first manned exploration of the Mid-Atlantic Ridge (Ballard et at., 1975), clearly demonstrated the complementary nature of manned and unmanned vehicle systems. The preliminary mapping and detection phase of the project made use of deep-towed acoustic and optical imaging vehicles like Deep-Tow, ANGUS, GLORIA, and LIBEC, and manned submersibles were used to conduct the subsequent detailed documentation and carry out sampling operations.

With the advent of high-bandwidth fiber-optic cables, the deep-submergence community began to develop and use highly sophisticated remotely operated vehicles, or ROVs, the first being the ARGO/JASON system at Woods Hole. It was the ARGO towed visual and acoustical mapping system that found the TITANIC in 1985 in 3,750 m of water and an early prototype of JASON, JASON Jr. or JJ, that explored the interior of TITANIC in 1986. Following the discovery of the TITANIC, ARGO was used to find the German Battleship BISMARCK in 5,000 m of water (Ballard and Archbold, 1990), followed by similar programs to find the aircraft carrier USS YORKTOWN, lost during the Battle of Midway in 5,500 m of water (Ballard and Archbold, 1999), PT-109 (Ballard, 2002), and a series of Australian, Japanese, and American warships lost in the Battle for Guadalcanal (Ballard and Archbold, 1993).

The first use of this new remotely operated vehicle technology to locate and document ancient shipwrecks in the deep sea began in 1988 when we used ARGO to locate ancient shipwrecks in the central Mediterranean Sea near a region at the northern end of the Straits of Sicily called Skerki Bank (Ballard et al. 2000). That effort was followed by a precision mapping effort of two warships from the War of 1812 in Lake Ontario that used the JASON ROV and the SHARPS navigation system to carry out a precision mapping program of the HAMILTON and SCOURGE. In subsequent years, our laboratory worked with a number of archaeologists in the Eastern Mediterranean and Black Sea as more and more ancient shipwrecks were found in the deep sea.

The driving force behind these initial programs was to determine if the ancient mariners used deep-water trade routes, venturing beyond sight of land with their ships, and traveling hundreds of miles out to sea. Finally, after several years of exploratory programs it became clear to the author that the deep sea is a large repository of human history having important historical, archaeological, and anthropological importance that required a concerted effort. But following the successful development of the ARGO/JASON vehicle systems in DSL, these assets were transferred to the National Deep Submergence Facilities at WHOI and came under the control of the National Science Foundation, which did not have a major interest at the time in having these assets used by social scientists.

From 1997 to 2002, the author continued his research in cultural history beneath the sea with the formation of the Institute for Exploration at Mystic Aquarium in Mystic, Connecticut. During that period of time, he began to develop a new series of underwater vehicles specifically for deep-water archaeological programs. It became clear to the author (Ballard, 2004), however, that for this newly emerging field of research to flourish, it needed to develop within an academic setting of both oceanographers and social scientists, and it needed its own vehicle technology that was designed specifically for this newly emerging field of Archaeological Oceanography. This realization has led the author to create the Institute for Archaeological Oceanography (IAO) at the University of the Rhode Island's Graduate School of Oceanography (GSO).

During this same period of time, other discoveries were being made in the deep sea that demonstrated the presence of ancient shipwrecks far from shore. Most of these programs, however, had other goals and the discovery of ancient shipwrecks was not their primary objective. For that reason, their results were not published in professional journals, so the full impact of these discoveries awaits further study.

In 1998, Odyssey Marine Exploration Inc. while searching for a British warship off Gibraltar that sank more than 300 years ago that was carrying a large cargo of coins, discovered a Phoenician shipwreck dating to the $5^t h$ century B.C. (Environmental News Network Staff, 1998). While statements have been made about returning to the site for subsequent scientific analysis, nothing has occurred as yet.

In the spring of 1999, the NAUTICOS Corp. discovered an ancient shipwreck in 10,000 feet of water south of Cyprus while searching for the lost Israel submarine DAKAR (Dettweiler et al., 2001). As before, statements have been made about returning to the sea to conduct an archaeological investigation of site; none has occurred to date.

The Norwegian Institute of Archaeology at the Norwegian University of Science and Technology (NTNU) has been conducting a number of underwater archaeological projects over the years and is now beginning to move into deeper water (Jasinski and Soreide, 2004).

The Institute for Nautical Archaeology (INA) has initiated a number of underwater search efforts using advanced technology, including one off Mt. Athos (Wachsmann, 2004), but none has yet located any ancient shipwrecks in the deep sea.

The purpose of this paper is to describe this new program, the technologies IAO has or is at present developing, and its recent field programs using this technology base.

2. IAO'S DEEP-SUBMERGENCE VEHICLE PROGRAM

2.1. Search Systems

A number of lessons that had been learned using DSL's undersea vehicles on deep-sea archaeological programs from 1988 to 2001 were taken into consideration when designing and developing a new family of undersea vehicles for IAO's new Archaeological Oceanography program at GSO. In the past, we had used the DSL-120 side-scan sonar as our primary acoustic search system and the ARGO towed sled for visual detection. But the DSL-120 was designed to map geologic features and to create bathymetric maps as well. With an operating frequency of 120 kHz, it had an acoustic swath-width of 600 m, but tuning the sonar to detect small archaeological targets resulted in the inability to measure the phase of the returning signal and therefore the loss of topographic information within the acoustic swath of returning signal.

Creating a topographic map of the ocean floor, however, was less important to the archaeological oceanography than the ability of the sonar to detect the subtle acoustic signal of an ancient shipwreck as well as its associated features.

2.1.1. ECHO

For the reasons given above, our engineers designed a five-channel side-scan sonar system called ECHO (Figure 1). Two channels are dedicated to looking side to side at a frequency of 100 kHz and a swath width of 1,000 m to detect ancient shipwrecks. Two additional channels are dedicated to looking side to

Figure 1. IAO's side-scan sonar vehicle ECHO (left), HERCULES ROV (middle), and LITTLE HERC ROV (right).

side at a higher frequency of 500 kHz to delineate shipwreck features and a fifth channel is dedicated to a chirp 2–7 kHz sub-bottom profiler to delineate buried objects. ECHO is towed behind a depressor weight that helps dampen out the heave of the towing ship so that ECHO follows a smoother path through the ocean, producing a clearer record of the bottom's features.

2.1.2. ARGUS

The primary visual search system used at DSL was the ARGO towed vehicle that was successful in finding both the TITANIC and BISMARCK as well as the first Roman trading ship on Skerki Bank in 1988. Its primary search sensor was an array of low light-level black and white SIT cameras. It also was outfitted with a 100-kHz side-scan sonar, but when towing it at the slow speed of 1 knot, ARGO commonly "crabbed" making its sonar ineffective as a search sensor.

When designing ARGUS (Figure 2), the traditional instability of ARGO was taken into consideration and lateral thrusters were added at both ends of the 3.5-m-long vehicle as well as an attitude package containing an electronic compass and angular rate sensor that made it possible for ARGUS to maintain a constant heading while being towed to depths of 6,000 m. A 675-kHz fan-beam scanning sonar replaced the side-scan sonar since ECHO now serves as the primary acoustic search system for IAO and ARGUS became less important as a search tool.

Figure 2. IAO's ARGUS optical and acoustical tow sled.

ARGUS fills the traditional role carried out by the MEDEA vehicle of DSL. MEDEA is the relay vehicle between JASON and the surface ship. When working on the high seas, the surface rises and falls with the open ocean swell. This motion is transmitted down the cable, moving MEDEA up and down with the swell. At times, this motion can be violent, causing MEDEA to rise and fall suddenly. The primary role of the MEDEA operator is to pay in and pay out cable so that the neutrally buoyant cable connecting MEDEA to JASON never comes up short, thereby transmitting its violent up-and-down motion to JASON and leading to snap load that might damage or sever the cable termination.

Such is the role ARGUS now plays, providing a valuable supporting role to IAO's ROVs HERCULES or LITTLE HERC. Several high-energy HMI lights are mounted on ARGUS' 1,800 kg stainless-steel frame. They can include two 1,200- and two 400-watt lights with a variety of reflector configurations. With these powerful lights, ARGUS is able to illuminate a large area of the ocean floor in which the ROVs can operate to provide outstanding imaging. ARGUS carries a high-definition video camera mounted on a motorized tilt platform. This tilting ability, coupled with its thruster controls, makes it possible for the ARGUS operator to keep the ROV constantly in its field of view as well as obtain high-resolution video footage and still images of the ROV working beneath it. This "eye-in-the-sky" vantage point makes it possible for the ROV operator to see the overall bottom setting in which he is working.

2.2. Remotely Operated Vehicles or ROVs

2.2.1. LITTLE HERC

LITTLE HERCULES (Figure 1), as was the earlier JASON JUNIOR vehicle that
DSL once had, was built so our scientists would have a highly maneuverable
ROV that could be deployed quickly to inspect an acoustic target detected by
ECHO using its high-definition video camera with paired parallel red lasers for
size reference. Designed to operate at a depth of 4,000 m, the 270-kg LITTLE
HERC is 1.2 m in length and carries two 250-watt piloting lights, a magnetic
compass, a simple heading rate sensor, a precision pressure/depth sensor, and
an altimeter. Its 675-kHz scanning sonar is used by the pilot to close on the
acoustic target until it comes into the view of the vehicle's high-definition
video camera. Its pressure/depth sensor, altimeter, and integrated navigational
software make it possible for the pilot to maintain a constant depth, heading,
or altitude while operating the vehicle. As a result, LITTLE HERC can be used
to create photo-mosaics of a wreck site. LITTLE HERC has been outfitted
with various crude sample devices and could carry a small manipulator in the
future, but its primary mission is as an imaging vehicle.

2.2.2. HERCULES

Once an ancient shipwreck has been located using either ECHO or ARGUS,
a visual inspection has been conducted using LITTLE HERC, and a determination is made the shipwreck merits further investigation, the HERCULES
(Figure 1) vehicle is then deployed,

HERCULES was designed by a team of archaeologists, oceanographers,
and engineers to excavate ancient shipwrecks in the deep sea to archaeological
standards. This concept includes the ability to carry out a variety of work
functions using a pair of highly sophisticated manipulators, one of which has
force-feed back sensors that enable the operator to "feel" objects in its grasp.

HERCULES can "fly" in any direction, like a helicopter. Unlike a
helicopter, it will gently float up to the surface if its thrusters stop turning.
Using its attitude package, which consists of a pressure/depth sensor, altimeter,
and Doppler, HERCULES can conduct a variety of precision documentation functions. These can include close-up visual inspection using its high-
definition video camera, surface sampling using its two manipulators that can
place recovered samples into a forward or starboard-side sample drawers or
place them in elevators dropped to the bottom from the surface ship above,
or precision acoustic and visual mapping of a site using its altimeter and/or
Pixelfly digital stereo still cameras.

Once an excavation effort gets underway, HERCULES uses its two manipulators to operate twin hydraulic suction/jetting pumps that are at the head
of two long tubes that move sediment through the vehicle's body to the
back of the vehicle and up its ARGUS/HERCULES neutrally buoyant tether

where it is released into the water column. By placing HERCULES' head into the bottom current and through careful tuning of the pumping and suction functions, the excavation effort can be conducted without loss of visibility. HERCULES also carries of variety of lights including two 400 W HMI for its primary lighting, up to four 250 W incandescent for auxiliary purposes, two 130 watt-sec strobes for the still cameras, and paired parallel red lasers for object-size determination.

2.3. Command/Control Center

All of our vehicles are operated 24 hours a day. This schedule means that there must be "watches", groups that trade off operating the system so the others can have time off to eat, sleep, and do other work. There are five watch stations. The Pilot operates HERCULES, controlling its thrusters, manipulator arms, and other functions. The Navigator monitors the work being done and communicates with the ship's crew to coordinate ship and vehicle motions (Figure 3). The Engineer controls the winch that moves ARGUS up and down, and controls the thrusters and other functions on ARGUS. Video and Data positions support the recording and documenting of the many streams of data that the vehicles generate.

A total of eight underwater video images from both ARGUS and HERCULES are transmitted via fiber-optic cable up to the command/control center (Figure 4) located within a containerized work area aboard ship

Figure 3. Ship's navigator and HERCULES pilot working at consoles inside Command/control Center aboard ship.

Figure 4. Artwork of HERCULES and ARGUS working together on an ancient shipwreck in the Black Sea.

where they were displayed on large plasma monitors. In addition to these optical images, five additional acoustic signals are also transmitted to the surface. These include forward-looking obstacle avoidance sonar on each vehicle, as well as a sub-bottom profiler and high-frequency scanning sonar on HERCULES. Also displayed within the command/control center aboard ship are various computer displays indicating the vehicles' performance as well as real-time navigational information.

This shipboard networking system also includes 24 channels of two-way audio communications permitting a series of separate conversations to occur among the deck force, vehicle operators, the bridge watch, and scientists throughout the ship while they are observing the real-time operations at various workstations. These signals are then routed to a shipboard satellite van with a tracking antenna able to maintain lock on a geosynchronous orbiting satellite while the ship rolls up to 15 degrees and spins on its axis.

Six video signals, including two high-definition images and three two-way audio channels, are transmitted off the ship via satellite. One of the video signals is a composite image containing eight separate video images including two images depicting activity within the command/control center and from the ship's after deck during the launch and recovery of vehicles.

2.4. Inner Space Center

Since one of the primary objectives of the Institute for Archaeological Oceanography is to delineate the ancient trade routes of the deep sea, it is important to be able to access experts ashore since it is impossible and impractical to have an array of experts aboard ship at any one time, waiting for a discovery to be made in their area of expertise. Internet2 (described below) and a high-bandwidth satellite link now make it possible for scientists to work on the ocean floor from the comfort of their University laboratory instead of actually going to sea on an archaeological exploratory expedition (Figure 5).

In such an exploration scenario, the survey team aboard ship consists primarily of engineers and technicians operating the exploration technology on a continuous 24-hour schedule with only a few scientists actually on the expedition. Once a new discovery is made during the cruise, the networking technology is used to network instantly the appropriate scientists from aboard ship via Internet2 so they might evaluate the significance of the discovery and to recommend immediate follow-up studies. This procedure greatly increases the overall effectiveness of this program. The scientists aboard ship include those who must be there to oversee the collection and processing of surface samples that are needed to determine the initial age and origin of a particular shipwreck.

Figure 5. Artwork showing the pathway of satellite signal from ship to shore and back to the Inner Space Center as well as other receiving sites via Internet2.

Figure 6. A team of archaeologists at the Inner Space Center participating in the 2004 TITANIC expedition.

The satellite signal is received by an antenna in the U.S. and placed on Internet2 (Figure 5). Internet2 is a consortium of 205 Universities working with industry and the government to develop and deploy an advanced Internet network that operates at 10 gigabits per second. Its future goal is to offer 100 megabits per second via its Abilene backbone to it use base (see www.internet2.edu/about). The primary Inernet2 site for our program is the newly created Inner Space Center (ISC) at the University of Rhode Island's Graduate School of Oceanography (Figure 6). During a recent expedition to the TITANIC a group of archaeologists and historians participated via the ISC. In the ISC, a series of plasma screens and other displays replicated the science workstation aboard ship. Scientists, students, and engineers at the Inner Space Center were able to participate in two-way conversations with the various people aboard ship. They could also request any of the images they saw on the composite display to be switched to a higher resolution screen for further evaluation.

3. IAO'S ON-GOING FIELD PROGRAMS

3.1. Thunder Bay Program

IAO is at present carrying out a number of programs around the world. For the last few years, IAO has been working with the Institute for Exploration in Mystic Aquarium and NOAA's Marine Sanctuary Program to conduct a long-term program in NOAA's Thunder Bay Marine Sanctuary. IAO has used its ECHO search system to carry out a systematic survey of the Sanctuary locating a series of known and unknown wreck sites (Figure 7). The eventual goal of this program is to use remotely operated cameras and Internet2 to provide "live" public access to some of its wreck sites.

Figure 7. ECHO side-sonar of new shipwreck discovered in Thunder Bay.

3.2. Skerki Bank Program

The Skeri Bank program began in 1988 (Ballard et al., 2000) and continued into the 2003 field season. During the course of the program, numerous shipwrecks, mostly those dating to the Roman period (Figure 8), have been located, documented, and sampled using a broad range of vehicle systems (McCann and Freed, 1994; McCann and Oleson, 2004).

3.3. Ashkelon Program

The Ashkelon program began in 1997 when the U.S. Navy nuclear research submarine NR-1 was searching for the lost Israel submarine DAKAR off the coast of Egypt and located, instead, what appeared to be two shipwrecks. Working with Dr. Lawrence Stager of Harvard University a follow-up expedition in 1999 confirmed the two shipwrecks to be Phoenician, dating to the 8th century B.C. resting below 400 m (Ballard et al., 2002). They were both carrying a large cargo of wine destined either for Egypt or Carthage (Figure 9). A second expedition was planned for 2003, but conflict in the Middle East resulted in its cancellation.

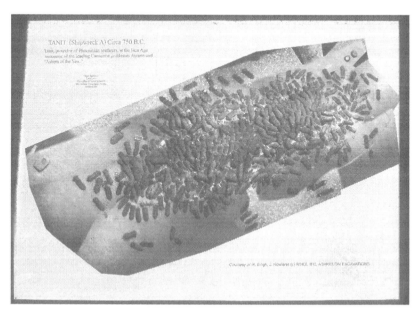

Figure 8. High-definition image of Skerki D Roman shipwreck of the 1st century A.D. taken by LITTLE HERC in 2003.

Figure 9. Mosaic of Phoenician wreck off Egypt.

3.4. Black Sea Program

The Black Sea program dates back many years when IAO's scientists began working with Dr. Fred Hiebert of the University of Pennsylvania in a long-range program to determine the ancient trade routes of the Black Sea. Following a series of preliminary coastal surveys using local fishing ships off the central Turkey seaport of Sinop, a major expedition was carried out in 2000 that located four shipwrecks of Byzantine age; three in the upper oxygen-rich layer and one in a lower anoxic layer (Ballard, 2001). In 2003, a follow-up expedition was carried out off Sinop when the HERCULES ROV was used successfully to excavate a portion of the anoxic shipwreck (Figure 10; Ward and Ballard, 2004).

Concurrent with the program off Turkey, IAO has also been conducting a long-term program off the coast of Bulgaria with the Bulgarian Institute of Oceanography (BIO). Working with Professor Petko Dimintrov, IAO scientists used the ECHO search system in 2001 that located a number of acoustic targets off the coast of Bulgaria and Romania thought to be ancient shipwrecks. IAO and BIO scientists returned in the summer of 2002 and using a BIO small submersible vehicle located a shipwreck of the 5th century B.C., which both organizations hope to investigate further in the near future.

3.5. TITANIC Program

The TITANIC program began in 1985 with the author's discovery of this historic shipwreck in 3,750 m of water. After numerous visits by

Figure 10. High-definition image of Byzantine shipwreck in anoxic waters of the Black Sea during excavation effort by HERCULES during the summer of 2003.

other countries using manned submersibles, the author returned in the summer of 2004 with the ARGUS and HERCULES vehicles to document the damage done to the site by those submersible operations (Figure 11). That most recent expedition led to the United States' joining the United Kingdom in signing a treaty to protect the site from further human induced damage.

Figure 11. High-definition image of TITANIC's bow taken by HERCULES during 2004 return expedition.

4. IAO'S PUBLIC OUTREACH PROGRAM

The new networking technology discussed above makes it possible for countless others to participate in remote field programs whether they are graduate or undergraduate students who have limited opportunities to go to sea or pre-college teachers, students, and the general public wanting to learn more about science and technology in an interactive and exciting way. It is also important the stress that use of this technology is not limited to marine studies as it has broad research and educational potential for scientists working in remote areas on land.

4.1. Future Programs

4.1.1. Educational Program

The Institute for Archaeological Oceanography was only two years old in 2004 when it admitted its first five students to the newly approved graduate program in Archaeological Oceanography. Students in this program will receive a Ph.D. in Oceanography and a Masters degree in Marine Archaeology. IAO will also continue working with the JASON Foundation, IFE, and National Geographic Society in their efforts to educate the public about cultural sites beneath the sea.

4.1.2. Development Program

AO's engineers are also continuing their development of additional undersea vehicles to support this newly emerging program. The next new vehicles at present under development are AUVs or Autonomous Undersea Vehicles that will operate in conjunction with the ROVs already being used in IAO's on-going field programs. Once a trade route has been located and the ROVs are being used to map or excavate a particular shipwreck, AUVs will be used to operate independently along the trade route using their side-scan sonars to locate additional shipwrecks.

4.1.3. Field Programs

IAO's field programs will continue to expand as more students and scientists join the Archaeological Oceanography program at the University of Rhode Island's Graduate School of Oceanography. Discussions are underway with Spain and Greece to begin long-term programs with those countries.

4.1.4. Conservation and Preservation of Submerged Cultural Sites

IAO's long-term program is not only to delineate the ancient trade routes of the world and excavate those of importance, but also to establish museums beneath the sea that can be visited through the use of tele-presence technology. It is estimated that as many as 1 million shipwrecks are lost in the world's oceans. It is impractical to think that all of those shipwrecks can be totally excavated. It is much more practical to think that only a limited number of the shipwrecks would be fully excavated with the rest remaining on the ocean floor for future generations to visit and appreciate.

4.1.5. Long-term Funding of Program

For Archaeological Oceanography to be successful, it needs to be accepted into the larger oceanographic community, because that community controls the major Federal funds sponsoring ocean research and the development and funding of the technology needed to work in the deep sea. The recently completed report by the President's Commission on Ocean Policy has recommended that Archaeological Oceanography be a part of mainstream oceanographic research, but only time will tell how fast that transformation takes place.

REFERENCES

Akal, T., Ballard, R.D., and Bass, G.F., eds., 2004, *The Application of Recent Advances in Underwater Detection and Survey Techniques to Underwater Archaeology. Conference Proceedings, Bodrum, Turkey, 3–7 May.* Ofset Basim Yayin A.S., Istanbul.

Ballard, R.D., 1987, *The Discovery of the TITANIC.* Warner/Madison Press Books, New York

Ballard, R.D., 1990, *The Lost Wreck of the ISIS.* Random House/Madison Press Books, New York.

Ballard, R.D., 1993, The MEDEA/JASON remotely operated vehicle system. *Deep-Sea Research Part I* 40(8):1673–1687.

Ballard, R.D., 2001, Maritime Archaeology. In *Encyclopedia of Ocean Sciences*, Vol. 3, edited by John H. Steele, Karl Turekian, and Steve A. Thorpe, pp. 1675–1681. Academic Press, New York.

Ballard, R.D., 2002, *Collision with History: The Search for John F. Kennedy's PT-109*, National Geographic Society, Washington, D.C.

Ballard, R.D., 2004, Technology, Oceanography, and Archaeology. In *The Applications of Recent Advances in Underwater Detection and Survey Techniques to Underwater Archaeology. Conference Proceedings, Bodrum, Turkey, 3–7 May.* edited by T. Akal, R.D. Ballard, and G.F. Bass, p. 7. Ofset Basim Yayin A.S., Istanbul.

Ballard, R.D., n.d., Deepwater Archaeology. In *Terra Marique: Studies in Honor of Anna Marguerite McCann.* Monograph Series of the Archaeological Institute of America, in press.

Ballard, R.D., and Archbold, R., 1990, *The Discovery of the BISMARCK*, Warner/Madison Press Books, New York.

Ballard, R.D., and Archbold, R., 1993, *The Lost Ships of Guadalcanal*, Warner/Madison Press Books, New York.

Ballard, R.D., and Archbold, R., 1999, *Return to Midway.* National Geographic Society/Madison Press, Washington, D.C.

Ballard, R.D., and Hively, W., 2000, *Internal Darkness.* Princeton University Press, Princeton, NJ.

Ballard, R.D., Stager, L.E., Master, D., Yoerger, D., Mindell, D., Whitcomb, L.,Singh, H., and Piechota, D., 2002, Iron Age Shipwrecks in Deep Water Off Ashkelon, Israel. *American Journal of Archaeology* 106(2):151–168.

Ballard, R.D., Hiebert, Fredrik T., Coleman, Dwight F., Ward, Cheryl, Smith, Jennifer, Willis, Kathryn, Foley, Brendan, Croff, Kathrine, Major, Candace, and Torre, Francesco, 2001, Deepwater Archaeology of the Black Sea: The 2000 Season at Sinop, Turkey. *American Journal of Archaeology* 105(4):607–623.

Ballard, R.D., Mc Cann, A.M., Yoerger, D., Whitcomb, L., Mindell, D., Oleson, J., Singh, H., Foley. B., Adams, J., Piechota, D., and Giangrande, C., 2000, The Discovery of Ancient History in the Deep Sea Using Advanced Deep Submergence Technology. *Deep-Sea Research Part I* 47(9):1591–1620.

Ballard, R.D., Yoerger, D.R., Stewart, and W.K., Bowen, A., 1991, Argo/Jason: A Remotely Operated Survey and Sampling System for Full-Ocean Depth. *IEEE, Oceans 1991 Proceedings*, vol 1, pp. 71–75.

Ballard, R.D., Bryan, W.B., Heirtzler, J.R., Keller, G., Moore, J.G., and van Andel, Tj.H., 1975, Manned submersible observations in the FAMOUS area: Mid-Atlantic Ridge. *Science* 190:103–108.

Dettweiller, T., Bethge, T. and Phaneuf B., 2001, Nauticos Discovers Ancient Shipwreck. *Meridian Passages* (publication of Nauticos Corp.) VIII(1, winter):1–3.

Environmental News Network Staff, 1998, Shipwreck of lost 'Sea People'found, http://edition.cnn.com/TECH/science/9810/16/shipwreck.yoto/

Jasinski, M.E., and Soreide, F., 2004, Deepwater Archaeology from a Norwegian Prespective, In *Proceedings of the 1st Conference on Deep Water Archaeological Exploration. Technology and Perspectives*, The Ephorate of Underwater Antiquities and the Hellenic Center for Marine Research, Athens, Greece.

McCann, A.M., and Freed, J., 1994, *Deep Water Archaeology: A Late-Roman ship from Carthage and an ancient trade route near Skerki bank off northwest Sicily. Journal of Roman Archaeology Supplement* 13,

McCann, A.M., and Oleson. J.P., 2004, *Deep-Water Shipwrecks off Skerki Bank; The 1997 Survey. Journal of Roman Archaeology Supplement* 58, Portsmouth, Rhode Island.

Wachsmann, S., 2004, The Persian War Shipwreck Survey 2003–2004: Preliminary Report. In *Proceedings of the 1st Conference on Deep Water Archaeological Exploration. Technology and Prespectives*, The Ephorate of Underwater Antiquities and the Hellenic Center for Marine Research, Athens, Greece.

Ward, C., and Ballard, R.D., 2004, Deep-water Archaeological Survey in the Black Sea: 2000 Season. *The International Journal of Nautical Archaeology* 33(1):2–13.

Chapter *20*

Precision Navigation and Remote Sensing for Underwater Archaeology

David A. Mindell

Abstract: As archaeological sites in very deep water have become accessible to the research community, remote sensing techniques have been developed to survey and document these sites to high degrees of precision and accuracy. Precision navigation has enabled closed-loop control of underwater vehicles for high-data-density multiple-pass surveys. The acoustic and optical data collected on these surveys have been used to produce centimeter-level of precision for three - dimensional maps of archaeological sites, within volumes of about $100\,m^3$. This paper reviews the work conducted by the author and collaborators in recent years, and presents the author's latest work in high-precision navigation.

1. INTRODUCTION

In the past twenty years or so, underwater vehicles have become adept at locating sites in the deep ocean for archaeological investigation. These capabilities are continuing apace, with the development of new vehicles and sensors for searching the deep oceans. Less publicized, however, have been the advances in technologies available to survey and analyze those sites once located. Remotely operated and human-occupied vehicles (ROVs and HOVs) still rely on the human eye and hand for detailed exploration, and autonomous underwater vehicles (AUVs) are still less capable of detailed site investigation. The key to archaeological-quality, remote-sensing surveys of

499

underwater sites is navigation: the ability to precisely, rapidly, and repeatably locate and position a vehicle in three dimensions above an underwater site. With precision navigation, remote sensing can be accomplished on fine scales in the ocean, allowing precise maps of archaeological sites.

This paper describes the development and state of the art in remote sensing using high-precision acoustic navigation systems for local use around underwater sites. It presents the basic techniques, data collected from recent expeditions, and paths for future development.

2. TECHNOLOGY AND ARCHAEOLOGY IN DEEP WATER

In 1989, in one of its first operational deployments, the JASON remotely-operated-vehicle system, a tethered, underwater robot built specifically for oceanography, imaged and explored a Roman shipwreck known as "Isis" near Skerki Bank in the Tyrrhenian Sea (McCann, 1994; Ballard et al., 2000). The following year, JASON investigated two sunken warships from the War of 1812, the *Hamilton* and the *Scourge* in Lake Ontario. In this case, the permit from the Canadian government forbade touching the site, so a variety of non-contact techniques were developed. These included instrumenting the site with precision acoustic navigation transducers (the SHARPS system) so the vehicle would be precisely tracked at all times, and using those data along with new techniques in computer graphics to create "photomosaics" and three-dimensional sonar images of the wrecks. These precision techniques also had applications in the ocean sciences, and the following year were further developed to map hydrothermal vent sites in the pacific, this time with a wireless version of SHARPS called EXACT (Yoerger and Mindell, 1992).

On these projects, the numerous interactions and discussions between archaeologists and engineers made it clear that scientific-quality archaeological practice required mapping the sites and recording the location and orientation of any artifacts as precisely as possible before any mechanical intervention like recovery or excavation. Spatial relationships are the cornerstone of archaeological investigation and interpretation (Muckelroy, 1978; Bass, 1966; Delgado, 1997). The position of each artifact must therefore be carefully documented prior to disturbance of the site, as well as throughout an excavation.

At the time, some naysayers asserted that such recording was impossible. Such critiques ironically came from opposite camps – both from archaeologists who did not believe scientific-quality work could be undertaken in deep water, and from treasure hunters who argued that since proper archaeology was impossible at great depths, the sites should be open to anyone with a steel cable and a clam bucket. Remote sensing proved both wrong, in dramatic, quantitative fashion. The development of these techniques has served not only to legitimize the science of archaeology in deep water, but also to protect cultural resources in deep water, as treasure hunters can no longer use difficulty as an excuse.

In 1997, the Roman and Carthaginian sites at Skerki Bank were revisited with an improved and updated JASON and detailed quantitative maps were made of the sites, producing dramatic images with spatial resolutions on the order of a few cubic centimeters. Precision, closed-loop control of the ROV allowed not only precise sonar surveys, but also detailed, stable and comprehensive photographic coverage with a digital camera (Whitcomb et al., 1998). Advances in image processing generated large, integrated digital photomosaics of the site, using both automated techniques (for single strips of images) and manual point-and-click feature identification (for digitally stitching strips together). These representations of the archaeological site were unlike anything that had been seen before. They not only delimited the precise outline, dimensions, and extents of the sites, but also revealed a variety of features that had been invisible through the video camera of the ROV, or even looking out the window of a submarine. A number of the amphorae, for example, were embedded in small craters that seemed to derive from sediment-scour patterns on the site. Similarly, the outline of the ship's buried hull may be present in the subtle change in the topography as it curves around the site. These images were so novel, in fact, that the engineering team had to educate the archaeological team to understand their importance as part of the scientific dataset (Singh et al., 2000). The engineers also learned to modify the images to present them in a way more useful to archaeologists, for example by adding specific artifact identifications to the photomosaic. These images have been important elements in archaeological analysis and publication (Ballard et al., 2000; Mindell et al., 2005).

These techniques were again employed and further refined off of Ashkelon, Israel. In 1997, the submarine NR-1 identified two ancient shipwrecks, and collected video footage that suggested they might be interesting and important. A group led by Dr. Robert Ballard and Dr. Lawrence Stager returned to the site with the ROV JASON in 1999, and the wrecks were confirmed to be Phoenician, from the 8th century B.C. Each wreck had hundreds of amphorae. When samples were recovered, they proved to be nearly exactly the same size (+/− about 3% in volume) (Ballard et al., 2002). As at Skerki Bank, JASON was put into closed-loop control, and precision photomosaics and microbathymetry were collected (Singh et al., 2000). Again, they revealed an overall picture of the site and precise dimensions and topography not available with the ROV alone—in this case showing a large, circular crater surrounding the entire wreck, analogous to the smaller craters on the Skerki bank wreck. Thus the precision mapping enabled interpretation of site-formation processes not otherwise visible.

The 1999 Ashkelon survey also tested a technology that had been inspired by prior work—a high-frequency, narrow-beam sub-bottom profiler. Using a 2–3 degree beam and a 150 kHz frequency with a 30 cm circular array, this device has the potential to acoustically peer inside buried shipwrecks without touching them (Mindell and Bingham, 2001). When combined with precision navigation and control, such narrow-beam profilers could potentially map the buried site in three dimensions. Such data could enhance the accuracy of, and reduce the need for, physical intervention. Archaeologists sometimes dig trenches through sites just to see what is there—this technology, while still at an early stage of development, could allow them to replace those trenches with sound waves.

3. MAPPING AND METHODOLOGY IN DEEP WATER

By themselves, *video and photography do not constitute archaeological-quality data*. They do not provide information about the size, shape, or topography of a site, and imagery by itself is not quantitative. The goal,then, is to produce a precise, three-dimensional map and image of an archaeological site in deep water, where no direct human presence is possible. Archaeologists on land spend a significant portion of their effort surveying and documenting a site. The same is true in shallow water, where divers routinely spend hundreds of dives taking measurements. In shallow water, the techniques used are generally manual and can consume a significant percentage of the total dive time on a site. The result is a site plan.

Archaeological ethics dictate that before any physical intervention is done on a site (i.e. sampling, recovery, or excavation), the site must be mapped with the highest precision possible. Because excavation inherently destroys information, that information must be preserved, not only for contemporary analysis but also to enable researchers in the future who may ask questions that we do not conceive of today. Such recording is particularly important for shipwrecks, where the location of the items (relative to the ship's structure and relative to other artifacts) holds important clues as to their origins and functions. (Bass, 1966; Delgado, 1997; Muckelroy, 1978)

4. PRECISION NAVIGATION AND CONTROL OF U/W ROBOTS

Making a site plan is difficult in the deep ocean which, as outer space does, epitomizes an uncontrolled environment. Achieving laboratory-like precision on a deep-water archaeological site presents formidable technical and operational obstacles. It is often said about the deep ocean that "Just knowing where you are is half the challenge," and indeed the problem of *navigation* is central to the enterprise. This problem can be broken down into the following two sub-problems.

1) *Global navigation*. Or, where is the site on the surface of the earth? Acceptable accuracy is about ten meters or so, enough to allow future researchers to return reliably to the site, and to allow conclusions to

be drawn from its geographical location and its location relative to other proximate sites.

2) *Local navigation.* Or, what is the precise map of the site? The precision required for local mapping must be sufficient to reconstruct the position and orientation of objects, even if those objects are quite small. Uncertainties of a few centimeters is sufficient for all but the smallest objects. Such high precision allows for three-dimensional images to be constructed from a site, and the higher the precision the higher the resolution. Better navigation makes sharper pictures.

Achieving the necessary precision for each of these requirements involves using technologies specifically designed for underwater navigation. On a land site, for example, the satellite-based global positioning system (GPS) provides sufficient accuracy for locating a site and many of its smaller features on the surface of the earth. This system depends on electromagnetic energy (i.e., radio waves) that do not penetrate the ocean, so it is unusable underwater. Instead, acoustic techniques are used to measure the travel time of sound waves over some distance.

On the Skerki Bank project, a technique known as "long baseline acoustic navigation," or LBL, was used. This technique is common to other oceanographic applications. First, acoustic transponders are laid in a "net" surrounding the area of interest. Then, a sonar on the vehicle sends out an audible "ping" of sound ($\sim$7–15 kHz, within the range of human hearing), which the transponders hear and respond to. The total travel time is measured, which is proportional to the range from each of the transponders, and then these ranges are triangulated to determine a position. This technique provides navigation accuracy of a few meters over an area several kilometers on a side (the actual range and accuracy depend on the topography, the acoustics of the water column, and other factors). Because it is described elsewhere in the literature, we will not dwell on it here (see Whitcomb et al., 1998 for quantitative comparison of navigation techniques).

More of interest for this paper are the techniques used for local navigation and precision imaging of the site. Actually, they are similar in structure to the larger LBL technique but operate with higher precision and in smaller areas. This system, called EXACT, also involves transponders, working at much higher frequencies of around 300kHz (well above human audible range). These devices are sent down on an elevator (essentially a basket lowered from the support ship), the same device used for recovery of artifacts. JASON's manipulator arm then grabs the transponders and deploys them next to the site, about 50–70 m apart. JASON has a similar device mounted on its frame, which emits a high-frequency coded ping. When the transponder detects a ping with a code that matches its own, it emits a similarly coded ping, which is then received by JASON. Again, the travel time of the sound is proportional to the distance—in this case resulting in an accuracy of under 1 cm. By measuring its distance to two known points, and combining that information with a precision depth measurement, JASON's computer receives

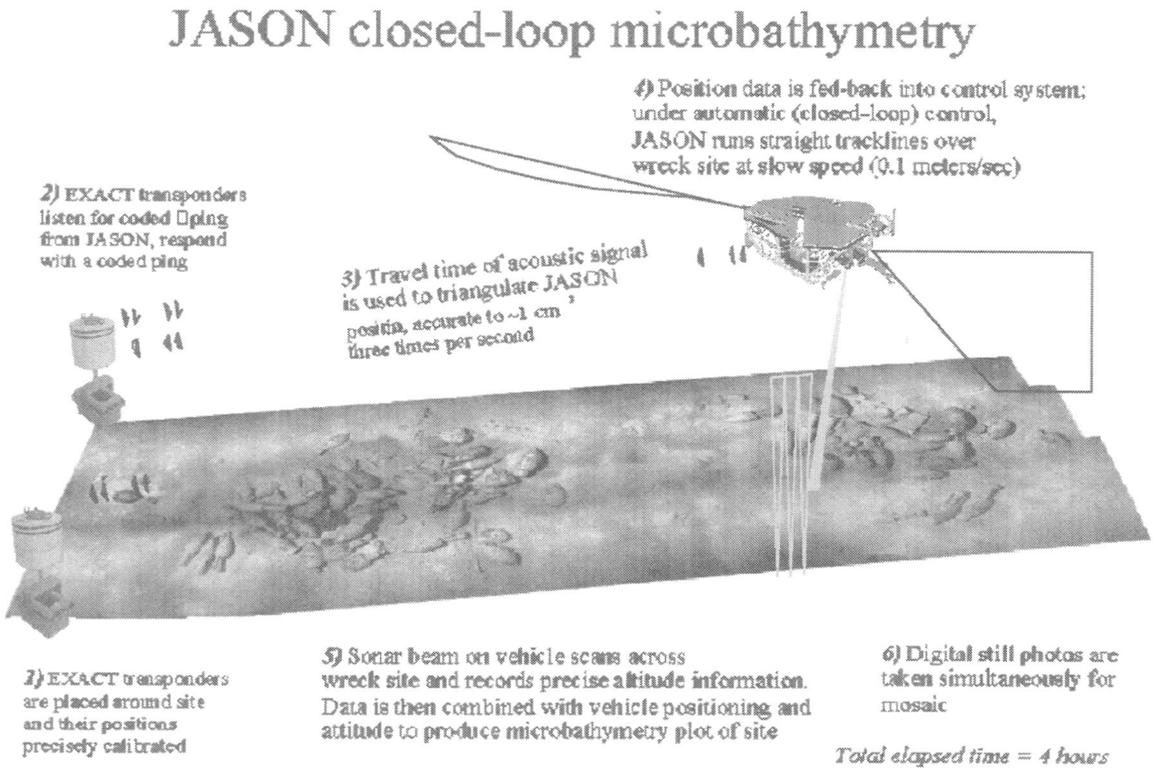

Figure 1. Jason closed-loop survey and precision navigation.

an X,Y,Z position reading a few times per second. (Yoerger and Mindell, 1992) JASON's closed-loop survey is displayed in Figure 1.

This position information can then be fed back into JASON's computer control system to create a digital autopilot. This feedback enables JASON to hover or to run precise track lines, independent of a human pilot. Thus the vehicle can "mow the lawn" above an archaeological site, all the while collecting data from other sensors (such as cameras and sonars). Precision control enables much straighter lines than a human pilot can follow. It also ensures that the entire site is completely covered, leaving no gaps in the imagery or measurement.

Some sensors (like a scanning sonar or a digital still camera) require relatively long time periods (i.e., a few seconds) to record their data; precision control allows the vehicle to move slowly ($\sim$10 cm/sec) across the site, thus increasing the resolution of the sonar survey. Precision control also allows the survey to be repeatable, so it can be run several times during the course of a project (a capability whose implications we explore in the final section). Once the survey is run, the sensor data can then be combined with the navigation data to create precise maps of the site. Precision navigation and control allow an underwater vehicle to make high-resolution, repeatable underwater surveys of an archaeological site, with various types of data.

5. PHOTOMOSAICS

One difficulty with imaging underwater is that the electromagnetic spectrum (including visible light) attenuates rapidly and nonlinearly. This means that large objects cannot be framed within a video or other optical camera's field of view. Obtaining a global perspective of a site of interest on the seafloor requires piecing together a series of images in a process called photomosaicing. This involves running a carefully planned survey over the site, collecting a series of overlapping images, identifying common features in the overlapping imagery, and then merging the images to form larger mosaics.

As the vehicle moves under closed-loop control, it collects data from a variety of sensors, including a digital still camera, similar in principle to those available to consumers today though optimized for undersea work and for producing quality images from the low-contrast conditions typical of the deep sea. The camera is pointed vertically downward and snaps an image every 13 seconds. Closed-loop control allows systematic coverage of the site, ensuring that digital images are captured for every point within the boundaries of the site; in fact, they overlap significantly. (Singh et al., 2000) (Figure 2). This mosaic provides an overall view of the site unavailable by other means, and, because it was produced early in the expedition, it allowed the archaeologists to orient themselves around the site as further investigations were conducted.

Several factors make producing such a mosaic a complex task. Physical constraints on the distance separating cameras and lights as well as constraints

Figure 2. Photomosaic of Skerki D shipwreck (from Singh et al., 2000), made of about 180 individual images.

on the energy available for operating the lights constitute major impediments. Most important, the unstructured nature of the underwater terrain introduces incremental distortions into the photomosaic as successive images are added in. Thus, while the mosaics produce an image of the site that is coherent and pleasing to the eye, the process introduces distortions that are often visible only on close inspection, or to an expert eye. Current research efforts are focusing on limiting the errors in the photomosaics to enable quantitative measurements, but the mosaics produced at Skerki Bank, while providing an excellent view of the site and the relative positions of artifacts, do not constitute a quantitative survey.

6. MICROBATHYMETRIC MAPPING

Bathymetry is the measurement of the depth of the ocean; when many points are assembled into a map, bathymetry generates data of the type found on nautical charts. On the Skerki Bank project, researchers coined the term "microbathymetry" to capture the idea of measuring depths on a grid of only a few centimeters. Then, rather than measuring the contours of a continental shelf, we measure the actual features of an archaeological site, right down to the shapes and orientations of the amphoras.

As mentioned in the previous section, once the JASON vehicle was under closed-loop control, then it could run slow, precise track lines over the site. A sonar mounted on the vehicle scanned a narrow-beam of high-frequency sound across the site, perpendicular to the vehicle's path of travel. Several times per second, this sonar returned the range to whatever object the sonar beam first encountered, and also the angle of the scanning beam at that moment. A single scan would typically cover about 45 degrees on either side of the vertical, and hence would return a 90-degree swath, or a line that indicated the profile of the site. As the scanning head moves the sonar beam across the target, the vehicle is moving forward, so the actual footprint is this "zipper" pattern. The slower the vehicle moves, then, the tighter the spacing of these scans, and the higher the resolution of the survey.

As these scans are recorded, JASON's computers also log the position (x,y, and z) of the vehicle at the point they were taken, as well as its attitude (pitch, roll, and heading). When these data are combined with the physical layout of the vehicle, the computer can solve for the actual 3-D point in space that the sonar indicates. Thus all of the scan-lines can be converted into real-world coordinates, with an accuracy of a few centimeters. When these real-world coordinates are plotted together as a surface, they reveal the "microbathymetric" image of the site, as shown in Figure 3 .

The detail in Figure 3 is fine enough to show individual amphorae, and the two piles of artifacts characteristic of the wreck. It also reveals features that were invisible to the naked eye (even for a human observer looking out the window of a submarine). The site is on a gradual slope (indicated by the gradual shading from red to blue across the site). Several of the amphoras,

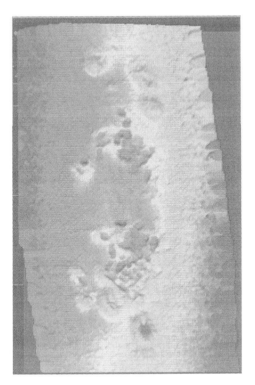

Figure 3. Microbathymetry of Skerki D shipwreck. Note individual amphorae and two separate piles. (Mindell et al., 2005).

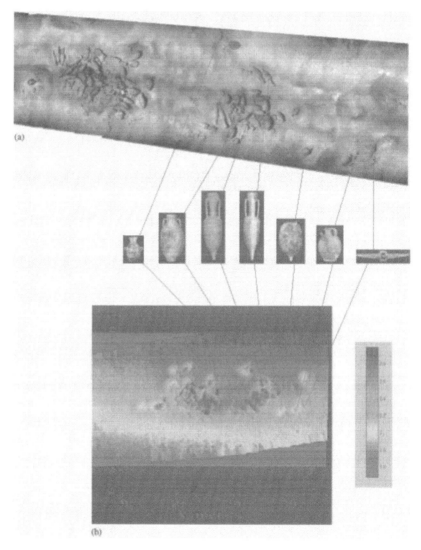

Figure 4. Correspondence between photomosaic (a) and microbathymetry (b). (Mindell et al., 2005)

or groups of amphoras, are in craters, probably created by some aspect of the currents' moving sediment across the bottom and leaving scour patterns around the amphora. Also, as the yellow to green transition indicates, one may be able to discern a rough hull pattern manifested in the topography. If one looks closely, one can also discern the residues of the sonar scans across the wreck, which have not been completely smoothed out by processing. Note the subtlety of the relief here: the entire vertical excursion, from the highest point to the lowest, is about one meter.

This microbathymetric map is spatially accurate to within a few centimeters of precision. When enlarged, it reveals both the size of the wreck, the size and orientation of the amphoras, as well as the relationship to surrounding topography. Figure 4 shows how the microbathymetry data correspond directly to the photomosaic. These datasets represent new tools for underwater archaeology, and may prove as useful on shallow-water sites as they are on deep-water sites.

7. TRANSPONDER-BASED PRESICION ACOUSTIC NAVIGATION

These imaging techniques are all based on precision navigation. The author has developed a high-precision, transponder-based acoustic navigation system, called EXACT, for absolute-referenced precision surveys of local sites. EXACT enables a precision mapping system by providing the fundamental ability to measure very precise travel times, with a resolution of one microsecond travel time (standard deviation). It builds on a similar system, called SHARPS, developed in the 1980s for archaeological sites, and on a previous version of EXACT developed by the author and Dr. Dana Yoerger at the Woods Hole Oceanographic Institution in the 1990s (Yoerger and Mindell, 1992). The present system uses updated signal processing electronics and spread-spectrum techniques around 200 kHz to achieve extended range, ping coding, and data transmission. The basic characteristics of EXACT are:

- Timing resolution μS
- Range resolution, 2–3 mm
- Position resolution $<1 \, cm^3$
- Range: $>200 \, m$
- Position updates: $\sim 1 \, Hz$

The high-precision advantages of this system are enabled by creating a structured environment—either by mounting transceivers in a tank or by deploying transponders in the ocean. Of course such operations have costs, both financial and logistical. The point should also be made that EXACT is not designed to be a replacement for other types of Doppler, LBL, or inertial navigation systems. Those systems are useful for telling "where am I" over a very large area, whereas this system is of at least an order of magnitude greater accuracy in range, is used for a local area, and is thus more properly classified as a "positioning" system rather than "navigation." That is, the data are useful for feeding back into a vehicle control system for precise, closed-loop control, and for correlating with other sensor data to make three dimensional maps. It is not generally a search system, and the capabilities really do not overlap with more traditional navigation systems—they are used for different purposes. Hence precision navigation is a specialized technology useful for the highest-accuracy three-dimensional mapping currently available underwater.

8. CONCLUSION

Previously, we have used this system to plot microbathymetry and some sub-bottom data, but other sensors could be used as well. *The importance of precision navigation is that it allows a vehicle to create a map of anything that can be measured.* One can easily imagine chemical maps of underwater sites, or salinity maps, or magnetic maps.

Absolute referencing is important because it allows precision surveys of underwater sites to be repeatable. For example, a vehicle can hover over a site for an indefinite amount of time, perhaps integrating data from an instrument to increase its signal-to-noise ratio. Or, if a transponder is left on a site for an extended period, then a precision survey can be conducted at regular intervals, allowing quantitative change detection on the order of cubic centimeters for an underwater site. Such techniques could measure changes in an archaeological site, such as decay and site formation, biological activity, or human intervention. A precision map could be produced not just once for a site, but every day during an excavation season, for example.

ACKNOWLEDGEMENTS

The author gratefully acknowledges the collaboration of Dr. Robert Ballard, Dr. Dana Yoerger, Dr. Hanu Singh, Dr. Louis Whitcomb, Jonathan Howland, and Dr. Anna McCann.

REFERENCES

Ballard, Robert D., Stager, L.E., Master, D., Yoerger, D.R., Mindell, D., Whitcomb, L.L., Singh, H., and Piechota, D., 2002, Iron age shipwrecks in deep water off Ashkelon, Israel. *American Journal of Archaeology* 106(2):151–168.

Ballard,Robert D., McCann, A.M., Yoerger, D.R., Whitcomb, L.L., Mindell, D.A., Oleson, J., Singh, H., Foley, B.P., Adams, J., and Piechota, D., 2000, The discovery of ancient history in the deep sea using advanced deep submergence technology. *Deep Sea Research I*, 41:1591–1620.

Bass, George F., 1966, *Archaeology Under Water*. Praeger, New York.

Delgado, James P., ed., 1997, *Encyclopedia of Underwater and Maritime Archaeology*. Yale University Press, New Haven.

McCann, Anna M., 1994, *Deep Water Archaeology: A Late Roman Ship from Carthage and an Ancient Trade route near Skerki Bank off Northwest Sicily. Journal of Roman Archaeology,* Supplementary Series 13 Ann Arbor.

Mindell, David A., and Bingham, B., 2001, A high-frequency, narrow-beam sub-bottom profiler for archaeological applications. *Oceans 2001 Conference Proceedings.* MTS/IEEE, Honolulu.

Mindell, D.A., Singh, H., Yoerger, D., Whitcomb, L., and Howland, J., 2005, Precision mapping and imaging of underwater sites at Skerki Bank using robotic vehicles. In *Deep-water Shipwrecks off Skerki Bank: the 1997 Survey,* edited by A.M. McCann and J.P. Oleson, *Journal of Roman Archaeology,* Supplementary Series 58.

Muckelroy, Keith. 1978. *Maritime Archaeology.* Cambridge University Press, Cambridge.

Singh, Hanu, Adams, J., Foley, B.P., and Mindell, D.A., 2000, Imaging for Underwater Archaeology. *Journal of Field Archaeology* 27(3):319–328.

Whitcomb, Louis, Yoerger, D.R., Singh, H., and Mindell, D.A., 1998, Toward precision robotic maneuvering, survey, and manipulation in unstructured undersea environments. In *Robotics Research—The Eighth International Symposium,* edited by Y. Shirai and S. Hirose, pp. 45–54. Springer-Verlag, Berlin.

Yoerger, Dana R. and Mindell, D.A., 1992, Precise navigation and control of an rov at 2200 meters Depth. In *Remotely Operated Vehicles* (ROV) '92. Marine Technology Society, San Diego.

Section VI

Cultural Resources and Heritage Management

Chapter *21*

Applications of Remote Sensing to the Understanding and Management of Cultural Heritage Sites

JOHN H. STUBBS AND KATHERINE L.R. MCKEE

Abstract: Remote Sensing imagery can be used as the baseline dataset in documenting
              and analyzing the historical and contemporary effects of human activities
              at cultural heritage sites. An integrated Remote Sensing (RS)/Geographic
              Information System (GIS) allows cultural resource managers, historians,
              planners, and engineers to catalogue and assess the organizational and structural
              patterns of such sites, and determine sustainable tourism and urban development
              within their regions. A regional dynamics survey can be used for analysis of
              cultural heritage resources by applying a multiscalar remote sensing approach to
              demonstrate patterns of land use and land cover. This method can be used to
              create an urban dynamic model that visualizes future consequences of environ-
              mental and man-made threats, thereby enabling mitigation and management of
              these non-renewable resources.

1.   UNDERSTANDING CULTURAL HERITAGE RESOURCES

S. A. Drury, a pioneer in remote sensing research, wrote "Humanity, its
economy, and all the attendant activities are inseparable from the natural
world" (Drury, 1998:16). This statement could not be any truer than when
we examine cultural heritage sites. Cultural heritage sites are immovable,

515

human-made, physical evidence fixed within the landscape that are insep-
arably connected to the natural environment. Like other forms of cultural
heritage, they are the present manifestation of the human past (Sullivan, 2000).
Whether these sites comprise one structure or several within a widespread
area, they indicate spatial patterns and provide clues to past activities relative
to human settlement on earth. Spatial and temporal investigations of these
sites provide information about a region's past and present socio-economic,
political, religious, and environmental aspects.

Physical cultural heritage resources are tangible artifacts representing a
defined time, place, and culture in and of the built environment. These fragile,
irreplaceable archives belong to humankind's collective identity and reveal its
history, artistic mastery, and technological advancement. Thus, interpretation
and protection of cultural heritage resources are as critical as conservation of
our natural resources. Cultural resource managers and other decision makers
who shape our world now realize that culture and heritage provide outstanding
value to humankind and our common identity, growth, and sustainability
(UNESCO, 2003a).

1.1. Characteristics of Cultural Heritage Conservation and Archaeology

Cultural heritage conservationists and archaeologists use similar reasoning
when examining heritage resources. Onsite, archaeologists focus on identifying
and understanding the people who lived there, how they lived, how they were
sustained, and what became of them, among other issues. Heritage conservationists apply information gathered by archaeologists when deciding how to
conserve, present, and manage selected sites for contemporary use. Fundamentally, both archaeologists and heritage conservationists are concerned with the
history and interpretation of sites, and understand that their most important
task is documentation of these extant tangible cultural resources, which are
finite and non-renewable.

Both specialists observe how human interaction with the physical
environment may have changed the social and historical contexts of a site and
its surrounding area. Each seeks to identify signatures, or imprints, on a site's
features—whether an individual building or an entire cultural landscape—in
order to determine societal patterns and meanings across time. Archaeologists and heritage conservationists aim to interpret how heritage sites reflect
the relationship between humans and their environments, and to present this
knowledge to the general public.

The difference is one of timeline: archaeologists investigate the past to
piece together how a site has arrived at its present condition, while heritage
conservationists work to conserve and protect the site for posterity. Conservation of cultural heritage resources requires an understanding of the past

and present values associated with the object, structure, or site in question in light of contemporary developments and possible future threats. In these respects, archaeologists and heritage conservationists share commonalities. When archaeologists and conservationists work together to realize these shared objectives they are able to interpret and represent heritage sites as spatial and temporal continuums.

1.2. Defining Cultural Heritage Sites in Context

This spatial and temporal analysis of a cultural heritage resource requires an assessment of its fundamental physical components combined with that of the larger environment surrounding the site. An operating cultural heritage site lies within a larger region that includes: 1) the designated historic site; 2) the city, town, village, or enclave where visitors are directed; 3) the hinterlands surrounding both the site and city where local residents live; and 4) the landscape enveloping the former three elements plus any conterminous outlying areas. In this instance landscape is defined as the spatial manifestation of the relations between humans and their environment (Marquardt and Crumley, 1987). Together these four parameters characterize *in situ* physical, cultural heritage resources. The built and natural environments of a cultural heritage site form a single synthetic and dynamic unit: one element cannot be sustained without the other.

An example of a *cultural heritage region* may be seen in Figure 1, the famous historic city of Angkor that served as the seat of the Khmer empire from the 9th through the 13th centuries A.D., and which today is the historic heart of Siem Reap province in north-central Cambodia. Modern Siem Reap is the administrative and economic seat of Siem Reap province serving the region's farms and villages while also hosting visitors to the Angkor archaeological park, termed by some "the Eco-site of Angkor" (World Monuments Fund, 1995).

Cultural heritage areas in developing countries can be defined as emerging urban environments when they 1) comprise a unique, densely populated, and environmentally distinct habitat; 2) exert significant environmental impacts on their immediate surroundings, their hinterlands, and the regions of which they are a part; and 3) are linked by transportation, trade, and population migration in an interacting system. Within this system, economic, demographic, and political decisions influence both the local and regional environments (Miller and Small, 2003).

Most cultural heritage sites are simultaneously tourism destinations and finite resources that call for both exploitation and protection. Tourism creates self-supporting, urbanized areas near heritage sites where spatial boundaries, commerce, and population concentrate to support the tourist industry, which in turn accommodates, promotes, and relies on the protected heritage area. If

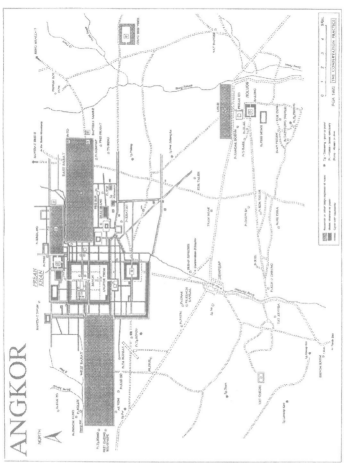

Figure 1. Map of Angkor Eco-site, which is characterized by its hydrological works (barays, canals, natural waterways, moats, and earthworks); key monuments, including five temple-cities built in succession in the Hindu-Buddhist period of the 9th–13th Centuries AD; and the modern provincial seat of Siem Reap (lower left of image). Map: WMF, 1992, drawn by Fred Alsworth, The Conservation Practice.

the heritage site is destroyed or seriously compromised, then the urban area ceases its functional priority.

1.3. Issues in Heritage Conservation

Tourism and urban growth are perhaps the two most pervasive mechanisms of change at heritage sites. Tourism creates growth and economic benefits for local communities, regions, and countries, and simultaneously makes both tangible and intangible cultural heritage vulnerable to irrevocable damage and loss. The economic modernization of developing countries accelerates this impact, because the heritage sites become more accessible to the global tourism industry.

Exploitation of heritage resources is often more expansive in developing countries, which experience rapid urban growth but have insufficient development regulations in place. To address such situations, UNESCO recently adopted principles of sustainable tourism that seek to devise strategies for long-term conservation of cultural and natural heritage areas (UNESCO, 2003b). A plethora of other institutions and organizations, including the World Bank, ICOMOS, the World Monuments Fund, regional associations of nations, and even tourism organizations themselves are also promoting sustainable development and tourism.

Myriad human and natural threats to cultural heritage sites are present irrespective of whether the place is deemed a UNESCO World Heritage Site, or is an obscure, undesignated, recently excavated archaeological site. While physical cultural resources are subjected to the same environmental degradation that commonly threaten natural landscapes, water resources, and air quality, it is human-induced changes that pose the greatest threat. Global climate and environmental changes require conservationists to "think globally" at heritage sites to manage the forces of attrition (Fitch, 1990) that have an impact on a site's local elements. Archaeological site conservation requires both macro and micro intervention and maintenance. At the micro-level, conservationists focus on stabilization and maintenance of the structural, architectural, and material integrity of the buildings or ruins. A macro approach addresses protection of the site, and its immediate landscape, from unregulated urban development, natural hazards and risks, looting, vandalism, and loss of tangible and intangible cultural values.

Within the Angkor Eco-site one can observe the effects of long-term exposure and neglect of archaeological fabric (Figure 2); vandalism and theft of bas-reliefs and sculptures, which are the very elements that give the temples and shrines of Angkor their meaning (Figure 3); and the negative results of uncontrolled tourism (Figure 4). These issues are common to many of the sixty-four key monuments at Angkor, and since the early 1990s have been addressed on a site-by-site basis by teams of architects,

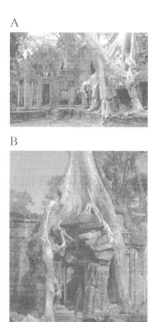

Figure 2. Neglect of monuments through much of Angkor's history as exemplified by ficus trees growing from Khmer temples. At some temples such as Angkor Wat and Banteay Srei all vegetation was removed by archaeologists without question, while at other temples, such as at Preah Khan and Ta Prom, this remarkable symbiosis of the natural and the human-made is being preserved where possible. Photos: J.H. Stubbs, WMF.

engineers, archaeologists, and conservation scientists through international projects sponsored by governments and NGOs in Japan, France, the United States, Germany, India, Italy, Hungary, Indonesia, Malaysia, and China.

Over the past decade the consortium of conservators and heritage resource managers working at Angkor under the aegis of the APSARA (site administrative) Authority have increasingly considered the Eco-site's larger perspective, and noted that its vast historic surroundings include numerous unprotected buildings and sites, earthworks, hydraulic features, roads, bridges, and natural features that also merit careful study and conservation. Sensitive historic areas beyond the core of the ancient capital are acutely vulnerable to threats from insensitive development, which are largely tourism-related (Figure 5). This is particularly alarming given that a recent analysis of remote sensing imagery found that the historic city of Angkor, including its supporting towns, villages, and agricultural lands, extended over 1,000

Figure 3. Vandalism and looting occurred at nearly every temple in Angkor especially during the 1990s before adequate heritage protection was put in place, and after Cambodia's civil war. In recent years the problem has been stemmed for the most part by Angkor's Heritage Police. Photo: P.J. Sanday, WMF.

sq km (Evans, 2003), which is well beyond UNESCO's World Heritage Site boundaries for Angkor.

An influx of tourists and a shift from agriculture to a service-based economy has increased Siem Reap's overall population and density, and created urban environmental problems. Sprawling development and congestion are immediately apparent, as are increasing strains on air quality, water supply, energy and food consumption, and waste removal and transportation systems. Conservationists and planners are examining how the Angkor Eco-site is affected by other changes in the natural and urban environments, including water sources, the hydrological cycle, deforestation, utilities, and farming practices. Such environmental changes not only affect the historic structures of Angkor and their immediate landscape features, but also have noticeable consequences at national, regional, and international levels. Tourism at Angkor plays such a critical role in the development of the entire country that Prime Minister Hun Sen stressed in a letter to the APSARA International

Figure 4. Throngs of visitors, some having arrived on elephants, atop 99-m-high Prasat Phnom Bakheng at day's end to view the sunset. Photo: P.J. Sanday, WMF.

Coordinating Committee in December 2002 that "Angkor will be the economic engine that will pull Cambodia into the future."

Understanding the dynamics of cultural heritage sites as specialized urban environments contributes to more effective management principles and procedures. The environmental issues affecting heritage sites cut across existing spheres of policy at local, national, and even international scales.

Figure 5. This aerial view from Sukhothai in Northern Thailand shows how insensitive development can harm cultural heritage: the diagonal line extending to the right is a road to the airport that was built directly through an historic temple complex. Reproduced by permission of UNESCO.

2. REMOTE SENSING APPLICATIONS FOR CULTURAL HERITAGE

Most issues confronting cultural heritage management can be mapped, measured, monitored, visualized, and managed using Earth Observation data. Spaceborne and airborne remote sensing provides cultural heritage managers the capability to discern spatial and temporal variations in the landscape over time. Remote sensing also facilitates informed scientific and quantitative analyses of resource monitoring and management, policy-making, and planning for sustainable development.

The application of remote sensing technology has been excellent for natural-resource management, hazards and risks mitigation, and analysis of global environmental change. The remote sensing techniques that contribute to a comprehensive understanding of these themes can be readily employed in the investigation of heritage resources, which possess amalgamations of these broad issues. Spaceborne and airborne radar, digital orthoimagery (rectified aerial photography), and multispectral satellite imaging have significantly enhanced the practice of archaeology, and hold great promise for the field of cultural heritage management. These data have proven to be an invaluable tool for the allied subject of cultural resource research and interpretation, in particular archaeological analysis, which is closely connected to cultural-heritage conservation. (See Moore, et al., in this volume.) Some of its most impressive applications to date have resulted in important discoveries of evidence of buried cities and other features of past civilizations: the Khmer capital of Angkor in Cambodia (Jacques, 1995; Moore and Freeman, 1997; Fletcher et al., 2002); Bronze Age, Edomite, Byzantine, and Crusader sites of Petra in Jordan (Comer, 1998); the ancient Greek colony of Chersonesos near Sebastopol, Ukraine (Carter and Morter, 1995); the long-lost key Silk Route town of Ubar in Oman (El-Baz, 1997); a pre-dynastic riverine civilization at Saf Saf Oasis in Southern Egypt (El-Baz, 2001), and Maya ruins in the jungles of Guatemala and Honduras (Sever, 1992).

Remotely sensed data acquired from heritage resources, in particular at archaeological sites and cultural landscapes, provides synoptic views over large areas and can detect spatial differentiations with quantitative measurements of physical conditions (Miller and Small, 2003). Determining variations in the heritage landscape provides a systematic means of mapping spatial extent and quantifying growth and morphology.

Perhaps the most promising application of remote sensing for heritage conservation is in detecting changes land use/land cover to quantify urban growth and land use over an extended length of time (Mesev et al., 1995; Stefanov et al., 2001; El-Baz, 2002; Small, 2002). "Land cover" refers to the physical materials on the surface of land (grass, trees, water, building surfaces, paving materials), while "land use" is the human activity that occurs on, or makes use of, that land (residential, commercial, industrial) (Barnsley et al., 2001). An underlying problem of remote sensing is that there is no direct way to discern land use based on spectral reflectance. Land use is a combination of cultural, socioeconomic, historical, and environmental factors that must be inferred using spatial and structural pattern recognition, as well as analysis of the textural and contextual arrangement of land cover.

Spectral characterization of heterogeneous land cover can, however, provide a basis for mapping the spatial extent of human land use with satellite imagery collected over the past thirty years (Small, 2002). Spectral mixture analysis, specifically Landsat-derived endmember fraction estimates (Small, 2002), can quantify the temporal and spatial variations in land cover over time, providing a systematic method of quantifying heterogeneous urban land cover and a means of consistent physical descriptions that quantifies the spatial extent of forest, agriculture, and built-up areas at different times and locations. For instance, a comparative analysis of vegetation abundance and population density in six U.S. cities demonstrates an inverse relationship between land cover and habitation patterns (Pozzi and Small, 2002) that enables calculation and prediction of urban growth.

3. REMOTE SENSING AT THE ANGKOR ECO-SITE

Since the early 1990s, applications of remote sensing to the vast Eco-site of Angkor have represented perhaps the most exemplary use to date of its benefits to archaeology and cultural heritage management.

Aerial photographs taken by archaeologists of the École Française Extreme d'Orient in the 1930s served as the basis for the first accurate maps of the historic city of Angkor. Subsequent mapping and low-level aerial surveys of northern Cambodia by French, American, and Vietnamese cartographers during and after the Vietnam War proved useful as well. However, it was the use of spaceborne multispectral imaging beginning in 1993 with infrared SPOT images that initiated a radically improved research method. Angkor's Zoning and Environmental Management Plan (1994), which was developed as a condition of Angkor's listing as a UNESCO World Heritage site in 1992, used SPOT[1] images to produce a baseline map, a GPS survey to accurately pinpoint elements, and a GIS for data integration and spatial analysis.

In late 1994, through the joint efforts of the World Monuments Fund (WMF) in New York and the Royal Angkor Foundation (RAF) in Budapest, the Jet Propulsion Laboratory of the United States National Space and Aeronautics Administration (JPL/NASA) applied a new kind of remote sensing technology in an effort to document, research, conserve, and present Angkor. In the early morning of October 4, 1994, the space shuttle *Endeavor,* flying at an altitude of approximately 256 km (160 miles) above the earth, acquired SIR-C/X-SAR radar images of Angkor from a northwest to southeast flight path. The mission also recorded nearly every stretch and bend of the Mekong River from central Laos to the river's delta in the South China Sea.

To analyze more efficiently the data acquired by this overflight, technical roundtables were held at Princeton University (February 1995) and the University of Florida (April 1996). They were organized by the World Monuments Fund with financial support from the J.M. Kaplan Fund (New York), and enabled international, multidisciplinary gatherings of archaeologists, historians, cultural landscape specialists, (Cambodian) administrators, architects, geologists, and JPL/NASA radar-imaging scientists to utilize the images for a greater understanding of Angkor, and to discuss applications of remote sensing at historic cultural sites in general. Of particular interest was new information about the evolution of the Angkorean landscape. For example Anthony Freeman of NASA/JPL and Elizabeth Moore of the School of Oriental and African Studies (SOAS), University of London, used Freeman's three-component scattering model to interpret curvilinear features from the pre-Angkorean period. This research led to interpretation on a variety of issues, such as: how the greater Angkor area may have supported its large population; its ancient land-settlement patterns, the extent and operation of its vast system of water management; and its ancient ecology. Also discussed were the modern needs for conserving and presenting the site (Moore and Freeman, 1997; see also Moore, Freeman, and Hensley, in this volume). These were urgent matters because Angkor was then beginning to re-open to tourism after ten years of isolation and civil unrest.

Building upon the start provided by the 1994 radar image, Moore, Freeman and Scott Hensley of NASA/JPL, with the help of Earnest Paylor of NASA and the Royal Governments of Cambodia and Thailand, were able to include Angkor within the 1996 AIRSAR-TOPSAR Airborne radar data acquisition. This second mission over the Angkor Eco-site acquired data from an altitude of approximately 9,144 meters (30,000 feet) in three swaths with each 10 km wide and 60 km long covering an area of 1800 sq km. These results enabled Moore to considerably advance her theories and research on the region's pre-Angkorian mound and moat culture. For example, one swath showed that circular land alterations extended north to the present-day border of Thailand.

It is now believed that these mound and moat settlements in the area served as exemplars of efficient prehistoric water management on a local scale for bolder water management schemes that followed. This development, plus the amenity of the nearby Tonle Sap (great lake) with its phenomenal seasonal size fluctuations, which may have supported even more ancient systems of aquaculture, served in their separate ways as *raison d'etre* for King Jayarvarman II's choice of the region as seat of his empire in ca. 802 A.D.

A second AIRSAR-TOPSAR overflight, and the third overall remote sensing undertaking of Angkor, was accomplished in September 2000 by NASA/JPL for the Greater Angkor Project, an international program concerned with Angkor's historic ecology and puzzling decline (Evans, 2003). The mission's principal investigator, Roland Fletcher of the University of Sydney, Australia, acquired contiguous swaths that were combined to form a highly accurate topographical map of 7,000 sq km (2,800 square miles) of the greater Angkor area. As with earlier overflights, Fletcher subsequently documented a number of new findings about Angkor, notably how the Khmer in the later phases of Angkor's historical period utilized the near and distant hinterlands.

The first decade of remote sensing imagery analysis at Angkor resulted in new documentation of the site's prehistory and a much greater understanding of the organization and extent of the historic capital. Equally as important as the various discoveries of archaeological features, which were derived from each of these remote sensing projects, is the new macro-perspective that is now assumed by those who manage the general Angkorian cultural landscape and region. Since the Angkor Eco-site's macro-planning operations began in the early 1990s, everyone involved in Angkor's documentation, analysis, and planning, and its environment, has been given the ability to view the vast site through "the other end of the telescope," as it were, with advanced visual techniques.

3.1. Radar Images of Angkor

The technological aspects of remote sensing allow researchers to observe large regions and detect landscape differentiations with precise measurements. As we mention in the section above, imaging radar has been acquired over the Angkor region three times. Imaging-radar systems are comprised of radar antennae that transmit and receive electromagnetic pulses to and from a remote surface, in this case Earth, at specific microwave wavelengths and polarizations (electromagnetic waves oriented in a single vertical or horizontal plane).[2]

The "classic" Spaceborne Imaging Radar-C/X-band Synthetic Aperture Radar (SIR-C/X-SAR) radar image of Angkor (Figure 6) shows an area approximately 55 km by 85 km (34 miles by 53 miles). Angkor Wat, the area's largest and most imposing temple, is the small, darkly outlined square towards the left center of the image. Angkor Thom, a fortified city containing the Bayon and Baphuon temples, is the bright square above Angkor Wat. Other temples are visible within square or rectangular outlines; the water sources are discernible as dark features: moats surround the temples, canals and dikes traverse the area in all directions, and the Western and Eastern Barays (reservoirs) show as large dark rectangles on two sides of Angkor Thom. The SIR-C/X-SAR radar imaging data produces multiband (varying wavelength ranges in quad polarization) images in resolutions between 10 and 25 m, enabling detailed information about surface geometrics, vegetation cover, and subsurface discontinuities. This system is particularly well suited for measuring vegetation type

Figure 6. The "classic" SIR-C/X-SAR image of Angkor. Image P-45156, Copyright JPL/NASA.

and extent, and soil-moisture content, which are important environmental factors at Angkor.

The Jet Propulsion Laboratory's airborne Synthetic Aperture System (SAR), known as AIRSAR/TOPSAR, flies on a NASA DC-8 jet and is capable of simultaneously collecting all four polarizations for three wavelength frequencies. In another mode of operation, the system collects all four polarizations for two frequencies while operating an interferometer simultaneously to generate topographic height data. Ancient and modern development is shown in the AIRSAR image (Figure 7). Angkor Wat (lower right) and Angkor Thom (upper right) are clearly noticeable; portions of the extensive

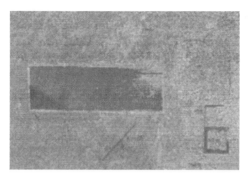

Figure 7. AIRSAR height data overlaid on a radar image of Angkor. Image P-49960, Copyright JPL/NASA.

water management system can be seen as the square moats surrounding the temples and large rectangular reservoir (Western Baray) in the left of the image. The dark line just below the Baray is the Siem Reap airport runway. Topographic height information is shown in color with one color cycle (yellow to yellow) representing 20 m of elevation change.

3.2. New Archaeological Discoveries through Remote Sensing at Angkor

The radar imagery at Angkor has been invaluable in allowing researchers to identify structural features previously hidden by dense tropical vegetation, or in areas that are inaccessible because of land mines. French historian and archaeologist Claude Jacques has noted several "new" features in the radar images that were not detected with SPOT, notably a huge wall between the Eastern Baray and Angkor Thom. He also detected what appear to be former canals, dikes, and roads in these images, and suggests that it may be possible to locate an older flow pattern of the Siem Reap River. Additionally, he noted another river within the Eastern Baray that is distinguishable by vegetation patterns that pass through the northern dike and continue to the south (World Monuments Fund, 1996).

Analysis of the AIRSAR imagery by Moore and Freeman (1997) has revealed circular prehistoric mounds and remains of previously undocumented temple structures, which suggest occupation as early as the 6th century A.D. The topographic maps created by these data show differences in the height of vegetation and significant features that can be seen in the perspective view of Figure 8. The AIRSAR data, which erroneously measured Angkor Wat's height at only 27 m (81 ft), reflects a rare shortcoming in the technology's recording of vertical topographical measurements. Its spatial resolution did not allow for an accurate resolution of the temple's steep and narrow canyon-like upper reaches, so that only the top stage of the multi-tiered structure was noted. In the radar image, Angkor Wat (whose accurate elevation is 65 m (213 feet) is central and surrounded by a moat. It is visible through the forested areas,

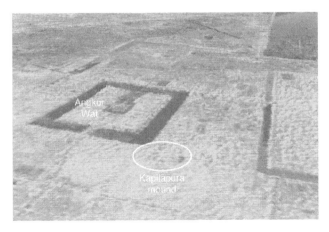

Figure 8. Three-dimensional perspective image showing moated Angkor Wat and the newly appraised Kapilapura (in circle) on left, and the southeast corner of Angkor Thom's moat on the right, created with radar interferometry data taken by NASA's AIRSAR. Image P-49602ac, Copyright JPL/NASA.

which appear yellow in the radar image. The moat of Angkor Thom is to the right, and Phnom Bakheng, the first temple mountain built at Angkor, stands at 99 meters high (324 feet) in the upper right. The height data on this image highlighted a small mound called Kapilapura in the Northeast area of Angkor. A subsequent field survey and further research led Moore, Freeman, and Hensley to include Kapilapura within the distribution of prehistoric circular mounds within the Angkor region (see Moore et al., in this volume).

Fletcher has used the AIRSAR images to reveal that the historic city of Angkor covered more than 1,000 sq km, a size that makes it the largest low-density, pre-industrial city in existence. At its apex, Angkor controlled a vast distribution of smaller cities and villages that spread throughout present-day Cambodia, most of Thailand, and into parts of Vietnam and Laos. Using radar images as a starting point for a comprehensive analysis, and incorporating ground-based examinations, low-level aerial surveys, and core-drilling, Roland Fletcher and the Greater Angkor Project, which is directed by Roland Fletcher, at the University of Sydney, seek to answer questions on the extent of the city, its duration, residential patterns, and reasons for its demise.

4. REMOTE SENSING FOR MANAGEMENT AND POLICY DECISIONS IN HERITAGE CONSERVATION

New archaeological discoveries are fascinating, and profoundly illuminate our understanding of a site such as Angkor. But what should be done now that the archaeologists have found new mounds reflecting early settlement, or previously undetected ruins, roadways, or canals? How do heritage conservationists protect the newly appreciated resource? Imperative questions need to be addressed: Is there modern settlement nearby, and if so, how close is it? What is the chance that this area will be chosen for development? How can the new appraisal of a site afforded by remote sensing inform governmental policies? How will it affect the local community? How can we best protect these newly appreciated features; and incorporate them into the larger ecological context? How and when do we inform the public of the new discoveries?

Protective measures need to be enacted swiftly so that newly discovered features are not compromised, or destroyed, when their knowledge becomes public. Archaeologists have given heritage conservationists an excellent historical perspective with which to begin formulating management plans and understanding the dynamics of change at heritage sites. It is now up to conservationists to ensure that all this work toward understanding the complete history of a place is safeguarded for posterity.

If heritage conservationists are to manage and plan for the mechanisms of cultural and environmental change, they must address the fundamental question of how these changes are encoded in, and inferred from, spatial variations (Marquardt and Crumley, 1987). In other words, heritage conservationists, planners, and others must use the archaeological evidence to affect decisions on contemporary urban growth, sustainable tourism, infrastructure development, environmental degradation, and other human-induced changes acting upon heritage sites.

As discussed earlier, tourism brings an assortment of environmental problems that affect a heritage site at multiple levels. In general, it increases population and urban growth for communities, regions, and countries; requires specific infrastructure to sustain itself; and consequently changes the natural and built environments. Since heritage resource dynamics function in a similar manner as urban environments, the remote sensing applications used in the study of urban environments should be easily adaptable to heritage conservation analysis.

At Angkor, information gathered by archaeologists can be applied toward policy, designation, and mitigation. Correlating archaeological data with conservation goals could, for example: increase the size of the designated World Heritage site; ensure compatible and sustainable development; provide cultural continuity of the landscape; monitor human-induced changes on the built and natural environments; track demographic change; and plan for, and mitigate, the effects of natural hazards on newly mapped historic structures.

Advances in remote sensing imagery facilitate the analysis of an array of spatial and temporal issues, and thus can affect policy and management decisions with quantitative, scientific data. Using remote sensing to quantify the environmental forces affecting cultural heritage resources will give heritage conservationists the ability to calibrate conservation principles and procedures with actual contemporary urban dynamics. Utilizing an integrated remote sensing/GIS is the optimal way to develop effective and sound comprehensive protection strategies.

4.1. Geographic Information Systems in Cultural Heritage Management

Remote sensing imagery provides an excellent baseline dataset for documenting and analyzing the contemporary effects of human activities at cultural heritage sites. When combined with a Geographic Information System (GIS) remote sensing delivers a tool to catalogue and assess organizational and structural patterns, quantify and monitor environmental conditions, and determine sustainable tourism and urban development of a cultural heritage site and its region. Integrating remote sensing and GIS generates a multi-scalar (multitemporal and multispatial) analysis that can be used to create an urban dynamics model that envisions future consequences of environmental and human-made threats. A Geographic Information System is capable of assembling, storing, manipulating, and displaying spatial information, for instance data identified according to location, and makes possible the linkage, or integration, of information that is otherwise difficult to associate. GIS applications for heritage conservation include automated cartographic display, property characterization and inventory, past landscape visualizations and viewsheds, impact assessment and prediction, large-scale synthesis, spatial sample design, and predictive modeling (Limp, 2000).

The United States National Park Service started using GIS as a landscape-management tool in the mid-1970s to document and manage Yosemite and the Great Smoky Mountains National Parks. Many National Park managers now use GIS to identify changes, patterns, and trends to solve resource management problems. GIS has also been implemented by the Cultural Resources Geographic Information Services (CRGIS) Facility of the United States National Park Service in their battlefield protection program. One project used GIS to plan for the preservation and development of Prairie Grove Battlefield State Park in Arkansas. The project assessed the visual integrity of the battlefield, identified important viewsheds, and used computer-imaging programs to model potential impacts of demographic changes on the integrity of the battlefield. The GIS demonstrated what a visitor would see from any location of the battlefield, the number of modern visual intrusions visible from current tour stops and viewing locations, how the view would change if a tour stop was moved to another location, and what kind of development potential a proposed interpretive location would have (Culpepper, 1997). In another project, the CRGIS created a GIS database for fifteen Civil War battlefields in the Shenandoah Valley, Virginia, to assess the integrity of the battlefields based on existing (1993) land use/land cover, and to use for further analysis and enhancement such as a Global Positioning Systems (GPS) survey, viewshed analysis, development scenario modeling, and conservation-priority analysis.

The use of GIS in archaeology and cultural resource management has increased in the last few years (Limp, 2000). GIS has been used as a management tool to integrate historical period archaeological sites into the urban planning process in Quebec City, Canada (Moss, 1998), and as the digital archiving system for a 3-D graphic and documentation assessment of the historic center of Thessaloniki, Greece (Patias, 2002). One of the most impressive GIS applications at an ancient site has been that of Pompeii where a 3-D model and rendering was made based on the latest site plan. It allows for the notation of building typologies, the ages of structures, morphological growth analyses, and the qualitative and quantitative analyses of conditions at the site. *Il Piano per Pompeii*, begun in 1996 (Longobardi and Mandara, 1998), has since become an invaluable tool for making conservation, maintenance, and display decisions at Pompeii, which is visited by approximately four million people every year. The Iraq Cultural Heritage Conservation Initiative is another a major undertaking of international importance. This collaborative program of the Getty Conservation Institute (GCI) and WMF in conjunction with the Iraq State Board of Antiquities and Heritage (SBAH) will develop the Iraq Cultural Heritage Sites Geographic Information Systems (GIS) Database, which will be used by SBAH to identify, analyze, and manage conservation priorities of the country's cultural heritage sites. The GIS will contain data on a site's or building's administrative status, management issues, topographic location and land use, period of occupation or construction, architectural features, disturbances and threats, physical conditions, and significance. Each record is linked to a shape representing the site and project area that is displayed on an interactive map, which supports satellite images and topographic maps. This comprehensive GIS will enable SBAH authorities to monitor development activities in at-risk areas, mitigate threats to sites and buildings, prioritize interventions and protection methods, and better manage conservation resources in coordination with other national and international agencies.

A formative use of GIS in cultural heritage management involves an integrated remote sensing and GIS utility that can be applied not only to mapping sites and urban environment issues, but also to a graphic representation (visualization) of time-space data. This type of analysis depicts changes such as urban development and policy decisions that could have a negative impact on a cultural heritage site.

4.2. Protection of Maya Sites in the Usumacinta River Basin

An example of a graphic model used in cultural heritage conservation to good effect for regional and industrial planning is the WMF-commissioned visual simulation model of potential flooding caused by a proposed dam on the Usumacinta River (Berendes, 2003). The Usumacinta River borders Guatemala and Mexico, and pierces a region that was a great cultural center of ancient

Maya civilization. The proposed dam threatened to flood significant ancient Maya sites at Piedras Negras, Yaxchilan, and at least 18 undocumented Maya located in the Usumacinta River basin. The model was developed with Landsat satellite imagery and digital elevation models (DEM) to conceptualize the potential impact of a dam 40 to 150 m high along the river at Boca Del Cerro, Mexico. Landsat, the United States' oldest land-surface satellite system, is particularly suitable for determining change in land use/land cover and served well the purposes of the Usamacinta River project. Landsat 7, the most recently deployed Landsat system, provides multispectral data at a resolution sharp enough to discern causes of land-surface change. Landsat 7 carries the Enhanced Thematic Mapper Plus (ETM+), a passive sensor calibrated for high accuracy that measures visible and reflected infrared radiation wavelengths in eight bands.

   A Landsat ETM+ image was used to spatially define the threatened region in the Usumacinta River basin and provide a visual frame of reference for the flooding analysis. Two Landsat ETM+ images were mosaicked together using ENVI 3.6 and Erdas Imagine 8.6 software to create a color-enhanced composite (Figure 9) in ENVI using the CN Spectral Sharpening algorithm. This refinement produced excellent graphical contrast while preserving the high spatial resolution (15 m) of the Landsat panchromatic band. Following this, a digital elevation model (DEM) that differentiated high and low elevations was required to demonstrate the effects of flooding. Data from three different sources were used to create one DEM that contained the entire study area. Two DEMs of the region in Mexico came from Instituto Nacional de Estadística Geografía e Informática (INEGI); the DEM of the Sierra del Lacandon region of Guatemala was obtained from the Guatemalan Consejo Nacional de Areas Protegidas (CONAP); and Digital Terrain Elevation Data (DTED®) derived from the Shuttle Radar Topography Mission (SRTM) was provided through a program co-sponsored by NASA and the National Geospatial Intelligence Agency (NGA). The composite DEM was then imported as a raster file into ArcGIS 8.2 software for use with the ArcGIS Spatial Analyst to detect the extent of land cover below 150 meters in elevation. The DEM showed that extensive overflowing of the reservoir would occur at dam heights greater than 90 m, and that the viable height for a dam at Boca del Cerro would be between 60 m and 90 m. The GIS was used to show the flooding that would occur if a dam 90 m high were built at Boca Del Cerro. The resulting flood layer was overlaid on top of the Landsat image to show that the extent of flooding over the area (Figure 10) would be over 5,600 sq km (2,100 sq miles), and would threaten the significant sites of Piedras Negras, El Porvenir, Yaxchilan, El Cayo, Macabilero, and La Pasadita.

   The Landsat imagery, DEM data, and GIS flood layers were then combined into a 3-D visualization software where the Landsat image was draped over the DEM to create a digital landscape. This scenario allowed the flood layers to be rendered at appropriate elevations to simulate flooding over the landscape,

Figure 9. Color Enhanced Landsat ETM+ image of the Usumacinta River basin from the proposed dam site at Boca Del Cerro in the upper left hand corner to Yaxchilan at lower right.

and showed the impact of flooding at varying dam heights. Figure 11 is a close-up view of the digital landscape near Piedras Negras with simulated flooding at dam levels of 30 m, 60 m, and 90 m. In this image it is evident that most of the structures at Piedras Negras and El Porvenir are inundated if the dam is built at 60 meters. They are almost all completely flooded by a dam of 90 m (Banderes, 2003).

The results presented in this visual simulation model clearly show that there would be enormous risk to several major Maya sites along the Usumacinta if a dam were built at Boca Del Cerro. Various public presentations of this simulation and its presentation to local authorities, plus a growing local opposition to the idea of a dam in general, have, according to the governor of the State of Tabasco, caused the proposed damming project to be canceled. As of June 2004, a new plan calls for Boca Del Cerro to be the center of a regional nature and culture reserve.

Figure 10. The GIS flood layer, which simulates the impact of a dam, 90 m high overlaid on the enhanced Landsat image.

5. FUTURE DIRECTIONS FOR REMOTE SENSING IN HERITAGE CONSERVATION

Progressing from the identification of remote sensing applications for heritage conservation to an established implementation of remote sensing/GIS for conservation management and policy-making is a sizeable movement. An integrated system of remote sensing/GIS must become a standard tool used by heritage conservationists for analysis, interpretation, management, and environmental monitoring of sizable heritage resources. Specifically, remote sensing/GIS technologies can provide systematic and quantitative measurements for land-use prediction, infrastructure placement, and development zoning and planning around cultural heritage sites.

Applying a comprehensive regional dynamics approach can serve as a good beginning for greater implementation of remote sensing/GIS in cultural heritage management. A regional dynamics survey is a broad-scale spatial

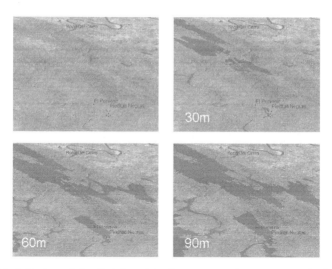

Figure 11. 3-D view of detailed digital landscape at Piedras Negras and El Porvenir with results of simulated flooding for dams 30 m, 60 m, and 90 m high at Boca Del Cerro.

analysis that allows the identification of the causes and effects of human activity over time. Examining a region for wide-ranging spatial discontinuities, rather than merely pinpointing localized anomalies, permits one to determine a region's mechanisms of change (Marquardt and Crumley, 1987). A regional dynamics survey establishes a multiscalar framework where initially ground data (field survey, documentary research) are collected and examined. This task is followed by a low-level aerial survey, which will verify the ground data, locate new sites, and document further stages of investigation. High-level vertical mapping photographs embed the compiled information within a wider regional context. These compiled data are then used to correlate digital classifications within the satellite or radar imagery. A regional settlement model that presents historical and contemporary habitation patterns will emerge from this data comparison.

A multiscalar, regional dynamics examination is straightforward when radar imagery is applied to additional remote sensing techniques to enhance and compare different types of information, such as vegetation, hydrologic, ecologic, geologic, and other environmental data. It also entails building a GIS to determine thematic classifications, which can then be extrapolated to a regional scale using topographic, elevation, soil, and vegetation maps, along with known physical, socioeconomic, organizational, and statutory boundaries. The GIS will generate themes, such as urban, water, barren areas, fields, crops, and vegetation that are assigned colors and compared with other maps and to obtain the percentage of a particular land use, enhance and quantify the existing classifications (Madry, 1987).

This approach is technically superior to the random-sample procedures currently used to analyze cultural heritage. It provides valuable information about past and present environments, land use, and settlement patterns, while also helping to target discovery efforts towards specific sites. This methodology has been successfully applied to a regional dynamics survey of southern Burgundy in France (Marquardt and Crumley, 1987; Madry, 1987), and is being employed, with some amazing results, in the Greater Angkor Project's historical ecology-based study of Angkor's collapse.

A regional dynamics investigation carries the archaeological inquiry into a contemporary timeframe, and highlights management issues for cultural heritage conservationists. An additional application of remote sensing of cultural heritage sites involves detection of urban growth and morphology, which can be quantified using the 30+ year global archive of Landsat imagery where spectral mixture analysis documents and calculates land-cover changes in heritage resource areas. This technique has been applied to both moderate resolution (30 m) Landsat imagery (Small, 2001; 2004) and high resolution (4 m) IKONOS imagery (Small, 2003). Multitemporal analysis of the resulting fraction maps can be accomplished using the method described by Small (2002) to produce maps of land cover change that subsequently reference specific types of land-use change (e.g., forest conversion to suburban development). These change analyses are best done at a regional scale (100–200 km) with Landsat imagery, and then followed up with more detailed land-cover mapping of critically changed areas using high-resolution IKONOS or Quickbird imagery.

The next phase in the use of remote sensing/GIS for heritage conservation will be to apply the analysis and known spatial and temporal relationships to predictive modeling. Urban dynamics theory uses time-series (temporal) analyses to project how a present city has been formed from the past, and employs many of the analytical techniques previously mentioned to visualize urban development. Current urban dynamics theory explores a concept of self-organized criticality on the premise that systems develop until they reach a critical state, which is a point of equilibrium that can accommodate most change, but can lead to dramatic changes involving a time-phase transition and alteration in the urban form. In this instance, change is attributed to local processes, which cause chain reactions that enact macro-level alterations at the global scale. These reactions, or relationships, exist across all spatial and temporal scales (Batty, 1996). Urban dynamics theory can demonstrate the space-time development of urban environments through computational models, such as those based on diffusion, cellular automata, genetic algorithms, probabilistic modeling, and fuzzy logic. Ultimately we hope all of these methodologies will become part of a larger system that enables conservationists to establish a theory explaining the process dynamics of cultural heritage resources.

6. CONCLUSION

Archaeologists primarily use remote sensing to locate previously undetected or unexcavated ruins, but cultural heritage conservationists, working closely with them and other specialists, can use this data to help conserve and interpret a known site and its surrounding area. The relatively new field of professional heritage conservation management is rapidly tackling many broad social-science issues, such as how environments and economies have shaped cultures. Cultural heritage projects that use remote sensing analysis have broken new ground. Remote sensing offers great promise for documentation, planning, and interpretation in cultural resource management practice. The strongest case can be made for its implementation on sites such as the Angkor Eco-site in Cambodia, where existing information is organized and enhanced through the baseline data that remote sensing applications offer.

The year 2004 marked a decade in the use of radar imaging at Angkor. Its three applications—beginning with the 'classic' SIR-C/X-SAR image—have prompted and defined the first uses of radar imagery in archaeology and cultural resource management in Southeast Asia. It has also offered important new tools for research, which in turn have spawned new methodologies and related training needs. The initial hope of the cultural heritage specialist who instigated the acquisition of the original 1994 'classic' image of Angkor was to learn more about the site, in particular at WMF's sole project at the time, Preah Khan, and its archaeology. As it turned out, far more was accomplished.

Indeed, the research potential at Angkor and its match with the powerful new tool of radar imaging has proved to be fortuitous. Among the many reasons for the success experienced thus far has been the initiative taken by those who sensed its possibilities at the beginning, their encouragement of others to join in this exciting and fruitful voyage of discovery, and the support all along of participating funders, organizations, and administrators. The beneficial outcome from the remote sensing work at Angkor is attributed to a mix of good timing, careful coordination, serious multidisciplinary research, and a fervent desire to understand and conserve an icon of the world's heritage. The example it offers can and should be repeated at other key cultural heritage sites in the future.

REFERENCES

Barnsley, M.J., Moller-Jensen, L., Barr, S.L., 2001, Inferring Urban Land Use by Spatial and Structural Pattern Recognition. In *Remote Sensing and Urban Analysis*, edited by Jean-Paul Donnay, M.J. Barnsley, and P.A. Longley, pp. 116–117. Taylor & Francis, London and New York.

Batty, M., 1996, Visualizing Urban Dynamics. In *Spatial Analysis: Modelling in a GIS Environment*, edited by P. A. Longley and Michael Batty, pp. 297–320. John Wiley & Sons, New York.

Berendes, T. , 2003, Visual Simulation of the Potential Impact of Damming the Usumacinta River. Final Project Report to World Monuments Fund, April 22.

Blom, R.G., and Crippen, R. E., 1997, Space Technology, Ancient Frankincense Trade Routes, and the Discovery of the Lost City of Ubar. Presentation at Geological Society of America, 1997 annual meeting. In *Abstracts with Programs*, 29, 6:237, Geological Society of America.

Carter, J., and Morter, J., 1996, How the Greek Countryside Was Divided in Southern Italy and Crimea: The Achievements and Potential of Remote Sensing. Paper presented at Symposium on New Technologies and Global Cultural Resource Management, University of Florida, April.

Comer, D. C., 1998, Discovering Archaeological Sites from Space: Using Space Shuttle Radar Data at Petra, Jordan. *CRM Bulletin*,21(5):9–11.

Culpepper, R.B., 1997, Better Planning Through GIS: Battlefield Management Efforts at CAST. *CRM Bulletin* 20(5):32–34.

Drury, S.A., 1998, *Images of the Earth: A Guide to Remote Sensing*, 2nd edition. Oxford University Press, Oxford and New York.

El-Baz, F., 2002, Change Detection of Urban Sprawl. Abstract in *Proceedings, 3rd International Symposium on Remote Sensing of Urban Areas, Istanbul, Turkey 11-13 June* (CD-ROM), edited by D. Makiav, C. Jürgens, F. Sunar-Erbek, and H. Akguen, p. 504. Istanbul, Turkey, Istanbul Technical University.

El-Baz, F., 2001, Gifts of the Desert. *Archaeology* 54 (2, March/April):42–45.

El-Baz, F., 1997, Space Age Archaeology. *Scientific American* 777 (2, August): 60–65.

Evans, D.H., 2003, Investigating the Decline of Angkor Using AIRSAR and GIS. *Proceedings, 30th International Symposium on Remote Sensing of Environment, Honolulu, Hawaii, November 10–14* (CD-ROM). Tucson, Arizona, ISRSE Office of the Secretariat, TS–43.4.

Fitch, J. M., 1990, *Historic Preservation: Curatorial Management of the Built World*. University Press of Virginia, Charlottesville, VA.

Fletcher, R., Evans, D.H., Tapley, I.J., 2002, AIRSAR's Contribution to Understanding the Angkor World Heritage Site, Cambodia—Objectives and Preliminary Findings from An Examination of PACRIM2 Datasets. *Proceedings of the 2002 AIRSAR Earth Science and Application Workshop, NASA/JPL* (http://airsar.jpl.nasa.gov/cgi-bin/sorkshop.x?tt=2). Pasadena, California.

Jacques, C., 1996, Description of Angkor and Overview of History: The Importance of Remote Sensing for the Knowledge of Angkorian Civilization. Paper presented at *Symposium on New Technologies and Global Cultural Resource Management*, April.

James, J., 1995, Shuttle Radar Maps Ancient Angkor. In *Science Magazine* 267 (no. 5200, February 17):965.

Jet Propulsion Laboratory (JPL), 1998, NASA Radar Reveals Hidden Remains at Ancient Angkor. California Institute of Technology, National Aeronautics and Space Administration, http://www.jpl.nasa.gov/releases/98/angkor98.html.

Limp, W. F., 2000, Geographic Information Systems in Historic Preservation. In *Science and Technology in Historic Preservation*, edited by Ray A. Williamson and Paul R. Nickens, pp. 231–248. Kluwer Academic/Plenum Publishers, New York.

Longobardi, G. and Mandara, A., 1998, *Il Piano per Pompeii*. World Monuments Fund for the Soprintendenza a Pompei, Rome.

Madry, S. L. H., 1987, A Multiscalar Approach to Remote Sensing in a Temperate Regional Archaeological Survey. In *Regional Dynamics: Burgundian Landscapes in Historical Perspective*, edited by C. L. Crumley and W. H. Marquardt, pp. 173–236. Academic Press, San Diego.

Marquardt, W. H., and Crumley., C. L., 1987, Theoretical Issues in the Analysis of Spatial Patterning. In *Regional Dynamics: Burgundian Landscapes in Historical Perspective*, edited by C. L. Crumley and W. H. Marquardt, pp. 1–18. Academic Press, San Diego.

McKee, K. R., 2003, Applying Remote Sensing to a Regional Landscape Survey of a Cultural Heritage Resource: A Dynamic Approach to Surveying and Modeling the Angkor Eco-Site. In *Proceedings, 30th International Symposium on Remote Sensing of Environment, Honolulu, Hawaii, November 10–14* (CD-ROM). Tucson, Arizona. ISRSE Office of the Secretariat, TS–43.2.

Mesev, T. V., Longley, P.S., Batty, M., and Xie, Y., 1995, Morphology from Imagery: Detecting and Measuring the Density of Urban Land Use. *Environmental Planning A* 27(5, May):759–780.

Miller, R. B., and Small, C., 2003. Cities from Space: Potential Applications of Remote Sensing in Urban Environmental Research and Policy."*Environmental Science & Policy* 6:129–137.

Moore, E., and Freeman, A., 1997, Circular sites at Angkor: A Radar Scattering Model. *The Journal of the Siam Society* 85, Parts 1 & 2 (April):107–119.

Moss, W., Simoneau, D., and Fiset, B., 1998, Archaeology, GIS, and Urban Planning in Quebec City. *CRM Bulletin* 21(5):19–20.

Patias, P., Karapostolou, G., Simeonidis, P., 2002, Documentation and Visualization of Historical City Centers: A Multi-Sensor Approach for a New Technological Paradigm. In *Proceedings, 3rd International Symposium on Remote Sensing of Urban Areas, Istanbul, Turkey 11–13 June,* edited by D. Makiav, C. Jürgens, F. Sunar-Erbek, and H. Akguen, pp. 609–619. Istanbul, Turkey, Istanbul Technical University. CD-ROM.

Pozzi, F., and Small, C., 2002, Vegetation and Population Density in Urban and Suburban Areas in the U.S.A. In *Proceedings, 3rd International Symposium on Remote Sensing of Urban Areas, Istanbul, Turkey 11–13 June* (CD-ROM), edited by D. Makiav, C. Jürgens, F. Sunar-Erbek, and H. Akguen. pp. 489–496. Istanbul, Turkey. Istanbul Technical University.

Sadler, G. J., and Barnsley, M. J., 1990, *Use of Population Density Data to Improve Classification Accuracies in Remotely-Sensed Images of Urban Areas.* South East Regional Research Laboratory, Birkbeck University, London.

Sever, T. L., 1992, Environmental Remote Sensing/GIS Analysis of the Peten area and Environs of Guatemala, Mexico, and Belize. Presentation at the American Anthropological Association meeting, December 2, San Francisco.

Small, C., 2001, Estimation of Urban Vegetation Abundance by Spectral Mixture Analysis. *International Journal of Remote Sensing* 21(7):1305–1334.

Small, C., 2002, Multitemporal Analysis of Urban Reflectance. *Remote Sensing of Environment* 81:427–442.

Small, C., 2003, High Resolution Spectral Mixture Analysis of Urban Reflectance. *Remote Sensing of Environment* 88:170–186.

Stefanov, W. L., Ramsey, M.S., and Christensen, P.R., 2001, Monitoring Urban Land Cover Change: An Expert System Approach to Land Cover Classification of Semiarid to Arid Urban Centers. *Remote Sensing of Environment* 77(2):173–185.

Sullivan, S., 2000, *Reader for Cultural Heritage Planning Workshop at Preah Khan.* Siem Reap, Cambodia, March.

United Nations Educational, Scientific and Cultural Organization (UNESCO), 1972, Convention Concerning the Protection of the World Cultural and Natural Heritage, 17th Session, Paris.

United Nations Educational, Scientific and Cultural Organization (UNESCO), 2003a, Ensuring Sustainable Development through Cultural Diversity. http://portal.unesco.org/en/ ev.php@ URL_ID=1219&URL_DO=DO_TOPIC&URL_SECTION=201.html. Updated 14-01-2003.

United Nations Educational, Scientific and Cultural Organization (UNESCO), 2003b, Towards More Responsible Tourism. http://portal.unesco.org/en/ev.php@URL_ID= 2481&URL_DO=DO_TOPIC&URL_SECTION=201.html. Updated 14-01-2003.

Wiseman, J., 1996, "Space Missions and Ground Truth, *Archaeology* 49(4, July/August):11–13.

World Monuments Fund and Royal Angkor Foundation (WMF/RAF), 1995, *Radar Imaging Survey of the Angkor Eco-Site; Report of the First Scientific Roundtable, Princeton University, February 1–2.* New York.

World Monuments Fund and Royal Angkor Foundation (WMF/RAF), 1996, *Workshop on Radar Imaging and Cultural Resource Management at the Angkor Eco-Site and Symposium on New Technologies and Global Cultural Resource Management. Report of the Second Scientific Roundtable, University of Florida, Gainesville, April 15–19.* New York.

Index

Made in the USA
Lexington, KY
30 July 2013